Environmental History *of the* Hudson River

Environmental History of the Hudson River

*Human Uses that Changed the Ecology,
Ecology that Changed Human Uses*

EDITED BY
Robert E. Henshaw

WITH A FOREWORD BY
Frances F. Dunwell

COVER: *Progress (The Advance of Civilization) 1853*, by Asher B. Durand, courtesy of the Westervelt Collection, Westervelt-Warner Museum of American Art in Tuscaloosa, AL. See legend for Fig. lntro.2.

INSIDE COVER IMAGE: *Novi Belgii Novaeque Angliae / Partis Virginiae tabula multis in locis emendate, 1685*, by Nicolaes Visscher with Schenk, Peter, Jr., courtesy of the Library of Congress. Based on a manuscript map by Adriaen Van der Donck, 1648. See Fig. lntro.1. Three generations of Visschers produced 27 versions based on this map.

Published by State University of New York Press, Albany

© 2011 State University of New York

All rights reserved

Printed in the United States of America

No part of this book may be used or reproduced in any manner whatsoever without written permission. No part of this book may be stored in a retrieval system or transmitted in any form or by any means including electronic, electrostatic, magnetic tape, mechanical, photocopying, recording, or otherwise without the prior permission in writing of the publisher.

For information, contact State University of New York Press, Albany, NY
www.sunypress.edu

Production by Ryan Morris
Marketing by Fran Keneston

Library of Congress Cataloging-in-Publication Data

Environmental history of the Hudson River : human uses that changed the ecology, ecology that changed human uses / edited by Robert E. Henshaw.
    p. cm.
  Includes bibliographical references and index.
  ISBN 978-1-4384-4026-2 (pbk. : alk. paper) — ISBN 978-1-4384-4027-9 (hardcover : alk. paper)
  1. Human ecology—Hudson River (N.Y. and N.J.)—History. 2. Nature—Effect of human beings on—Hudson River (N.Y. and N.J.)—History. 3. Natural history—Hudson River (N.Y. and N.J.) 4. Environmentalism—Hudson River (N.Y. and N.J.)—History. 5. Hudson River (N.Y. and N.J.)—Environmental conditions. I. Henshaw, Robert E.
   GF504.N7E68 2011
   304.209747'3—dc22
                                                                                    2011014090

10 9 8 7 6 5 4 3 2 1

This book is dedicated to my father, Dr. Paul S. Henshaw, nuclear biophysicist, from whom I learned the unity of physical and biological sciences with the social sciences; and to Dr. G. Edgar Folk Jr., environmental physiologist at the University of Iowa, from whom I learned professional persistence.

Contents

Foreword ix
Frances F. Dunwell

Acknowledgments xiii

Introduction xv
Robert E. Henshaw

The Hudson River Watershed:
An Abbreviated Geography xxi
Robert E. Henshaw

PART I
History and Biology: Providing Explanations 1
Robert E. Henshaw

CHAPTER 1
Historical Facts/Biological Questions 3
Robert E. Henshaw

CHAPTER 2
Linkages between People and Ecosystems:
How Did We Get from Separate to Equal? 7
Stuart Findlay

CHAPTER 3
Symbioses between Biologists and
Social Scientists 13
Lucille Lewis Johnson

PART II
River of Resources 23
Robert E. Henshaw

CHAPTER 4
Hudson River Fisheries: Once Robust,
Now Reduced 27
*Robert A. Daniels, Robert E. Schmidt,
and Karin E. Limburg*

CHAPTER 5
Herpetofauna of the Hudson River
Watershed: A Short History 41
Alvin R. Breisch

CHAPTER 6
Human Impacts on Hudson River
Morphology and Sediments: A Result
of Changing Uses and Interests 53
*Frank O. Nitsche, Angela L. Slagle, William
B. F. Ryan, Suzanne Carbotte, Robin Bell,
Timothy C. Kenna, and Roger D. Flood*

CHAPTER 7
The Earliest Thirteen Millennia of Cultural
Adaptation along the Hudson River Estuary 65
Christopher R. Lindner

CHAPTER 8
Archaeological Indices of Environmental
Change and Colonial Ethnobotany in
Seventeenth-Century Dutch New Amsterdam   77
Joel W. Grossman

CHAPTER 9
Linking Uplands to the Hudson River:
Lake to Marsh Records of Climate Change
and Human Impact over Millennia   123
*Dorothy M. Peteet, Elizabeth Markgraf,
Dee C. Pederson, and Sanpisa Sritrairat*

CHAPTER 10
Vegetation Dynamics in the Northern
Shawangunk Mountains: The Last Three
Hundred Years   135
John E. Thompson and Paul C. Huth

CHAPTER 11
Agriculture in the Hudson Basin Since 1609   153
Simon Litten

CHAPTER 12
Ecology in the Field of Time: Two Centuries
of Interaction between Agriculture and Native
Species in Columbia County, New York   165
Conrad Vispo and Claudia Knab-Vispo

CHAPTER 13
The Introduction and Naturalization of
Exotic Ornamental Plants in New York's
Hudson River Valley   183
Chelsea Teale

PART III
River of Commerce   195
Robert E. Henshaw

CHAPTER 14
The Rise and Demise of the Hudson River
Ice Harvesting Industry: Urban Needs and
Rural Responses   201
Wendy E. Harris and Arnold Pickman

CHAPTER 15
Human Sanitary Wastes and Waste Treatment
in New York City   219
*David J. Tonjes, Christine A. O'Connell,
Omkar Aphale, and R. L. Swanson*

CHAPTER 16
Foundry Cove: Icon of the Interaction of
Industry with Aquatic Life   233
Jeffrey S. Levinton

CHAPTER 17
River City: Transporting Commerce
and Culture   247
Roger Panetta

CHAPTER 18
Out of the Fray: Scientific Legacy of
Environmental Regulation of Electric
Generating Stations in the Hudson
River Valley   261
John R. Young and William P. Dey

PART IV
River of Inspiration   275
Robert E. Henshaw

CHAPTER 19
Birth of the Environmental Movement
in the Hudson River Valley   279
Albert K. Butzel

CHAPTER 20
The Influence of the Hudson River School
of Art in the Preservation of the River,
Its Natural and Cultural Landscape, and
the Evolution of Environmental Law   291
Harvey K. Flad

CHAPTER 21
"Thy Fate and Mine Are Not Repose":
The Hudson and Its Influence   313
Geoffrey L. Brackett

CHAPTER 22
The Past as Guide to a Successful Future   325
Robert E. Henshaw

Afterword   335
Robert E. Henshaw

Contributors   337

Web Addresses of Cited and Key Agencies,
Not-for-Profit Organizations, and Academic
Institutions in the Hudson River Basin   341

Index   343

Foreword

THE HUDSON IS A RIVER of dreams. Human dreams have transformed this body of water and recreated it. They have explained and interpreted it. It is a river that has been sculpted by the ideas of a people. In its waters and on its shores are written the changing thoughts of Americans over the great sweep of history.

Many of our nation's rivers have come to embody an idea or a moment in time in our history. The Mississippi will always be the river of Huck Finn and steamboats and jazz. The Columbia tells the story of Louis and Clark, while the Rio Grande echoes with memories of ancient canyons, water wars, and the human heartbreak of border crossings. George Washington will forever be crossing the Delaware and dwelling on the Potomac, the seat of our national government and our monuments and shrines. The Saco we think of as wild and free, and the Red River Valley a place of goodbyes.

The Hudson is a different kind of river, because it tells not one but many stories. It was the river of the frontier, a battleground for freedom, and the creative inspiration for a generation of American poets and painters. Here, the civil engineers' visions of possibility bore fruit, and so did the dreams of entrepreneurs and captains of industry. It has been the gateway to America for millions of immigrants who aspired for a new life. From its harbor the Statue of Liberty sends forth her beacon of light. As rivers go, the Hudson may be short—it is a mere 315 miles in length—but its connection with our country's history is long and deep.

This environmental history of the Hudson, compiled by the Hudson River Environmental Society (HRES), begins, as it must, with the river's unique geography, but it also weaves in the human element, exploring the role of ideas, innovation, and passion. It shows how science can unravel the mysteries of our past. It illustrates the deep divide of values that forced legal showdowns, as well as the attitudes and practices that allowed the river to become polluted. It shows how the emergence of new ideas inspired a later generation to focus on restoring the estuary and its ecosystem.

Nature blessed the Hudson with a deep harbor that doesn't freeze, a pleasant climate and good soils for agriculture, a long estuary that provides habitat for abundant fish and wildlife, plus a geologic storehouse of metals and minerals. Its port is one of the best in the world, a function of its size, shape, location, and geologic history. The river also radiates breathtaking natural beauty.

For centuries—long before the arrival of European colonists—these natural assets have attracted people who seek a better life. Nature set the stage for prosperity and entrepreneurship that is best reflected in the great city on Manhattan Island at the mouth of the Hudson. The accumulation of power and wealth in New York City can be directly traced to the river's ecosystem. In turn, the city shaped the future of the river and changed many aspects of its ecology. It is through this lens that HRES has asked the authors to explore the river's history.

The way the city and the river co-evolved reflects not only the unique geography of the Hudson but also its place in world history, its mix of ethnic groups, and its power to inspire human imagining. The Age of Enlightenment, the Romantic Era, the transportation revolution, and the landing of a man on the moon all colored the vision of those who sought to arrange the Hudson to their own designs. Advances in technology have also been critical to this story. Inventions such as the Mercator map, the steam engine, the Bessemer process for making steel, the use of dynamite, and the harnessing of electricity from water power all shaped the future of the river in profound ways. Laws and policies have similarly been important, all influenced by the people who settled here. Native People, Dutch, Africans, and English in particular established concepts of governance, trade, commerce, and land use that echo throughout the ages of river history. Later, the French, Irish, Germans, Italians, Eastern Europeans, and Chinese left their mark. You name it: the Hudson has been the quintessential melting pot of world cultures, each influencing the next with their notions of the role a river should play in meeting human needs.

Like many American rivers, the Hudson has been dammed, filled, channeled, and polluted, but it has also been a success story for cleanup and a model for protection of scenery. Precedents for conservation of the environment have spread from here across the nation to other places in the world. Fundamentally, this history comes back to ideas and how we relate to nature.

Rivers have always been a window into the deeper and sometimes hidden emotions, and there is an essential spiritual element to the river's history. "My soul runs deep like rivers," poet Langston Hughes once wrote. Taking the journey to the source, finding the hidden headwaters, the "heart of darkness," is part of the Hudson's story as well as that of many other rivers, yet this river, more than any other in America is populated by fairies, heroes, and scoundrels. Here myth and reality are blended. Legends and literature have been born from such things as the rolling thunder in its Catskill mountain shoreline and odd occurrences, such as the rare white whale that swam into the fresh tidewaters of Albany and Troy. The Hudson has been used as a metaphor for madness, for death, and for life.

Not surprisingly, the river's stories and dreams are intertwined. The artist, funded by the entrepreneur, painted works that inspired the conservationist. The engineer remodeled the river and then designed the mechanics of its recovery. The fabulously wealthy became the philanthropists whose treasure has preserved a natural and historic heritage. Immigrants who carried out the transformation of the Hudson's shores raised children and grandchildren who fought to save the river from destruction. Politicians whose childhood was spent on the banks of this storied river drew lessons from their childhood ramblings and applied these experiences to state and national policies that reverberated here and everywhere.

Having grown up on the river and studied its history, I have concluded that the story of the Hudson is really about passion. Among the ranks of those who have made a difference in the history of the river are governors, journalists, bankers, surveyors, singers, aristocrats, fishermen, congressmen, lawyers, scientists, mothers, tree farmers, businessmen, teachers, and Presidents. Their voices, their energy have profoundly affected how civilization has proceeded across the river valley and how it spread from the Hudson to the nation and the world.

The one thing all those individuals have in common is the power of their imagination. The Hudson inspires big dreams and energizes the people who can fulfill them. Most of the people who have made a difference on this river have been steeped in personal experience of it. They swim in it, they study its rocks, and they listen to the songs of its birds and observe the habits of its fishes. They smell the fragrance of the sweet flag growing in its shallows or contemplate the scenery in quiet meditation. They are moved by it, as am I.

My own personal experience of the river grew over a period of years. When I was a child, in the '50s and '60s, the river was at its worst, a stinking sewer that was hard to love. I remember having to get shots to go out on a boat with a friend, in case I fell in. Then Earth Day came along when I was in college, and the Clearwater Sloop began having festivals on the waterfront, spreading a message of both anger and hope. Like many young people of my generation, I was inspired to do something about the pollution of the river. The environmental movement coincided with the women's movement and the civil rights movement. I was lucky to get an internship that launched what has become a career in conservation. Now, with more than thirty-five years of experience in protecting the river's water quality, historic sites, fisheries, habitats, and scenery, I am one of a number of women who have made a profound difference for the river as we know it today, and I have been blessed to know many of the people who played key roles in its recovery.

Among those who have made a great contribution to the future of the river are the scientists, engineers, and historians who make up the membership of the Hudson River Environmental Society. This fine book is a collection of essays from people who have worked in the trenches, bringing a depth of personal experience, scientific knowledge, and historical perspective that shines a light on our understanding of the river and its people.

Frances F. Dunwell,
author of *The Hudson: America's River*
New Paltz, NY
November, 2010

Acknowledgments

Robert E. Henshaw

FUNDING WAS RECEIVED from several sources which we gratefully acknowledge. We thank: former Albany County Historian and now NYS Assemblyman John J. "Jack" McEneny, for a Legislative Member Item in support of this volume; NYS Senator Neil D. Breslin, for a Legislative Member Item in support of the conference; the Hudson River Foundation for grants from the Hudson River Improvement Fund; the Hudson River Valley National Heritage Area in partnership with the National Park Service and Congressman Maurice D. Hinchey, whose grant funds are administered by the Hudson River Greenway Council; The Lucy Maynard Salmon Research Fund of Vassar College; Ms. Hollee H. Haswell, a dedicated Hudson River Valley botanist; Henningson, Durham, and Richardson Architecture and Engineering PC, a long time supporter of environmental work for the Hudson such as the present volume; and the Environmental Consortium of Hudson Valley Colleges and Universities. We especially thank the Hudson River Estuarine Program of the NYS Department of Environmental Conservation directed by Frances Dunwell. We also thank the State University of New York Press for production of the resulting text you now hold.

Believing that readers wish to connect with the Hudson River system, we have provided a list of useful Web addresses sequestered from many sources, and with the assistance of many people; in particular, we thank Manna Jo Greene of Hudson River Clearwater Inc. and Emilie Hauser of NYS Department of Environmental Conservation.

This conference and book project were possible because of arduous work by several key individuals, all recognized experts in their respective disciplines. I thank fisheries ecologist Dr. Robert A. Daniels, New York State Museum, Albany, NY; aquatic ecologist Dr. Stuart E. G. Findlay, Cary Institute for Ecosystem Studies, Millbrook, NY; archeologist/anthropologist Dr. Lucille L. Johnson, Vassar College, Poughkeepsie, NY; industrial historian Dr. Roger Panetta, Fordham University, New York, NY; ornithologist Dr. Kathryn J. Schneider, Hudson Valley Community College, Troy, NY; and then Executive Director of Hudson River Environmental Society Mr. Stephen O. Wilson, Albany, NY. My own background as an environmental analyst with the NYS Department of Environmental Conservation, Albany, NY, served me well in the present comparative study.

INTRODUCTION

Robert E. Henshaw

> We found a pleasant place below steep little hills. And from among those hills a mighty deep-mouthed river ran into the sea.
> —Giovanni da Verrazano, Italian explorer, 1524 for "his most serene and Christian Majesty," Francis I of France, his patron

> This is very good land to fall in with, and a pleasant land to see.
> —Robert Juet, Henry Hudson's historiographer, 2 Sep 1609

MUH-HE-KUN-NE-TUK—the River That Flows Both Ways. Native Americans revered the river and defined themselves by it for at least twelve millennia. Early European explorers and colonists renamed it many times: *Mauritius; River of the Mountains; North River; Hudson River*. It has been called America's Rhine. Those who have lived or traveled along it, worked upon it, defended it, or simply beheld it, have valued this river out of all proportion to its meager 315 mile length. Others, too, who built on its shores, conducted commerce along it, transported upon it, and expelled industrial and municipal wastes into it, also relied on the river. Over the years the people who settled along it deforested its shores and built vast businesses. Communities and industries sullied it, causing the people to shun it for many years. Today they rediscover it. This volume attempts to explain the enigma of this, the queen of America's rivers, the Hudson River.

In 2009 New York State celebrated the quadricentennial anniversary of Henry Hudson's 1609 voyage into the river that came to bear his name. Most public recognition celebrated the last four hundred years of human presence on the Hudson River. Less attention was given to the many environmental influences of the human presence since the coming of the Europeans, and even less to the preceding thirteen millennia of human presence on the river. The Hudson River Environmental Society participated in that anniversary celebration by convening a conference to examine the ways that human activities affected the ecosystems of the Hudson River watershed, and how once those ecosystems had been altered, subsequent human activities had to be modified. More than a historical accounting of events, the speakers, each an expert in his or her respective subject, sought to explain the interplay of human activities with the Hudson River watershed. Because the audience comprised many disciplines, discussions ranged widely. This volume is based on the speakers' original presentations, now modified based on those discussions.

Seemingly widely diverse, these chapters all consider, directly or by implication, a single concept: the reciprocal effects of human uses and ecosystem responses. Some authors look for causation in preceding events or ramifications in later events. Others consider ecological conditions and the implications in preceding or subsequent animal or plant communities. Still others focus directly on human activities implying ecological effects. If one looks only at specific details and events, the history of the Hudson River appears unique. If, on the other hand, one considers patterns of change and general mechanisms, then the Hudson is seen to model any river.

This volume is divided into four parts, each examining particular aspects of the human use/ecological response feedback relationship. Each part begins with a short introductory narrative history

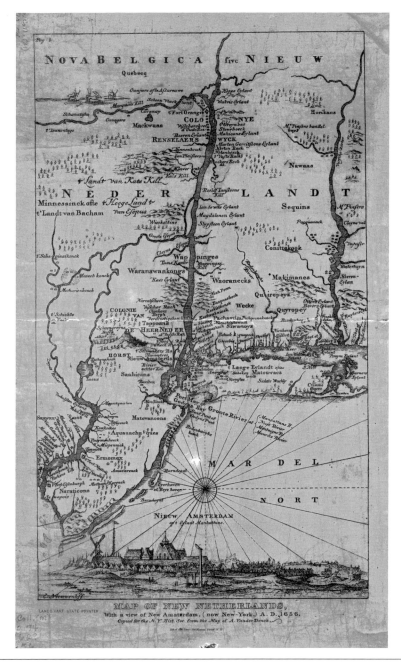

FIG. INTRO.1. Map of New Netherlands with a View of New Amsterdam 1656.

Earliest map of the Hudson River (Courtesy of the New York State Historical Association Library, Cooperstown, NY). From the second edition of "A Description of the New Netherlands," by Adriaen Van der Donck, published by Evert Nieuwenhof, Bookseller in 1656, copyrighted for fifteen years in 1653. In a foreword Nieuwenhof states his intention for the book: "Comprehending the fruitfulness and natural advantages of the country, within itself, and from Abroad, for the subsistence of man…" Preceding Van der Donck's text, Nieuwenhof also offered this poem:

ON THE PATRONS AND THE HISTORY OF NEW NETHERLANDS
Still Amstel's faithful Burger-Lords do live
Who East and West extend their faithful care;
To lands and men good laws they wisely give,
That like the beasts ran wild in open air.,
With aged care Holland's gardens still they save—
And in New Netherlands their men will ne'er be slaves.

Why mourn about Brazil, full of base Portugese? [sic]
When Van der Donck shows so far much better fare;
Where wheat fills golden ears, and grapes abound in trees;
Where fruit and kine are good with little care;
Men may mourn a loss, when vain would be their voice,
But when their loss brings gain, they also may rejoice.
Then, reader, if you will, go freely there to live,
We name it Netherland, though it excels it far;
If you dislike the voyage, pray due attention give,
To Van der Donck, his book, which, as a leading star,
Directs toward the land where many people are,
Where lowland Love and Laws all may freely share.

(Evert Nieuwenhof)

Every line provides clues to life, conditions and authority in New Netherlands; many are discussed throughout this volume.

FIG. INTRO.2. Asher B. Durand, *Progress (The Advance of Civilization), 1853,* oil on canvas. (Generously provided by the Westervelt Collection; original is displayed in the Westervelt-Warner Museum of American Art in Tuscaloosa, AL)

Asher B. Durand was the second prominent artist in the Hudson River School of Landscape Painting. A former respected banknote engraver, he was noted for his accurate and realistic depictions. This painting, done on commission from Jay Gould, the railroad magnate, is thought to be a conflation of many scenes, rather than a specific location (Ferber 2010). We may consider it a composite snapshot of the Hudson River Valley in the mid-nineteenth century. While some in the Hudson River School, especially Thomas Cole, its founder, deplored encroachment on the wildness, Durand deals unapologetically, perhaps even approvingly, with the notion of "continued upward progress" of American society which was much debated in mid-nineteenth-century learned society. In this one scene Durand encapsulates many ways that humans were changing the Hudson River ecosystems, and how those changes both enabled and forced changes in the ways humans continued using those ecosystems. Some art historians believe that two years later, Durand began to criticize "progress."

In the foreground are remnants of the primeval forest that so attracted arriving Europeans, though shown here in dark tones that contrast with the scene below. Durand features "blasted" (storm-damaged) trees—iconic of Hudson River School paintings. Three Native Americans (nostalgically?) view the scene from the cliff top. A tributary stream to the Hudson has been canalized to facilitate flatboat traffic but impeding fish access to headwaters for spawning. The floodplain is now cleared and farmed, reducing biodiversity. Poles along the road suggest the arrival of the telegraph, a recent invention of fellow resident of the Hudson River Valley, Samuel F. B. Morse. Villages line the shoreline, attesting to their continued reliance on the river for both water source and waste disposal. A railroad connects river communities to far off places, but creates a linear slice separating people from the river, and dividing properties and communities. The tracks have required excavation of the rock bluffs and are supported on a causeway that restricts the river's circulation through a former embayment. Beyond, the river sweeps around a point where industries and brick factories send plumes of smoke and steam skyward. In the far distance, mountains, probably based on the Hudson Highlands, rise into the clouds, but unlike the present course of the river, in this scene the river turns to the right beyond the distant prominence. A church and a large institution occupy commanding views of the river, suggesting they valued landscape vistas. If the large institution in the middle ground is meant to suggest Sing Sing Prison, it was not placed there for a commanding view, but rather because, as they explained at the time, that is where the stone used in its construction was in abundance. Thus, some construction was river-dependent, drawing on the river for resources or transportation, or simply for aesthetic joy, while other developments clearly were not. Lighthouses may be in the river channel attesting to the amount of boat traffic. Durand shows one steamboat, one oceangoing sailing ship, and one Hudson River freight-hauling sloop. Although many HRS paintings portray robust river traffic, the artist has not done so in this scene—perhaps to force the viewer to focus on riverside development.

The eye is inevitably drawn to the center of the painting where the sky is bright at the very point where the sun would be setting in shades of red if this view were northwestward. Was this Durand's optimism for "progress?" The viewer is left to ponder what is gained and what is lost when civilization advances.

FIG. INTRO.3A. *Progress, 2010* Panorama of Lower Hudson River Valley at Rivermile 41 (looking toward south and west). (Scene continues in Fig. Intro. 3b)

At this point the Hudson River threads through the Allegheny Mountains. During the last ice age an ice plug blocked southward river flow and diverted the flow northward into the St. Lawrence River. To the south, Peekskill Solid Waste Facility is exactly centered in the focal point of the southern vista. Easily seen on a clear day, the large structure is out of character with the regional natural landscape. This plant might have been sited 1,000 m to the east and have been completely out of sight, but during reviews of plans New York State regulators concentrated on protecting fish from discharges and not on protecting scenery. Directly across the river, Iona Island is the site of a former U.S. Navy munitions storage depot with underground ammunition storage vaults. The island is ostensibly open to the public, however access is difficult. Reintroduced bald eagles frequent the island. Beyond is Iona Marsh, ca. 80 ha of cattails where once the river flowed freely. This is one of four tidally flushed wetlands in the Hudson estuary managed by the National Estuarine Research Reserve System. The marsh resulted from restriction of river flushing of the former embayment when the Hudson River Line Rail Road (today the West Shore RR) was constructed in the mid-nineteenth century without regard to environmental effects. This railroad is still the principal route to deliver freight northward to the crossover near Albany bound for New York City south of here. On the far shore is the site of Doodletown, a colonial riverside community, now vanished leaving virtually no signs of its previous existence. Above rise the forested low mountains of Bear Mountain State Park; a gift to the state from a wealthy river family. (Photo by the author)

of events and context surrounding the subjects of chapters that follow it. These introductions may be read in sequence for an unfolding of the environmental history of this phenomenal river we call home—how we got to our present state and how we leave it to future generations.

In Part I, "History and Biology: Providing Explanations," the convening authors discuss general concepts of the use/response interaction. They seek to demonstrate the valuable and necessary synergism between biologists and historians—one that is too often ignored by both. We hope the result of the present volume will be that biologists and historians in the future will consider collaboration essential to complete their work.

Part II, "River of Resources," concentrates on the resource base exploited by Native Americans and colonists. Native American tribes varied in their uses of the Hudson River and the surrounding forest ecosystems. Archeologists believe that because of their low population density and their lifestyles, they lived sustainably within the productive capacity of the Hudson region—as surely they must have since they occupied the Hudson region for at least twelve millennia. This volume examines evidence of early Native American uses of resources, but it concentrates on the most recent four hundred years since the arrival of European colonists, on their increasing presence and accumulating impacts on the Hudson River and its surrounding terrestrial communities, and on their (our) failure to live sustainably within the productive capacity of the Hudson region.

For millennia the Native American tribes arrayed up and down the Hudson exploited the river for trade and transportation—uses that were ecologically benign. Part III, "River of Commerce," demonstrates that from the first viewing, Europeans sought to exploit the river and its surroundings with little or no regard to ecological impact. Arriving from stressed and crowded Europe, the new lands appeared almost magically productive and available. Many assumptions were made; buffalo and unicorns were hypothesized in the remote north, and

FIG. INTRO.3B. *Progress, 2010* Panorama of Lower Hudson River Valley at Rivermile 41 cont. (looking toward west and north). (continued from Fig. Intro. 3a)

Beyond the northern tip of Iona Island is the Hudson River Line RR trestle that restricts water flow through the Doodletown embayment (see Fig. Intro. 3a.). Between the island and the trestle are mute swans, a Eurasian exotic species introduced during the 1800s on large Hudson River estates. Unlike most invasive species, the swan population is stable. Above the shoreline, Bear Mountain rises 300 m (900 ft.), accessible to the public as a state park. At the north end is Bear Mountain Bridge. When constructed in 1924 it was the longest suspended bridge as well as the largest privately owned bridge in the world. It remained in private ownership until 1940 when the state bought it. Just upstream of the bridge are the sites of the former Revolutionary War Fort Clinton and Fort Montgomery. Here the first Great Chain was suspended across the river during the Revolutionary War to prevent British incursion. The British dismantled and salvaged the chain. The cliff on the east side is 300 m (900 ft.) high Anthony's Nose named to celebrate the prodigious proboscis of Anthony Van Corlaer, trumpeter for early governor Petrus Stuyvesant. The mountains above the picture site (out of view) are Camp Smith, a former New York State militia training ground. Although ostensibly off-limits, the peaks are popular with day hikers. Immediately below the camera's location the eastside railroad tracks carry passengers and freight to New York City. As with the 1853 Durand version of "Progress" (Fig. Intro. 2.), the viewer of this modern-day vista may ponder what is gained and what is lost each time decisions are made for human uses that affect ecosystems. (Photo by the author)

the earliest map of the region (Fig. Intro. 1) indicated that the today's Rockland County and New Jersey were an island.

The authors in the present volume describe effects of forestry, agriculture, ice, sanitary, chemical, and power industries on the river and its surroundings. Many other industries might be examined with respect to the use/change feedback model but space limitations prevent inclusion in this volume. Perhaps the best summary of nineteenth-century effects of human uses on ecosystems in the Hudson River Valley was created by the artist Asher B. Durand (cover and Fig. Intro. 2). This 1853 painting all but encapsulates the content of this volume. That scene may be compared to a present-day vista from the Scenic Overlook opposite Bear Mountain State Park. Here, compressed into one glorious view, one may see the pervasive effects of prior development decisions imposed on stunning, otherwise pristine scenery as well as present-day uses of the region (Figs. Intro. 3a and 3b). To provide a geographic context for all that follows, we begin with an abbreviated geography of the Hudson River Basin.

Reactions to growth and development of the Hudson were varied. Many residents observed environmental deterioration and despaired, forsaking the river. Others reveled in growth, arguing that "Progress" (see Fig. Intro. 2) was inevitable and even "Manifest Destiny." A remarkable number of Hudson citizens were stimulated to strike out in new directions, creating new literature, art, law, and regulatory procedures. Even the form of the new federal government following the Revolutionary War was shaped by study of the Iroquois Confederacy. The authors in Part IV, River of Inspiration, discuss the uniquely large spiritual impact this river has had on its inhabitants generation after generation.

It should come as no surprise that the world's environmental movement began in the Hudson River Valley, initiated specifically to protect its unique environmental qualities. Similarly, it was along the Hudson River, at Ft. Edward in 1986, that one of the first negotiated agreements between the federal and state governments and the Native American community (coordinated by Joel Grossman, author of chapter 8 in this volume), established a new protocol for archaeological investigations which respects Native American values through on-site nondestructive documentation and in-place reburial under Native American supervision.

This diminutive river system continues to hold a disproportionate level of public interest. A computer search on "Hudson River" returns more hits per river mile than any other river in the world. The Hudson River has had a differentially large impact on American history and culture up to now. We must believe that it will continue to provide lessons, guidance, and inspiration in the future. May it spur *your* curiosity, imagination, and participation, that you too may become a part of the inspiring spiraling history of the Hudson River.

REFERENCES CITED

Durand, Asher B. 1853. *Progress (The Advance of Civilization) 1853.* Painting, oil on canvas, Westerfelt-Warner Museum of American Art.

Juet, R. 2006. *Juet's journal of Hudson's 1609 voyage, from Collections of New York Historical Society,* Second series, 1841. Transcribed by Brea Barthel, Albany: New Netherlands Museum. http://halfmoon.mus.ny.us/Juets-Journal.pdf. 2 Sep.

Van der Donck, Adriaen, J.U.D. 1656. *A description of the New Netherlands.* Tr. from the original Dutch by Hon. Jeremiah Johnson. Evert Nieuwenhof, Bookseller, Amsterdam

Verrazano, Giovanni da. 1524. In Carl Carmer, The Lordly Hudson: Over 350 years a mighty pageant of history has moved through the myth-haunted valley of the Great River of the Mountains. *American Heritage Magazine.* December 1958, 10(1). http://beta2.americanheritage.com/articles/magazine/ah/1958/1/1958_1_4.shtml. accessed Dec 2010.

The Hudson River Watershed

An Abbreviated Geography

Robert E. Henshaw

HUDSON RIVER WATERSHED

By comparison with America's other great rivers, the Hudson River is small, yet because of its location, topography, and ecology, it is one of America's most interesting rivers. By virtue of its location, it is one of America's most important rivers. It originates in the Adirondack Mountains and flows 507 km (315 miles) to the Atlantic Ocean. The northern half brings the river down from the side of New York's highest mountain peak to sea level; the southern half is a single long estuary emptying through its all-season harbor into the Atlantic Ocean. En route it traverses two mountain ranges, bisects a broad pastoral valley, receives sixty-five tributaries draining eleven sub-watersheds, and merges with branches of the sea isolating Manhattan and Long Island. It creates scenic landscapes so striking that they are protected by federal law. The reader should take this incredible journey in the fifteen-minute virtual airplane tour of the entire length of the river from the southern point of discharge into the Atlantic Ocean northward to the source of the river in the Adirondack Mountains at the Hudson River Environmental Society's website, www.hres.org.

The watershed of the Hudson River comprises an area of ca. 3.5×10^6 hec (13,600 mi^2), 93 percent of which is in New York; the rest is in Vermont, Massachusetts, Connecticut, and New Jersey (Fig. G.1 and on the CD). It lies strategically between New England and the interior of the continent and saw key battles to control it during the American Revolutionary War. The watershed is 65 percent forested, 25 percent agricultural, and 8 percent urban. The Hudson River system comprises three legs: the Upper Hudson, the Mohawk River, and the Lower Hudson. Each leg is distinct in hydrology, biology, terrestrial setting, and history of human habitation, uses, and impacts (Fig. G.2).

UPPER HUDSON RIVER

The Upper Hudson River begins its 277 km (172 mi.) descent to sea level near the 1,460 m (4,800 ft.) level of New York State's tallest mountain, Mt. Marcy. It issues from the diminutive, and romantically named, Lake Tear of the Clouds, as Feldspar Brook, which empties into the Opalescent River before becoming the mainstem Upper Hudson River. For the first three-fourths of the way it tumbles through the granitic Adirondack Mountains. As a result it has very low turbidity and is highly oxygenated. It receives little allochthonous (in-washed) nutrient material, and therefore is oligotrophic. Total aquatic biomass and biological productivity are low. When it reaches Troy, New York, still 246 km (153 mi.) inland from the ocean, the river is virtually down to sea level. The Upper Hudson is virtually free flowing even as it passes through more than twenty dams. It takes but a few days for a drop of water leaving Lake Tear of the Clouds to arrive at Troy.

Precipitation is 1.0–1.2 m/year (40–48 in./year), distributed fairly uniformly from month to month. Average surface runoff is 46–61 cm/yr (18–24

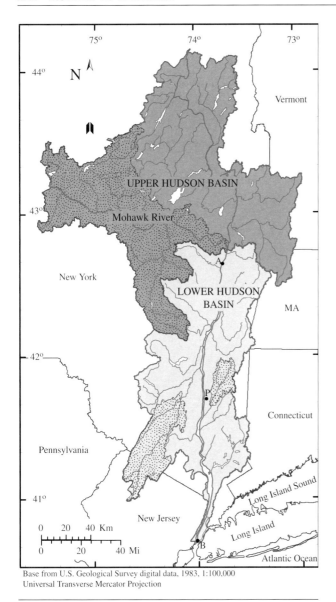

FIG. G.1. Watershed of the Hudson River Basin. Most of the 165 tributaries are shown. (Courtesy of Wall 2010)

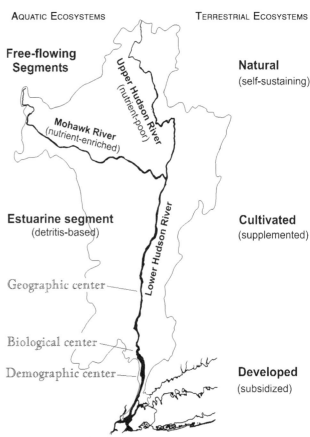

FIG. G.2. Limnological map of the Hudson River Basin. The distinctly different productivity of the three main legs is characterized. Based on the relative biological productivity and surrounding demographics, the "biological center" and "demographic center" are characterized with respect to the "geographical center." The productivity of terrestrial ecosystems surrounding the river is generalized. Fundamentally these terrestrial ecosystems are second-growth forest in the north, agrarian in the middle, and urbanized in the south.

in./yr). However, surface runoff varies greatly with season because virtually all of the winter precipitation is held in the Adirondack Mountains as snow cover. During April and May the snow melts, rapidly releasing all of the winter-accumulated water into the river in just a few weeks producing a dramatic and short "spring high flow." By August, surface runoff declines so greatly that the "summer low flow" is equally dramatic. The spring high flow supports world-class whitewater kayak and canoe events, but by late summer a child can wade across the river. For the lower sixty-four km (forty mi.), flow of the Upper Hudson River is streamlined and uninterrupted until it arrives at Troy.

MOHAWK RIVER

The principal tributary to the Hudson River is the Mohawk River. It enters the mainstem Hudson from the west near Troy. The Mohawk River is 225 km (140 mi.) long, flowing mostly through agricultural lands. As a result the water is turbid due to in-washed silt and nutrients from surface runoff. It supports a mesotrophic food web although many tributaries are oligotrophic. The annual flow of the Mohawk River is seasonal, although spring high flow is less dramatic than in the Upper Hudson because it extends over a period of months. Situated between the Adirondack Mountains and the

Catskill Mountains and Allegheny Plateau, the Mohawk River provided a route along which the Erie Canal was constructed—the first reliable commercial transportation route to the west through the Allegheny Mountains.

LOWER HUDSON RIVER

Hydrology

The Lower Hudson (Fig. G.3), often referred to as the Hudson River Valley, flows in a nearly straight path from Troy southward to the southern tip of Manhattan ("River Mile 0"), a distance of 246 km (153 mi.). It then continues between Staten Island and Long Island to the Verrazano Narrows on the Atlantic Ocean south of New York City. Unless specified otherwise most of the chapters in this volume concentrate on the Lower Hudson River.

This segment of the Hudson River is commonly considered to be a drowned river because the main channel for the Lower Hudson formed at a time when sea level was about 140 m (430 ft.) lower than today. As the Wisconsinan glacier retreated around 18,000 years ago, the ocean inundated the channel. Subsequently, in-washed materials have filled much of the river channel. Today the Lower Hudson averages ca. 6 m (21 ft.) deep for about 80 km (50 mi.) south of Troy. Where it breaches the Allegheny Mountains (see Fig. G.3), often referred to as "the Gorge," the river narrows as it rounds West Point. Here the increased water velocity flushes out settleable materials, so that the river maintains a depth of as much as 59 m (194 ft.). South of The Gorge the river broadens to 5.6 km (3.5 mi.) wide to become the Tappan Zee (Tappan Sea) that is as little as 1 m (3 ft.) deep. Water here warms in the summer and is the nursery area for many species of fish. Because ocean-going freight vessels commute to the Port of Albany, a navigation channel is maintained by the US Army Corp of Engineers of 10 m (32 ft. to just south of Albany and 9 m (27 ft.) at Albany (USACE 2009, 1).

Mean annual flow in the Lower Hudson River is ca. 385 m³/sec (13,600 ft.³/sec), but because the inflow from the Upper Hudson and the Mohawk are seasonal, flow in the Lower Hudson varies from ca. 1300 m³/sec (46,200 ft.³/sec) during spring high flow to ca. 245

FIG. G.3. Satellite map of the Hudson River Valley in false color.
South of Albany and Troy the Hudson River flows southward along an ancient crease in the earth's surface. For the first ca. 80 km (50 mi.) it flows through a wide pastoral valley. Where the Allegheny Mountains cross from the west into New England, the Hudson must pass through a narrow deep gorge between mountains ca. 300 m (1,000 ft.) high. Following the recent Wisconsinan glaciation, as an ice plug in this gorge melted, the ocean inundated the channel all the way to Troy, N.Y., 240 km (150 mi.) inland. South of the gorge is the broad and shallow Tappan Zee (Tappan Sea). For the last ca. 80 km (50 mi.) south, the river channel is deep, creating an all-season port for deep draft oceangoing vessels. (Courtesy of NYS Department of Environmental Conservation)

m³/sec (8700 ft.³/sec) during summer low flow. Flushing rate is ca. 126 days. Although the average residence time for a drop of water in the estuary is ca. 4 days, during summer low flow it can lengthen to as long as three weeks for a drop of water entering at Troy to reach the ocean.

The entire Lower Hudson River is an estuary with mean tidal height of ca. 1.5 m (4 ft.), and tidal flux varying between 6 and 13,026 m³/sec (200 and 460,000 ft.³/sec) all the way to Troy. Peak flood and ebb tide water velocities are ca. 1.8 m/sec (4 mph). Thus, tidal circulation is about ten times the river flow. Tidal movements, combined with river flow and variable winds that predominately blow north and south, present sailors with interesting challenges.

The northern two-thirds of the Lower Hudson River is freshwater. The southern third is brackish (Table G.1). As freshwater pushes toward the ocean, sea water attempts to enter the river channel creating an interface of sea water and freshwater, the halocline, or more commonly, the "salt front." As the tide pulses in and out, the salt front moves up and down the river twice a day. During spring high flow of freshwater the salt front is pushed down to about "River Mile 30" (48 km). During summer low flow the salt front may advance as far north as about "River Mile 80" (128 km) (Fig. G.4). This "front" is not a vertical wall. Sea water contains about 3 percent salt, and therefore is more dense than freshwater. Thus, as the sea water attempts to flow in, the freshwater tends to "float" on top of the sea water for a distance before they fully mix. The salt front is more correctly viewed as a "salt wedge" (Fig. G.4). The salt wedge is extremely important to larval aquatic animals because they can maintain their location in the river simply as a result of their normal vertical migrations up into freshwater moving out, then down into sea water moving in.

Biology

The Lower Hudson holds a rich abundance of aquatic plants and animals. Overall biodiversity is great because there are freshwater, salt water, and brackish water species, and there are salt marshes, tidal saltwater wetlands, and tidally flushed freshwater wetlands. Total biological productivity is very high because there are two foundations for the food web, the usual algae-based primary production as well as bacterial and fungal decomposition of in-washed organic matter (detritus) such as leaves. In essence the Lower Hudson is a long trough within which the tidal flux sloshes the contents back and forth for up to months while the freshwater moves slowly from Troy to the Atlantic Ocean. This means that in-washed organic matter remains in the trough long enough for bacteria and fungi to break it down into a rich broth available to support a robust mix of ecological communities of organisms. In the Hudson River detrital decomposition provides twice as much chemical energy to the trophic steps above as algal production (Cole and Caraco 2001, 101; Caraco and Cole 2004, 308). The sum of both foundations for the food webs above makes the Hudson River estuary an unusually productive one.

Overall river productivity has been affected by several important anthropogenic modifications

TABLE G.1. Distribution of Salinity

River Mile	River Km	High Flow	Low Flow
100	161	0.0	0.0
90	144	0.0	0.1
80	128	0.1	0.2
70	112	0.2	0.7
60	96	0.5	1.0
50	80	0.6	1.8
40	64	1.1	3.0
30	48	2.3	6.1
20	32	6.0	10.5
10	16	12.3	17.0
0	0	23.5	25.0
-1	-2.5	25.0	25.0

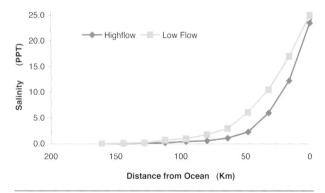

FIG. G.4. Salinity of the Hudson River.
Spring high flow pushes the halocline, or "salt front," downriver. Summer low flow allows sea water to penetrate farther, pushing the salt front upriver. (Based on Oak Ridge National Laboratory 1977)

through the last century. The center-river channel has been dredged from the harbor to Albany to accommodate oceangoing freight ships and barges. Dredge spoils were placed in channels among islands effectively removing the braided channel. Railroad embankments tended to linearize the shorelines and simplify them with riprap. Where railroads encountered meandering shorelines and embayments, the tracks were often laid on elevated gravel beds and trestles, semi-enclosing the embayments. All of these affected aquatic productivity (Strayer and Findlay 2010; Findlay et al. 2002).

These conditions prevailed until about 1993 when the exotic zebra mussel arrived in the Hudson River. With no natural parasites present in the Hudson, it quickly produced a huge population. It is such an effective filter feeder of suspended nutrients that algae and detritus were greatly reduced. As a result, abundance of most animal groups in the trophic web changed dramatically. Biologists continue to monitor populations to determine how the balance of processes will eventually stabilize.

In excess of two hundred species of fish occur in the Lower Hudson River. Spawning, egg laying, and early larval growth concentrate in the brackish region of the salt front. For this reason this region is characterized in Fig. G.2 as the "biological center" of the Lower Hudson. A number of species of fish are anadromous, relying on the freshwater and brackish parts of the river for spawning and early growth before spending the bulk of their lives at sea. Seasonal migration runs, when the mature fish commuted into and out of the river, historically supported robust commercial fisheries. A few species of fish are catadromous, such as the eel, which spends its life in the river but travels to the mid-ocean Sargasso Sea for reproduction. During colonial times, the nonmigratory short nosed sturgeon that could grow to more than 2 m (7 ft.) and a body weight in excess of 60 kg (150 lbs.) was taken as "Albany beef."

For many years wastes from the metropolitan areas along the Hudson's shores affected the biological integrity of the river. Until the installation of secondary sewage treatment in 1975, waste released from Albany and Troy caused a severe reduction of oxygen in the water downstream for ca. 30 km (24 mi.). This "oxygen block" precluded virtually all species of fish from reaching Troy. Within two years of installation of secondary treatment most species of fish again could be found as far north as the federal dam at Troy. New York City also released vast quantities of raw sewage until 1991 although it did not seem to cause the degree of impact as occurred at the north end of the estuary.

Demography

The Lower Hudson passes several large cities and many smaller cities and villages. Albany and Troy are at the northern end. Poughkeepsie is at about the midpoint. By far, the largest and densest metropolitan area is at the southern end, Yonkers to Manhattan on the east side and Jersey City to Elizabeth on the west side. For this reason, Fig. G.2 suggests the "demographic center" of the Lower Hudson is south of the "geographic center." The smaller cities and villages differ from highly urban to nearly rural in character, and provide a wide variety of community quality of living, cost of living, and opportunity to their citizens (hudson-river-valley.com 2010). Some of those communities have concentrations of certain ethnic groups or economic levels. As a loose generalization, mean household income, home value, cost of living, and cost of education increase from south to north. Cost of housing in New York City (up to five hundred times higher than in Albany) accounts for the bulk of the high cost of living there. Demographics of the five boroughs of Greater New York City differ markedly; household income, home value, and all costs of living are highest in Manhattan. Environmental quality parameters do not follow such a trend from south to north but rather reflect local conditions. New York City's drinking water, world famous for its high quality, is delivered through two massive aqueducts by gravity flow from nineteen reservoirs in the Catskill Mountains more than 160 km (one hundred miles) northwest of the city. The reader may go online to find the fascinating story of the creation in the nineteenth century of "Greater New York City" from its five separate boroughs to increase political and purchasing power in order to gain access to mountain water resources.

Industry

Throughout the nineteenth and twentieth centuries large industries located along the shores, capitalized on the river to transport products to market. Without regulation, industrial wastes, including chemical wastes, were increasingly disposed of in the receptacle of greatest convenience, the river. Fish populations bore the brunt of those impacts. Many organic chemicals, notably the infamous polychlorinated biphenyls (PCBs), were carcinogens and hormone mimetics. Sport fishing for most fish species and commercial fishing for striped bass and shad finally were closed at the end of the twentieth century because of chemical pollution. In the new century, public, technical, and regulatory attention is now focusing on contamination from pharmaceutical residues in municipal waste water and pesticides in agricultural runoff.

During the early part of the twentieth century large power plants were constructed along the shores of the Lower Hudson, some in the region of the salt front, that is, unknowingly in the "biological center" of the river. These large power plants, being only about 30 percent efficient, produce enormous quantities of excess heat that must be dissipated by passing millions of cubic meters of water per day through the plant. In this region where fish spawning and early larval growth predominate, tiny fish with limited mobility cannot avoid the cooling water intakes. Nearly all are killed. The particularly large impact on fish populations due to power plants became the subject of litigation that culminated in a historic settlement in 1980 modifying the operation of the power plants during fish spawning season (Barnthouse et al. 1988).

READINGS IN GEOGRAPHY

In this short geography facts are assembled from many sources and are presented mostly without attribution. The reader is encouraged to explore the unusually robust literature on the Hudson River. As a starting place one might review the following respected general discussions: "An Atlas of the Biologic Resources of the Hudson Estuary" (L. H. Weinstein, ed. 1977), "The Hudson River Ecosystem" (Limburg et al. 1986), and "The Hudson: An Illustrated Guide to the Living River" (Stanne et al. 1996). The Internet provides access to a vast amount of data. Of particular importance are excellent summaries such as "The Hudson River Watershed Alliance" (U.S. Fish and Wildlife Service 2009); "Significant Habitats and Habitat Complexes of the New York Bight Watershed" (2009); and a compendium of data from the United States Geological Survey (USGS 2009). Also, it is possible to actually experience ongoing limnological research in real time via the Internet from in-river monitoring stations (USGS) and even free-floating data logging devices (Stevens Institute of Technology 2010; HRECOS 2010). This present cursory description of the geography of the Hudson River Watershed must serve only to whet the reader's appetite.

Geography is an inclusive subject. Description demands functional analysis, which in turn facilitates interpretation and interconnections. The authors of the papers in the present volume attempt to describe and to interpret some of the functional relationships of humans and the Hudson River Basin. They cannot disguise their personal involvement and love for the Hudson River. We can be sure that the remarkable beauty, diversity, abundance, and importance will continue to captivate and inspire the observer whether he or she has not (yet) traveled the length of the river or has spent a lifetime upon it.

REFERENCES CITED

Barnthouse, L. W., R. J. Klauda, D. S. Vaughan, and R. L. Kendall. 1988. *Science, law, and Hudson River power plants. A case study in environmental impact assessment. American Fisheries Society Monograph* 4.

Caraco, N. J., and J. J. Cole. 2004. When terrestrial matter is sent down the river: Importance of allochthonous carbon to the metabolism of lakes and rivers. In *Food webs at the landscape level,* ed. A. Polis and M. E. Power, 301–16. Chicago: University of Chicago Press.

Cole, J. J., and N. F. Caraco. 2001. Carbon in catchments: Connecting terrestrial carbon losses

with aquatic metabolism. *Marine and Freshwater Research* 52: 101–10.
Findlay, S. E. G., E. Kiviat, W. C. Nieder, and E. A. Blair. 2002. Functional assessment of a reference wetland set as a tool for science, management, and restoration. *Aquatic Science* 64: 107–17.
Hudson River Environmental Conditions Observing System (HRECOS). 2010. http://www.hrecos.org/joomla.
Hudson-river-valley. 2010. http://www.hudson-river-valley.com/htm/zip/demographics/town.html; accessed Aug. 2010.
Hudson River Watershed Alliance. http://www.hudsonwatershed.org/; accessed Dec. 9, 2009.
Limburg, K. E., M. A. Moran, and W. H. McDowell. 1986. *The Hudson River ecosystem.* New York: Springer-Verlag, 331.
Oak Ridge National Laboratories. 1977. General considerations of the Hudson River Estuary. In *A selective analysis of power plant operation on the Hudson River with emphasis on the Bowline Point Station* 1: 2.1–2.19.
Stanne, S. P., R. G. Panetta, and B. E. Florist. 1996. *The Hudson: An illustrated guide to the living river.* New Brunswick: Rutgers University Press, 1–25.
Stevens Technological Inst. 2010. http://hudson.dll-stephens-tech.edu/maritimeforecast/PRESENT/data.html.
Strayer, D., and S. E. G. Findlay. 2010. Ecology of freshwater shore zones. *Aquatic Science* 72: 127–63.
US Army Corps of Engineers (USACE). 2009. Public Notice: HR-AFO-09. http://www.nan.usace.army.mil/business/buslinks/navig/cntldpth/albany.pdf; accessed Aug. 2, 2010).
US Fish and Wildlife Service, Coastal Ecosystems Program. 1997. *Significant habitats and habitat complexes of the New York Bight Watershed.* http://library.fws.gov/pubs5/web_link/text/toc.htm; accessed Dec. 10, 2009.
US Geological Service (USGS). New York Water Science Center.http://ny.water.usgs.gov/projects/hdsn/fctsht/su.html; accessed Dec. 10, 2009.
Wall, G. R., K. Riva-Murray, and P. J. Phillips. 1998. Water quality in the Hudson River Basin, New York and adjacent states, 1992–1995. US Geological Survey Circular 1165. http://pubs.usgs.gov/circ/circ1165/.
Weinstein, L. H. (ed.). 1977. *An atlas of the biologic resources of the Hudson Estuary.* Yonkers: Boyce Thompson Institute for Plant Research, 1–9.

PART I

HISTORY AND BIOLOGY

Providing Explanations

Robert E. Henshaw

[The] river is really the summation of the whole valley. To think of it as nothing but water is to ignore the greater part.
—Hal Borland, *This Hill, This Valley*

[H]istory in its broadest aspect is a record of man's migrations from one environment to another.
—Huntington Ellsworth, *The Red Man's Continent: A Chronicle of Aboriginal America* (1919)

BIOLOGISTS ARE MOST COMFORTABLE describing present-day ecological conditions or population size, i.e. parameters that can be directly measured. But of what importance are these specific conditions? Are populations up or down? By how much? And is that change important to the organism, the ecosystem, or the human community dependent on it? It is easy for a present-day researcher to believe that because the present sample is well quantified that it somehow may be more than simply one "grab-sample" in the long life history of a species or ecosystem. But the population abundance of a plant or animal at any given time results as that species exploits the habitat available and its nutrient supply while contending with predation, disease, and parasitism, as well as the effects of human actions. Establishing a timeline, or trend, for a parameter over time can facilitate interpretation of single samples. Thus, the population of a species results from preceding population levels. Earlier conditions may affect a species in ways that impact future population size. The present volume concentrates on effects of human activities on the River's ecosystems, and how ecological conditions determine the possible human activities (Fig. 2.1; Findlay develops this concept in chapter 2 in this volume). Because the feedback relationship derives from functional processes, components may be systematically examined, quantified, and predicted. The feedback loop can look forward suggesting effects of human activities on the present population, possibly lasting, in the future. Likewise, the feedback relationship can permit predictions backward

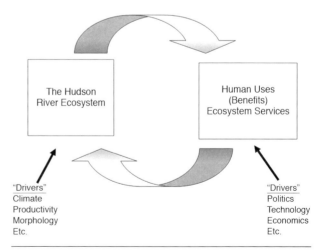

FIG. 2.1. (from Findlay, ch. 2) Diagram of feedbacks between human society and ecosystem attributes. Drivers may affect either humans or ecosystems separately but eventually all are connected by the proposed feedbacks.

in time and suggest a likely earlier population condition due at least in part to preceding human uses. The feedback loop, then, really is more accurately viewed as a spiral of events through time. No matter where you enter the diagram, it suggests earlier and later conditions, and infers drivers and mechanisms that might be fruitfully examined. The authors in this first section examine the functional and predictive value of this approach for biological and physical sciences.

Historians have documented past events, but their descriptions leave us with questions. How important were those preceding events in determining subsequent events? Why did the historically documented activity occur? Is it explained by simply describing it? Was it due only to other human activities, or might the resources available to the people have driven their choices in their activities? The historian who simply documents *that* historical events occurred clearly misses important explanations of those historical events. What might the historian, then, learn from consideration of the biological resources that were available to the study population? How did the availability of key natural resources determine what, how, and to what extent a particular human activity could be practiced? What preceding human activities might have influenced the biological resources available at the time? Can knowledge of key human activities permit extrapolation of, or at least hint at, changes in floral or faunal populations or ecosystems? Might this information permit the biologist to estimate trends in biota? And might establishing a pattern provide greater confidence to extrapolate biological conditions when no direct biological data exist? Historical descriptions and accountings may imply or demonstrate biological conditions even if they may not have been written to directly describe extant species or ecosystems or they were written before naming of species was standardized or ecological processes well understood. Authenticated historical accounts do provide firmly established points in time, something that may be very weak in biological trend analysis. Thus, historical literature represents a potential valuable resource to biologists as they attempt to describe earlier biological communities. Likewise, establishing biological conditions existing at earlier times can provide historians with valuable inferential information into why earlier human activities were undertaken.

This first section explores the power, implications, and mutual benefit of this approach to both historians and biologists. The convening authors seek to demonstrate the value of history to biological interpretation of existing conditions, and the equally important value of biology to historians when they wish to understand the reasons for many past activities. In chapter 1, Henshaw, drawing on established events from colonial Hudson River history, demonstrates the value of historical analysis to biologists. Findlay, in chapter 2, analyzes the components of the feedback relationship between human uses and ecological response. In chapter 3, Johnson finds relationships among the subsequent chapters in this volume and demonstrates how the reader may make productive associations among the authors' analyses of human activities and ecology. These conceptual chapters then are followed in subsequent sections by specific discussions of the Hudson River's resources used by humans, the growth of Hudson River commerce and industry, the emergence of environmental consciousness, conservation concern, and environmental law in the Hudson River Valley, together with lessons to be learned by looking backward to better see the future.

REFERENCES

Borland, Hal. 1990. *This Hill, This Valley.* Baltimore: Johns Hopkins University Press, 314.

Ellsworth, Huntington. 1919. *The Red Man's Continent: A Chronicle of America Series.* Edited by Allen Johnson. New Haven: Yale University Press, 183.

CHAPTER 1

HISTORICAL FACTS/BIOLOGICAL QUESTIONS

Robert E. Henshaw

ABSTRACT

For millennia, the impact of humans on the ecosystems of the Hudson River Valley was minimal. That changed with the arrival of Europeans in 1609. Native Americans inhabited the valley and utilized resources concomitant with their needs. Europeans, although initially few in number, marketed the Hudson's resources to a vastly larger population in the home continent. The fur trade, a good example, began in 1610, but declined beginning only a dozen years later. The Native Americans who trapped and dried beaver skins were exceedingly skillful, and thousands of skins were shipped in the first few years. If the decline was due to extirpation of the beaver in the region, major changes in the ecosystems surely resulted. The high biological productivity and species richness of the wet communities would have declined as drier ecosystems developed. Biologists seek to describe animal abundances during colonial times and before. Although direct contemporaneous observations may not exist for species abundances, it may be possible to reconstruct abundances based on review of historical documents written on entirely different subjects. Subsequent chapters in the present volume propose a feedback relationship between human uses and changes in the ecosystems that human communities relied on. Reciprocally, changes made in keystone species such as the beaver must have required changes in uses that human communities made of those ecosystems as well as changes in the methods humans used to exploit their ecosystem resources.

INTRODUCTION

For thirteen thousand years the Native American groups lived along the Hudson River; Lenape in the southern and coastal portion, and Mahicans north of the Hudson Highlands. These groups fished the waters, hunted the forests, and created river-related cultures. When Europeans came to the area, they too created a culture that relied on the river and its surrounding ecosystems. Each of these cultures surely influenced the region's ecosystems. Biologists have tried to determine the character of forest and river ecosystems during the precolonial and early part of the colonial period, but few data are available. However it may be possible to reconstruct those early ecological conditions using residual evidence of earlier ecosystems, for example, pollen deposits or historical documents left by persons who had no formal knowledge of biology.

Europeans first traveled the Hudson River when Dutch businessmen in 1609, desiring to compete with Portugal in the spice trade, sought a northern route to the Orient. They engaged one of the many available English navigators, Henry Hudson, who hired a crew of Dutch and English sailors. On his third voyage Hudson, his way blocked by sea ice and facing a threatened mutiny by his crew, violated his contract with his employers and turned due west. He avoided fur trading European settlements in New France on the north as well as the religious English colonies in New England and Virginia. In September he entered the river that came to bear his name and sailed as far north as then navigable to just south of present-day Albany. His crew's first impressions of the Hudson River, noted in the ship's log, were of the rich forests and likely productivity of the land (Juet 1610, Sept. 2nd).

FUR TRADE

Other Dutch explorers, Hendrick Christiaensen and Adriaen Block, investigated the lower reaches of the Hudson River from 1610 to 1614 (Lewis 2005). Block established a trading post on the island of Mannahatta (later Manhattan.). His report spurred Dutch businessmen to establish the New Netherlands Company and begin vigorous fur trade in the Hudson River region relying on active collaboration of the Native American population to do the trapping and drying of beaver skins.

Beaver trade in the Hudson River Valley, however, reportedly declined beginning in 1624 (Figler 2009), and was apparently commercially ended in the region by 1640 (Leach 1966). One may ask why fur trade from the Hudson River declined forty years before fur trade from adjacent regions. In those same years around eighty ships filled with pelts returned from New France on the St. Lawrence River (Ray 1978), demonstrating that the European market for beaver fur remained strong during the seventeenth century (Feinstein-Johnson 2009). Fur trade from New England also was active until the end of the century. European settlement of the Hudson River with consequent land clearance started so slowly that that could not have accounted for reduced beaver trade. Two possible explanations for the early decline seem probable. Either beaver trapping declined or beaver populations gave out, or trapping continued unabated but the pelts were sold to the French because the French provided better trade goods (see Bradley 2007, 61).

All trapping of furbearing animals and drying of the pelts was carried on by Native American hunters because Europeans recognized their superior skills as hunters and trappers. This meant that Europeans were entirely dependent on the Native Americans for their success in the fur trade. On the St. Lawrence River, the French respected the Natives and treated them as equals (Fischer 2008). However, from the beginning in the Hudson River, the Dutch and English explorers, and then the fur traders, treated the Natives with distrust and disrespect. Henry Hudson's scribe, Robert Juet, whose journal is the only surviving record of the voyage, noted this distrust even on the first day in the river. "This day many of the people came aboard, some in Mantles of Feathers, and some in Skins of divers sorts of good furres. At night they went on Land againe, so we rode very quiet, *but durst not trust them*" (Juet 1610, Sep 5; emphasis added). Juet repeated the crew's distrust of these new people several times in his journal.

By the second day, conflict erupted and one of their crew was killed by an arrow. Hudson sailed north away from the coastal Lenape Indian tribes. Above the Hudson Highlands he encountered warmer acceptance by the Mahicans, but following further conflict, he hurriedly returned to Europe. In the following years, this early disrespect for, and conflict with, the Hudson River Native Americans became pervasive. Clearly settlers did not understand Native values and customs. They referred to them as "lazy" "sauvages" and "wilder" (wildmen) (Frey 2001; Lewis 2005). This distrust surely was known to the Native fur trappers. Trading with the Europeans may have declined as a result. An alternative explanation for the early decline in Hudson River fur trading might credit the Natives with trapping so efficiently that beaver in the region were effectively trapped out within a couple of decades. Either beaver populations remained high due to reduced taking or trapping exceeded replacement potential and the beaver was locally depleted. Because the Native Americans quickly became addicted to the Europeans' trade items, knives, axes, and beads especially, they likely continued to grudgingly trap to the point of commercial extirpation of the beaver from the Hudson River region.

The biological implications of the decline in beaver trade are profound. The biologist interested in environmental change might be able to examine records of the beaver harvest and infer conditions that could explain change. Actual field surveys of animal abundances did not begin until the nineteenth century, so knowledge of abundances in prior times must be mostly conjectural in lieu of definitive contemporaneous reports of beaver populations. However, given the keystone importance of the beaver in controlling its ecosystem, the biologist might fruitfully examine historical records for indications of other ecological changes that may correlate with beaver abundance. Such changes would be predicted if trapping significantly reduced beaver populations.

The proclivity of the beaver to dam streams and create water impoundments that support extensive wet forest and swamp ecosystems is well known. As beaver populations were reduced, the number of water impoundments would have been reduced. The ecological effects would have cascaded through wetland and nearby dry ecosystems. Forest type should have shifted toward dry habitat communities. Cronon (2003) reported an increase in dry forest species occurred in New England in the late seventeenth and early eighteenth centuries as beaver trapping declined there. The abundance of wet habitat–dependent amphibians and birds, and possibly certain mammals, surely declined in response. Overall ecological productivity as well as species richness probably declined. For the biologist attempting to estimate abundances of species during this early settlement period when direct contemporaneous data are lacking, it may be possible to extrapolate those abundances and changing ecosystems by analysis of the beaver trade.

USE OF FIRE

Humans also effect large-scale change to their surroundings with predictable environmental consequences. An example is use of fire. Prior to arrival of the Europeans the Lenape people in the southern part of the Hudson River Valley used fire to clear patches in the forest for agriculture (Juet 1610, Sep. 2, Oct. 2; Williams 2001; Cronon 2003). In northern areas, where the Mahicans relied primarily on hunting and gathering, fire may have been used only occasionally to flush deer from cover (Cronon 2003). Fire, regularly used, causes profound effects on local ecosystems. Fire-tolerant tree species survive while others are lost. Cronon reported that during colonial times, fire used to open forests in southern New England caused fire-tolerant tree species to be favored. Similar changes in forest types in the lower Hudson River Valley and coastal region may be conjectured. Cronon also reported that the shift in forest type in New England favored large mammals, as well as turkey, grouse, and raptors. Similar changes could have occurred in the Hudson River Valley if forests were modified or reduced.

COLONIAL LAND USE

Beginning with the New Netherlands Company the overarching motivation of the Dutch businessmen in the Hudson River region was commercial exploitation. Commerce in furs soon led to commerce in commodities and eventually to extractive industries.

Following their arrival European settlers did not adopt the Native peoples' hunting and fishing lifestyle. Rather, they imported their familiar European customs including land ownership, land clearing for homestead and pastures, and keeping of livestock. As land became exhausted, they cleared more forest. As livestock increased in number, more forest needed to be cleared for pasturing. Shift in the relative abundance of pollens from forest species to grass species in lake bottom soil cores correlate with the arrival of Europeans in the Hudson region (Peteet, 2010).

VALUE OF FUNCTIONAL ECOSYSTEM ANALYSIS

Historians have documented events and actions. They have given less attention to the patterns of ecological change over time, or to the possible underlying reasons for the changes. We can assume that colonial enterprises and activities had ecological effects; however, the induced changes in the ecosystems have been poorly described in the Hudson River region so that actual abundances of floral and faunal species at earlier times are not well established. Even so, it should be possible to reconstruct estimates of those plant and animal populations by relying on the interplay between human uses of the ecosystem (documented by historians) and the expected ecological effects (researched by biologists). It is expected that such a collaborative approach may establish the character of the ecosystems during presettlement and early colonial periods.

REFERENCES CITED

Bradley, J. W. 2007. *Before Albany: An archaeology of Native-Dutch relations in the Capital Region 1600–1664.* Albany: New York State Museum.

Cronon, W. 2003. *Changes in the land: Indians, colonists, and the ecology of New England.* New York: Farrar, Straus, and Giroux, 46–51, 62, 99–101, 108–26, 130–38, 257.

Feinstein-Johnson, K. A brief history of the beaver trade. University of California at Santa Cruz. Http://people.ucsc.edu/~feirste/furtrade.html; accessed Dec. 22, 2009.

Figler, G. 2009. Environmental impacts of the Hudson Valley fur trade in regard to beavers, Hudson River Valley Institute, Interim Report, Fall 2009, 1–11.

Fischer, D. H. 2008. *Champlain's dream.* New York: Simon and Shuster, 130–134, 154, 238.

Frey, C. H. 2001. For(e)knowledge of youth: Malaeska: The Indian wife of the white hunter. *The ALAN Review* 28(3): 19.

Juet, R. 2006. Juet's journal of Hudson's 1609 voyage, from *Collections of New York Historical Society,* Second Series, 1841. Transcribed by Brea Barthel, Albany: New Netherlands Museum. Http://halfmoon.mus.ny.us/Juets-Journal.pdf. (see July 19, Sept. 2, Oct. 2).

Leach, D. E. 1966. *The northern colonial frontier 1607–1763.* New York: Holt, Rinehart, and Winston, 91, 98.

Lewis, T. 2005. *The Hudson: A history.* New Haven: Yale University Press, 42.

Ray, A. J. 1978. History and archeology of the northern fur trade. *American Antiquity* (Soc. for Amer. Archeology) 43(1): 26–34. Http://www.jstort.org/stable/270628); accessed Dec. 21, 2009.

Williams, G. 2001. References on the American Indian use of fire in ecosystems. Http://www.wildlandfire.com; accessed Dec. 21, 2009.

CHAPTER 2

Linkages between People and Ecosystems

How Did We Get from Separate to Equal?

Stuart Findlay

ABSTRACT

It is well established that humans derive multiple benefits from the ecosystems they inhabit as well as causing a certain degree of alteration, generally but not exclusively damaging natural features and processes. What has not been fully explored is the network of interactions and feedbacks between ecosystem services and impacts and how these interactions affect future human actions and ecosystem attributes. Most attention has been paid to negative effects of human activities on ecosystems (and the Hudson River has certainly not been spared), but this chapter argues that ecosystem attributes have also caused significant changes in human behavior and then these new behaviors cause further changes in ecosystem attributes. Thus, there may often be two-directional influences of humans on ecosystems and ecosystems on humans and an explicit recognition of these feedbacks yields a clearer understanding of how human uses and ecosystem attributes have changed over time. This chapter reviews how human-ecosystem interactions have become explicitly recognized and lays out some examples from the Hudson River. These examples will highlight the long-lasting effects of these feedbacks, demonstrating how their consequences may change over time. Understanding the environmental history of the Hudson Valley requires that we consider how human uses have affected and been affected by the provision of natural resources from the Hudson River ecosystem.

RECOGNIZING FEEDBACK RELATIONS

The science of ecology deals with factors affecting the abundance and distribution of organisms and human society has obviously been a major factor driving increases in some species, declines in others, and greatly changing how species are distributed around the planet. Concurrently, there is no doubt that human societies rely on ecosystems to provide goods and services including timber, water supplies, opportunities for recreation, etc. In hindsight it would seem obvious that natural scientists could not understand the world around them without some consideration of what humans might be doing (or have done) to make the ecosystem appear and function the way it presently does. By the same token, much of human behavior, history, culture is driven by the availability of (or lack of, or conflict over) natural resources including food, fuel, water, minerals, etc. and so historians or sociologists would need to understand how these resources become available. Despite the apparent two-way need for those studying nature to include people as a factor and for those studying people to consider nature, it is only within the past two decades that there has been a widespread and explicit recognition of these feedbacks and approaches for their study. As few as twenty years ago there was a view that one could not study an ecosystem that included a significant human component since the natural functioning would somehow be masked or

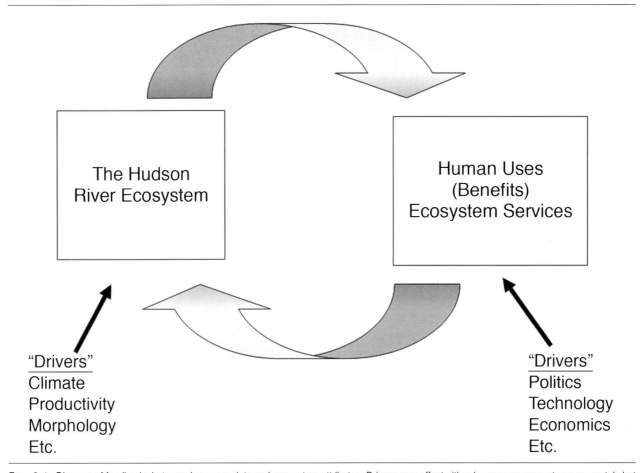

FIG. 2.1. Diagram of feedbacks between human society and ecosystem attributes. Drivers may affect either humans or ecosystems separately but eventually all are connected by the proposed feedbacks.

fundamentally altered (Cronon 1993). Change has been swift, with an almost complete acceptance of the idea that human effects do not remove an ecosystem from the pool of systems worthy of study. One may argue persuasively that there is no ecosystem on the planet that has not been altered to some degree by one of many far-reaching human activities (Likens 1991). In fact, there are now two Long-Term Ecological Research sites located in cities of the United States (Baltimore and Phoenix) and a scholarly journal entitled *Urban Ecology*. This theme of bidirectional human-ecosystem interaction represented as a feedback loop (Fig. 2.1) is proposed to be helpful in drawing together the social science and natural science perspectives on particular subjects represented in this volume. Such a diagram reminds and encourages all researchers to consider how their findings have been altered by these feedbacks. I would argue that these feedbacks are essential to understanding the human-ecosystem interactions we observe around us today and the current state of the ecosystem is partially due to prior patterns in human-ecosystem interaction.

I will use two Hudson River Valley examples to illustrate how these ecosystem-human feedbacks have, and continue to, operate locally. Firstly, the aquatic biology, natural history, and particularly the ecology of native fishes in the Hudson have shaped the way people in the Hudson River Valley have used these resources over the past four hundred years (derived from Daniels et al., ch. 4, this volume). In the first two hundred years, as during the preceding twelve millennia, fishing was primarily for consumption within the community and so impacts on the fisheries resources would be dispersed and relatively minor due to the low human population. With the Industrial Revolution came the need for a vastly increased food supply delivered to the growing urban centers. Shad, with their timed migrations were relatively easy to harvest and were the target species for the first fifty-plus years of commercial fishing (late 1800s). As Daniels et al. recount, eventual overfishing, possibly compounded by habitat destruction and water quality problems, led to the first collapse of the shad population and a shift to the next most feasible species—sturgeon. The locations for best results and necessary gear were quite different and so the characteristics of the resource base caused changes in human behavior. Clearly with this example, humans have altered the nature of the ecosystem by changing the abundance of important top consumers. Daniels et al. (ch. 4, this volume) argue that the most recent change is a shift to management intended to increase the abundance and size of piscivorous sportfish, which may well lead to altered abundances of other species. Concurrently, the nature of the available resources has affected how human society meets food demands or where and how they target harvest efforts. Thus, the two-way interaction between ecological attributes and human society has fed back and forth to change the components of the Hudson River ecosystem as well as the functioning of human society.

Human occupation and use of the Hudson's shorelines is another case illustrating two-way effects. Clearly, the suitability of the Hudson as a transportation corridor has led to a concentration of human development along its shores. In addition to communities and facilities intended to service travelers and the transportation infrastructure, vast ice-harvesting operations (see Harris and Pickman, ch. 14, this volume) and brick factories were concentrated along the shore. The proximity to raw materials (ice and clay) and easy transportation drew these operations to the near-river areas and so again the nature of the resource influenced human behavior. While the nature of the Hudson has altered how people occupy and use the near-shore area, human modifications of the Hudson have been occurring for centuries, with large and persistent effects. Humans have stabilized the shoreline and altered near-shore habitats by dredging and filling to improve the navigation channel, vastly changing the flow regimen and abundance of many habitat types. These examples or numerous others illustrate the pervasive nature of the two-way interactions between the Hudson River and the people in the valley. Understanding the history of natural resource use or the status of the ecosystem will be less complete than is possible without a broadening of views about how ecosystems shape people's lives with long-lasting effects.

There are two aspects of the human-ecosystem feedback loop relevant to consideration of environmental history: (1) How did this blending of social and natural sciences develop? and (2) How might this blending improve conservation of ecosystems while allowing human use of natural resources? In perhaps an overly pessimistic view of how this merger has progressed, I would argue that the realization that all parts of the globe have somehow been altered by human changes in global biogeochemical cycles (atmospheric CO_2, widespread nitrogen release), species introductions/extirpations, and demand for resources, which have forced ecologists to think more directly about human actions. From a more intellectually satisfying perspective the merger has progressed because practitioners of both social and natural science have realized they could expand their domain of inquiry by applying new tools and concepts that may ultimately allow a seamless view of how humans and ecosystems interact. McDonnell and Pickett (1993) suggest (somewhat facetiously) that human effects on ecosystems may be divided into "The good, the bad and the subtle." "The good" are far too often overlooked and underappreciated and would include broad, far-reaching, and long-lasting management actions such as passage of the Clean Water Act and the Endangered Species Act and creating a system of National Parks, National Forests, Wild and Scenic Rivers, and a legal standing for citizens to advocate in favor of natural resource protection. In this volume on the environmental history of the Hudson River Valley there are many good examples of human actions with positive effects on the ecosystems as they exist today that increase the likelihood these ecosystems will help sustain future generations. In fact, the landmark court case centered in the Hudson River Valley (see Suszkowski and D'Elia 2006) and described by Butzel (ch. 19, this volume) established a voice for the public in decisions about natural resource management.

Unfortunately, there are also many examples of "the bad," which can be shortsighted (although frequently unknowing) human actions that have had large, long-lasting negative consequences. Extinction of species, introduction of exotic species (see Teale, ch. 13, this volume), and discharge of contaminants (PCBs, cadmium) (Levinton, ch. 16, this volume; Levinton and Waldman 2006, chs. 23–28) must all lie near the top of the list. The subtle effects may include historical legacies, displaced effects, and indirect effects (see McDonnell and Pickett 1993 for details). Combined storm and sanitary sewers serve as a good example of an historical legacy effect. When originally designed to remove animal waste from city streets (see Tonjes et al., ch. 15, this volume) they were clearly a step forward in waste handling and beneficial for water quality. Presently, they primarily act to overwhelm WWTP and are detrimental since the problem they were intended to address is no longer present. Therefore, their legacy is a negative and difficult management problem, although they undoubtedly improved living conditions when first constructed. Cumulative effects might also be considered among subtle effects. The first power plant on the Hudson with once-through cooling (Danskammer, operational 1951—water withdrawal <5 percent of annual freshwater flow) probably had a tiny impact on larval fish survival, but together with five other large plants (with a cumulative demand of as much as one-third of freshwater flow) has been found to have a significant negative effect (Young and Dey, ch. 18, this volume). Similarly, a small amount of impervious surface in a drainage basin probably has little effect on groundwater recharge or delivery of pollutants to surface waters, but once a threshold has been passed even small additions of impervious surface can result in disproportionate effects (Cunningham et al. 2009).

In contrast to the fairly well-documented record of human effects on ecosystems, the record of how ecosystem attributes affect human society and individual behavior is less robust. There have long been beliefs that our surroundings influence our individual behavior (consider the underlying themes in Thomas Mann's *Death in Venice*; Joseph Conrad's *Heart of Darkness*) and, perhaps the most egregious in early writings, the view that "savages" are less than human due to ther rustic way of life dictated by their untamed surroundings. A controversial debate exists over which came first, cities or agriculture (see Ali 2009, ch. 3). Some argue that the suitability of an area for high-yielding agriculture allows the higher population density of cities. The converse argument is that the need to live in cities drove the development of agriculture. In either case, the bidirectional influences of people on place and place on

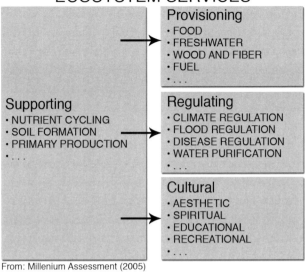

FIG. 2.2. Representation of classes of Ecosystem Services as used in the Millennium Assessment. The four types of Services (described in text) are subject to differing schemes for valuation and provide fundamentally different benefits to human society.

people are recognized and few would suggest we can understand early stages of civilization without considering these feedbacks. Furthermore, many of the early environmental protections were put in place following the recognition that environmental conditions could lead to greater risk of human disease transmission, so there was a clear appreciation that humans were not isolated from or somehow "above" their surroundings.

DEVELOPMENT OF HUMAN-ECOSYSTEM FEEDBACKS

In the last few years there has been a formalization of the various ways ecosystems provide services to human society. Development of a solid conceptual basis for ecosystem services grew out of a desire to assess the status of ecosystems around the world *including* some measures of how well ecosystems were supporting human demands and whether these might be sustainable. The Millennium Assessment required standardized concepts, terminology, and measures of how humans relied on ecosystems, and this framework has come into widespread use (Fig. 2.2) (MEA 2005).

This scheme describes four broad classes of ecosystem services: Supporting, Provisioning, Regulating, and Cultural. The first, Supporting, might be loosely considered to be the capital or infrastructure the ecosystem requires for maintenance. For instance, allowing soil formation and suitable nutrient cycling pathways is essential to the future of an ecosystem, and any human action that impedes or degrades these components will ultimately cause degradation of the ecosystem. Biodiversity may be considered a Supporting service since there is substantial (although not universal) evidence that more diverse systems are more productive and resilient (see Hooper et al. 2005). Provisioning services include the fairly well-known and traditional extraction of natural resources (fish, fur, fuel), and these are commonly perceived as valuable and can be readily converted to economic terms if necessary.

Regulating services include the capacity of wetlands to store flood waters or forested catchments to remove pollutants derived from atmospheric deposition. For example, forested catchments currently remove well over 50 percent of nitrate delivered from above (Lovett et al. 2002) and this service essentially displaces the need for any engineering approach to ameliorate nitrogen loading. Cultural services include the value placed on natural entities by various religions, as well as opportunities for education and recreation. Several of the services in this category are quite difficult to convert to economic terms yet are sacrosanct to many groups, and their continued availability may be legally required by treaty.

The ecosystem services framework has been very helpful in showing individuals, managers and policymakers how functioning ecosystems are essential to viable human societies, and there has been tremendous progress in recognizing, quantifying, and applying "value" to a wide array of ecosystem services (see Bennett et al. 2009 for brief review). However, there is not yet a robust scheme to translate this knowledge into management action. Several problems are evident, perhaps the most fundamental being that any ecosystem provides multiple services that may be negatively correlated, so maximizing one service may be associated with decline in another. There is not a generally accepted method for resolving this issue of trade-offs among ecosystem services. Layered on this are differences in perceived value among services and stakeholder groups. For instance, it may be easy to justify actions that protect human health even though there is some cost to another species or some biogeochemical process. While it is understandable that value is in the eye of the beholder the risk is a ratcheting down of some of the supporting services that allow the ecosystem to persist but perhaps are less directly associated with immediate benefits to human well-being. At present we do not have an objective way to ensure that supporting services or those harder to value in economic terms receive the requisite attention in management plans. Quite recently there have been proposals for a circular feedback between ecosystem service valuation and management tools (institutions, incentives, etc.) to sustain those services (Daily et al. 2009). I feel there is a reasonable expectation that we will eventually have mechanisms tightly connected to tracking and maintaining (and improving) ecosystem services for all segments of the global population.

What might be done to help sustain and even expand delivery and management of ecosystem services with the ultimate goal of broadening access? I believe that a more explicit recognition of the feedbacks between people and the ecosystems around them would increase public awareness and support for novel management approaches. Explicit description of any feedbacks would generate a much greater appreciation of the two-way effects of people on ecosystems and of ecosystems on people. These interactions can cause long-lasting effects such that knowledge of history is essential for understanding how societies and ecosystems came to function as they do today. Conversely, if people realize how tightly dependent they are on services from their ecosystems they should (must?) have a greater interest in understanding how those ecosystems work and what is needed for their long-term viability. The concept of ecosystem services can therefore act as the icebreaker for realizations of how people and ecosystems depend on and influence each other.

REFERENCES CITED

Ali, S. 2009. *Treasures of the Earth*. New Haven: Yale University Press.

Bennett, E. M., G. D. Peterson, and L. J. Gordon. 2009. Understanding relationships among multiple ecosystem services. *Ecology Letters* 12: 1394–1404.

Cronon, W. J. 1993. Foreword: The turn toward history. In *Humans as components of ecosystems,* ed. M. J. McDonnell and S. T. A. Pickett, vii–x. New York: Springer-Verlag.

Cunningham, M. A., C. M. O'Reilly, K. M. Menking et al. 2009. The suburban stream syndrome: Evaluating land use and stream impairments in the suburbs. *Physical Geography* 30: 269–84.

Daily, G. C., S. Polasky, J. Goldstein, P. M. Kareiva, H. A. Mooney et al. 2009. Ecosystem services in decision making: Time to deliver. *Frontiers in Ecology and the Environment* 7: 21–28.

Hooper, D. U., F. S. Chapin, J. J. Ewel et al. 2005. Effects of biodiversity on ecosystem functioning: A consensus of current knowledge. *Ecological Monographs* 75: 3–35.

Levinton, J. S., and J. R. Waldman. 2006. *The Hudson River Estuary.* New York: Cambridge University Press.

Likens, G. E. 1991. Human-accelerated environmental change. *BioScience* 41: 130.

Lovett, G. M., K. C. Weathers, and M. A. Arthur. 2002. Control of nitrogen loss from forested watersheds by soil carbon: Nitrogen ratio and tree species composition. *Ecosystems* 5: 712–18.

McDonnell, M. J. and S. T. A. Pickett. 1993. Introduction: Scope and need for an ecology of subtle human effects and populated areas. In *Humans as components of ecosystems,* ed. M. J. McDonnell and S. T. A. Pickett, 1–5. New York: Springer-Verlag.

Millennium Ecosystem Assessment. 2005. www.millenniumassessment.org/en/index.aspx.

Suszkowski, D. J., and C. F. D'Elia. 2006. The history and science of managing the Hudson River. In *The Hudson River Estuary,* ed. J. S. Levinton and J. R. Waldman, 313–34. New York: Cambridge University Press.

CHAPTER 3

SYMBIOSES BETWEEN BIOLOGISTS AND SOCIAL SCIENTISTS

Lucille Lewis Johnson

ABSTRACT

For at least one million years, humans and our immediate ancestors have interacted with the environments within which they operate through the use of technology, which can be seen as humanity's species-specific adaptive mechanism. In order to survive and thrive, humans create technologies for living, which include tools, the rules for making and using them, and the social and political systems within which they are enmeshed. Thus, much of our adaptation as individuals is to the sociotechnical world we have created for ourselves. On the other hand, for individuals and societies to survive, we also have to adapt to the natural environments within which we find ourselves. While our technologies and the social mystique with which we surround them often make us feel independent from nature, the choices we make and patterns we create will not succeed if we too blatantly ignore the constraints imposed by the environments in which we find ourselves. Feedbacks between cultural and natural processes can be seen with particular clarity in the Hudson Valley, where, for millennia and particularly for the last four hundred years, humans have intentionally and accidentally created changes in the watershed that have influenced the ecology of the valley, necessitating additional human interventions which attempt to correct problems created by the earlier cultural interventions. To understand the environmental history of the valley, it is necessary to study the natural environment; the geographical facts and climatic conditions that have shaped the valley and its species, and the historical factors that have led humans to act in certain ways toward the valley; and thus change the valley in ways that are often detrimental and require further intervention to correct.

AT A CONFERENCE entitled *State of the Hudson* in the fall of 2009, it was clear that scholars studying the river and its valley as it is today clearly recognized the interrelations and feedback between natural and cultural processes. For example, Burns (2009) discussed ways in which human land use and disturbance can completely overwhelm climate effects, which were exemplified in Wall's paper (2009) showing that the differing amounts of sediment in the Mohawk drainage in comparison to the upper Hudson were due to the heavy agricultural use of the Mohawk Valley as opposed to the forested upper Hudson. So, the history of land use in the two drainages determined the sediment flows, which then influenced growth conditions for aquatic plants and animals in the estuary.

These papers and others framed the Hudson River Environmental Society's quadricentennial conference and this book, with a series of questions: How have these intersecting processes operated in the past? How can we enhance our historical understanding by combining scientific and cultural/historical approaches? How can this enhanced historical understanding help us move forward into the future in ways that will preserve both natural and human health in the Hudson Valley?

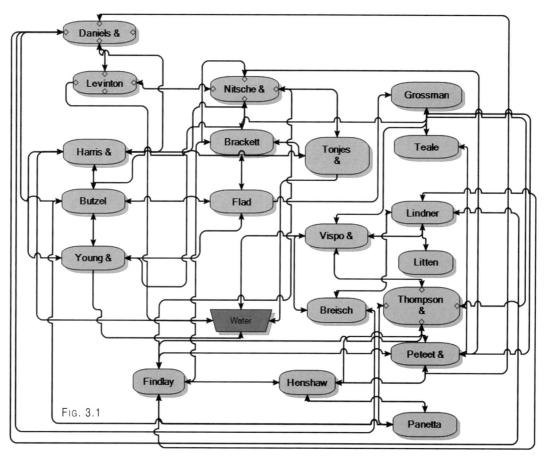
Fig. 3.1

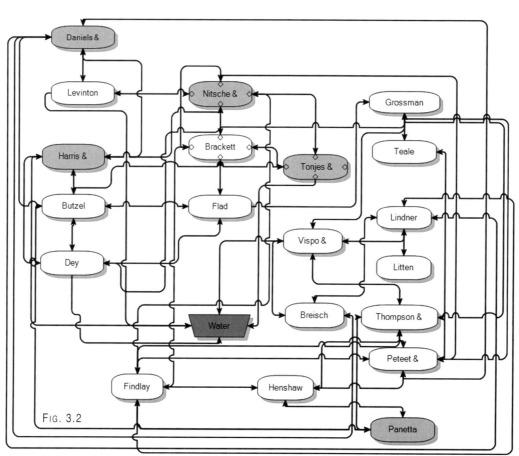
Fig. 3.2

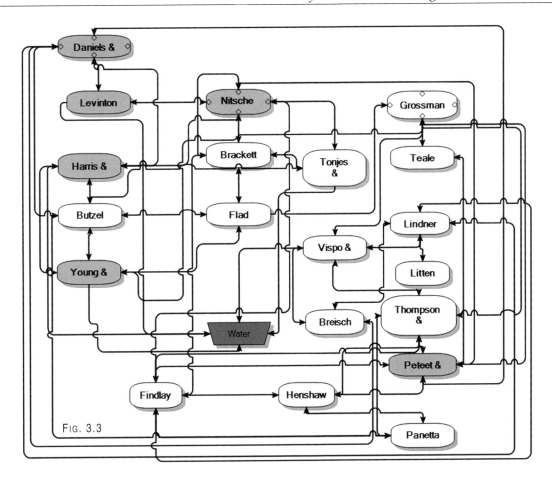

FIG. 3.3

The papers included in this volume have been arranged by the editor in one logical linear structure, but the interactions and intersections among these papers (Fig. 3.1) are much richer than a linear order might suggest. In this chapter I will attempt to provide a guide to some of the interactions and feedbacks between these papers and the phenomena they study as I see them from an anthropological perspective. While I am sure there are additional relationships that I have not considered, I have found seven rubrics under which to look at interconnections: changes in the river's shape and their effects, changes in the river itself, changes in the valley lowlands and margins, changes in land use and their effects, changes in the vegetation of the valley, changes in human use of the shorelines, changes in valley transportation.

The book and my discussion begin by looking at how humans have purposely changed the shape of the river during the historic period in order to enhance transportation (Fig. 3.2; see Panetta, ch. 17; all references to chapters are to those in this volume). Nitsche et al. (ch. 6), consider the historic reasons why people purposely changed the river's shape, how they went about doing it, and then analyze the intended and unintended results and consider what can be done to mitigate the latter. Other changes in the river's course were caused by such activities as ice harvesting and clay and sand mining for bricks (Harris and Pickman, ch. 14), while in the lower Hudson the river was both dredged, to improve shipping, and narrowed, to create more land for Manhattan (Tonjes et al., ch. 15). All of these changes affected the fish and the fisheries, (Daniels, Schmidt, and Limburg, ch. 4).

A major change in the shape of the river was also created by the railroads (Panetta, ch. 17), whose tracks now line both sides of the river. The railroad tracks and trellises separated bays from the river, creating swamps and backwaters where they had not been before, and this affected the fish spawning grounds (Daniels, Schmidt, and Limburg, ch. 4). The railroad tracks also had a number of major effects on the human occupation of the valley, both bringing people and goods into and out of the valley and separating river towns from their riverfronts,

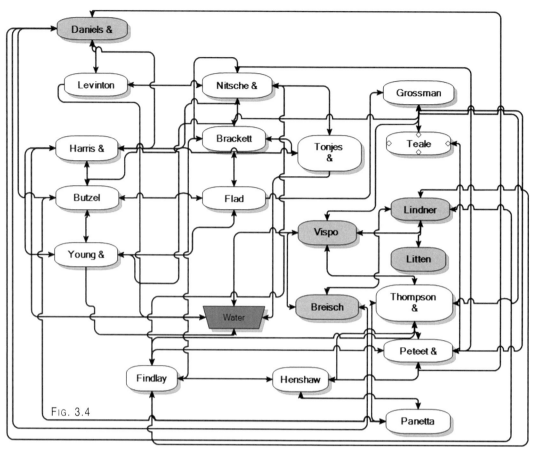

Fig. 3.4

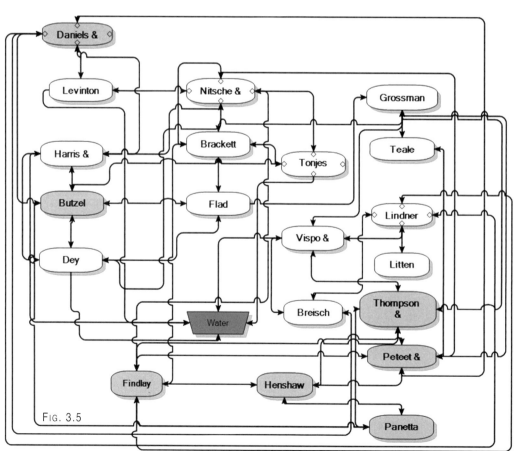

Fig. 3.5

turning what we, today, think of as prime residential land on the river into industrial zones, and this also exacerbated the pollution in the river (Tonjes et al., ch. 15).

In focusing on the river itself (Fig. 3.3), Nitsche et al.(ch. 6) examine the changes in bathymetry due to dredging and filling of the river, Daniels, Schmidt, and Limburg (ch. 4) look at the overall health of and changes in the fishery, while Levinton (ch. 16) looks at the effect of one particular pollution event and its cleanup to understand how organisms reacted to human degradation and subsequent cleanup of the environment. The issue of river health is also treated in Peteet et al.'s chapter (ch. 9) on river marshes. Both Levinton and Peteet et al. consider not only the human attempts to clean up humanly degraded environments, but also the natural processes by which the environment tries to clean itself up. The ice harvesting industry (Harris and Pickman, ch. 14) used the river water itself as its resource, cutting it up and taking it away during the dormant period of the year for biological organisms, while power plants (Young and Dey, ch. 18) use the river water for cooling during all seasons of the year, seriously affecting the marine biota during their peak activity.

While this first set of papers focuses on the river itself, the valley includes the watershed as a whole, including its valley bottoms and uplands (Fig. 3.4). In order to understand how the valley has been used and changed by its human residents over the last twelve thousand years, we turn to archaeology (Lindner, ch. 7) and history (Litten, ch. 11). These overviews set the stage for a detailed study of changes in a particular region over a three century time span (Vispo and Knab-Vispo, ch. 12), considering again how land forms and water influence people, and human activities change land forms and water. A detailed study of herpetofauna (Breisch, ch. 5) and how they have been affected by human activities compliments Daniels et al.'s (ch. 4) study of fish, and further enhances our understanding of human action and environmental resilience.

Looking at changes in the land and its uses and how these changes affect the waterways, the waters and the biota forms another major focus of the papers included in this volume (Fig. 3.5). Henshaw (ch. 1) looks particularly at beavers and pigs, how the hunting out of the former changed the quantity and distribution of wetlands in the valley and how the latter modified the forests. Findlay (ch. 2) indicates how the building of dams by humans for hydropower reversed the earlier decline in wetlands that was partially caused by the extirpation of the beaver noted by Henshaw. Additional factors of land use include the farm ponds (one of which, now swamp, sits outside my window as I write) as important humanly created wetlands and, conversely, the filling of wetlands during construction as important human destruction of wetlands. Thompson and Huth (ch. 10) show how the vegetation of the Shawangunk Ridge, which forms the western margin of the watershed in the mid-Hudson region, was decimated by human activity and has recovered, but not to the state it was in before. Many of the activities that they detail resulted in increased runoff into the streams and other wetlands of the valley. Some of these are detailed by Peteet et al. (ch. 9), who, through the examination of deep cores taken from uplands and coastal marshes, both in the mid and lower Hudson Valley, show the effects of changes in land use on riverine deposits, which then, of course, affect the fish and fisheries (Daniels et al., ch. 4). Changing valley transportation systems have also had a major effect on waterflow and waterways (Panetta, ch. 17). Finally, one major aspect of the Storm King Con Ed plant was its potential impact on the striped bass (Butzel, ch. 19).

The landscape of the valley has also been affected by vegetation change, as Grossman (ch. 8) and Teale (ch. 13) discuss in two very different historical contexts (Fig. 3.6). Grossman looks at evidence for native/Dutch interaction in the Colonial Period New Amsterdam, particularly as it concerned food and medicinal uses for both native and European plants, while Teale looks at plants which were introduced into the valley in the nineteenth century to enhance the beauty of the valley landscapes as perceived by American Romantics, and which, in many cases have had devastating effects on native vegetation. Flad (ch. 20) and Brackett (ch. 21) examine other effects of this romantic vision in terms of the creation of a Hudson River aesthetic and the effects of this aesthetic on subsequent human interactions with the valley. Peteet et al. (ch. 9) also consider introduced and now invasive species such as reeds, while Thompson and Huth (ch. 10) look at the changing vegetation of the Shawangunk Ridge as people cut the trees for various industrial uses,

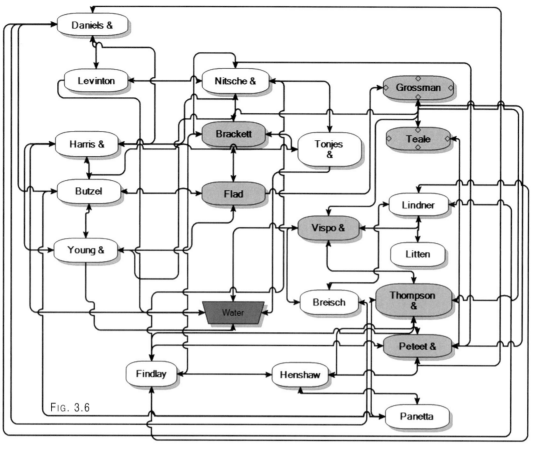

FIG. 3.6

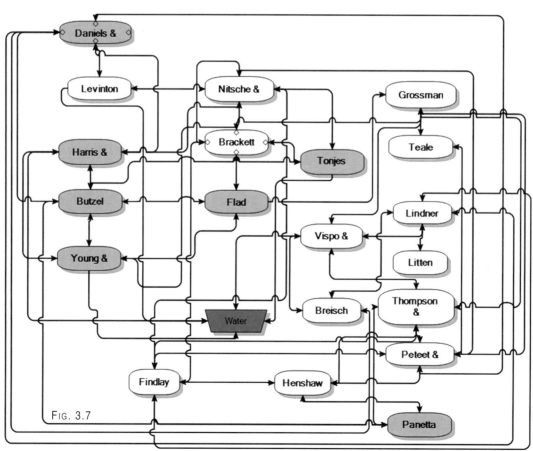

FIG. 3.7

burned the trees to enhance blueberry meadows, and then, as human uses of the ridge changed, allowed the vegetation to grow back as it would, though meadows are still subject to controlled burns to retain the historic farm landscape in the lowland areas of the preserve. In an interesting twist, Vispo and Knab-Vispo (ch. 12) examine how the humanly created landscape of Columbia County mimics, in many ways, the natural landscape it replaces.

While there have been dramatic changes in the river's shape due to transportation concerns, and to valley lands, both above the ground as plant communities have changed and changed again, and structural, as mining has resculpted much of the valley, human uses of the shoreline have effected even greater changes, as the next set of papers demonstrates (Fig. 3.7). Most of these changes are strongly related to the pull of the city at the mouth of the river. Harris and Pickman (ch. 14) discuss the major social and environmental effects of the ice harvesting industry, which, in the mid to late nineteenth century developed to provide city residents with the refrigeration necessary for the population to grow and thrive. At the same time, the Great Fire of 1835 in the city led to the blossoming of the Hudson Valley brick-making industry, which completely altered the shape of the estuary south of the canalization zone as well as having major social and population effects in the valley. In the same period, the railroads along the Hudson were being built, creating more environmental change and allowing more city people to visit the valley and enjoy their romantic visions, as will be mentioned later by Panetta (ch. 17), and Flad (ch. 20). Tonjes et al. (ch. 15) ask, What effect did the explosive growth of the city, which led to such huge changes in the valley, have on its own shores and water? Looking at the long spiral of decline in water quality and New Yorkers' relationship to their harbor and river, they ask whether the desired climb back can be fully achieved given political realities.

Whereas ice harvesting removed water from the river, brick making used water for processing, and the lower Hudson was sullied by human waste, another city-driven industry, the power industry, has had the most dramatic effect on the upriver fauna of any of these industries. Young and Dey (ch. 18) discuss the major research that has occurred to understand the effects that the use of water for power generation has had on river life and Butzel (ch. 19) takes up the topic in discussing the creation of environmental legislation designed to mitigate those effects over the past forty years. Brick making, railroad construction, sewage and power plants have all had a major effect on Hudson River fisheries (Daniels, Schmidt, and Limburg, ch. 4).

Turning to the final set of papers looking at transportation and the valley (Fig. 3.8), Roger Panetta (ch. 17) considers transportation within the watershed, showing that major routes, until the early twentieth century, went north and south up the valley, with only a few east-west routes made possible by canoes (Lindner, ch. 7) and ferries. The modifications of the river's course considered by Nitsche et al. (ch. 6) were, of course, directly motivated by transportation, as continued dredging of the river, and winter ice clearing continue to be. For much of its course, the Hudson has lowlands on the east shore and highlands on the west, which resulted in great differences in human use until tunnels and bridges brought the two sides together—to an extent. Harvey Flad (ch. 20) and Geoffrey Brackett (ch. 21) show how historical perspectives on the valley changed over time, but also how these perspectives shaped the ways in which people interacted, and continue to interact, with the valley and the "other people"—the plants and animals—who also inhabit it.

This examination of the numerous ways in which the chapters in this volume interact with and complement each other to enrich our understanding of human/environment interactions over the past 12,000 years has in no way exhausted the richness of the data. For example, one might focus on the water of the river itself, how the water flows from the tributaries (Fig. 3.9) (Vispo and Knab-Vispo, ch. 12) through riverside marshes (Peteet et al., ch. 9) and through ground water sources, and how it has been polluted by sewage (Tonjes et al., ch. 15) and by heat (Young and Dey, ch. 18).

As obligately technological creatures, we must and do change our world, but should we, and how should we, and how far should we go? In order to chart our future course we must understand what we have done in the past and what the effects of these actions have been: we need history and

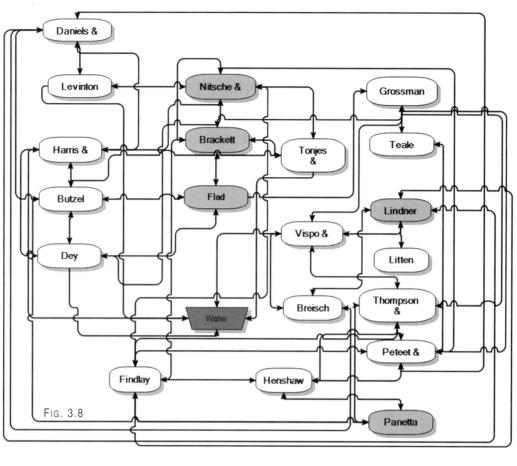

Fig. 3.8

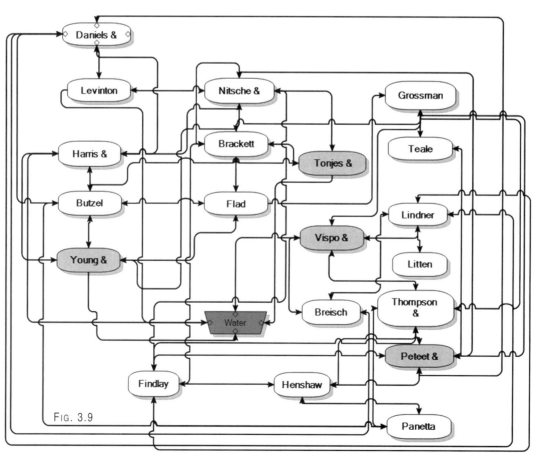

Fig. 3.9

science, to understand the results of our past actions and to chart a future based on strong scientific evidence and historical understanding. As Gerda Lerner has said:

> We can learn from history how past generations thought and acted, how they responded to the demands of their time and how they solved their problems. We can learn by analogy, not by example, for our circumstances will always be different than theirs were. The main thing history can teach us is that human actions have consequences and that certain choices, once made, cannot be undone. They foreclose the possibility of making other choices and thus they determine future events. (http://womenshistory.about.com/cs/quotes/a/gerda_lerner.htm; accessed March 11, 10)

I hope that as you read this book, you will encounter additional connections that will enrich both your understanding of the valley and your future research.

REFERENCES CITED

Burns, D. 2009 Trends in ecologically significant flows. Paper presented at the State of the Hudson Watershed conference, September 29, Hyde Park, New York.

Lerner, G. Womenshistory. About.com Gerda Lerner Quotes http://womenshistory.about.com/cs/quotes/a/gerda_lerner.htm; accessed March 11, 10.

Wall, G. 2009 Trends in Mohawk River hydrology and implications for sediment delivery to the tidal Hudson River. Paper presented at the State of the Hudson Watershed conference, September 29, Hyde Park, New York.

PART II

RIVER OF RESOURCES

Robert E. Henshaw

> We . . . saw many salmons, and mullets, and rays very great. (2 Sep.)
>
> [Our crew] went up into the woods (on the east side) and saw a great store of very goodly oaks, and some currents . . . which were sweet and good. (5 Sep.)
>
> The land . . . were as pleasant with grass and flowers, and goodly trees, as they had seen. (6 Sep.)
>
> We went on land on the west side of the river, and found good ground for corn, and other garden herbs, with great store of goodly oaks, and walnut trees, and chestnut trees, yew trees, and trees of sweet wood in great abundance, and a great store of slate for houses, and other good stones. (25 Sep.)
> —Robert Juet, *Journal of Henry Hudson's historiographer,* 1609

PRISTINE WILDERNESS

The first European explorers to enter the Hudson River commented on the abundance of natural resources. Sponsored by businessmen, European voyages around the world were strictly utilitarian. By the sixteenth century Dutch merchants traded in salt and slaves from South America, sugar from the Caribbean, and fish from the Outer Banks of North America. They desperately wanted in on the immense profits possible with spices from the Orient. They hired Henry Hudson to find a direct northern route. When sea ice blocked his way, and his crew threatened mutiny, Hudson diverted to the New World to seek a western route. Upon entering the great bay, Hudson's historiographer, Robert Juet, recorded their immediate awareness that they were in a great river and not a possible sea route to the Orient, and that they set about documenting the immense resources to be exploited there. Similar accounts by Dutch and English explorers also extolled the majestic forests, extensive grasslands and marshlands, and abundant aquatic life in the Hudson River and terrestrial fauna in the neighboring areas (Sanderson and Brown 2007).

Thirteen millennia of occupation by Native Americans seem to have left the aquatic and terrestrial ecosystems throughout the Hudson River region rich and viable. In the southern part the agrarian Lenape harvested the abundant shellfish, birds, and mammals, and also cleared patches of forest to accommodate their crops of corn, beans, and squash (Cronon 2003). North of the Hudson River Gorge (see "An Abbreviated Geography," in this volume), the Mohawk relied entirely on hunting, fishing, and gathering. No Native American group concentrated on the Adirondack Mountains because of the low biological productivity there. Throughout the entire region the Native American custom of taking only the number of game that they wished to consume at the time led to their living entirely sustainably. Hudson's crew commented on the abundance of game even near Native villages, confirming that game was not overexploited even there. The high abundance of beaver was readily apparent and noted. By 1611 the Dutch East Indies Company and its immediate successor, the New Netherlands Company, began active trade for beaver and other furbearing species (Lewis 2005).

Juet described, and modern botanists have confirmed, that the entire northeastern region was densely forested except for the patches in the southern coastal regions cleared by Native Americans for agriculture. One can surmise that those forests contained a high percentage of wet forest communities, given the high abundance of beaver. As the Dutch West Indies Company established patroonships beginning in 1624, homesteads became more numerous and pasturelands and orchards increasingly replaced the forests. In the following century commercial lumbering, tanning bark harvest, and coke production further reduced those dense forests until the entire Hudson River Valley was virtually deforested by around 1800 (Gurth 1935; NYS Department of Environmental Conservation 2009).

Direct contemporary quantitative assessments of early abundance of terrestrial animal species are virtually lacking. Reconstruction of the abundance of the pre-European ecosystems must be by extrapolation from human activities established by historians to have occurred at known times, as discussed in the preceding chapters. Homesteading, keeping of livestock, the freeing of pigs to forage in salt marshes or the forests, and the later extractive industries all took their toll. By the end of the seventeenth century game was no longer the principal source of protein in the colonists' diet (Cronon 2003). This may have been due to the increased numbers of settlers hunting beyond sustainable levels, or to the increasing numbers of livestock supplying consumptive needs.

Early colonists also exploited the fish and shellfish in the river and its tributaries. Commercial fishing began immediately, both for local consumption and later for shipment to the sponsoring Dutch merchants. It later declined as channelizing of the river (Nitsche et al., ch. 6) removed spawning and nursery habitat and the increasing population and nineteenth-century industries sullied the river with wastes. Commercial fishing for striped bass ended, due to PCB contamination from shoreline industries, and for shad in 2010, due probably to a combination of pelagic and in-river overharvest (Streeter 2009; Mylod and Nack 2010).

HUMAN ACTIVITIES THAT AFFECTED THE RIVER

In the present volume, the authors detail many ecological changes due to human activities in the Hudson River. Modifications in the shorelines are documented by Nitsche in chapter 6. In chapter 7, Lindner discusses archeological evidence from Native American occupation for thousands of years. Peteet et al., in chapter 9, examine pollen counts in soil cores and confirm the drastic reduction in forest density as Europeans settled in. Thompson and Huth, in chapter 10, demonstrate further reduction in forests as early settlement increased. The changing forest density may be correlated to evolving agricultural practices throughout the Hudson River Valley as described by Litten in chapter 11. Uses of plants in the growing village of New Amsterdam on Manahatta Island is described by Grossman in chapter 8 and throughout the Hudson River Valley by Teale, in chapter 13.

RESPONSES TO HUMAN ACTIVITIES

The Hudson River and its surroundings are remarkably rich and resilient. The Hudson provided resources to the human occupants for fourteen millennia without known deterioration. During the last four hundred years the increasing presence of humans and their activities has greatly affected the river. Hudson River fauna responded to the anthropogenic changes in forest cover and river channel. Daniels et al., in chapter 4, detail the declines in productivity of fish populations during four hundred years of changes in fishing pressure as human populations grew and fishing methods changed. Breisch, in chapter 5, discusses likely changes in amphibian and reptile populations as these species faced on-land pressures similar to those faced by fish populations. Conspicuously absent in this volume are similar analyses of responses of other flora and fauna, especially birds and mammals, in the face of increasing human settlement and development. Likewise absent is a thorough consideration of the unusually high biological productivity of the aquatic food web due to its autotrophic and heterotrophic foundations (see "An Abbreviated Geography" in this volume). Papers on these topics were not offered in response to the call for papers. We hope the present chapters will stimulate the reader to construct similar analyses of the long-term interplay of human activities and other components of the aquatic and terrestrial flora and fauna. The present chapters in Part II focus on the ecological component of the feedback relationship between ecosystems and human activities. Today a thorough analysis of any of the physical and biological resources provided by the river demands careful attention to the reciprocal feedback effects of humans on their resources and of those ecological resources on the human users. This volume fosters that consideration.

REFERENCES CITED

Cronon, W. 2003. *Changes in the land: Indians, colonists, and the ecology of New England.* New Haven: Yale University Press, 51.

Division of Lands and Forests. History of state forest program. http://www.dec.ny.gov/lands/4982.html; accessed Dec. 15, 2009.

Gurth, W. 1935. *A history of half a century of management of the natural resources of the Empire State, 1885–1935.* Albany: Conservation Department and New York State College of Forestry, 1–10.

Juet, R. 2006. *Juet's journal of Hudson's 1609 voyage*, from *Collections of New York Historical Society,* Second series, 1841. Transcribed by Brea Barthel, Albany: New Netherlands Museum. http://www.halfmoonreplica.org/Juets-modified.pdf, 2 Sep., 5 Sep., 6 Sep., 25 Sep.

Lewis, T. 2005. *The Hudson: a history.* New Haven: Yale University Press, 42–53.

Mylod, J., and S. Nack. 2010. A net gain for the Hudson? In New York State Department of Environmental Conservation. Albany *Times Union,* Jan. 4, 2010.

Sanderson, E. W. and M. Brown. 2007. Mannahatta: An ecological first look at the Manhattan landscape prior to Henry Hudson. *Northeastern Naturalist* 14(4): 545–70.

Streeter, R. 2009. Shad fishing in doubt. Albany *Times Union,* Sep. 24, 2009.

CHAPTER 4

HUDSON RIVER FISHERIES

Once Robust, Now Reduced

Robert A. Daniels, Robert E. Schmidt, and Karin E. Limburg

ABSTRACT

The Europeans that entered the Hudson River in 1609 and the years immediately following entered a system with a fish assemblage that included almost a dozen diadromous species, several estuarine species, and many freshwater species. Actual numbers for population size of any given species are not available, but every early account noted the abundance of fish present in the river and its tidal tributaries. Reconstructions of life on the pre-contact river also suggest that fish were exceptionally abundant. Over the last four hundred years, species composition and overall and relative abundance of the fish assemblage has changed. The current assemblage is also made up of diadromous, estuarine, and freshwater species, many of them introduced within the last two hundred years. However, abundance of most, if not all species is arguably less than that noted in the earliest reports. In the intervening four hundred years, fish stocks have been harvested, in some cases, to a point where they were no longer commercially viable. They have been subjected to changes in river morphology and have responded to effects of changes in landscape use throughout the drainage. They also have been affected by increases in pollutants, both chemical and biotic, in the water. Fish resource use by humans has focused not on protecting individual species as sustainable resources, but on protecting the commercial catch as a whole, with an interesting exception for the striped bass fishery. The trend since the mid-nineteenth century has been to harvest an individual species until fished to extremely low numbers and then switch to another, more abundant species. Here we examine the composition of the assemblage in 1609. We examine trends in fishery resource use beginning in the early seventeenth century using archaeological inferences, early written reports, and fisheries statistics published by government agencies. We link these trends to changes in human attitude toward the use of these resources.

INTRODUCTION

When Henry Hudson sailed up the North River in September 1609 he entered a river that, by all accounts, was teeming with fish. The established indigenous populations had recognized these fishes as important resources and the arriving Europeans, such as Juet (1909) and Van der Donck (2008), quickly learned that a little effort in fishing provided ample returns. Although the native populations used effective harvest methods (Brumbach and Bender 2002), their numbers were too low to affect fish abundance (Dunn 1994). For almost two centuries after 1609, the number of Europeans that arrived in the Hudson River valley remained low as well (Swaney et al. 2006), and their arrival coincided with declines in the populations of Native Americans (e.g., Snow and Starna 1989). In fact, fishing was a household effort, where individuals caught fish primarily for their own use (Cheney 1896). Few were commercial fishers. That changed with the Industrial Revolution when the number of people increased and fish became a resource to the market economy. In the nineteenth century commercial fishers dominated the resource, a status that was retained until the late twentieth century when commercial fishing declined and recreational anglers became dominant (Hudson River Estuary Management Program 1996).

Here we examine changes in fisheries resources in the Hudson River from 1609 to the present. We speculate on the composition of the fish assemblage present during the early years of European settlement and trace changes in commercial harvests. We focus on three species as examples of exploitation of a natural resource and human responses to changes in abundance. Patterns of use and response are similar and suggest that the individual species were relatively unimportant. In each case, the focus had been on protecting the overall market value of the resource and not the component species that make up the total resource. Only in the last few decades has this pattern been challenged by resource managers (Hattala and Kahnle 1997; Limburg et al. 2006).

HUDSON RIVER FISHERY RESOURCES IN THE SEVENTEENTH AND EIGHTEENTH CENTURIES

It is impossible to delineate with absolute certainty the component species of the Hudson River fish assemblage that existed prior to the early nineteenth century. Lake (2009) suggested that fish were bountiful and easily harvested. Brumbach and Bender (2002) noted that prior to AD 800 fish were the dominant component of the diet of people living in the Hudson Valley. After AD 800 maize became dominant but fish continued to be an important food. Rostlund (1952) estimated that the fish harvest in the Lower Hudson River was 29,000–35,000 kg/year. Brumbach (1978) estimated pre-contact fishery stocks at 10.63 million kg of anadromous fish alone. Either number supports the contention that fish were abundant in the river when considering that the population of Mahicans living in the lower valley at contact was estimated between 4,000–8,000 individuals (Brasser 1978; Dunn 1994).

Historical records support the assessment that fish were abundant in the river during the first century of European exploration and settlement. These reports also offer clues to the species present. The problem with identifying species based on early reports is that the authors rarely used names that are easily interpreted by modern readers. Typically the early reporters used familiar European names. Take Robert Juet's report of the 1609 Hudson voyage up the North River. He corroborates that fish were abundant: "The River is full of fish." He also names the fish caught by crew. During the twenty-two days that the *Half Moon* was in the river, Juet mentions that the crew fished on three days and that they caught "mullets, breames, bases and barbils" [*sic*]. He also mentioned catching mullets and a ray at the mouth of the river and in seeing many "salmons" at the mouth and upriver. The only additional information provided was that the mullets exceeded 1.5 feet in length. The identification of salmon as Atlantic salmon (*Salmo salar*), a species that these European mariners should have recognized, is controversial and has been largely discredited (Webster 1982) because this species has not been found in the river in the last several centuries. In fairness, Juet only claimed to observe this species and did not report catching one. Identifying the other four species is problematic and depends on which characteristics of the fish are used in making the identification. Most have made the assumption that these names were given because the fish caught was taxonomically close to a European counterpart: breame, then, could be a golden shiner (*Notemigonus crysoleucas*), a fish similar to the common bream (*Abramis brama*) and in the family Cyprinidae. Juet, however, may not have focused on taxonomic similarity but instead on the deep body and laterally flattened appearance of the fish and the term *bream* denotes a variety of freshwater and marine fish today. Juet may have applied the name to sunfish (Centrarchidae: *Lepomis*) or temperate bass (Moronidae: *Morone*) or both. So "mullets, breame, bases and barbils" could be interpreted as striped mullet (*Mugil cephalus*), golden shiner, white perch

TABLE 4.1. Fishes present in the Hudson River, seventeenth century, see Juet (1909), De Vries (1909), Megapolenus (1909) and van der Donck (2008).

Juet, 1609	De Vries, 1633–43	Megapolenus, 1644	van der Donck, 1656	Possible Identification
		lamprey	lamprey	2 lamprey species
	sturgeon		sturgeon	2 *Acipenser* spp.
			aal	Young *A. rostrata*
	eel	eel	paling	Adult *A. rostrata*
	elft	shad	elft	3–4 *Alosa* spp.
barbils		catfish		2 *Ameiurus* spp.
	dirteinen		dirteinen	2–5 sucker species, marine strays
		sucker	sucker	2–3 sucker species, lamprey
breames	roach, carp		carp	3–5 species of large cyprinid
			minnow	8–18 species of small fish in several families
	salmon		salmon	*Salmo salar*
			trout	*Salvelinus fontinalis*
			silverfish	*Menidia* spp., *Alosa* spp., *Osmerus mordax*?
	pike	pike	pike	2 *Esox* spp.
			frostfish	*Microgadus tomcod*
bases		bass	bass	*Morone americanus*?
mullets	twalift		twalift	*Morone saxatilis*
		sunfish	sunfish	*Lepomis gibbosus*, other *Lepomis* spp.
	perch	perch	perch	*Perca flavescens*
			flounder	*Trinectes maculatus*, marine strays
			brikken	?
4	10	9	20	38–55

(*Morone americana*), and fallfish (*Semotilus corporalis*); these are fish present in the river that have closely related European counterparts. Or, Juet may have focused on a single characteristic most similar to the European species so that barbils refers to white catfish (*Ameiurus catus*), which have prominent barbels, and mullet to striped bass (*Morone saxatilis*), because both species have prominent lateral stripes. Juet unfortunately offered no clues to his naming process so that any later interpretation is problematic.

Similar interpretation problems arise with early Dutch reports and are compounded because English-language researchers must rely on interpretations. Nonetheless, these reports offer clues to the composition of the seventeenth-century Hudson River fish assemblage, and interpretations arising from the New Netherlands project, led by Charles Gehring, have yielded a wealth of detail of the era. Writing in 1644, Johannes Megapolensis described fisheries of the upper Hudson (or possibly Mohawk River, below Cohoes Falls) in Mohawk Country:

> In this river is a great plenty of all kinds of fish—pike, eels, perch, lampreys, suckers, cat fish, sun fish, shad, bass, etc. In the spring, in May, the perch are so plenty, that one man with a hook and line will catch in one hour as many as ten or twelve men can eat. My boys have caught in an hour fifty, each a foot long. They have three hooks on the instrument with which they fish, and draw up frequently two or three perch at once. There is also in the river a great plenty of sturgeon, which we Christians do not like, but the Indians eat them greedily.... This river ebbs and flows at ordinary low water as far as this place, although it is thirty-six leagues inland from the sea. (Megapolensis 1909)

In addition to the report of Reverend Megapolensis, Captain David Pieterszoon de Vries, chronicling 1633–1643, also listed and described fish present in the drainage (Table 4.1). The lists show some consistency and suggest that at least several dozen species of fish were the basis of the fish assemblage at that time.

The most extensive coverage of seventeenth-century resource use is detailed by Adriaen Van der Donck in 1656 (2008). He notes the presence of twenty types of fish found in the Hudson River and its tributaries (Table 4.1). He often used European names and, for most species, included little descriptive information. Several names are relatively easy to interpret. The European sturgeon, bass, pike, trout, perch, and lamprey have North American counterparts that closely resemble and are taxonomically close to their Old World cousins. Although Van der Donck's name may not be attributable to a specific species, it does provide evidence of the presence of a group of organisms with similar characteristics. Sturgeon, for example, probably covered both species native to the Hudson River, and pike probably refers to two species of pickerel. Van der Donck also noted the presence of aal and paling, which refer respectively to the young and adult of American eel (*Anguilla rostrata*), another species with an obvious European counterpart.

Two species are listed with sufficient descriptive information to link to a Hudson River species. The sunfish, with its "spotty skin of brightly flecked scales" is the pumpkinseed (*Lepomis gibbosus*). The frostfish, which comes up from the sea in winter and resembles the silver hake, is clearly the Atlantic tomcod (*Microgadus tomcod*). Several names are more difficult to assess either because they could represent several species or because there is no translation of the name: carp, minnow, silverfish, flounder, and brikken. The only name in Van der Donck's list that is associated with an obvious North American family without a European counterpart is sucker, family Catostomidae. However, this name may be a descriptor rather than reference to a taxonomic group of fish and if so, may refer to lamprey or remora rather than one of the catostomids now present in the river.

An interesting aspect of Van der Donck's list, and that of others (Table 4.1), is the quaint habit of naming unknown fishes sequentially. Consequently the Dutch settlers caught fish named elft (eleven), twalift (twelve), and dirtienen (thirteen). Elft and twalift are described well by both Van der Donck (2008) and De Vries (1909) and refer to American shad (*Alosa sapidissima* and possibly the two species of river herrings) and striped bass respectively. Dirtienen is much less obvious. The Dutch word used to name this fish is *lipfis*, which is translated as wrasse (C. Gehring, New Netherlands Project, personal communication). Wrasses are common shore fishes in Europe and should have been recognized by the early settlers. Van der Donck noted that dirtienen is not comparable to any known fish. De Vries provided other clues to its identity: it is yellow, shaped and with scales like a carp, bigger than a cod, and runs the river after the twalift. Because there is no obvious candidate, this fish is often thought to be a marine fish, such as red drum (*Sciaenops ocellatus*) or one of the two species of wrasse present in the area (Limburg and Waldman 2009 for discussion of marine species in the river). This interpretation relies heavily on the translation of *lipfis* as wrasse and less on De Vries's description. If the term *lipfish* is taken as a descriptor rather than a name, then there are other Hudson River fishes that qualify as dirtienen. White sucker (*Catostomus commersonii*), creek chubsucker (*Erimyzon oblongus*), and shorthead redhorse (*Moxostoma macrolepidotum*) are present in the Hudson River drainage today. They are members of the Catostomidae, a group of fish notable for their distinctive lips. They have scales like carp and are shaped like carp and, although they do not run the river from the sea, they probably do run the tributaries from the river, and the redhorse, at least, would make the spawning run after striped bass. One of these species, or all three, may be the dirtienen.

A second way to determine the components of the fish assemblage is to back calculate from the assemblage present today. The species composition of the Hudson River assemblage changes because of introductions and increased sampling; there are approximately 217 species reported from the drainage (e.g., Daniels et al. 2005). Of these, 104 are marine

TABLE 4.2. Piscivorous fish in the Hudson River and year of establishment. See Daniels et al. (2005) for additional information.

Species	Year
Brook trout, *Salvelinus fontinalis*	Native
Striped bass, *Morone saxatilis*	Native
Chain pickerel, *Esox niger*	Native
Redfin pickerel, *Esox americanus*	Native
White catfish, *Ameiurus catus*	Native
Largemouth bass, *Micropterus salmoides*	1830s
Smallmouth bass, *Micropterus dolomieu*	1830s
Rock bass, *Ambloplites rupestris*	1840s
Northern pike, *Esox lucius*	1840s
Walleye, *Sander vitreus*	1893
White crappie, *Pomoxis annularis*	1900s
Black crappie, *Pomoxis nigromaculatus*	1900s
White bass, *Morone chrysops*	1975
Channel catfish, *Ictalurus punctatus*	1976
Snakehead, *Channa argus*	2007

fish, 44 are exotic, and 14 are reported only from the Mohawk River or upper Hudson River. The number of native freshwater, estuarine, and diadromous species in the lower Hudson River is roughly fifty-five, a number similar to that obtained by examining historical records. Of these, approximately twenty-five species would be commercially desirable because they are large species, migratory and therefore easy to catch, or abundant. One interesting aspect of the historic assemblage is that few of the species are fish-eating predators. Of the possible species in the native assemblage, approximately five are largely piscivorous (Table 4.2).

THE NINETEENTH CENTURY AND THE DEVELOPMENT OF COMMERCIAL FISHERIES

The origin of commercial fishing, that is, harvesting fish for sale rather than personal use, probably dates to a time soon after the arrival of Europeans; however, large-scale commercial fishing developed in the Hudson River in the nineteenth century (Cheney 1896). Statistics on commercial harvests do not become routine until mid-century, but it is likely that fishers harvested and marketed most of the species in the river and did not necessarily concentrate on a single-species catch. For example, Gill (1856) noted harvests of striped bass, white sucker, and common carp. Mearns (1898) reported that redbreast sunfish and rock bass taken in fyke nets were commonly sold at fish markets. Nonetheless, as markets developed, the value of some species increased over that of others and, with use of selective gear and careful attention to fish behavior, commercial catches could and probably did concentrate on single-species catches.

American shad had been a mainstay in the diets of Native Americans and European arrivals (see McPhee 2002 for a popular account). The shad fishery probably dates to the arrival of humans in the valley, and American shad was clearly part of the Native American diet (Lake 2009). Cheney (1896) noted that, in the early part of the nineteenth century, farmers traveled to fishing camps at natural barriers to harvest and salt this species and "use them for home consumption." Harvesting for family use probably was the pattern until the Industrial Revolution and the influx of large numbers of immigrants, when commercial harvesting of fish in general and American shad in particular became common in the Hudson River (Limburg et al. 2006). American shad and the two smaller, related species called river herring (blueback herring *Alosa aestivalis*, and alewife *A. pseudoharengus*) are anadromous species and easily caught in large numbers because their spawning runs are predictable in time and space (Schmidt et al. 1988). Initially, these species supported the commercial fishing effort in the river.

Landing records for American shad in the Hudson River were episodic until the early part of the twentieth century, but by 1880, more than 1.2 million kg of shad were caught by commercial fishers and sold in cities along the river (Fig. 4.1). It is likely that there remained a personal use harvest as well. Commercial harvests were impressive during this period; in the 1890s, commercial landings reached nearly two million kg.

Catches were a response to the need for inexpensive food for recent immigrants. The recognition that the fishery was an important economic mainstay must have occurred early in its development because the first laws protecting shad were passed in New England in the early nineteenth century and on the Hudson River, the first restriction on fishing passed was in 1870 with net lift regulations (Laws of New York 1870). It was not until 1895 that the first assessment of the fishery was undertaken (Cheney 1896). In that year, 5,520 nets were fished in the river from Bay Ridge (Brooklyn)

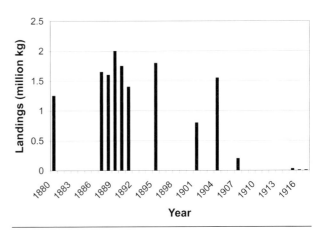

FIG. 4.1. American shad (*Alosa sapidissima*) landings in the Hudson River, 1880–2005. See Limburg et al. (2006).

to Castleton and 1,868,547 fish were caught weighing more than 1.8 million kg—staggering statistics by today's measure. The reason an assessment was funded was that the catches had begun to decline (Cheney 1896). Beginning in 1882, American shad young-of-year were stocked in the Hudson River (Fig. 4.2). Each year, until 1902, shad from New York State and federal hatcheries was stocked. As is true for any stocking effort, it is assumed that the stocked fish enhance the fishery, but no assessment was ever undertaken to demonstrate this. By the beginning of the twentieth century, it was clear that the American shad fishery was in deep decline, and it collapsed completely by 1910 (Blackford 1916).

Overfishing may have been the sole reason for the collapse of this fishery, but other factors may have affected abundance as well. During the mid-nineteenth century, in-stream environmental conditions changed with added pollution and exotic species, damming, dredging, and channelization (Stephenson 1899). Lossing (1866) specifically noted that the construction of a railroad bridge across the river at Troy and the ensuing modification of islands in the river at that site led to the loss of both the shad and sturgeon fishery in that area. Land-use practices in the surrounding valley changed. The railroads, in particular, altered the riparian zone of the river (Swaney et al. 2006) by eliminating riparian vegetation and the associated biological community, stabilizing the shoreline with the actual tracks and with riprap, isolating embayments or minimizing tidal entry into embayment by restricting entry through culverts, and restricting access to tributary streams by trestle construction and the creation of bridge pools at the mouths of streams. Changing conditions may have affected the ability of American shad and the river herrings to successfully reproduce, but there is little information directly linking environmental conditions and fish abundance during this period, whereas there is ample evidence that the fishing effort was extremely high.

American shad were important because they provided an inexpensive food source for an increasing human population. This led to an increase in the fishing effort and a change in fishing methods, coupled with environmental changes that were not favorable to the shad. The response was twofold: regulations that provided some protection for the

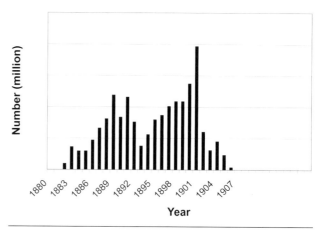

Fig. 4.2. Stocking effort of American shad (Alosa sapidissima), Hudson River, New York, 1882–1906. See Cheney (1896).

shad and an intensive stocking program, neither of which were either assessed or successful. The final response to the declining shad fishery was to switch to a new fishery. Having failed to protect American shad as a resource, the ultimate tactic was to switch to a new resource, which kept markets open but required consumers to accept and buy a different fish.

ALBANY BEEF AND A SECOND LATE-NINETEENTH-CENTURY MARKET RESOURCE

American shad and the river herrings constituted an important fishery in the nineteenth century along the Hudson River but other fishery resources were as important to fishers in the valley. Although not highly regarded by either Dutch (Megapolensis 1909; Van der Donck 2008) or English (Thacher 1854), sturgeon was present, available, and abundant. Saffron (2004) linked the inception of the commercial sturgeon fishery to the increase in price of halibut (*Hippoglossus hippoglossus*) from the Grand Banks in the mid 1800s. Albany beef (smoked sturgeon) was an inexpensive food alternative for newly arrived immigrants in New York City.

The Hudson River is home to two species of sturgeon, the anadromous Atlantic (*Acipenser oxyrinchus*) and the estuarine shortnose (*A. brevirostrum*) sturgeon (Smith 1985). Nineteenth-century reports suggested that Atlantic sturgeon grew to 5 m and 350 kg and, like the American shad and river herrings, was harvested as the fish moved upriver to spawn. Albany beef, initially a term of derision, became an important and inexpensive replacement resource as the marine halibut fishery declined. By the 1850s, sturgeon had eclipsed other fishes as an inexpensive staple. Sturgeon was caught from the Battery to Troy, although Hyde Park and Low Point, just downriver of Newburgh, were identified as the most productive areas (Lossing 1856). Munsell (1869) noted that approximately 2,500 individual sturgeon with an average weight of 100 kg came to market in Albany during the May through August season and had a value of $18,750 (or about $4.5 million in today's dollars). In addition to Albany beef, sturgeon provided other products—its offal was rendered into oil for lamps and medicine and sturgeon roe is caviar. Oil was valuable; one hundred barrels of oil were produced in Albany from the 2,500 individuals caught (Munsell 1869).

Reports of declining sturgeon catches began in the 1880s (Anonymous 1881a). By this time, Albany beef was no longer an inexpensive protein source for arriving immigrants. In 1880, Albany marketed on average 450 kg of smoked sturgeon per week at about 40 cents per kg, but on occasion more than two thousand kg per week was sold (Anonymous 1881a). This same article noted that sturgeon from the river was becoming increasingly rare and the much of the sturgeon marketed in Albany and other river cities came from Maine, Florida, and the Great Lakes. Finally, the article mentions that the catch from the Hudson River could not meet local demand.

The response to the decline in the sturgeon fishery was met with increasing importation of sturgeon from other areas within the country and with attempts to save the fishery through legislation. In 1894, the New York State Assemblyman Howard Thorton of Newburgh introduced legislation that was designed to protect the sturgeon oil industry, which was important to that city (Anonymous 1894). His bill would close sturgeon fishing from September 1 to June 1, which would make harvest illegal during the early part of the spawning season (Smith 1985). The bill was enacted into law May 10, 1894, as part of a major overhaul of the Fish and Game regulations (Laws of New York 1894). In addition to the closure of the fishing season, the law required that nets used to harvest sturgeon have bar mesh greater than 17.8 cm. In 1902, legislation requiring that sturgeon from the Hudson have a minimum size of 91.5 cm was passed (Laws of New York 1902).

As was the case for American shad, when Atlantic and shortnose sturgeon declined in the catch, legislation was enacted to protect the resource while minimizing major harm to the commercial fishery.

The sturgeon fishery on the east coast of the United States peaked late in the nineteenth century and by the beginning of the twentieth century the sturgeon fishery had been reduced to harvest from a few populations (Saffron 2004). Unfortunately, newspaper reports indicated that the Hudson River sturgeon fishery peaked as early as the 1870s (e.g., Anonymous 1881a) and was already on a precipitous decline before declines were noted along the rest of the east coast. Legislation was too late to save sturgeon and shad as resources. In the case of shad, the stocking program also failed to save the river fishery, but was successful in introducing American shad and alewife into other bodies of water where they were not native, particularly in western North America (cf. Boyle 1979). As both American shad and sturgeon declined in the Hudson River another species was quickly found as a replacement so that neither the commercial fishery nor availability of fish at market was seriously affected.

THE IMPORTANCE OF AN EXOTIC—COMMON CARP

The actual date of introduction of common carp (*Cyprinus carpio*) is controversial. DeKay (1842) quotes extensively a letter from Henry Robinson, Esq., of Newburgh, Orange County, in which Robinson states that he brought six or seven dozen carp from France in 1831 and 1832, and introduced them into his ponds. DeKay's description of the fish is inconclusive, noting characteristics that are clearly carp-like and others that are not. He reported that Robinson released one to two dozen individuals into the Hudson River each year, where they grew larger than the pond fish and were caught by fishers in nets. In 1877, Spencer Baird (cited in Zeisel 1995) argued that the first common carp were imported to North America by the U.S. Fish Commission in 1872 and that on his trips to the New York City markets, he had found no genuine carp but rather "the common gold-fish, reverted to its original normal condition." He also stated that market netters took "precisely similar specimens of white, red, and all intermediate conditions" from the Hudson. It seems to us that there is little reason to believe that DeKay was in error, but early records of carp catches from the Hudson are probably a mixture of carp and goldfish.

Common carp were readily available from Hudson waters by the 1880s and were sold at half the price of American shad (Zeisel 1995). Smith (1896) reported that commercial fishers in the Illinois River in 1893 had an offer from New York [City?] dealers who would take all that the fishers could catch at $0.08 per kg rough. Smith (1896) also reports carp being sent from the federal hatcheries to forty states, the District of Columbia, and the "Indian Territories." As American shad, sturgeon, and striped bass declined in the commercial catches in the late 1800s, commercial fishers switched to common carp. Initially, common carp was sold as an ethnic food and marketed under names such as "Great Lakes Salmon" (Zeisel 1995), but as other species became exhausted at the turn of the century, common carp began to dominate the market. By 1908, for example, 1,425 metric tons were sold at the Fulton Market for 25 cents per kg (Zeisel 1995).

The technology of catching common carp from the river did not differ from that used to capture the native fishes such as American shad, although the switch did entail capital investment. Exotic carp was hardy and had been reared for aquaculture for millennia (Balon 1995) and could be treated differently than the native species after capture. Carp was shipped alive to New York City on commercial steamers (e.g., *The Redfield*) and sold at the Fulton Fish Market. Several commercial operators built holding ponds for carp and fattened them with grain before market. One fisher built a barge with six live wells. The barge, named the *Live Carp*, was 15 x 5 m and could hold more than three metric tons of fish. By adapting aquaculture techniques to the harvest, transport became easier, the product arrived at market fresher and, by holding the fish in ponds, the market could be manipulated to the benefit of the fishers.

American shad, the river herrings, and the sturgeons were commercially extinct by the beginning of the twentieth century and common carp and related fishes, such as the native white sucker, which had been harvested for an ethnic market since the 1880s, immediately assumed dominance in the overall market. Despite substantial local catches, the Hudson River fishers could not keep up with demand. In 1901, 3.1 million kg of carp were imported from the Great Lakes and the Mississippi River (Illinois) and were shipped alive in specially designed railroad cars. By 1922, 135,000 kg of carp was harvested from the Hudson River and 58,500 kg of white sucker was caught. In 1924 the carp and sucker catch totaled almost 230,000 kg. Common carp and related species became the fish of choice for millions in New York City and protected the freshwater market when other species were no longer abundant enough to support a fishery.

Then in the 1930s, harvest of carp declined with a resurgence of the American shad population and the resultant in-river shad fishery. Carp have languished in relative fisheries obscurity ever since (Zeisel 1995). However, catches of common carp exceeded forty metric tons per year until after 1947. Sucker catches were always less than ten metric tons except in 1939. Common carp and suckers were worth the same or more than shad (on a per pound basis) until the mid-1940s (Fig. 4.3). The fall in total catch of common carp in the Hudson (Fig. 4.4) is correlated with a substantial increase in value of American shad (Fig. 4.3) and therefore a relative devaluation of common carp. The price for suckers remained comparable to shad in the 1950s, but catches were never large. The carp fishery allowed the freshwater fishing industry and its market to continue during a period when other fishes were unavailable for harvest. As one of them, American shad, rebounded, common carp was replaced. Its fall from favor did not seem to be solely related to a decline in numbers but to the rebound of a more popular and recently readily available species. Common carp as a resource was abandoned, because the market was secured by the increase in American shad numbers.

SWITCH IN TACTICS—COMMERCIAL FISHING TO SPORT FISHING

Prior to the 1840s, fishing on the Hudson River was limited primarily to harvests for personal use

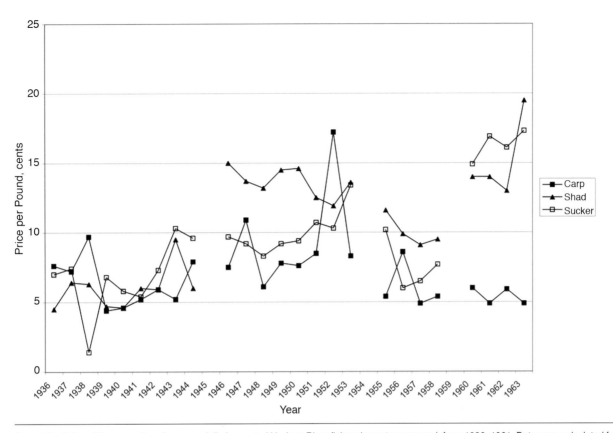

FIG. 4.3. Value of three species of commercially harvested Hudson River fishes, in cents per pound, from 1936–1961. Data were calculated from Annual Reports of the State of New York Conservation Department, 1936–1961. There are no data for 1945, 1954, and 1959.

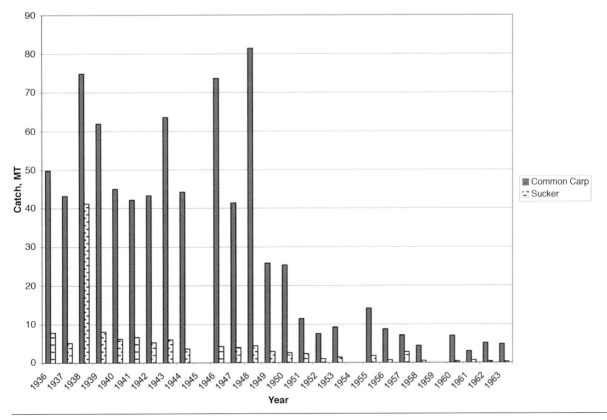

FIG. 4.4. Commercial catch of common carp (*Cyprinus carpio*) and suckers (*Catostomidae*) from the Hudson River, 1936–1961. Data are from Annual Reports of the State of New York Conservation Department, 1936–1961. There are no data for 1945, 1954, and 1959.

(Cheney 1896). From the beginning of commercial market fishing in the 1840s to the 1960s, the general pattern was to harvest a species until it was commercially extinct and then harvest a different species until it became commercially extinct. Regardless of collapses in fish stocks, the market demand for fish was consistently high. Importation of fish from the Great Lakes, the Mississippi River, and farther abroad filled some of the demand, but local fishers were best positioned to respond. They did so by shifting exploitation to more common species. In the three cases examined here, harvests were intensive and, although there were efforts to enhance certain fisheries through stocking and laws were enacted to limit catches by season and size, the mitigating measures were often late and insufficient to protect the individual fisheries.

A possible exception to harvesting to commercial extinction is the striped bass. Commercial exploitation of striped bass has a long history in the Hudson River. Ackerly (1818) reported that striped bass were for sale in considerable quantity in February and they were shipped frozen to Albany and "the interior of the country." The market price for striped bass in New York was between forty and sixty-five cents per kilogram from 1879 to 1882 (Anonymous 1879, 1880, 1881b, 1882), often sold as fish from the Delaware River, North Carolina, or "the provinces." Hornaday (1904) mentioned a catch of 10,164 lbs of striped bass on Fire Island in 1879 (single seine haul), which suggests that local stocks were harvested for market in this time period. These prices indicate that the consumers of striped bass were from a different social stratum than the consumers of American shad or carp.

Striped bass was managed differently from the other species we have discussed. Many efforts were made to protect the stock including creel limits and size limits imposed on sport fishers and fishing seasons, lift periods, size limits, and gear restrictions placed on commercial fishers (McLaren et al. 1988). There is some evidence that this single species management began early in the fishery. For example, in 1879, it was illegal to possess striped bass less than 0.25 kg (Anonymous 1879). Because of its economic importance, striped bass was one of several species that were the focus of intensive study during the 1970s and 1980s. These studies primarily examined the effect of water use in power generation along the river. Boyle (1979) popularized the evidence on the effect power generation had on fish, particularly striped bass, with graphic photographs. Others (e.g., McDowell 1986) reviewed the extensive literature and noted that environmental changes resulting from shoreline power generation clearly and negatively affected fish populations and the fish assemblage in general and that there eventually developed a consensus on the extent of the impact that led to mitigation measures and changes in management for these species.

Striped bass aside, mitigation measures aimed at protecting the individual fishery were weak and often inadequate, in part because there was always an alternative, underutilized stock that could be marketed instead. Clearly during the 120 years of intense commercial fishing, overfishing was not the only factor that affected catches. This period coincides with increases in the human population in the valley and with changes in land use practices and massive in-stream modifications (Swaney et al. 2006). Whether it was overfishing or environmental change that affected riverine fish stocks, the response was consistent: utilize a new resource.

In the 1960s, when the American shad population collapsed a second time, the number of commercial fishers on the river began to decline (Limburg et al. 2006). Although some commercial fishing continues on the river, although not for American shad, in the last several decades Hudson River fish resources have become increasingly a game fishery. This has led to the introduction, often unsanctioned, and repeated stocking of fish that are popular game fish (Table 4.2). The effort to establish game fish in the river markedly increased the number of species that are piscivorous, from about six species to fifteen. Two of these, largemouth and smallmouth bass, have affected fish and invertebrate assemblages and are important economically (Mills et al. 1996). At present, it is impossible to assess the extent of the change to the fish assemblage that has occurred or may occur due to this increase. Daniels et al. (2005) reported declines in forage fish and several of the smaller anadromous species.

These declines may not result solely from the presence of more exotic predators in the system, because pollutants, increasing temperatures (which, in the case of rainbow smelt [*Osmerus mordax*], may have led to extirpation), the introduction of other exotics, harvesting, and habitat modification may also affect numbers. For example, pollutants, such as polychlorinated biphenyl (PCB), contaminate fish, although there is high variability among species and even individuals (Zlokovitz and Secor 1999). Because of the contamination, catch restrictions and consumption advisories were implemented, which have reduced catches and led to increased numbers of fish that are nonetheless contaminated (Barnthouse 2000). The introduction of exotic zebra mussel and water chestnut has had profound effects on ecosystem processes (e.g., Caraco et al. 2000; Caraco and Cole 2002). Zebra mussels' removal of phytoplankton and microzooplankton has been enough to demonstrably alter fish community structure (Strayer et al. 2004). Habitat complexity that was probably important for juvenile life stages of many species was largely removed by repeated dredging of the up-estuary shallows and filling of side channels (Nitsche et al., ch. 6 in this volume; Daniel Miller, New York State Department of Environmental Conservation, personal communication). The effect of the loss of shallow side channels, embayments, and shoreline wetlands to riverine fishes is well documented (e.g., Hughes et al. 2005) and their loss in the Hudson River probably has affected community structure, species diversity, and abundance.

In addition to exotic predatory fish, the native piscivores, particularly striped bass, and several marine species that utilize the river affect the composition of the Hudson River fish assemblage. Striped bass feed when in the river and, at least the larger ones, are largely piscivorous (Gardinier and Hoff 1982). With the resurgence in numbers in the last few decades, the predatory demand of this species in the river could account for observed declines in some species (Heimbuch 2008). Marine fish, such as bluefish (*Pomatomus saltatrix*), can account for large declines in river fish as well (Buckel et al. 1999). Nevertheless, the number of predatory species now in the river that were not present previously, increasing populations of native predatory fish resulting from management efforts to increase numbers, and the presence of marine species all affect the survivability of many of the smaller species and young-of-year of larger species, particularly the anadromous American shad, river herrings, and striped bass. Sport fisheries in the river have the potential to be sustainable, in contrast to the history of commercial fisheries. In a sport fishery, there are lower catches, effective regulation, and both single-species and ecosystem management of the resources (Hudson River Estuary Management Program 1996). Conversely, the effort to successfully develop a sport fishery with several exotic predators may stymie efforts to protect and manage the several native species that serve as prey.

By developing the river as a sport fishery, the pattern established with previous declines was only slightly modified: as each riverine fishery collapsed, a new resource was tapped. However, the development of the game fishery resulted in the introduction of exotic fishes and the enhancement of native predatory fishes, which has altered the assemblage by increasing the number of top predators in the system.

CONCLUSION

Hudson River fish compose a resource that was recognized as important by Native Americans, early European settlers, commercial fishers, management agencies, and decision makers. However, since the development of a market economy almost two centuries ago, individual fish species were not recognized as resources; instead, the catch, with serial replacement of the dominant species, was treated as the resource. Fishing in the Hudson River displayed a pattern typical of commercially driven resource extraction; there were few if any provisions for conservation or setting sustainable quotas (see Richards 2003). As individual species declined in number, possibly due to overfishing, environmental change, or a combination of both, protective regulations, enhancement efforts, and new technologies were mustered in an effort to protect the individual species, but ultimately and consistently, the effort rapidly moved to switch to a new species and protection of the overall catch. This culminated in a switch from commercial fishing to sport fishing in the last several decades and an effort to establish several exotic top carnivores in the Hudson River and enhance through management native predatory species. As exotic game fish were promoted there has been a change in attitude regarding individual species. Recent efforts to recognize each species as a separate resource have resulted in management plans that specifically protect and enhance sturgeon, American shad, river herrings, striped bass, and other species (Hudson River Estuary Management Program 1996). It is ironic that now, after belatedly recognizing the importance of each species as a resource worthy of individual protection and management, the introduction of several exotic species to the Hudson River has altered the system (Mills et al. 1996), which may make it difficult to restore effectively the anadromous fisheries that were prominent in the last two centuries.

REFERENCES CITED

Ackerly, S. 1818. Economical history of the fishes sold in the markets of the City of New York. *American Monthly Magazine and Critical Review* 2: 370–71.

Anonymous. 1879. The household; Features of the markets. The supply of fish. Prices asked for other provisions. The *New York Times,* November 16, 1879, 9.

Anonymous. 1880. The household; Features of the markets. The supply of fish. Prices asked for other provisions. The *New York Times,* August 29, 1880.

Anonymous. 1881a. The supply of "Albany beef." The *New York Times,* August 19, 1881.

Anonymous. 1881b. The household; Features of the markets. The supply of fish. Prices asked for other provisions. The *New York Times,* April 3, 1881.

Anonymous. 1882. The household; Features of the markets. The supply of fish. Prices asked for other provisions. The *New York Times,* August 27, 1882.

Anonymous. 1894. Caring for a Newburgh industry. The *New York Times,* January 14, 1894.

Balon, E. K. 1995. Origin and domestication of the wild carp, *Cyprinus carpio*: From Roman gourmets to the swimming flowers. *Aquaculture* 129: 3–48.

Barnthouse, L. W. 2000. Impacts of power-plant cooling systems on estuarine fish populations: The Hudson River after 25 years. *Environmental Science and Policy* 3: S341–S348.

Blackford, C. M. 1916. The shad—A national problem. *Transactions of the American Fisheries Society* 46: 5–14.

Boyle, R. H. 1979. *The Hudson River: A natural and unnatural history.* Expanded edition. W. W. Norton, New York.

Brasser, T. J. 1978. Mahican. In *Handbook of North American Indians*, Volume 15, Northeast, ed. B. G. Trigger, 198–212. Washington, DC: Smithsonian Institution Press.

Brumbach, H. J. 1978. Middle woodland fishing economics: The Upper Hudson River drainage. PhD dissertation, Department of Anthropology, State University of New York at Albany.

———, and S. Bender. 2002. Woodland period settlement and subsistence change in the upper Hudson River Valley. In *Northeast Subsistence-Settlement Change A.D. 700–1300,* ed. J. P. Hart and C. B. Rieth, 227–39. NYS Museum Bulletin 496.

Buckel, J. A., D. O. Conover, N. D. Steinberg, and K. A. McKown. 1999. Impact of age-0 bluefish (*Pomatomus saltatrix*) predation on age-0 fishes in the Hudson River estuary: Evidence for density-dependent loss of juvenile striped bass (*Morone saxatilis*). *Canadian Journal of Fisheries and Aquatic Sciences* 56: 275–87.

Caraco, N. F., J. J. Cole, S. E. G. Findlay, D. T. Fischer, G. G. Lampman, M. L. Pace, and D. L. Strayer. 2000. Dissolved oxygen declines in the Hudson River associated with the invasion of the zebra mussel (*Dreissena polymorpha*). *Environmental Science and Technology* 34: 1204–10.

Caraco, N. F., and J. J. Cole. 2002. Contrasting impacts of a native and alien macrophyte on dissolved oxygen in a large river. *Ecological Applications* 12: 1496–1509.

Cheney, A. N. 1896. Shad of the Hudson River. First Annual Report of the Commissioners of Fisheries, Game and Forests, 1895. Albany, NY, 125–34.

Daniels, R. A., K. E. Limburg, R. E. Schmidt, D. L. Strayer, and R. C. Chambers. 2005. Changes in fish assemblages in the tidal Hudson River, New York. In *Historical changes in large river fish assemblages of the Americas,* ed. J. N. Rinne, R. M. Hughes, and B. Calamusso, 471–503. American Fisheries Society, Symposium 45, Bethesda, MD.

DeKay, J. E. 1842. *Zoology of New York. Part IV. Fishes.* Albany: W. & A. White and J. Visscher.

De Vries, D. P. 1909. From the "Korte Historiael Ende Journaels Aenteyckeninge" 1655. In *Narratives of New Netherland, 1609–1664,* ed. J. F. Jameson, 181–234. New York: Charles Scribner's Sons.

Dunn, S. W. 1994. *The Mohicans and their land: 1609–1730.* Fleischmanns, NY: Purple Mountain Press.

Gardinier, M. N., and T. B. Hoff. 1982. Diet of striped bass in the Hudson River Estuary. *New York Fish and Game Journal* 29: 152–65.

Gill, T. 1856. On the fishes of New York. Report of the Smithsonian Institution 1857, 253–69.

Hattala, K. A., and A. W. Kahnle. 1997. Stock status and definition of over-fishing rate for American shad of the Hudson River estuary. Report to the Atlantic States Marine Fisheries Commission. New York State Department of Environmental Conservation, New Paltz, NY.

Heimbuch, D. G. 2008. Potential effects of striped bass predation on juvenile fish in the Hudson River. *Transactions of the American Fisheries Society* 137: 1591–1605.

Hornaday, W. T. 1904. *The American natural history.* New York: Charles Scribner's Sons.

Hudson River Estuary Management Program. 1996. *Final Hudson River Estuary Management Plan.* Albany: New York State Department of Environmental Conservation.

Hughes, R. M., R. C. Wildman, and S. V. Gregory. 2005. Changes in fish assemblage structure in the main-stem Willamette River, Oregon. *In Historical changes in large river fish assemblages of the Americas,* ed. J. N. Rinne, R. M. Hughes, and B. Calamusso, 61–74. American Fisheries Society, Symposium 45, Bethesda, MD.

Juet, R. 1909. From "The Third Voyage of Master Henry Hudson," by Robert Juet, 1609. In *Narratives of New Netherland, 1609–1664,* ed. J. F. Jameson, 16–29. New York: Charles Scribner's Sons.

Lake, T. R. 2009. The ancestral lure of the Hudson Estuary: Predictable aquatic resources. In *Mohican Seminar 3, The journey—An Algonquian peoples seminar,* 16-29. Albany: New York State Museum Bulletin 511.

Laws of New York. 1870. Chapter 567, An act to amend the game law. § 1, § 3, § 4.

Laws of New York. 1894. Chapter 627, An act to amend the game law. § 136.

Laws of New York. 1902. Chapter 361, An act to amend the forest, fish and game law in relation to sturgeon. § 1.

Limburg, K. E., K. A. Hattala, A. W. Kahnle, and J. R. Waldman. 2006. Fisheries of the Hudson River. In *The Hudson River estuary,* ed. J. Levinton and J. R. Waldman, 189–204. New York: Cambridge University Press.

Limburg, K. E., and J. R Waldman. 2009. Dramatic declines in North Atlantic diadromous fishes. *BioScience* 59: 955–65.

Lossing, B. J. 1866. *The Hudson, from the wilderness to the sea.* Troy: H. B. Nims.

McDowell, W. H. 1986. Power plant operation on the Hudson River. In *The Hudson River ecosystem,* ed. K. E. Limburg, M. A. Moran, and W. H. McDowell, 40–82. New York: Springer-Verlag.

McLaren, J. B., R. J. Klauda, T. B. Hoff, and M. Gardinier. 1988. Commercial fishery for striped bass in the Hudson River, 1931–1980. In *Fisheries research in the Hudson River,* ed. C. L. Smith, 89–123. Albany: State University of New York Press.

McPhee, J. 2002. *The founding fish.* New York: Farrar, Strauss, and Giroux.

Mearns, E. A. 1898. A study of the vertebrate fauna of the Hudson Highlands, with observations on the Mollusca, Crustacea, Lepidoptera, and flora of the region. *Bulletin of the American Museum of Natural History* 10: 303–52.

Megapolensis, J., Jr. 1909. A short account of the Mohawk Indians, 1644. In *Narratives of New Netherland, 1609–1664,* ed. J. F. Jameson, 163–80. New York: Charles Scribner's Sons.

Mills, E. L., D. L. Strayer, M. D. Scheuerell, and J. T. Carlson. 1996. Exotic species in the Hudson River Basin: A history of invasions and introductions. *Estuaries* 19: 814–23.

Munsell, J. 1869. *Annals of Albany.* Albany: J. Munsell.

Richards, J. F. 2003. *The unending frontier. An environmental history of the early modern world.* Berkeley: University of California Press.

Rostlund, E. 1952. *Freshwater fish and fishing in Native North America.* Berkeley: University of California Press.

Saffron, I. 2004. Introduction: The decline of the North American species. In *Sturgeons and paddlefish of North America,* G. T. O. LeBreton, F. W. H. Beamish, and R. S. McKinley, 1–21. Dordrecht: Kluwer Academic Publishers.

Schmidt, R. E., R. J. Klauda, and J. M. Bartels. 1988. Distributions and movements of the early life stages of three species of *Alosa* in the Hudson River, with comments on mechanisms to reduce interspecific competition. In *Fisheries research in the Hudson River,* ed. C. L. Smith, 193–215. Albany: State University of New York Press.

Smith, C. L. 1985. *Inland Fishes of New York State.* Albany: New York State Department of Environmental Conservation.

Smith, H. H. 1896. Report of the Division of Statistics and Methods of the Fisheries. In U.S. Commission of Fish and Fisheries, *Report of the commissioner for the year ending June 30, 1894,* 115–75.

Snow, D. R., and W. A. Starna. 1989. Sixteenth-Century Depopulation. A View from the Mohawk Valley. *American Anthropologist* 91: 142–149.

Stevenson, C. H. 1899. The shad fisheries of the Atlantic coast of the United States. In U.S. Commission of Fish and Fisheries, Part XXIV, *Report of the commissioner for the year ending June 30, 1898,* 101–269. , Washington, DC: Government Printing Office.

Strayer, D. L., K. Hattala, and A. Kahnle. 2004. Effects of an invasive bivalve (*Dreissena polymorpha*) on fish populations in the Hudson River estuary. *Canadian Journal of Fisheries and Aquatic Sciences* 61: 924–41.

Swaney, D. P., K. E. Limburg, and K. M. Stainbrook. 2006. Some historical changes in the patterns of population and land use in the Hudson River watershed. American Fisheries Society Symposium 51: 75–112.

Thacher, J. 1854. *Military journal, during the American revolutionary war, from 1775 to 1783; describing the events and transactions of this period, with numerous historical facts and anecdotes. To which is added, an appendix, containing biographical sketches of several general officers.* Hartford: Silas Andrus and Son.

Van der Donck, A. 2008. *A Description of New Netherland.* Ed. C. T. Gehring and W. A. Starna. Trans. D. W. Goedhuys. Lincoln: University of Nebraska Press.

Webster, D. A. 1982. Early history of the Atlantic salmon in New York. *New York Fish and Game Journal* 29: 26–44.

Zeisel, W. 1995. Angling on a changing estuary: The Hudson River, 1609–1995. Report to the Hudson River Foundation, New York.

Zlokovitz, E. R., and D. H. Secor. 1999. Effect of habitat use on PCB body burden in Hudson River striped bass (*Morone saxatilis*). *Canadian Journal of Fisheries and Aquatic Sciences* 56: 86–93.

CHAPTER 5

HERPETOFAUNA OF THE HUDSON RIVER WATERSHED

A Short History

Alvin R. Breisch

ABSTRACT

The amphibian and reptile species found in the Hudson River watershed today are the result of a series of invasions that followed the retreat of the Laurentide Ice Sheet about 11,000 years ago. Although Native Americans utilized these species in a variety of ways, little of this use is documented in early literature or archaeological digs. European settlement of the Hudson River Valley has resulted in significant changes to the native populations of amphibians and reptiles, some through direct action but much through the unintentional consequences of other actions. The earliest written reports of amphibians and reptiles in what is now New York came from the Dutch shortly after Henry Hudson sailed up the river now bearing his name. Amphibians and reptiles were originally exploited or persecuted by the Dutch settlers and those that followed. Recognition of New York's herp diversity peaked during the "Naturalist Period," which extended from the early 1700s to the late 1800s. This transitioned into the modern era around the beginning of the twentieth century when popularizing, scientifically studying, and protecting amphibians and reptiles became the standard. Although still threats today, exploiting and persecuting herps have to a large degree been replaced by acceptance, appreciating, and understanding.

THE EARLY YEARS

Native flora and fauna found in New York today have had a relatively short period to colonize the state since the entire state, except for a small area in the western Allegany Plateau, was buried under ice during the most recent glacial period, the Wisconsinan, when the Laurentide Ice Sheet reached its maximum about 21,750 years ago (Isachsen et al. 2000). Approximately 11,000 years ago the ice front had retreated from what is now the Hudson River watershed.

The retreat of the ice 12,000 to 10,000 BP created land forms that can still be seen today and influenced this recolonization. Whereas mammals, birds, and fish can invade new habitats relatively quickly, most amphibians and reptiles (i.e., herpetofauna, herps for short) would have been slower to recolonize the state. Little is known of this invasion, but Holman (1992) suggests that the Wood Frog (*Lithobates sylvaticus*) was the first species of herp invader followed by about two dozen other primary invaders as tundra gave way to boreal forest. Secondary and tertiary invaders would have followed as mixed conifer-hardwood forests and finally deciduous forests developed by about 7500 BP resulting in a herpetofaunal assemblage similar to what we have today. Few archeological sites have been uncovered in New York containing significant evidence of amphibians and reptiles except for species that were used in ceremonies or eaten. Most Native American tribes in New York at the time of European contact revered turtles and adopted the turtle as one of their clan symbols. Turtle rattles used in ceremonies were considered particularly sacred.

The earliest written record of amphibians and reptiles in the state comes from Adriaen Van der Donck (1656) less than fifty years after Henry Hudson sailed up the river now bearing his name. Two sections of his book *A Description of New Netherlands* are of particular interest to herpetologists.

The first section, titled "Of the Fishes," deals with aquatic species of all sorts, from true fishes and sharks to invertebrates such as lobsters, shrimps, oysters, and crabs but also includes whales, porpoises, and turtles. Van der Donck notes, "There are shrimps and tortoise in the water and on the land. Some persons prepare delicious dishes from the water-terrapin, which is luscious food. There are also sea-spiders, and various other products of the ocean, which are unknown in Holland, and are of little consideration, as they contribute little to the wants of human society." He does not further describe "tortoises in the water and on the land" or "water-terrapin" but since his concern was clearly for animals that could be eaten he was most likely referring to the Northern Diamond-backed Terrapin *(Malaclemmys terrapin)* and the Common Snapping Turtle *(Chelydra serpentina),* although it is possible that several species of sea turtles that enter New York waters during the warmer months might have been included. And we cannot rule out the Eastern Box Turtle (*Terrapene carolina*), which was eaten by coal miners in the nineteenth century, or the Wood Turtle (*Glyptemys insculpta*) which was collected and sold as food about the same time (Babcock 1971), as both species are still frequently found in the lower Hudson River watershed.

Understandably, food was a major concern to these early European settlers, but perhaps equally important were the things that could cause harm, which Van der Donck treated in a section titled "Of the Poisons." In this section he relates some of the lore from the Native American tribes he encountered along with observations he claims to have made himself. Van der Donck lists "Several kinds of black, speckled and striped snakes" that are "found in the country" and "are said to have connections with the eels." He notes that "Snakes of those kinds do no damage except destroying young birds" so at least he did not consider all snakes to be a threat but he also note that "Unless they escape from travelers and farmers, they are usually put to death. The Indians do not fear snakes of this kind, for they will run after and take them by the tails, then take hold of them behind the heads and bite them in the neck: thus they kill them."

Van der Donck spent considerable space discussing rattlesnakes. His statement that "[r]attlesnakes like those of Brazil, are found in the country" indicates that in the mid-seventeenth century Brazilian rattlesnakes were better known to the Dutch than the Timber Rattlesnake (*Crotalus horridus*) found in New York. And although the "yellow, black and purple colours" can be seen in modern Timber Rattlesnakes, his description that they had "four sharp teeth in the front of their mouth, which the Indians use for lancets" makes one wonder how he determined that until one realizes that since the snakes were dead when the Indians extracted the fangs, they may have also removed the smaller replacement fangs that lay next to or behind the two front fangs that are normally used to inject venom into the snake's prey (Greene 1997).

Van der Donck continues with a description of the rattlesnake: "at the end of the tail it has a hard, dry, horny substance . . . with which these snakes can rattle so loud that the noise can be heard several rods; but they never rattle unless they intend to bite." "When persons are bitten by those serpents and the poison enters the wound, their lives are in a great danger." His observation that "fortunately the rattlesnakes are not numerous" indicates that he did not travel extensively in the Hudson Highlands. He does report rattlesnakes on Long Island, an area from which rattlesnakes were extirpated about 1920 (Welch 1995). Van der Donck cites "an experiment made on Long-Island with snake-wort, on a large rattlesnake, when a person chewed a quantity of the green plant, and spit some of the juice on the stick, which was then put to the nose of the snake, it caused the snake to thrill and die instantly. The Indians hold this plant in such high estimation, that many of them carry some of it, well dried, with them to cure the bites of those serpents." However, he does note that when bitten by

TABLE 5.1. Amphibians and reptiles first collected and described for science from the Hudson River watershed.

Common name	Species	Notes	Source
Eastern Red-backed Salamander	Plethodon cinereus	Originally reported by Rafinesque 1818	Smith (1963)
Gray Treefrog	Hyla versicolor	As reported in Schmidt 1953	McCoy (1982)
Wood Frog	Lithobates sylvaticus	Originally reported by LeConte 1825	Martof (1970)
Snapping Turtle	Chelydra serpentina	Originally reported by Linnaeus	Gibbons, et al (1988)
Painted Turtle	Chrysemys picta	Originally reported by Schneider 1783	Ernst (1971)
Wood Turtle	Glyptemys insculpta	As reported in Schmidt 1953	McCoy (1982)
Northern Watersnake	Nerodia sipedon	As reported in Schmidt 1953	McCoy (1982)
Eastern Milksnake	Lampropeltis triangulum	Originally reported by Lacepede 1788	Williams (1994)
Timber Rattlesnake	Crotalus horridus	Originally reported by Linnaeus 1758	Collins and Knight (1980)

a rattlesnake, Indians frequently die of the bite, so the snake-wort cure apparently did not always work. Modern observations have shown that as many as 25 percent of rattlesnake bites are "dry," that is, no venom is injected. This could account for the observation by these early settlers of an occasional "cure" after being bit.

Rattlesnakes were not the only species of herpetofauna believed to threaten early humans. "Lizards . . . which have pale bluish tails . . . are much feared by the Indians, because (as they say) this kind will crawl up into their fundamentals, when they lay asleep on the ground in the woods, and cause them to die in great misery." This blue-tailed lizard is the juvenile of the Common Five-lined Skink *(Plestiodon fasciatus)*. Gibbons (1983) warns that one bite of a fried blue-tailed lizard will make a person immediately ill, but there is no documentation that one crawling into your "fundamentals" or any other location will cause great misery let alone death.

"Toads" were included in Van der Donck's account as one "of the poisonous reptiles" he "discovered in the country." It was several centuries until a clear distinction was made between amphibians and reptiles, but the poisonous nature of toads, an amphibian, has long been known.

THE NATURALIST PERIOD

After Van der Donck's detailed history there followed a period of exploration by early naturalists searching for new species of plants and animals to collect and describe. In the eastern United States this "Naturalist Period" (Adler 1979) extended from about 1725 to the mid-nineteenth century. In New York, interest in the native herpetofauna began about 1750 with the advent of the Linnaean system of classification of living organisms, peaking in the early 1800s. As a result, the Hudson River watershed, including portions of New York City, is credited as being the "type locality" (the first place a previously unknown species is collected and described for science) for nine species of amphibians and reptiles (Table 5.1).

The first attempt to list all the species of amphibians and reptiles found in New York was published by James Macauley in 1829. His list placed all species in the class "Amphibia" which was divided into two orders: "Reptilia," which included fourteen species of turtles, frogs, toads, salamanders, and lizards, and "Serpentes," which included thirteen species of snakes. His imprecise descriptions make it impossible to determine exactly which species he was referring to. He referred to salamanders as "lizards" and he recognized four species of toads mainly by color (only three species of toads occur in New York) which he placed in the genus *Rana* (now *Lithobates*), thereby grouping toads with the true frogs. He also considered a number of harmless snakes to be venomous.

Macauley did, however, make a number of observations on Timber Rattlesnakes that are valid. He listed known occurrences from across the state but noted that they do not inhabit all parts of the state. He identified six areas in the Hudson River watershed where rattlesnakes could be found including parts of the Highlands, the Nose in Montgomery County, several places along Schoharie Creek, Glenville in Schenectady County, the Helderbergs in Albany County, and Snake Hill near Newburgh. Rattlesnakes are still found at the former three locations but have been extirpated at the latter three. He noted that they have a preference for oak forests, they can swim across rivers and lakes and may make excursions of four to five miles, all observations that have been confirmed in recent studies (Brown 1993). Macauley noted that rattlesnakes were slaughtered in great numbers directly by man, swine, and by "conflagrations incidental to clearing land." Even in the first half of the nineteenth century, habitat loss was recognized as a reason for the decline of a species. But direct persecution can be credited with causing the biggest losses. In some instances several hundred would be killed in the spring at a single den. In the area around Lake George 1,500 Timber Rattlesnakes were killed in a single year!

James E. DeKay, as Head of Zoology Section of the New York Geological and Natural History Survey, published a five-volume treatise, *The Zoology of New York* (1842–44). Part III, the reptiles and amphibians, was the first comprehensive treatment of these species for New York and was considered a model for a state faunal survey at that time. DeKay described all species known from the state but also included species found in surrounding states that he expected would be found in New York. The volume also included colored illustrations, some more realistic than others, with most of the reptiles appearing more lifelike than the amphibians. Quite advanced for its time, DeKay's *Zoology* corrected previous errors such as clearly distinguishing amphibians from reptiles and salamanders from lizards. But it was not without errors. DeKay described what we now call the Red-spotted Newt (*Notophthalmus viridescens*) as three distinct species in two separate genera: the Yellow-bellied Salamander (*Salamandra symmetrica*), the Scarlet Salamander (*Salamandra coccinea*), and the Crimson-spotted Triton (*Triton punctatus*).

James Eights, a contemporary of DeKay's, documented a number of unusual records from the Hudson Valley. It was Eights who first reported to DeKay that he had found the "Brown Swift" (Eastern Fence Lizard, *Sceloporus undulatus*) in New York, at a site near Fishkill, Dutchess County. And when DeKay (1842) reported that the range of the "Muhlenburgh Tortoise" (Bog Turtle, *Glyptemys muhlenbergii*) had at last been extended into New York, Eights (1853) responded with a short note that for the last thirty years he "had been in the habit of obtaining them from a morass in the county of Rensselaer." Bog Turtles have not been found in Rensselaer County since the time of Eights. In a series of articles in a monthly periodical published in 1835 in Albany, "The Zodiak," Eights (McKinley 2005) reported Red Salamanders (*Pseudotriton ruber*) in swamps around Albany and a Soft-shell Turtle (*Apalone spinifera*) under the Cohoes Falls. He also reported receiving two specimens of Tiger Salamanders (*Ambystoma tigrinum*) from the vicinity of Albany, which are the only records of this species north of Rockland County. And at that time Spotted Turtles (*Clemmys guttata*) were "every where seen in the pools of clear water about the pine plains," an area currently known as the Albany Pine Bush where Spotted Turtles are now extremely rare.

Toward the end of the nineteenth century, and the end of the Naturalist Period, naturalists were revisiting lands that had previously been explored, reporting new species or extending their ranges, even in areas as well known as the Hudson Valley. Around the start of the twentieth century, herpetofauna were championed by three distinct movements: to popularize them, to scientifically study them, and to provide some level of protection to them.

POPULARIZING HERPS

The first person to popularize herps was Raymond L. Ditmars, Curator of Reptiles at the Bronx Zoo, who published numerous books and articles on reptiles. Most significantly for us is his pamphlet on the reptiles in the vicinity of New York City (Ditmars 1905). His observation that "the Spotted Turtle rivals the Painted Turtle in being the most common of the local chelonians" is reminiscent of Eights's earlier statement about Spotted Turtles near Albany. Sadly today neither of them is true. Ditmars also noted that "within fifty miles of New York City, the rattlesnake is now scarce." That statement sounds a lot like Van der Donck's assessment, but today, although far less numerous than they once were, the greatest concentration of the state's rattlesnakes do occur in the Hudson Highlands within about fifty miles of New York City.

Keeping with the tradition of zoo curators reaching out to the public, Carl Kaulfield, Curator of the Staten Island Zoo, published two books

(Kaulfield 1957, 1969) about snakes. Although he included a plea for conservation of these animals, his description of where and how to find them unintentionally led to the decimation of populations of rattlesnakes in Dutchess County and other locations. The late John Behler, Curator of Herpetology at the Bronx Zoo from 1973 to 2006, was also a champion of promoting reptiles and amphibians to the general public. In addition to his international scientific and conservation work, primarily on turtles and crocodilians, Behler found time to undertake long-term detailed studies of the ecology and life history of Bog, Spotted, and Wood Turtles in Westchester, Putnam, and Dutchess counties. He, often with his wife Deborah, produced a number of popular books and field guides on reptiles and amphibians, most notably the field guide to amphibian and reptiles as part of the Audubon Field Guide Series (Behler and King 1979). John was also one of the coauthors of the recently published book on the herpetofauna of New York State (Gibbs et al. 2007).

THE SCIENTIFIC PERIOD

The change from the Naturalist Period of the eighteenth and nineteenth centuries to what I call the Scientific Period of the twentieth century is best exemplified in New York by a team of researchers from Cornell University, Professor Albert Hazen Wright and his wife Anna Allen Wright. The Wrights produced two handbooks (Wright and Wright 1949, 1957) that are still cited as key references for anyone working on frogs or snakes. Earlier, in 1932 A. H. Wright (reprint 2002) published a book on the frogs of Okefenokee Swamp, Georgia, and in this book included life history studies of

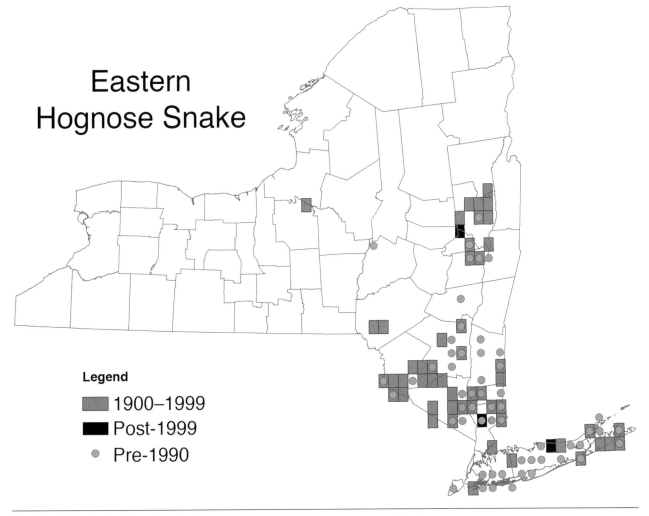

FIG. 5.1. Preliminary map showing current and historical distribution of Eastern Hognosed Snake.

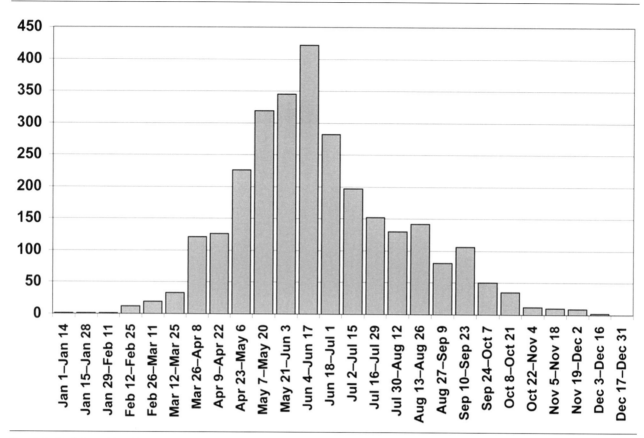

FIG. 5.2. Preliminary bi-weekly activity reports for Painted Turtle.

frogs he conducted far from Georgia and even on species not found in Georgia at all. Hidden in this book is a twenty-two-page treatise on the Mink Frog, *Lithobates septentrionalis*, based on studies he made largely within the Upper Hudson River drainage in the Adirondacks. The information provided on the life history and seasonal activity patterns of this species was by far the most detailed account produced at the time. Wright suggested, based on the Mink Frog's range and habitats, it might be found in the Catskill Mountains. He was apparently not familiar with the earlier report by Mearns (1898) of Mink Frogs from the Hudson Highlands, but based on work conducted since then it is unlikely Mink Frogs are in either the Catskills or the Hudson Highlands. Of all species of amphibians in New York, the Mink Frog, a cold climate–adapted species, may be the species most detrimentally affected by global warming (Popescu 2007) and could possibly disappear from the state's herpetofauna before the end of this century.

Sherman C. Bishop was a contemporary of A. H. and A. A. Wright. Although Bishop published on fishes, birds, mammals, and spiders as well as reptiles and amphibians, he is best known for two publications (Grobman 1952). His "Salamanders of New York" (1941) is still considered one of the most comprehensive treatments of this group produced for any state. Based at the New York State Museum in Albany for several years, he conducted life history studies of all the salamanders associated with the Helderberg Escarpment and associated wetlands in Albany County but also traveled, observed, and collected throughout the Hudson Valley and the rest of the state. In 1947, he produced the "Handbook of Salamanders," which provided in-depth summaries of what was then known about all salamanders in the United States and Canada.

Many other researchers have chosen the Hudson River watershed for their herpetofaunal studies. Space does not allow mention of all of them, but a few stand out as being leaders in the field. Daniel Smiley, whose family established the Mohonk Preserve on the Shawangunk Ridge in Ulster County, was one of them. Dan was one of the pioneers in monitoring acid precipitation and its possible effects on vernal pool amphibians. Perhaps longer than anyone else in the state, Dan recorded the sea-

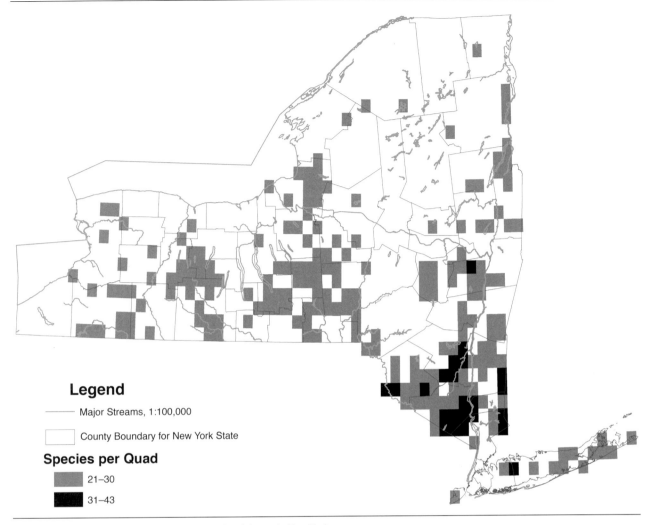

FIG. 5.3. Preliminary map of herpetofaunal species richness in New York.

sonal activity patterns of vernal pool amphibians, notably the ambystomid salamanders, beginning in 1929 and continuing until his death in 1988. The work he began is being continued today under the direction of Paul Huth and John Thompson (see Thompson and Huth, ch. 10 in this volume).

Dr. Margaret M. Stewart, known as "the frog professor" on campus at the State University of New York at Albany, was well known in the Hudson River Valley, both as a conservationist and researcher (Brown and Breisch 2005). Her long-term studies of the amphibians and reptiles in the Albany Pine Bush (Stewart and Rossi 1975), and continued work with numerous undergraduate and graduate students until her death in 2006 (Brown and Breisch 2006), has provided a basis for measuring changes and setting priorities for conservation actions. Stewart also played many roles in her work with the Eastern New York Chapter of The Nature Conservancy, notably as chair of the Conservation Committee. Her commitment is memorialized at both the Discovery Center in the Albany Pine Bush and the Environmental Education Center at Sam's Point in the Shawangunks.

Perhaps the researcher most closely associated with the herpetofauna of the Hudson River itself, is Erik Kiviat of Hudsonia, Ltd., and Bard College. Kiviat has conducted extensive research and inventory of the region's natural resources, always with great consideration for the amphibians and reptiles (Kiviat 1997, 1998). Most notably, he is the expert on the Blanding's Turtle (*Emydoidea blandingii*) in Dutchess County. When asked to assist with developing a wetland mitigation plan to protect the habitat of the Blanding's Turtle, which was required when Arlington High School expanded their facilities, Kiviat and his colleagues incorporated the management strategy into a long-term research project to test the effectiveness of the action (Kiviat et al. 2000). This project serves as an example of how difficult, and expensive, it is to recreate habitat for a threatened turtle species such as the Blanding's Turtle. The project was given the Society for Ecological Restoration Award in 1997.

PROTECTING HERPS

New York was the first state to regulate the collection of any reptile or amphibian (Breisch 1997) when, in 1905, the "Forest, Fish and Game Law" was amended to prohibit the "taking, killing, or exposing for sale of all land turtles or tortoises, including the box and wood turtle." The law was again amended in 1912 setting seasons on the "bullfrog, green frog and spring frog [leopard frog]." What seems like an enlightened beginning of the twentieth century was followed by decades of continued persecution and unregulated collecting of all other species as well as illegal collection of the protected species. The New York State Conservation Department (1939) encouraged people to "control and utilize" Snapping Turtles because of their perceived threat to fish and young waterfowl. And the state authorized counties to establish bounties on "undesirable" species. In three counties in northern New York (Warren, Washington, and Essex) people could collect bounties on Timber Rattlesnakes until the bounty system was outlawed in 1973. Some of the bounty collectors admitted that they collected rattlesnakes in counties that did not have a bounty and turned them in to the county clerk in counties where they could collect the bounty (Furman 2007).

In 1974, New York adopted an endangered species law and the Bog Turtle was the first herp to be placed on the list. With almost all of the known Bog Turtle sites in New York occurring within the Hudson River watershed, this became a useful tool in protecting wetlands within this region of the state. The law was amended in 1983 to include species that could be designated as threatened. Five additional herp species found in the Hudson River watershed have now been added to the lists of endangered and threatened species: Eastern Cricket Frog (*Acris crepitans*) and Eastern Mud Turtle (*Kinosternon subrubrum*) as endangered, and Blanding's Turtle, Timber Rattlesnake, and Eastern Fence Lizard as threatened. Completing the protection picture, in 2006, a bill was enacted giving the New York State Department of Environmental Conservation (DEC) authority to regulate the collection and possession of all native amphibian and reptile species in New York.

As strong as the Endangered Species Law is, a landmark court decision added even more strength. In 1990, a hard-rock mining application near Fishkill, Dutchess County, was proposed by Sour Mountain Realty, Inc. Subsequent review showed that the proposed mine would have a significant impact on a Timber Rattlesnake population by modifying the habitat the snakes used and by making a significant portion of the habitat inaccessible to them. The rattlesnakes from this population overwintered in a den that was located only 260 feet from the Sour Mountain Realty property. Subsequently, a second den was found on their property. The court determined that the actions of Sour Mountain Realty to modify the habitat could be considered a "taking" of a threatened species under the law even if direct killing of the snakes could not be demonstrated, and that DEC could deny the permit (Amato and Rosenthal 2001).

A 240-unit residential housing project below the ridge of Schunnemunk Mountain, Orange County, also involved Timber Rattlesnakes. In this case the den was initially believed to be about a half mile from the proposed development. An eight thousand foot long fence four feet high, constructed of half-inch hardware cloth was erected as a snake barrier to keep the rattlesnakes and the residents separated. A three-year study demonstrated that the fence was quite ineffective in keeping the snakes out of the development and that snakes that managed to find their way past the fence and into the development often were killed (Breisch, et al. 2005).

A three-year undercover investigation, dubbed "Operation Shellshock," by DEC into the poaching, smuggling, and illegal sale of protected reptiles and amphibians led in early 2009 to charges against eighteen individuals, eight of whom illegally collected wood turtles, box turtles, rattlesnakes, copperheads, hognosed snakes, and salamanders from the Hudson River Valley, often from state park lands or lands owned by not-for-profit conservation organizations. More than 2,400 individual turtles, snakes, and salamanders were involved in the documented crimes with several hundred of these being confiscated during the investigation. The investigation was coordinated through DEC's Bureau of Environmental Crimes Investigation. Undercover investigators spent hundreds of hours afield with poachers and at commercial herp shows where the under-the-table sale of protected species was occurring. Operation Shellshock should serve not only as a deterrent to those who illegally trade in our native herpetofauna but also should raise public awareness about the need to protect all exploitable wildlife. During the summer of 2009, six of the confiscated copperheads were released back into the wild at Mohonk Preserve following a year of quarantine to assure that the snakes were healthy and did not carry any exotic diseases. In July 2010, the Operation Shellshock team members were awarded the first annual Peter A. A. Berle Memorial Award in recognition of their outstanding contribution in furtherance of the DEC's mission of environmental stewardship.

THE HERP ATLAS

In 1990, DEC began a ten-year project to map the distribution of all species of amphibians and reptiles that occur in the wild in New York. This project, involving more than 1,900 volunteers, collected data from nearly a thousand atlas blocks across the state. More than 65,000 individual species reports were received documenting the distribution of sixty-nine native and three introduced species. Of these seventy-two species, twenty-seven species of amphibians and twenty-eight species of reptiles occur in the Hudson River watershed, making it the most diverse watershed in New York for herpetofauna. Preliminary data have been used to create maps for a book on New York's herpetofauna (Gibbs et al. 2007). A publication is now in preparation providing more detailed distribution maps which also include historic and museum records before 1990 and new locations documented since 1999. Accompanying each species account will be a map showing these data (Fig. 5.1). A chart showing number of records received by date will also be included to highlight seasonal activity peaks for each species (Fig. 5.2). When all the distribution information is combined, the atlas blocks with the highest species richness for herps are seen to be primarily in the Hudson River watershed (Fig. 5.3). These data can be used in project reviews for proposed developments, conservation planning on public and private lands, or to identify "Important Herp Areas," similar to a program now being used to identify Important Bird Areas.

ACKNOWLEDGMENTS

I thank Dennis Wischman and John Ozard for help creating the maps. Ozard also designed the database used to analyze the Herp Atlas data. I thank Zack Cava for sharing historical information on herps of New York. This manuscript benefited from comments by Thomas Pauley, Mark Fitzsimmons, and an anonymous reviewer. Scientific and common names of amphibians and reptiles follow that of Crother (2008).

REFERENCES CITED

Adler, K. 1979. A brief history of herpetology in North America before 1900. Society for the Study of Amphibians and Reptiles. *Herpetological Circular* No. 8.

Amato, C. A., and R. Rosenthal. 2001. Endangered Species protection in New York after *State v. Sour Mountain Realty, Inc. New York University Environmental Law Journal* 10(1): 117–45.

Babcock, H. 1971. *Turtles of Northeastern United States.* New York: Dover.

Behler, J. L., and F. W. King. 1997. *National Audubon Society field guide to North American amphibians and reptiles.* New York: Alfred A. Knopf.

Bishop, S. C. 1941. The salamanders of New York. New York State Museum Bulletin No. 324.

———. 1947. *Handbook of salamanders.* Ithaca: Comstock Press.

Breisch, A. R. 1997. The status and management of turtles in New York. In *The status and conservation of turtles in the northeast,* ed. T. Tyning. Massachusetts Audubon Society.

———, T. Tear, and R. Stechert. 2005. Rattlesnake exclusion fences: Do they really work? Joint Meeting of Ichthyologists and Herpetologists. Norman, Oklahoma.

Brown, W. S. 1993. Biology, status, and management of the timber rattlesnake (*Crotalus horridus*): A guide for conservation. *SSAR Herpetological Circular* 22:1–78.

———, and A. R. Breisch. 2005. Margaret McBride Stewart. *Copeia 2005* (3): 701–708.

———. 2006. Margaret McBride Stewart (1927–2006). *Herpetological Review* 37(4): 396–98.

Collins, J. T., and J. L. Knight. 1980. *Crotalus horridus. Catalogue of American Amphibians and Reptiles* 253: 1–2.

Crother, B. I. 2008. Scientific and standard English names of amphibians and reptiles of North America north of Mexico, with confidence in our understanding. *Herpetological Circular* No. 37.

DeKay, J. E. 1842. *Zoology of New-York; Part III. Reptiles and Amphibia.* Albany.

Ditmars, R. L. 1905. The reptiles of the vicinity of New York City. *Journal of the American Museum of Natural History* 5: 93–140.

Doyle, R. F. 1995. The timber rattlesnake on Long Island. *Long Island Forum* (Spring): 18–21.

Eights, J. 1853. Scraps from a naturalist's note book. No. 15. *Country Gentleman.* Albany.

Ernst, C. H. 1971. *Chrysemys picta. Catalogue of American Amphibians and Reptiles* 106: 1–4.

Furman, J. 2007. *Timber rattlesnakes in Vermont and New York: Biology, history, and the fate of an endangered species.* Lebanon, NH University Press of New England.

Gibbons, W. 1983. *Their blood runs cold.* Tuscaloosa: The University of Alabama Press.

Gibbons, J. W., S. S. Novak, and C. H. Ernst. 1988. *Chelydra serpentina. Catalogue of American Amphibians and Reptiles* 420: 1–4.

Gibbs, J. P., A. R. Breisch, P. K. Ducey, G. Johnson, J. L. Behler, and R. A. Bothner. 2007. *The amphibians and reptiles of New York State: Identification, life history, and conservation.* New York: Oxford University Press.

Greene, H. W. 1997. *Snakes: The evolution of mystery in nature.* Berkeley and Los Angeles: University of California Press.

Grobman, A. B. 1952. Sherman C. Bishop 1887–1951. *Copeia 1952* (3): 127–28.

Holman, J. 1992. Patterns of herpetological re-occupation of postglacial Michigan: Amphibians and reptiles come home. *Michigan Academician* 24: 453–66.

Isachsen, Y. W., E. Landing, J. M. Lauber, L. V. Rickard, W. B. Rogers, eds. 2000. Geology of New York: A simplified account. 2nd ed. New York State Museum Educational Leaflet 28.

Kaulfield, C. 1957. *Snakes and snake hunting.* Garden City: Hanover House.

———. 1969. *Snakes, the keeper and the kept.* Garden City: Doubleday.

Kiviat, E. 1997. Where are the reptiles and amphibians of the Hudson River? Part 1. *News from Hudsonia* 12(2–3): 1, 3–5.

———. 1998. Where are the reptiles and amphibians of the Hudson River? Part 2. *News from Hudsonia* 13(3): 1–7.

———, G. Stevens, R. Brauman, S. Hoeger, P. J. Petokas, and G. G. Hollands. 2000. Restoration of wetland and upland habitat for Blanding's turtle, *Emydoidea blandingii*. *Chelonian Conservation Biology* 3: 650–57.

Macauley, J. 1829. *The natural, statistical, and civil history of the State of New York in three volumes.* Volume 1: 430–520.

Martof, B. S. 1970. *Rana sylvatica. Catalogue of American Amphibians and Reptiles* 86: 1–4.

McCoy, C. J. 1982. Amphibians and reptiles in Pennsylvania: checklist, bibliography and atlas distribution. Carnegie Museum of Natural History Special Publication Number 6.

McKinley, D. L. 2005. James Eights; 1798–1882. Antarctic explorer, Albany naturalist, his life, his times, his works. New York State Museum Bulletin 505. Albany: University of the State of New York.

Mearns, E. A. 1898. A study of the vertebrate fauna of the Hudson Highlands, with observations on the Mollusca, Crustacea, Lepidoptera, and the Flora of the region. *Bulletin of the American Museum of Natural History* X, Article XVI: 302–52.

New York State Conservation Department. 1939. Snapping turtle control and utilization.

Popescu, D. V. 2007. Complex interactions shaping Mink Frog (*Rana septentrionalis*) distribution in New York State: pond factors, landscape connectivity and climate change. MS Thesis. SUNY College of Environmental Science and Forestry, Syracuse.

Smith, P. W. 1963. *Plethodon cinereus. Catalogue of American Amphibians and Reptiles* 5: 1–3.

Stewart, M. M., and J. Rossi. 1981. The Albany Pine Bush: A northern outpost for southern species of amphibians and reptiles in New York. *American Midland Naturalist* 106: 282–92.

Van der Donck, A. 1656. *Description of the New Netherlands.* Amsterdam.

Williams, K. L. 1994. *Lampropeltis triangulum.* Catalogue of American Amphibians and Reptiles 594: 1–10.

Wright, A. H. 2002. *Life-histories of the Frogs of Okefinokee Swamp, Georgia.* Ithaca: Comstock Publishing Associates, Cornell University Press.

———, and A. A. Wright. 1949. *Handbook of frogs and toads.* Ithaca: Comstock Publishing Associates.

———. 1957. *Handbook of snakes of the United States and Canada.* Vol. 1. Ithaca: Comstock Publishing Associates.

CHAPTER 6

HUMAN IMPACTS ON HUDSON RIVER MORPHOLOGY AND SEDIMENTS

A Result of Changing Uses and Interests

Frank O. Nitsche, Angela Slagle, William B. F. Ryan, Suzanne Carbotte,
Robin Bell, Timothy C. Kenna, and Roger D. Flood

ABSTRACT

Since the beginning of European colonization, the Hudson River Estuary has been a focus of commerce and transportation in the northeastern United States. Varying demands caused by changing technology and increasing population have resulted in significant modifications of large sections of the Hudson River shoreline including channel constriction, filling, relocating, stabilization, and the construction of piers and docks. These modifications resulted in substantial changes of estuarine flow, sediment transport, and deposition patterns in the Hudson River. In this chapter we compare historic documents with modern data from the Hudson River Estuary. Using examples from various parts of the estuary we demonstrate the impact of human modifications on bottom morphology and sediment distribution and how these modifications reflect the changing priorities in the use of the Hudson River.

INTRODUCTION

The Hudson River is one of the major rivers in North America and has been used by humans since long before Europeans arrived. But the exploration of the river by Henry Hudson in 1609 and the European colonization that followed marks a transition to more direct modification and major impacts on river morphology and sediment transport. As described by other papers in this volume, during the eighteenth and nineteenth centuries the Hudson River became the dominant route connecting Europe via New York City to the northern Midwest and the Great Lakes region.

Over the last four centuries, changes in technologies, economic interests, and aesthetic values have influenced the way people used and viewed the Hudson River. The first European settlers used the river mostly for trade, connecting early settlements, and as a food source. Therefore, a major concern of these settlers and tradesman was easy access to the river. On the other hand, the first ships, especially the early Hudson River sloops with their wide flat hulls, were small enough to reach most shores and most of the Hudson River was naturally deep enough for shipping at this time.

With time, the population in the Hudson Valley increased and so did the need for transportation capacity and reliability on the river. Larger boats with deeper drafts required better access to the water, such as larger piers, and improved navigation of the waters, especially after the introduction of the steamboats during the nineteenth century. Further improvements in technology and engineering during the nineteenth and twentieth centuries brought about the building of railway and highway systems, as well as the construction of bridges to further improve transportation along and across the Hudson River.

Although the Hudson River remains an important transportation corridor today, industrialization during the twentieth century shifted the focus of use from transportation toward a resource for the growing industries along the river including power plants and factories. More recently the focus shifted again toward recreational use of the river including parks, sportfishing, and riverview housing developments (Dunwell 2008).

Several sections of the Hudson River in its natural stage did not meet these increasing demands for navigation and usage and, as a result, were significantly modified by humans. These various human activities not only affected the landscape, flora, and fauna along the shores of Hudson River, but also had an impact on the river itself by changing its shape, hydrodynamics, and the nature and distribution of sediments at the bottom of the river.

In this chapter we discuss several examples of human modifications of the Hudson River Estuary during the last centuries and analyze their effects on morphology, sediment distribution, and hydrodynamics.

THE HUDSON RIVER ESTUARY

Our analysis focuses on the ~240 km long Hudson River Estuary, also called the Lower Hudson River, between the Battery in New York City and the Federal Dam in Troy (Fig. 6.1). The Lower Hudson River is a partially mixed, meso-tidal estuary with a tidal range of 1–2 m all the way to Troy (Abood 1974). The estuary has strong seasonal variations in river flow with discharge varying between 100 and >1500 m^3/s at low and peak flow conditions respectively and current velocities up to 1 m/s (Geyer et al., 2000; Geyer and Chant 2005). Maximum water depths (natural and dredged) vary along the estuary with an average of ~15 m in the main channel and a maximum depth of 55 m in the Hudson Highlands. Along some sections of the estuary the main channel is bound by extensive shallow flats with depths between 1 and 3 m (Nitsche et al. 2007). The extent of salt intrusion typically reaches the area of Haverstraw Bay and the Hudson Highlands, but it moves farther northward to Newburgh Bay during extreme dry conditions or southward into the Tappan Zee during extreme fresh water flow events (Geyer and Chant 2005).

The general trends in bottom morphology and sediment distribution of the Hudson River Estuary have been described by Sanders (1974) and Coch (1986) and have recently been refined through comprehensive acoustic mapping and sediment sampling (Bell et al. 2004; Bell et al. 2006; Nitsche et al. 2007).

METHOD AND DATA

To identify different modifications over time and to analyze their effects on riverbed morphology, river flow, and sediment processes we analyze data from the present Hudson River Estuary as well as from historic documents and maps.

We obtained information about the present morphology and sediment distribution of the estuary mainly from the Hudson River Benthic Mapping Project (HRBMP; Nitsche et al. 2005; Bell et al. 2006) and from NOAA's National Ocean Services estuarine datasets (http://estuarinebathymetry.noaa.gov/). These datasets include high-resolution bathymetry, sidescan sonar, seismic

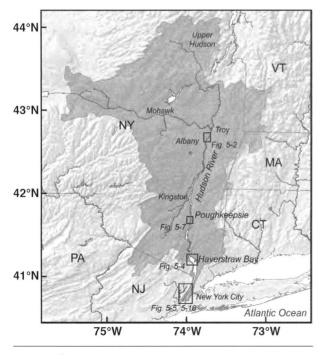

FIG. 6.1. Overview map of Hudson River Valley with the watershed area outlined in dark gray and black rectangles showing the locations of examples discussed in this chapter.

profiles, and sediment information from cores and grabs that allow characterization of the bottom of the river in great detail (Nitsche et al. 2007). These data provide a reference point for determining changes compared to historical data. In addition, past modifications and their effects are in some cases directly visible in the present morphology or in sub-bottom data.

We compare the present morphology, shoreline location, and sediment distribution with historic documents from various sources including historical charts and soundings from NOAA (http://www.nauticalcharts.noaa.gov/), maps from the Library of Congress (http://memory.loc.gov/ammem/gmdhtml/), and historic maps provided by the New York State Department of Environmental Conservation (NYSDEC). Because the Hudson River has always been intensively used for navigation, many of the historic maps include bathymetric information that we used to identify changes in the bottom morphology of the river. We loaded digitized version of these maps into a Geographic Information System (ArcGIS), where we geo-referenced the maps based on known fix points and unchanged segments of the shoreline. Using ArcGIS we digitized shorelines and depth soundings, and converted the depth sounding from fathoms and feet to meter following established procedures (Nitsche et al., 2002). In addition, we used historical shoreline data provided by NYSDEC for the Albany-Troy section of the Hudson River and the 1865 shoreline by the New York State Office of General Services (http://www.nysgis.state.ny.us/gis9/hrshorline.zip).

EXAMPLES OF MODIFICATIONS AND THEIR IMPACT ON THE RIVER

Navigation Enhancements

Since the beginning of the European colonization, the Hudson River has been a major conduit for transporting goods and people connecting the immigration hub of New York City to upstate New York, the Great Lakes region, and the northern Midwest. An increase in traffic and size of ships with time, especially in the second half of the nineteenth century, required improvements of the waterway (Dunwell 2008). Modifications to the river were made to accommodate this growing need.

One of the biggest modifications was the channelization of the Hudson River between New Baltimore and Troy. Originally this stretch of the river was characterized by a system of primary and secondary channels with frequent sandbars and islands. Very likely this was a dynamic system, where depth, location, and shape of channels and sand bars changed over time. This became a major obstacle to increasing river traffic with bigger and deeper vessels. Thus, at the end of the nineteenth century it was decided to close off the side channels and narrow the main channel by constructing longitudinal dikes (Collins and Miller 2009).

In addition to the loss of wetlands, this channelization also altered the morphology and sediment composition of the river bottom. Figure 6.2 shows the present river bottom together with the historic and present shorelines of a small section of the altered part. The reduction in surface area of the channels is clearly visible, but it also shows that the

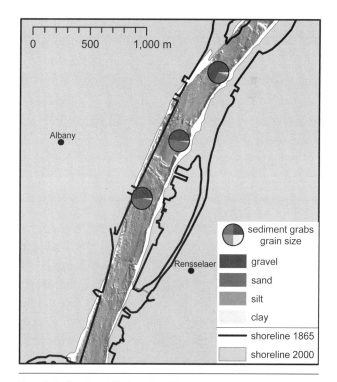

FIG. 6.2. Section of Hudson River Estuary between Albany and Troy, N.Y., (Fig. 6.1) with the present land showing in light gray and the shoreline of 1865 showing as thick black line (http://www.nysgis.state.ny.us/gis9/hrshorline.zip). Pie graphs show the results of grain size analysis of sediment grab samples. They show that gravel and sand dominate the bottom sediment in this stretch. River bottom is shown as sun-illuminated bathymetry derived from Hudson River Benthic Mapping data (Bell et al. 2006).

bottom is now dominated by gravelly sediments where older maps indicate a softer bottom dominated by sand and mud. This change in bottom sediments is most likely caused by increased flow velocities resulting from the reduced cross-section in the narrower channel. The faster flow is moving finer particles from this section of the river and leaving mostly coarser gravel and sand behind.

This change in substrate from sand and mud toward coarser gravel also changed the habitat conditions for species living on the bottom and might even have increased the changes of invasive species to establish themselves. For example, the invasive zebra mussels prefer a harder substrate for colonization and the increased amount of gravel might have helped the migration and establishment of zebra mussels in this section of the Hudson River (Strayer et al. 1996).

Dredging

Before the navigation was influenced by introducing channel constrictions and re-locating the shoreline through filling, different parts of the estuary were frequently dredged to make them passable for larger ships.

Although the channelization improved the navigation significantly, it was not sufficient for the continuing increase in size and draft of ships operating on the Hudson River. Larger vessels required a deeper main channel. The Rivers and Harbors Act of 1910 authorized the U.S. Army Corps of Engineers to create and maintain a navigational channel of 3.6 m (12 ft) up to Albany. The mandatory depth was then subsequently increased to 8.2 m (27 ft) and 9.75 m (32 ft) in 1925 and 1954 respectively (USACE Public Notice No. HR-AFO-09). Although large sections of the Hudson River were at this depth naturally, several other sections required dredging to achieve this depth.

Figure 6.3 shows that large sections of the estuarine Hudson River have been dredged throughout the twentieth century to establish and maintain a continuous navigation channel at the required depth. Most dredging activities took place north of Kingston, N.Y., in the Manhattan section of the river south of the George Washington Bridge and in Haverstraw Bay.

Records of the U.S. Army Corps of Engineers, who manage the dredging operations, show that between 1930 and 2000 ~57,000,000 m³ of material were removed from the Hudson River by dredging (Fountain 2003). This sediment volume is comparable to the estimated sediment input to the lower Hudson River, which ranges from 300,000–1,000,000 metric ton per year (Olsen

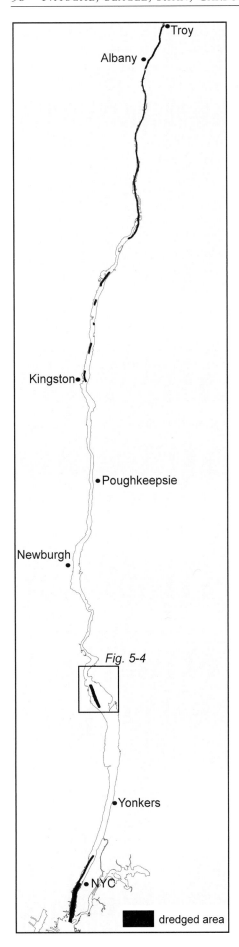

FIG. 6.3. Dredged areas in the northern (left) and southern (right) section of the Hudson River based on information provided by the U.S. Army Corps of Engineers (after Fountain 2003). Dredged areas in the New York harbor are not included.

1979; Bokuniewicz 2006; Wall et al. 2008). Thus, it is likely that the removal of this sediment had a substantial impact on the sediment budget and processes of the system.

One of the smaller dredged areas is located in Haverstraw Bay (Fig. 6.1). Haverstraw Bay consists of a wide, shallow embayment with an ~1 km wide main channel on the western side (Fig. 6.4). The central section of the main channel is naturally less then the required 32 feet (9.75 m) deep. To provide this depth, a central part of the channel was dredged in 1961 and again in 1987 to maintain the channel (Houston et al. 1989).

Haverstraw Bay is one of the major depositional areas of the Hudson River (Nitsche and Kenna 2007). Sediment accumulation during the twentieth century is concentrated within the main channel area and ranges from 0 to >2 m (Fig. 6.4). The highest deposition rates are found in the dredged portion of the main channel. Since this section was dredged in 1961 and 1987, the observed sediment accumulation occurred over a much shorter time (~13 years) while accumulation in the remaining areas took place over a period of eighty years. This large difference in accumulation rates (0.01–0.02 m/y versus 0.15–0.2 m/y) indicates that the dredged channel has a focusing effect on local deposition (Nitsche and Kenna 2007).

It is likely that these high, local accumulation rates are a direct effect of the dredging (Fountain 2003). Similar focusing effects in dredged areas have been observed in other places such as the Elbe River (Johnston 1981; Kerner 2007). In some cases, dredging might even cause erosion along nearby flanks (Johnston 1981). To date, dredging-related erosion has not been identified in the Hudson River itself, but it is suspected for the nearby Jamaica Bay (Hartig et al. 2009).

In addition to these large-scale dredging projects, smaller areas along the edge of the river are dredged to maintain shore access in places such as between piers and marinas. These small-scale operations have only local effects on sediment distribution and flow.

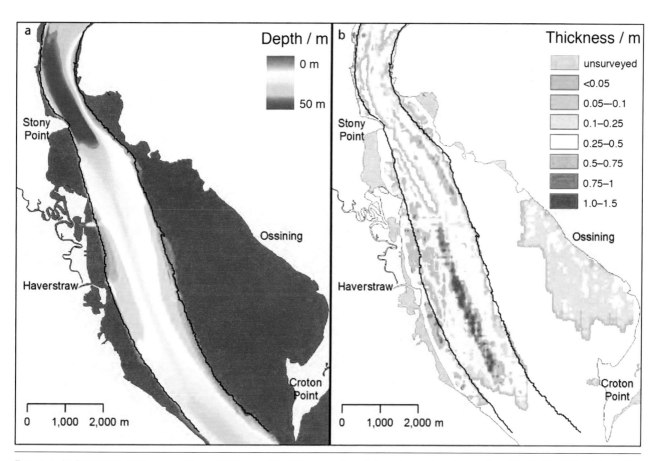

FIG. 6.4. (a) Bathymetry of Haverstraw Bay and (b) interpolated thickness of twentieth-century sediment deposition. Note that the thickest deposition occurs in the area where a narrow band was dredged in the center of the main channel.

Shoreline Development

The increase of transportation as a result of population and industrial growth in the Hudson River Valley and the use of larger vessels had also affected the shoreline. Piers and quays were needed to connect these deeper draft vessels to the shore and allow for rapid unloading and loading of goods and people.

These shoreline developments are most obvious in the New York City area of the Hudson River. Long before the population of the New York metropolitan area reached its current numbers, extension of piers toward deeper waters caused major shoreline changes on the Manhattan and New Jersey shores. Slagle (2007) analyzed the shoreline changes of the Hudson River near New York City area using historical nautical charts. Analysis of the historic maps shows that over time the shoreline developments significantly narrowed the Hudson River in the Manhattan section (Fig. 6.5). Comparing the bathymetry data reveals a successive deepening of the main channel as the cross-section becomes narrower (Klingbeil and Sommerfield 2005; Slagle 2007). It is likely that this deepening is a direct effect of the narrowing as the river system is trying to maintain its cross-sectional area and preserve its dynamic equilibrium surface (e.g., Olsen et al. 1993; McHugh et al. 2004).

The last major shoreline alteration was the creation of Battery Park City in the 1970s, an extension of the lower tip of Manhattan into the Hudson River (Fig. 6.5b). Although other shoreline extensions often filled the shallow water areas, the Battery Park City extension was built into the main channel, where it provided an obstacle to the flow. Comparison of more recent bathymetry data shows a shift of dominant erosion from the deep channel near the east shore toward the middle of the Hudson River (Fig. 6.5b). This could indicate a shift of the main channel flow from the east to the center of the river (Slagle, 2007).

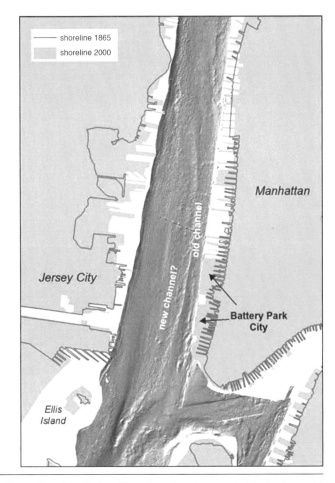

FIG. 6.5. (a) Changes in shoreline between 1865 and 2000 in the Manhattan section of the Hudson River show the enlargement of the land area; (b) Shoreline and bottom morphology of the Hudson River near Battery Park City. River bottom is shown as sun-illuminated bathymetry derived from Hudson River Benthic Mapping data (Bell et al. 2006). Battery Park City is built into the old channel path and a new channel is forming closer to the center of the river.

Railroads

The introduction of the railway in the middle of the nineteenth century provided an additional mode of transportation within the Hudson Valley. Although the waterway continued to be an important pathway for the transportation of goods, especially bulk materials, trains became more and more important (Dunwell 2008).

The shore of the Hudson River is close to sea level all the way to Troy, New York, and the relief along the Mohawk River valley between Troy and the Great Lakes is relatively modest compared to other crossings of the Appalachian Mountains. This low-relief route is one of the preferred paths across the Appalachian Mountains to the Great Lakes region and the Chicago area.

To take advantage of this natural, low-relief route, the railroad tracks have been built directly along or close to the shoreline of the Hudson River. Where the tracks run directly along the shore they were stabilized with beds and causeways constructed of boulders, sometimes called riprap (Fig. 6.6), which created a harder, erosion-resistant shoreline. The railroad tracks also cut off many smaller inlets and creeks to keep the tracks straight, which limited the connection and exchange of water and sediments between these wetlands and smaller tributaries with the main stream of the river.

The importance of the railroad beds and causeways and associated hardening on river flow and sedimentation of the entire system is unknown.

FIG. 6.6. Hudson River shoreline with railroad dam hardened with riprap and a narrow bridge providing limited exchange with small tributary.

Large segments of the Hudson River shoreline naturally consists of hard bedrock. On the other hand, Ellsworth (1986) estimated that exchange with shoals and bank erosion could be a significant component of the Hudson River sediment budget. Today more than 90 percent of the Lower Hudson shoreline is composed of rocks or artificially hardened, and, thus, only a fraction of the shoreline is available for erosion. As a result the contribution from shoreline erosion to the sediment load of the river is only on the order of six thousand metric tons per year (Ellsworth 1986).

The disrupted or reduced physical connection between wetlands and the main river has reduced sediment exchange between the river and the wetlands. This probably reduced sediment contributions from the river to the wetlands, while also reducing the loss of sediments from the wetlands to the main river.

Bridges

The first railroads ran along the river shore. Crossing the Hudson River using ferries was time consuming, but the width of the river, which varies between 400 m near Albany to >4,800 m in Haverstraw Bay, was a challenge for bridge construction. During the second half of the nineteenth century growing traffic and population increased pressure to replace ferries with bridges. In 1888 the Poughkeepsie Railroad Bridge was the first major bridge south of Albany to cross the Hudson. Several more bridges followed, including the Bear Mountain Bridge (1924), the George Washington Bridge (1931), and the Tappan Zee Bridge (1955), as the pressure for bridging the Hudson River increased as result of the use of automobiles (Adams 1996; Dunwell 2008).

Although bridges can have a great impact on the economy and society of a wider region their impact on the riverbed is local and usually small compared to dredging and shoreline alterations. Some bridges, such as the George Washington Bridge, do not have pillars in the river bed and, thus, do not have an effect on morphology or sediment distribution. Conversely, bridges constructed with pillars in the river influence the local flow, and affect riverbed morphology.

Fig. 6.7. Perspective image of the river bottom morphology near Poughkeepsie. Two bridges span the Hudson River with pillars in the river bed for support. The bottom topography shows a series of sediment ridges that originate from the bridge pillars.

Figure 6.7 shows a perspective view of the river bottom near Poughkeepsie. The local morphology around the bridge is dominated by ridges and scours running parallel to the river flow. The ridges usually form up- and downstream of the bridge pillars whereas the troughs are located between the pillars. This pattern reflects the changes in the flow dynamics created by the bridge pillars (Melville and Coleman 2000). The focusing of the flow between the pillars generates higher flow velocities and thus erosion, whereas reduced flow behind the pillars allows sediments to settle and form the observed ridges. Since the Hudson River is a tidal estuary with water flowing both ways, the ridges and troughs are observed up- and downriver of the bridge.

The change in flow dynamic also affects the sediment distribution. The faster flow between the pillars sweeps finer particles away and leaves a coat of coarser sediments behind whereas the ridges often consist of more fine-grained sediments. However, the observed differences in Hudson River sediments are small and, like the effect on bottom morphology, local, reaching only up to one thousand meters up- and downstream of the bridge (Fig. 6.7).

INFLUENCES ON SEDIMENT SUPPLY

The sediment budget of the Hudson River varies annually between 300,000 and 1,000,000 metric tons per year (Olsen 1979; Ellsworth 1986; Bokuniewicz 2006; Wall et al. 2008). Beside the direct changes to the Hudson River caused by modifying the shoreline and bottom by dredging and filling, human activity also altered the sediment supply to the river.

The New York State Department of Environmental Conservation dam inventory of 2008 lists more than two thousand dams in the Hudson River watershed (Fig. 6.8; http://www.nysgis.state.ny.us/gisdata). The total effect of these dams on the sediment flux is uncertain. Monitoring of suspended sediment loads shows that much of the suspended sediment flows over these dams, especially the Green Island Dam in Troy (Wall et al. 2008). On the other hand, these dams are likely to interrupt bedload transport of sediment, and it has been shown that the cumulative effect of smaller dams can significantly alter sediment load in a system (Walter and Merritts 2008). Although a model of the sediment input of twenty tributaries in the lower estuary estimates a contribution of eighty to one hundred thousand metric tons per year (Lodge 1997), the actual sediment load of most tributaries remain uncertain.

Dams might have reduced the effects of land use changes with time. Clearing original forest for agriculture and settlements after colonization and replanting of forests again in the twentieth century should have resulted in significant changes of sediment availability and runoff pattern. The dams would have acted as buffers for the increased sediment supply when the original forest was cleared. On the other hand, the dams would have also reduced the sediment signal of the subsequent reduction of sediment delivery during twentieth-century reforestation. Unfortunately, the necessary data to reconstruct the impact of these changes on sediment delivery are not available.

Although dams might have reduced sediment input from tributaries, the increasing number of people living in the Hudson River Valley became a significant source of sediment. Systematic treatment of sewage including the removal of large particles did not begin until the second half of the twentieth century (Brosnan et al. 2006). For most of prior centuries, sewage and other waste was simply released into the Hudson River. Gross (1974) analyzed various sources that contribute sediment to the Hudson River before major upgrades of the sewage treatment plants were installed. He estimated the input of solids from sewage into the Hudson River, Newark Bay, and Upper Bay to be ~270,000 metric tons per year, which is about a third of the estimated annual load of ~800,000–1,000,000 metric tons per year of waterborne "natural" sediments. A later estimate from 1986 shows the significantly lower contribution of ~52,000 metric tons per year of sediment from sewage and waste sources (Ellsworth 1986), which demonstrate the effect of systematic sewage treatment since the Clean Water Act of 1972.

It is unclear whether much of this anthropogenic sediment remained in the Hudson River. Much of these sediments were probably deposited in the New York Harbor and removed as part of the dredging operations there. On the other hand, some studies identified changes in sedimentation rates in neighboring areas such as Raritan Bay that could be related to increased anthropogenic sediment supply (Gaswirth et al. 2002).

On a much smaller scale people also dumped a variety of objects and waste into the Hudson River—sometimes intentionally, sometimes unintentionally. At many sites we find shipwrecks and other remnants of human and industrial activity, which are a byproduct of the intense use of the Hudson River as a waterway. In some cases these objects cause local scouring of sediments (Fig. 6.9). In other cases, when they are located in shallow water they could create navigation obstacles.

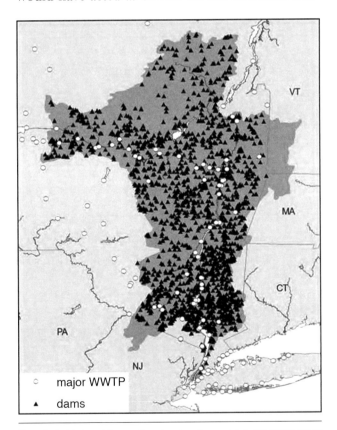

FIG. 6.8. Hudson River valley dam locations and major waste water treatment plants (WWTP) (http://www.nysgis.state.ny.us/from State Pollutant Discharge Elimination System-New York State (NYSDEC) 2007 and NYS DEC dam inventory 2008).

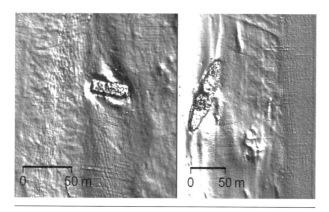

FIG. 6.9. Examples of shipwreck on the bottom of the Hudson River showing local scouring (left) and a combination of scour and deposition (right). Images are artificial sun-illuminated bathymetry data (after Nitsche et al. 2005).

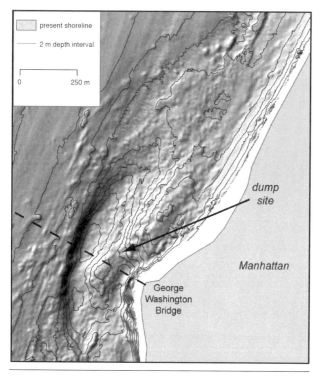

FIG. 6.10. Mound of material dumped underneath and north of the George Washington Bridge.

One specific example is a mound in the main channel underneath and north of the George Washington Bridge (Fig. 6.10). During the early twentieth century the municipal garbage was dumped on the continental shelf in New York Bay. During World War II the threat of enemy submarines caused the city to temporarily abandon this practice and fifty million cubic yards of waste were dumped in the harbor and the Hudson River. This included eleven million metric tons that were put under the George Washington Bridge (Gross 1974).

SUMMARY AND DISCUSSION

The different examples described in this chapter demonstrate that the Hudson River Estuary was subject to significant human alterations during the last four hundred years. After colonizing the Hudson Valley people modified the river according to their changing needs and uses, which often followed technological advances and population increase.

The largest direct alterations, which had the biggest impact on bottom morphology and sediment processes, are shoreline changes, dredging and filling operations that were made to enhance navigation on the river and to gain additional land.

Relocating and hardening shorelines appear to have the most permanent effects, especially since these modifications often prevent natural readjustments. In addition, these shoreline changes can have consequences in terms of unwanted erosion and permanent alterations of the river flow system.

Dredging also alters the riverbed morphology and can affect sediment processes by focusing sediment flow into the dredged areas, starving other areas of sediment supply, and occasionally causing erosion in nearby areas. On the other hand, the focusing of sediment accumulation in dredged areas tends to restore natural equilibrium conditions to the river bottom, if left to itself.

Smaller, local alterations such as bridges with pillars have only local effects on the bottom morphology and sediment distribution, although locally the effects can be large.

Indirect impacts of human activities on sediment and morphology of the Hudson River are related to alterations of sediment supply. Changes in land use, the construction of dams, and the release of sewage and other waste into the river modified the amount and type of sediment delivered to the system. In this chapter we do not discuss the impact of land use, although it could have had a stronger impact on the sediment budget, because the necessary detailed data on this topic are still insufficient.

The alterations of the river by dredging, channelization, and stabilization often caused additional problems, such as unwanted erosion, shifts in sedimentation pattern and changes of habitats. A common response to these consequences was further alterations of the channel, landfill, or additional dredging.

This analysis shows that the present Hudson River is a result of many human modifications. Although some sections are close to their natural state most others have been significantly altered. The available information allows us to identify and quantify the impacts some modifications had on the river. However, the existing data are not sufficient yet to quantify the impact of other, watershed-wide modifications such as dams or land use changes.

ACKNOWLEDGMENTS

This study used data acquired primarily as part of the Hudson River Benthic Mapping Project managed by the Hudson River National Estuarine Research Reserve and funded by the New York State Department of Environmental Conservation with funds from the Environmental Protection Fund through the Hudson River Estuary. Further funds were provided by the Hudson River Foundation grant 008/05A. Many people contributed to the work with their discussion and insights into the history of the Hudson River. We thank especially Margie Turin, Cecilia McHugh, Dan Miller, John Ladd, Steve Chillrud, Rocky Geyer, and others for their input and contribution to this work. This is Lamont contribution 7340 and contribution 1405 of the School of Marine and Atmospheric Sciences.

REFERENCES CITED

Abood, K. A., 1974. Circulation in the Hudson Estuary. *Annals of the New York Academy of Sciences* 250(1): 39–111.

Adams, A., 1996. *The Hudson River guidebook.* New York: Fordham University Press.

Bell, R. E., R. D. Flood, S. M. Carbotte, W. B. F. Ryan, C. M. G. McHugh, M. Cormier, R. Versteeg, H. Bokuniewicz, V. Ferrini, J. Thissen, J. W. Laad, and E. A. Blair. 2006. Benthic habitat mapping in the Hudson River Estuary. In *The Hudson River Estuary,* ed. J. S. Levinton and J. R. Waldman. New York: Cambridge University Press, 51–64.

Bell, R. E., R. D. Flood, S. M. Carbotte, W. B. F. Ryan, F. O. Nitsche, S. Chillrud, R. Arko, V. Ferrini, A. Slagle, C. Bertinato, and M. Turrin. 2004. Hudson River Estuary Program Benthic Mapping Project New York State Department of Environmental Conservation Phase II—Final Report, Lamont-Doherty Earth Observatory.

Bokuniewicz, H., 2006. Sedimentary processes in the Hudson River Estuary. In *The Hudson River Estuary,* ed. J. S. Levinton and J. R. Waldman. New York: Cambridge University Press, 39–50.

Brosnan, T., A. Stoddard, L. Hetling, J. Levinton, and J. Waldman. 2006. Hudson River sewage inputs and impacts: Past and present. In *The Hudson River Estuary,* ed. J. S. Levinton and J. R. Waldman. New York: Cambridge University Press, 335–48.

Coch, N. K., 1986. Sediment characteristics and facies distributions. *Northeastern Geology* 8: 109–29.

Collins, M. J. and D. Miller. 2009. Hudson River floodplain change over the 20th century, Geological Society of America, Northeastern Section—44th Annual Meeting. Abstract volume 41.

Dunwell, F., 2008. *The Hudson: America's river.* New York: Columbia University Press.

Ellsworth, J. M., 1986. Sources and sinks for fine-grained sediment in the lower Hudson River. *Northeastern Geology* 8: 141–55.

Fountain, K., 2003. The concept of an equilibrium profile and the effect of dredging on sedimentation in two reaches of the Hudson River Estuary. MA Thesis, Columbia University, New York.

Gaswirth, S. B., G. M. Ashley, and R. E. Sheridan. 2002. Use of seismic stratigraphy to identify conduits for saltwater intrusion in the vicinity of Raritan Bay, New Jersey. *Environmental and Engineering Geoscience* 8(3): 209–18.

Geyer, W. R., and R. Chant. 2005. The physical oceanography processes in the Hudson River Estuary. In *The Hudson River Estuary,* ed. J. S. Levington and J. R. Waldman. New York: Cambridge University Press, 24–38.

Geyer, W. R., J. H. Trowbridge, and M. M. Bowen. 2000. The dynamics of a partially mixed estuary. *Journal of Physical Oceanography* 30: 2035–48.

Gross, M. G., 1974. Sediment and waste deposits in New York Harbor. *Annals of the New York Academy of Sciences* 250(1): 112–28.

Hartig, E. K., V. Gornitz, A. Kolker, F. Mushacke, and D. Fallon. 2009. Anthropogenic and climate-change impacts on salt marshes of Jamaica Bay, New York City. *Wetlands* 22(1): 71–89.

Houston, L., M. W. LaSalle, J. D. Lunz, and A. Vicksburg. 1989. Predicting and monitoring dredge-induced dissolved oxygen reduction. Environmental effects of dredging, Technical note. Army engineer waterways experiment station Vicksburg, MS, environmental lab.

Johnston, S. A., 1981. Estuarine dredge and fill activities: A review of impacts. *Environmental Management* 5(5): 427–40.

Kerner, M., 2007. Effects of deepening the Elbe Estuary on sediment regime and water quality. *Estuarine, Coastal and Shelf Science* 75(4): 492–500.

Klingbeil, A. D., and C. K. Sommerfield. 2005. Latest Holocene evolution and human disturbance of a channel segment in the Hudson River Estuary. *Marine Geology* 218(1–4): 135–53.

Lodge, M. J., 1997. A model of tributary sediment input to the tidal Hudson River. MS Thesis, State University of New York at Stony Brook, Stony Brook.

McHugh, C. M. G., S. F. Pekar, N. Christie-Blick, W. B. F. Ryan, S. Carbotte, and R. Bell. 2004. Spatial variations in a condensed interval between estuarine and open marine settings: Holocene Hudson River Estuary and adjacent continental shelf. *Geology* 32: 169–72.

Melville, B., and S. Coleman. 2000. Bridge scour. Water Resources Publication, Highlands Ranch, Colorado.

Nitsche, F. O., R. E. Bell, C. Bertinato, G. Fails, S. M. Carbotte, and W. B. F. Ryan. 2002. Sediment dynamics derived from historic bathymetry data of the Hudson River. Eos Trans. AGU, 83(47): Fall Meet. Suppl. Abstract OS21D-02.

Nitsche, F. O., R. E. Bell, S. M. Carbotte, W. B. F. Ryan, R. D. Flood, V. Ferrini, A. Slagle, C. M. G. McHugh, S. Chillrud, T. Kenna, D. L. Strayer, and R. M. Cerrato. 2005. Integrative acoustic mapping reveals Hudson River sediment processes and habitats. EOS Trans. AGU, 86(24): 225, 229.

Nitsche, F. O., and T. C. Kenna. 2007. Distribution, timing, and dynamic of deposition in the Hudson River Estuary. Hudson River Foundation, Final report 008/05A, Lamont-Doherty Earth Observatory of Columbia University, Palisades.

Nitsche, F. O., W. B. F. Ryan, S. M. Carbotte, R. E. Bell, A. Slagle, C. Bertinado, R. Flood, T. Kenna, and C. McHugh. 2007. Regional patterns and local variations of sediment distribution in the Hudson River Estuary. *Estuarine, Coastal, and Shelf Science* 71(1–2): 259–77.

Olsen, C. R., 1979. Radionuclides, sedimentation and the accumulation of pollutants in the Hudson estuary. PhD Thesis, Columbia University, New York.

Olsen, C. R., I. L. Larsen, P. J. Mulholland, D. K. L. Von, J. M. Grebmeier, L. C. Schaffner, R. J. Diaz, and M. M. Nichols. 1993. The concept of an equilibrium surface applied to particle sources and contaminant distributions in estuarine sediments. *Estuaries* 16(3): 683–96.

Sanders, J. E., 1974. Geomorphology of the Hudson Estuary. *Annals of the New York Academy of Sciences.* New York Academy of Sciences 250(1), 5–38.

Slagle, A., 2007. Spatial and temporal variability of sedimentary processes in the Hudson River Estuary. PhD Thesis, Columbia University.

Strayer, D., J. Powell, P. Ambrose, L. Smith, M. Pace, and D. Fischer. 1996. Arrival, spread, and early dynamics of a zebra mussel (Dreissena polymorpha) population in the Hudson River estuary. *Canadian Journal of Fisheries and Aquatic Sciences* 53(5): 1143–49.

Wall, G., E. Nystrom, and S. Litten. 2008. Suspended sediment transport in the freshwater reach of the Hudson River Estuary in eastern New York. *Estuaries and Coasts* 31(3): 542–53.

Walter, R. C., and D. J. Merritts. 2008. Natural streams and the legacy of water-powered mills. *Science* 319(5861): 299–304.

CHAPTER 7

THE EARLIEST THIRTEEN MILLENNIA OF CULTURAL ADAPTATION ALONG THE HUDSON RIVER ESTUARY

Christopher R. Lindner

ABSTRACT

Here we explore through time the shifting balance between ancient people's activity in contrastive habitats, either close to the Hudson River estuary or in the backcountry. Our lenses are new techniques of analysis, many of them microscopic, which focus on relatively recent information about subsistence. We consider five shifts in settlement patterns that may have occurred due to changes in food procurement strategies.

OVERVIEW

During the last quarter century, most professional fieldwork in eastern New York, as elsewhere in this country, has taken place in response to proposals for development. Municipal application of the New York State Environmental Quality Review Act (SEQRA) increasingly mandates that cultural resource management (CRM) archaeologists do research ahead of ground disturbance. Such archaeology has produced a meaningful accumulation of new data, particularly on the physiographic settings of sites, as fieldworkers investigate a broad spectrum of riverine and backcountry habitats. Meanwhile, academic and museum researchers have led in the examination of microscopic evidence for the uses of plants and animals. With recent discoveries of buried camps in diverse biomes and new techniques of inquiry into their contents, we have fresh glimpses of ecological relationships that may change our views of ancient lifeways in the Hudson watershed. By virtue of a greater familiarity with the middle reaches of the estuary, there is likely a bias in my emphasis on sites from Nutten Hook to the Hudson Highlands, even though I include discoveries up and down the river, and in neighboring drainages.

Despite the probability that canine companions were with ancient people throughout their stay in the valley, I'll avoid "arch bark," the archaeological jargon that is endemic to scientists and lay enthusiasts alike. I won't use the technical names of evolutionary stages, or cultural phases and periods, such as the new "South Hill complex of the early Late Archaic period." Instead, I will discuss lengthy spans of the past in terms of adaptation. Dates will express calendar years before 2010 and will approximate times given originally according to the radiocarbon clock, which were more recent by centuries in the earliest four-fifths of the thirteen millennia in question.

Based on recent models and evidence from various sources, we'll consider whether the estuary and its valley saw five major shifts in subsistence-settlement systems that *may* have taken place in the seasonal balance between the Hudson and inland habitats. The major subsistence-settlement pattern shifts are as follows:

1. From starting places along the river, near migration routes of caribou, people relocated

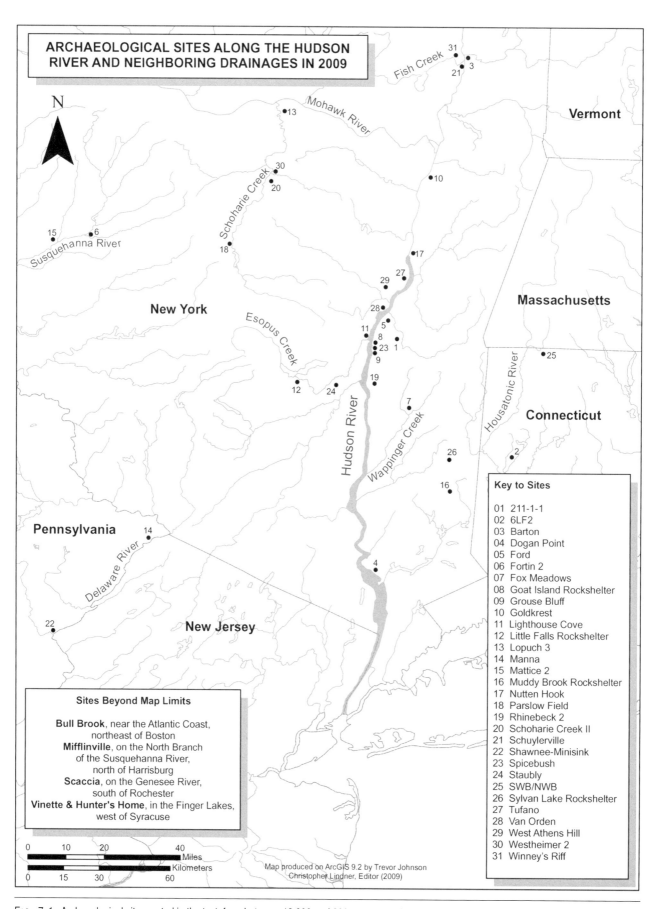

FIG. 7.1. Archaeological sites, noted in the text, from between 13,000 and 300 years ago along the Hudson River and neighboring drainages.

their principal camps for greater proximity to backcountry wetland resources in glacial lake basins.
2. Eventually, settlement took on an even more widely dispersed pattern, along interior streams for access to game and nuts in the increasingly productive forests.
3. Gradually people, greatly diminished in population, developed a focus on floodplains so as to intensify small seed harvesting and begin cultivation of domesticated plants.
4. In a relatively sudden shift, more people located their camps on the Hudson for its anadromous fish.
5. Finally, certain groups renewed their dispersal onto the floodplains of tributary streams to practice plant cultivation.

The estuary was always part of the seasonal round, even though its weight in the balance of ancient adaptations varied through time. One should maintain awareness that even when there was a focus on particular resources, people continued to use a broad array of foods and moved camp frequently through diverse habitats.

William A. Ritchie (1958) wrote the first fairly comprehensive prehistory of the Hudson Valley although four of its seven main sites were upstream of the estuary, including its only site not located on the river itself. The study compared artifact styles with those of central New York in an attempt to define the distinctiveness of the more eastern region. Bert Salwen's (1975, 65) article "Post-glacial Environments and Cultural Change in the Hudson River Basin" provided a general treatment of climatic and adaptive shifts while it gave rough estimates of population sizes with the interpretation that "each society, like any other biological population, has tended to live up to the carrying capacity of its habitat, using the technologies at its command. Until relatively recently, the limits of these technologies limited, in turn, the rate of population growth and the rate and extent of modification of the habitat."

In a final synthetic comparison of his research on estuarine, coastal, and interior riverine environments, the late New York State Archaeologist Emeritus, Robert E. Funk (1992), contrasts his early "back-country versus riverine model of settlement" for the Hudson Valley (1976), with his more recent "valley-floor versus upland model" for the upper eastern branch of the Susquehanna. We should keep in mind that the latter research effort, partly done as CRM testing for the extension of Interstate 88, was heavily skewed to discovery of riverine sites sealed in alluvial deposits. Yet Funk's (1993, 294) survey reveals a pattern of site location closer than 100 m to the Susquehanna River in floodplains and outwash terraces on the valley floor.

While the Hudson has tidal flats above Nutten Hook, riverine floodplains are rare southward. Adjacent to the estuary's middle reaches is a relatively level, broad valley. Its hills are mainly low until they attain higher elevations in the Catskills, Helderberg Escarpment, and the Taconic Mountain Range. The Hudson River farther downstream, in the Highlands and below, resembles a fjord. The oak-chestnut-forested ridge and valley terrain of the Hudson-Champlain lowland contrasts markedly with the Allegheny Plateau of high uplands covered with hemlock-white pine-northern hardwoods that border the Susquehanna River floodplains and alluvial terraces. In Fig. 7.1 I provide a list of archaeological sites discussed.

STARTING PLACES (13,000 TO 10,000 YEARS AGO): CARIBOU TO HUNT, PLANTS TO FORAGE, AND FISH TO CATCH

Funk (2004) revisited West Athens Hill, a few kilometers back from the Hudson, to update his earlier findings (Ritchie and Funk 1973) of a flint quarry, tool manufacture workshop, and multipurpose campsite that could date as far back as thirteen millennia. The high ridge top has expansive views of valleys to both sides. The waterways through the valleys were likely the migration routes of caribou, the principal game animal for several thousand years, until the herds moved northward permanently. Funk's (2004, 128) analysis of debris reduces the number of habitation clusters from fourteen to five, about which he comments, "Even in the absence of features and postmolds, it is difficult to escape the notion that the clusters were family dwelling and working areas." He had searched without success for features, such as hearths or firepits with burned earth, and looked in vain for postmolds, usually stains where saplings had been set in the ground, until they rotted away or were pulled out and their holes were filled with earth.

The largest known aggregation place in North America during this era, where groups from various areas gathered nearly 13,000 years ago, is the Bull Brook site in northeastern Massachusetts (Robinson et al. 2009). Possibly twenty-eight families camped there in a circular configuration for manufacture of tools from Hudson Valley chert (flint) to hunt and process migrating caribou. It would be at least six millennia before a large, intensively occupied site was deposited in the Hudson Valley, the most intact one being the newly documented Rhinebeck 2 site in northwest Dutchess County (Lindner 2008).

GIS expert Susan Winchell-Sweeney and Bard College master's student Patterson Schackne (2005) analyzed the four counties of the Mid-Hudson Valley to delineate the high-water level of the post-glacial river, known to geology as the Lake Fort Ann stage of Lake Albany. Their map plots the proximity of West Athens Hill, and several Mid-Hudson find-spots of tools from the earliest several millennia of human habitation in the Northeast, to within several hundred meters of the former shoreline, strongly suggestive that this habitat had a particular attraction, possibly as a pathway for caribou migration.

Michael Gramly (1988, 267–68) observes that certain artifacts, made of jasper quarried ca. two hundred km away but found in the earliest sites along the Hudson, may be evidence that human bands "had a well-developed cycle of movement up and down the Lowland into New Jersey and eastern Pennsylvania." Microscopy of biotic remains from stratified floodplain camp deposits (McNett 1985) indicates that people near the Watergap at Stroudsburg brought a variety of wild plants and fish from the Delaware River to the Shawnee-Minisink site almost 13,000 years ago (Gingerich 2007). These resources included hawthorn plum (*Crataegus* sp.), bramble berry (*Rubus* sp.), and wild spinach (*Chenopodium* sp.). Such plants were likely procured at this time by groups in the Hudson Basin, too. There is not, however, any sign of plants or fish having been consumed along the estuary during the earliest millennia, although Funk (2004, 95) notes that people at the 6LF2 or Templeton site nearby in Connecticut "were collecting acorns, reflecting the presence, though probably not abundance, of oak trees" in the area at the end of this span. Evidence is also lacking in the Hudson Lowlands for human interaction with megafauna; the numerous mastodon bones found there thus far have had neither cut marks from stone tools, nor artifacts near enough in the ground to be considered the result of exploitation. The Hiscock site in New York's far western lake plains (Laub et al. 1988) provides the first example of artifacts in association with mastodont in the Northeast. Funk (2004, 94) mentions the extinction of megafauna prior to 10,000 years ago but does not implicate human predation. He notes the high likelihood that sites along the coast, now deeply submerged by sea level rise, were the scenes of significant early human activity in the region.

THE FIRST SHIFT (10,000 TO 6,000 YEARS AGO): REGIONAL ADAPTIVE DIVERSIFICATION—GLACIAL LAKE BASINS AND THE MAST FOREST

George Peter Nicholas (1988) proposes a "glacial lake basin mosaic model" for clusters of camps that exhibit recurrent use around eleven millennia ago. He theorizes that diverse and high biomass environments, near the SWB/NWB sites in the Robbins Swamp along the Housatonic River in northwestern Connecticut, may have attracted more settlement after the caribou herds moved north permanently. Funk (1992, 25) agrees, but he makes the focus at that time appear less remarkable when he notes, "Examination of available site location data indicates that archaeological sites of all periods tend to be more frequently concentrated around wetlands than in other areas of the landscape." The upland forests eventually grew in productivity, and settlement increased there. Funk (1993, 283) regards a rise in oak on regional pollen charts as indication that by 8,000 years ago acorns had become "reasonably abundant." He notes an increase in hazel and walnuts in the pollen assemblage of an oak-hemlock forest zone that he calls "essentially modern" except for the lack of significant amounts of hickory (Funk 1993, 264).

Funk's (1976) Hudson Valley monograph charts the distribution of datable projectile points and counts the sites on which they occur, in relation to six environmental settings. The earliest projectile point that he can plot in meaningful frequency is typical of the second several millennia in the basin. The thin distribution of these characteristic bifurcated-base projectiles is notably spread across open camps in the backcountry, in contrast to sites in riverine habitats. For example, bones of deer and other mammals from this era were found at the Muddy Brook Rockshelter, far in the backcountry of Putnam County (Funk 1993, 264).

At least a few such projectile points have occurred on bluffs and low-lying landforms along the Hudson. Several bifurcated-base points are documented by avocationalist finds on a Hudson River beach at the Spicebush site along Tivoli South Bay in Red Hook, Dutchess County (Lindner 2002). Riverine habitation evidence from this era may well exist underwater. Waterman (1992) depicts two scenarios for the submergence of the Tivoli Bays as the Hudson rose, at first rapidly and then at a rate of one meter per millennium over the last 6,000 years. The embayment may originally have been dry land, later received tidal flow, and last held permanently the waters of its tributaries and the river. Or the embayment may have been for a long time much deeper but eventually filled with sediment, possibly in historical times, until its waters became quite shallow.

THE SECOND SHIFT (6,000 TO 3,000 YEARS AGO): FOREST SETTLEMENT AND POPULATION PEAK

The best faunal subsistence information for eastern New York during this span comes from the Sylvan Lake Rockshelter in the Taconic foothills of eastern Dutchess County, where Funk (1976) identifies bones from several datable strata. The deepest of these layers at the Sylvan Lake site has diverse fauna that include mostly deer, but also elk, woodchuck, turkey, grouse, goose, and turtle. The next layer up contains deer, elk, bear, raccoon, turkey, turtles, clams, and the earliest fish bones recorded in the Hudson basin. Assay of charcoal from a hearth in this stratum yields the radiocarbon age of 4,730 +/- 80 years (Lab #Y-1535), which calibrates to 5,500 calendar years ago. Faunal materials in a deposit from a few centuries later at Sylvan Lake Rockshelter comprise 90 percent deer remains, with raccoon also fairly abundant.

Ritchie (1980, 167) considers the increase of shellfish remains on Long Island as potential evidence of a shift in orientation "toward the sea" and away from an interior forest adaptation, as "a possible correlation with a growing scarcity of large game." In his research report on a western inlet of Narragansett Bay, Bernstein (1993, 50) allows the possibility of demographic pressures in the intensification of shellfish harvesting but emphasizes more the effect of "stabilization of the coast and the concomitant development of habitats, such as marshes and clam flats, that can support substantial mollusc populations" 5,000 to 4,000 years ago.

Along the lower estuary in the same era, the Dogan Point shell midden excavations reveal the start of the most intensive ancient use of oysters on the Atlantic coast (Claassen 1996). The shell deposit preserved the bones of a variety of finfish, but not sturgeon. On a high bluff along the river, at the Roeliff Jansen Kill confluence, a Ford site firepit dates approximately to this time, and contains the oldest known sturgeon remains in the Hudson watershed (Funk 1976).

Funk's (1993) decade-long excavations on the upper Susquehanna produced enough data for him to draw a model of seasonal mobility of several bands and to propose a model of springtime aggregation of groups to exploit fish migrations and perform ceremonies of annual renewal. He interprets a line of hearths as evidence of the oldest known house in the region, at the Mattice 2 site near Oneonta, datable to approximately 4,500 years ago. Mast foods such as acorns, hickory, and butternuts were also important to the occupants' diet, along with animals from the hunt.

Hoffman (1998) argues that researchers should accept the oldest radiocarbon dates for ceramic pots in the Northeast, ca. 5,000 years ago, rather than ca. 3,500 years ago as conventionally held. Instances of crushed nutshells in association with ceramic sherds, particularly during the third millennium before present in the Delaware Valley, suggest to Hummer (1994) extraction of oil from nuts and acorns. Innovative pyrolysis gas chromatography/mass spectrometry analysis of residues on ceramic containers along the St. Lawrence River by Taché et al. (2008) suggests that pots made between 3,200 and 2,700 years ago were used to process fish and nuts, possibly for their oils in ceremonial feasts.

Preparation of special foods for feasting may have been the function of soapstone vessels in the Northeast for the millennium of their abundance after 4,000 years ago (Sassaman 2006, 146). Using trace element analysis, Truncer (2004; 2006) presents an interpretation for steatite vessel proliferation that emphasizes prolonged cooking of red oak (*Quercus rubra*) acorns. Hart et al. (2008) challenge his hypothesis with analysis of phytoliths and fatty acids on soapstone sherds from the 3,700-year-old Hunter's Home site in the Finger Lakes. They identify residues of meat, pine resin, grass, and a legume, but not mast foods.

Along the upper Delaware, numerous netsinkers and large beds of burned rock suggest to Kraft (2001, 155) the drying and smoking of fish harvests during this time span. Custer (1996) notes unpublished experiments that indicate such "platform hearths" were used for extraction of fish oil. Jacoby (2000) uses protein analysis to detect the residues of fish and mammals on tools used to process them at the Mifflinville sites, along the Susquehanna River's North Branch in Pennsylvania. He supports the identifications by comparison of tools found on the bottomland that exhibit residues of fall migrating eel, in contrast to artifacts on a higher terrace, above the reach of the spring flood, that have residues of vernal spawning shad. These finds support Funk's emphasis on the importance of riverine fishing that started more than four millennia ago.

THE THIRD SHIFT (3,000 TO 1,500 YEARS AGO): INTENSIFICATION OF SEED PROCUREMENT AND INCEPTION OF HORTICULTURE DURING A MILLENNIUM OF APPARENT LOW POPULATION

For almost a thousand years, centered about two millennia ago, there appears to have been a population decline, down to 7 percent of its previous levels in the Hudson Valley and adjacent drainages, according to Funk's (1993) graphs of numbers of sites per century and occurrence of temporally diagnostic projectile points. There remains a lack of causal evidence, although Fiedel (2001) surveys a variety of climatic and epidemiological factors, suggesting that they took effect successively or in concatenation. In the Hudson and Susquehanna valleys at least, population levels significantly rebounded by 1,500 years ago, back to within 30 percent of their apex two millennia earlier (Funk 1993). Funk and Ritchie (1973, 369) note the hiatus and speculate that plant cultivation began soon afterward, but evidence to fill the gap and to factor in horticulture has begun to accrue just recently.

During the population low, burials with exotic items such as ocean shell beads and copper tools were made at the Van Orden site south of Catskill and the Barton site above Albany (Funk 1976). These components evidence participation in an interaction sphere of subcontinental range, with Ohio at its center. That the rise in funerary elaboration in the Hudson Valley occurs during a time of apparent population collapse suggests ritual behavior to counter severe stress. At the Lopuch 3 site in the lower Schoharie Creek valley, a buried layer dates to approximately 2,500 years ago where nutshell fragments rest beside a hearth, while in and around an adjacent firepit are thirteen fragmentary drill bits (Lindner and Folb 1998). Such "microdrills" of flint could have made the holes in shell beads found as offerings in roughly contemporaneous burials along the Mohawk River nearby.

Funk's (1976, 278) distributional data support his generalization of "a decided preference for the Hudson and its major tributaries in all seasons" during this interval. In the middle of the third millennium before today there was wild rice (*Zinzania aquatica*) at the Parslow Field site in Schoharie County (Hart et al. 2003, 635) and wild spinach seeds at the Schoharie Creek II site (Rieth 2008). Both locations are near the largest tributary to the Mohawk River. At the Vinette site in the Finger Lakes, Hart et al. (2007) have identified opal phytoliths of maize (*Zea mays*) in pottery encrustations that they cautiously date by accelerator mass spectrometry (AMS) at 2,300 years ago, and more definitely between 1,900 and 2,000 years ago. They anticipate skepticism about the older reading, as it currently constitutes the earliest date on maize in the Eastern Woodlands. Their two later readings on Vinette ceramic residues are only slightly younger than a site in Illinois where researchers identify maize by a massive screening project to find the oldest known macroremains of maize in the subcontinent.

Through the same process of phytolith identification and AMS dating, Hart et al. (2007) document maize and squash (*Cucurbita* sp.) approximately 1,600 years ago at the Westheimer 2 site near the confluence of Fox and Schoharie creeks. Ritchie and Funk (1973) interpret the same deposit as a recurrently occupied fall-winter camp for hunting, with fishing a minor activity. Similarly, Hart et al. (2007) date to 1,500 years ago maize, squash, and sedge (*Cyperus* sp.) in occupation zone 3 at the Fortin 2 site near the confluence of Charlotte Creek and the Susquehanna River. Funk (1998) offers butternut (*Juglans cinerea*) in support of his (1993) interpretation of Fortin as an intensively occupied (probably recurrent) fall camp where hunting and fishing took place. Finally, Hart et al. (2007) date squash phytoliths on potsherds from the Scaccia site on the Genesee River to 3,000 years ago (cf., Ritchie and Funk 1973, 114).

The earliest seeds in the Hudson Valley proper are currently those of arrowwood (*Viburnum dentatum*) in an earth oven datable to 2,150 years ago at the Grouse Bluff site on Tivoli Bays in Dutchess County (Lindner 1992; 2002). At the nearby Lighthouse Cove site numerous lobate-stemmed projectile points indicate an intensive occupation at this time (Lindner and Folb 1998). Flotation of approximately twelve liters of platform hearth sediments yields fifty-two carbonized shells of hickory (*Carya* sp.), one of acorn (*Quercus* sp.), and three bony fragments similar to turtle carapace/plastron (Largy personal communication 2005). Labrador's (2008) microscopic analysis of pottery temper composition supports the interpretation that people occupied Lighthouse Cove mainly about two millennia ago, even though some artifacts from earlier and later times have become interspersed during tillage, a depositional situation typical of most sites in the region.

THE FOURTH SHIFT (1,500 TO 1,100 YEARS AGO): A SUDDEN FOCUS ON FISHING

There is abundant evidence for sturgeon procurement for several centuries around 1,300 years ago in the Mid-Hudson Valley (Lindner 2009). At the Spicebush site, less than a meter above mean high tide today, Bard College students have found fragments of sturgeon scutes in deposits of similar age. In the Goat Island rockshelter at Tivoli North Bay, deposits possibly of this time contained bones of sturgeon and three other species of fish, as well as several species of waterfowl (Chilton 1992). Funk's theory that large flint blades had been used for the removal of scutes or bony plates from sturgeon receives support from Reifler's (2004) replicative experiments at Bard College that also identify wear traces from butchering deer on the archaeological specimens at the Tufano site. There, on the bank of the Hudson, Funk (1976) records hearths and shallow pits, possibly for storage, in which many of the large blades occurred together with sturgeon scutes and deer bones. The pits were adjacent to graves that contained a total of two dozen human burials.

Brumbach and Bender (2002) theorize a shift away from weirs and traps to catch migratory herring around two millennia ago, at the Schulyerville site near a Hudson confluence. At the ca. seven hundred-year-old Winney's Riff site along Fish Creek in Saratoga County nine km farther up this major tributary, people may have used gill nets and spears, as evidenced by deposits of numerous small fish vertebrae. The adjacent floodplains were potentially the scene of horticultural activity. Ceramic technology changed in tandem with settlement location and procurement strategies. The pottery at the later site along Fish Creek was more dense in composition than before and had thinner walls, which would have improved the cooking of maize or wild seeds (Brumbach and Bender 2002, 236).

THE FIFTH SHIFT (1,100 TO 400 YEARS AGO): MAIZE CULTIVATION AND RESOURCE DIVERSIFICATION

Cassedy and Webb (1999) report pits datable to 1,100 years ago that contain the earliest evidence of corn or maize along the Hudson River estuary. At the 211-1-1 site, on the Roeliff Jansen Kill, flotation was used to process one hundred liters of soil for each individual bit of maize. I used the same technique of "flotation," half-millimeter-mesh sifting of measured samples of soil dissolved in water, to process a few liters of earth from each of a dozen hearths or dark-stained pits at the Grouse Bluff site on Tivoli South Bay (Lindner 2002). Paleoethnobotanist Tonya Largy studied the flotation residue and identified charred shells of nuts and acorns, and a carbonized seed similar to pond weed (*Potamogeton* sp.), but no maize, potentially due to the small sizes of the samples or preservation biases. Two firepits datable to an average of 1,000 years ago yielded huckleberry seeds (*Gaylussacia* sp.).

Another approach to detection of cultigens, roots, and tubers is the study of residues on tools, such as mortars and pestles, possibly used to process plants. At the Manna site near Milford, Pennsylvania, on the upper Delaware River, Messner and Dickau (2005) identified starch grains of maize on stone mortars from the last millennium.

At the Fox Meadows site on Wappingers Creek (Iroquois Pipeline site 230-3-1), radiocarbon assay roughly corroborates the stylistic dating of its ceramics as late in the span 650–500 years ago (Cassedy 1998). Its excavator, Jonathan C. Lothrop, discerns the postmold pattern of a house, ten by five m in extent, which surrounded four hearths in a line. At least one of these hearths contained maize, as did several pits for storage and/or refuse disposal. Some of the hearths and pits lay outside, but near the house. They date earlier and later than the house by a few centuries, suggestive of more permanence than usually envisaged. Projectile points reflected hunting and netsinkers indicated fishing, although no animal bone was in evidence, probably due to acidic soils. Mast food remains were acorns, hickory, walnut (*Juglans nigra*), butternut, and hazelnuts (*Corylus* sp.). Seeds were from smartweed (*Polygonum* sp.), bramble berry, and sumac (*Rhus* sp.).

Native adaptations in the Hudson basin had no extensive impacts on the environment. Patterson and Sassaman (1988) find evidence of the widespread practice of land clearance by burning in the Northeast for horticulture and intensification of terrestrial wild food procurement, but only in the last few centuries of ancient times. Gramly (1977) argues that conflicts arose during the last half millennium to protect deer hunting territories valued for hide procurement to clothe populations, burgeoning from increased production of food, particularly in central New York and southern Ontario.

Diamond (1996, 103) summarizes his excavations at the Staubly site near Esopus Creek in Ulster County, where he identifies glass beads from "perhaps 1580–1620." Bones of deer and bear are present there, with walnut, hickory, and carbonized maize. Diamond also reports beans (*Phaseolus* sp.) from Staubly, a first for the Hudson Valley. Hart and Scarry (1999) demonstrate the total lack of evidence that beans entered the Northeast at the same time as maize, contrary to previous interpretations. Using accelerator mass spectrometry, they directly date beans in the Susquehanna Valley to about seven centuries ago. Hart (2008) has recently explored the "changing histories" of the Three Sisters: squash, maize, and bean in the Northeast.

At the Goldkrest site, on Papscanee Island in the Hudson River across from Albany, archaeologists and paleoethnobotanists (Lavin et al. 1996; Largy et al. 1999) associate two houses, probably from the early 1600s, with a variety of floral and faunal remains that include sturgeon. From sediments in a firepit between the dwellings, Largy identifies poisonous buttercup seeds (*Ranunculus* sp.) and comments, "Assuming that the presence of this species represents intentional use, some medicinal or ritual (charm) purpose likely accounts for its presence in the hearth" (Largy et al. 1999, 77). She suggests that buttercup may have been a protective measure against the spread of European diseases that had begun to decimate the Mohican populations. This firepit contained two pieces of sheet brass that possibly were cut from kettles traded from Fort Nassau or Fort Orange on the opposite side of the river.

Native camps have come to light in the Catskills. I excavated a hearth at the Little Falls rockshelter along Esopus Creek, immediately below the Ashokan Reservoir (Lindner 1998). From a radiocarbon assay, I estimate its date as ca. 1700, not long after the Dutch founding of Kingston. The hearth contains a mixture of Native ceramics, chert flakes, a bone tool possibly for sewing, a white clay pipe bowl, and a sheet brass arrowhead with its tip rolled, potentially as a symbolic ornament. Fish bones are present, along with deer, raccoon, and other small mammals, according to faunal analyst David Steadman (personal communication 1998). Paleoethnobotanist Roger Moeller (personal communication 1998) identifies charred walnuts and bramble berry seeds from flotation of thirty-six liters of hearth sediment.

I compare Little Falls and seven other Ashokan Catskill rockshelters with clusters of eight such sites in the lower Hudson basin's Bear Mountain area and eleven in southwestern Connecticut. The frequency of temporally diagnostic tools at these sites reveals two peaks in occupation, for several centuries around 5,000 years ago, during the millennium of highest population density, and again around five centuries ago. If one makes an adjustment for significant proportion of mountainous terrain in the range of the Mohican and mid-Hudson Lenape groups, their population densities would approach that of the Lenape farther down the estuary, which Salwen (1975, 55) estimates at about one person per square kilometer at the start of colonial times.

CONCLUSION: ANCIENT PEOPLES OF THE HUDSON VALLEY AND THEIR IDENTIFICATION WITH THE ESTUARY

Four hundred years ago when Henry Hudson's voyage laid claim to the region around Manhattan, the estuary was known as *Muhheakunnuk* (Brasser 1978, 211), "the river that ebbs and flows." Descendants of the large group of Native Americans who inhabited its upper reaches now have a reservation in central Wisconsin. Yet they still call themselves Mohicans after the estuary, and maintain an active interest in the valley. They are officially the Stockbridge-Munsee Band, Mohican Nation (Davids 2004, 7). The Munsee part of their name refers to the language of Algonquian groups downstream from the Mohicans' ancestral lands. Toward the end of the seventeenth century, English colonists referred to Native groups who resided near the Hudson as "River Indians" (Brasser 1978, 204).

Rather than an expression of Native identity, does the name River Indians better reflect the colonists' experience with the various groups who lived along the estuary? Many of the River Indians may have self-identified as *Lenape*, or "people" in their language. The colonists called the Lenape groups "Delaware" after the bay at the southern extent of their range, which had been named for the first governor of Virginia, Sir Thomas West, Lord de la Warr (Goddard 1978, 235). Descendents of the Lenape groups have tribal lands today in Oklahoma and Ontario.

The colonists of New Netherland, later New York, concentrated their settlements along the estuary, but had Native ancestors done the same? Could the primary reason for Mohican self-identification with the river be a heritage of seasonal ceremonial gatherings that combined abundant fishing and feasting with annual rituals of group solidarity? Does the estuary's seasonal bounty in migratory fish loom too large in our reconstruction of past adaptive patterns, as we continue to learn more about sites along tributary streams and elsewhere in the backcountry?

This sketch of shifts in adaptation along the Hudson River gives the outline of sustained swings in emphasis between the estuary and its hinterland. While the Hudson's role appears not to have always been dominant in the seasonal round, its part was usually influential. Now we need to contrast the Hudson Valley with comparable environments—estuaries of the Delaware, Susquehanna, and Connecticut—to distinguish their adaptive particularities.

The five shifts in habitat are drawn from evidence at various levels of supportive quality. The model of an initial shift, from a focus on riverine migration routes to a diversified adaptation to fit glacial lake basins, applies thus far only to relocation along the Housatonic River in Connecticut. The second major adjustment, to concentrate on backcountry forest resources of nuts or acorns, and animals that feed on them, appears fairly demonstrable. The third shift, to diversify by intensification of seed harvesting and inception of horticulture, may be due to differential application of field sampling and laboratory analysis techniques. After all, the use of flotation at the oldest dated site in the region demonstrates this technique's effectiveness on deposits of even the most remote times. The fourth shift, to migratory fish exploitation, appears quite dramatic and its suddenness perhaps particular to the Hudson estuary. Finally, the adoption of horticulture in the region has a growing body of evidence for its inception as the fifth and final shift before Colonial times. The sparse use of cultivation in the Hudson Valley and the slowness of its increase, however, pose questions about the overall subsistence pattern in the basin during the last millennium before people from the Old World arrive. As new analytical techniques proliferate and more sites come to light, these interpretations will likely change or become refined, by additional data and wiser theoretical frameworks.

REFERENCES CITED

Bernstein, D. J. 1993. *Prehistoric subsistence on the southern New England coast: The record from Narragansett Bay.* San Diego: Academic.

Brasser, T. J. 1978. Mahican. In *Handbook of North American Indians, Volume 15, Northeast*, ed. Bruce G. Trigger, 199–212. Washington, DC: Smithsonian.

Brumbach, H. J., and S. J. Bender. 2002. Woodland period settlement and subsistence change in the upper Hudson Valley. In *Northeast subsistence-settlement change: A.D. 700–1300*, ed. John P. Hart and Christina B. Rieth, 227–39. Albany: New York State Museum Bulletin 496, University of the State of New York.

Cassedy, D. 1998. From the Erie Canal to Long Island Sound: Technical synthesis of the Iroquois Pipeline Project, 1989–1993. Unpublished report by Garrow and Associates, on file at the New York State Historic Preservation Office.

———, and Paul Webb. 1999. New data on the chronology of maize horticulture in eastern New York and southern New England. In *Current Northeast Paleoethnobotany*, ed. John P. Hart, 85–100. Albany: New York State Museum Bulletin 494, University of the State of New York.

Chilton, E. S. 1992. Archaeological investigations at the Goat Island Rockshelter: New light on old legacies. *Hudson Valley Regional Review* 9: 47–75.

Claassen, C. 1996. The shell matrix at Dogan Point. In *A Golden chronograph for Robert E. Funk*, ed. Christopher Lindner and Edward V. Curtin, 99–107. Bethlehem, CT: Occasional Publications in Northeastern Anthropology 15.

Custer, J. F. 1996. *Prehistoric cultures of eastern Pennsylvania.* Harrisburg: Pennsylvania Historical and Museum Commission Anthropological Series 7.

Davids, D. 2004. *A brief history of the Mohican Nation: Stockbridge-Munsee band.* Bowler, WI: Stockbridge-Munsee Historical Committee.

Diamond, J. P. 1996. Terminal Late Woodland/Contact period settlement patterns in the Mid-Hudson Valley. In *A northeastern millennium: History and archaeology for Robert E. Funk*, ed. Christopher Lindner and Edward V. Curtin, 95–111. *Journal of Middle Atlantic Archaeology* 12.

Funk, R. E. 1976. *Recent contributions to Hudson Valley prehistory.* Albany: New York State Museum Memoir 22, University of the State of New York.

———. 1992a. Some major wetlands in New York State: A preliminary assessment of their biological and cultural potential. *Man in the Northeast* 43: 25–41.

———. 1992b. The Tivoli Bays as a middle-scale setting for cultural-ecological research. *Hudson Valley Regional Review* 9: 1–22.

———. 1993. *Archaeological investigations in the Upper Susquehanna Valley, New York State,* Volume 1. Buffalo: Persimmon.

———. 1998. *Archaeological investigations in the Upper Susquehanna Valley, New York State,* Volume 2. Buffalo: Persimmon.

———. 2004. *An Ice Age quarry-workshop: The West Athens Hill site revisited.* Albany: New York State Museum Bulletin 504, University of the State of New York.

Gingerich, J. A. 2007. Picking up the pieces: New Paleoindian research in the Upper Delaware Valley. *Archaeology of Eastern North America* 35: 117–24.

Goddard, I. 1978. Delaware. In *Handbook of North American Indians, Volume 15, Northeast*, ed. Bruce G. Trigger, 213–39. Washington, DC: Smithsonian.

Gramly, R. M. 1977. Deerskins and hunting territories: Competition for a scarce resource of the northeastern woodlands. *American Antiquity* 42: 601–605.

———. 1988. Paleo-Indian sites south of Lake Ontario, western and central New York State. In *Late Pleistocene and Early Holocene paleoecology and archeology of the eastern Great Lakes region*, ed. Richard S. Laub, Norton G. Miller, and David W. Steadman, 265–80. Buffalo: Buffalo

Society of Natural Sciences Bulletin 33.

Hart, J. P. 2008. Evolving the Three Sisters: The changing histories of maize, bean, and squash in New York and the greater Northeast. In *Current Northeast Paleoethnobotany II*, ed. John P. Hart, 87–99. Albany: New York State Museum Bulletin 512, University of the State of New York.

———, and C. M. Scarry. 1999. The age of common beans (*Phaseolus vulgaris*) in the northeastern United States. *American Antiquity* 64: 653–58.

Hart, J. P., R. G. Thompson, and H. J. Brumbach. 2003. Phytolith evidence for early maize (*Zea Mays*) in the northern Finger Lakes region of New York. *American Antiquity* 68: 619–40.

Hart, J. P., H. J. Brumbach, and R. Lusteck. 2007. Extending the phytolith evidence for early maize (*Zea mays ssp. mays*) and squash (*Cucurbita sp.*) in central New York. *American Antiquity* 72: 563–83.

Hart, J. P., E. A. Reber, R. G. Thompson, and R. Lusteck. 2008. Taking variation seriously: Testing the steatite mast-processing hypothesis with microbotanical data from the Hunter's Home site, New York. *American Antiquity* 73: 729–41.

Hoffman, C. 1998. Pottery and steatite in the Northeast: A reconsideration of origins. *Northeast Anthropology* 56: 43–68.

Hummer, C. C. 1994. Defining Early Woodland in the Delaware Valley: The view from the Williamson site, Hunterdown County, New Jersey. *Journal of Middle Atlantic Archaeology* 10: 141–51.

Jacoby, R. M. 2000. Prey selection and prehistoric settlement in the Susquehanna Valley: A test of protein residue analysis. *Journal of Middle Atlantic Archaeology* 16: 97–115.

Kraft, H. C. 2001. *The Lenape-Delaware Indian heritage: 10,000 B.C.–A.D. 2000*. Stanhope, NJ: Lenape Books.

Labrador, A. M. 2008. Building tools for identifying local variability and cultural patterns: A digital ceramic attribute analysis. *North American Archaeologist* 29: 287–96.

Largy, T. B., L. Lavin, M. E. Mozzi, and K. Furgerson. 1999. Concobs and buttercups: Plant remains from the Goldkrest site. In *Current Northeast Paleoethnobotany*, ed. John P. Hart, 69–84. Albany: New York State Museum Bulletin 494, University of the State of New York.

Laub, R. S., N. G. Miller, and D. W. Steadman. 1988. *Late Pleistocene and Early Holocene paleoecology and the archeology of the eastern Great Lakes region*. Buffalo, NY: Bulletin 33 of the Buffalo Society of Natural Sciences.

Lavin, L., M. E. Mozzi, J. W. Bouchard, and K. Hartgen. 1996. The Goldkrest site: An undisturbed, multi-component site in the heart of Mahikan territory. In *A northeastern millennium: History and archaeology for Robert E. Funk*, ed. Christopher Lindner and Edward V. Curtin, 113–29. *Journal of Middle Atlantic Archaeology* 12.

Lindner, C. 1992. Grouse Bluff: An archaeological introduction. *Hudson Valley Regional Review* 9: 25–46.

———. 1998. Eight rockshelters in the Ashokan Catskills and comparison with site clusters in the Hudson Highlands and Connecticut. *Bulletin of the Archaeological Society of Connecticut* 61: 39–59.

———. 2002. Paleoethnobotanical assessment of prehistoric facilities at the Grouse Bluff site, Tivoli Bays, Annandale-on-Hudson, Town of Red Hook, Dutchess County, New York. Unpublished report on file at Hudsonia Limited, Annandale, NY.

———. 2003. Archaeology of the Spicebush site, on Tivoli South Bay, Annandale-on-Hudson, Town of Red Hook, Dutchess County, New York. Unpublished report on file at Hudsonia Limited, Annandale, NY.

———. 2008. Site evaluation archaeology for the Thomas Thompson/Sally Mazzarella Park: Phase 2 report for the Rhinebeck Town Board and Park Committee. Unpublished report on file at the New York State Historic Preservation Office.

———. 2009. *The archaeology of fishing along the estuary*. Permanent exhibit at Bard College; web pages at http://inside.bard.edu/archaeology/spice-bush/index.shtml (December 29, 2009).

———, and L. Folb. 1998. Lopuch 3 and microdrills: Site report and use-wear analysis. *Archaeology of Eastern North America* 26: 107–32.

McNett, C. W. 1985. *Shawnee-Minisink: A stratified paleoindian-archaic site in the Upper*

Delaware Valley of Pennsylvania. Orlando: Academic.

Messner, T., and R. Dickau. 2001. New directions, new interpretations: Paleoethnobotany in the Upper Delaware Valley and the utility of starch grain research in the Middle Atlantic. *Journal of Middle Atlantic Archaeology* 21: 71–82.

Nicholas, G. P. 1988. Ecological leveling: The archaeology and environmental dynamics of early postglacial land use. In *Holocene human ecology in northeastern North America*, ed. George P. Nicholas, 257–96. New York: Plenum.

Patterson, W. A., and K. E. Sassaman. 1988. Indian fires in the prehistory of New England. In *Holocene human ecology in northeastern North America*, ed. George P. Nicholas, 107–36. New York: Plenum.

Reifler, A. R. 2004. Prehistoric Hudson Valley riverine adaptation: A use-wear analysis of Petalas blades to explore their proposed role as sturgeon-processing tools. Unpublished MS thesis, on file at the Bard College Library, Annandale-on-Hudson.

Rieth, C. B. 2008. Early Woodland settlement and land use in eastern New York. *Journal of Middle Atlantic Archaeology* 24: 153–66.

Ritchie, W. A. 1958. *An introduction to Hudson Valley prehistory.* Albany: New York State Museum and Science Service Bulletin 367, University of the State of New York.

———. 1980. *The archaeology of New York State* (Revised Edition). Harrison, NY: Harbor Hill.

———, and R. E. Funk. 1973. *Aboriginal settlement patterns in the Northeast.* Albany: New York State Museum Memoir 20, University of the State of New York.

Robinson, B. S., J. C. Ort, W. A. Eldridge, A. L. Burke, and B. G. Pelletier. 2009. Paleoindian aggregation and social context at Bull Brook. *American Antiquity* 74: 423–47.

Salwen, B. 1975. Post-glacial environments and cultural change in the Hudson River Basin. *Man in the Northeast* 10: 43–70.

Sassaman, K. E. 2006. Dating and explaining soapstone vessels: A comment on Truncer. *American Antiquity* 71: 141–56.

Swigert, E. K. 1974. *The prehistory of the Indians of western Connecticut: Part I, 9000–1000 B.C.* Washington, CT: American Indian Archaeological Institute.

Taché, K., D. White, and S. Seelen. 2008. Potential functions of Vinette I pottery: Complementary use of archaeological and pyrolysis GC/MC data. *Archaeology of Eastern North America* 36: 63–90.

Truncer, J. 2004. Steatite vessel age and occurrence in temperate eastern North America. *American Antiquity* 69: 487–513.

———. 2006. Taking variation seriously: The case of steatite vessel manufacture. *American Antiquity* 71: 157–63.

Waterman, B. 1992. Searching for clues to prehistoric human interaction with the environment at Tivoli Bays. *Hudson Valley Regional Review* 9: 77–92.

Winchell-Sweeney, S., and P. Schackne. 2005. *Glacial Lake Fort Ann.* http://www.cemml.colostate.edu/paleo/mapgallery3.htm (December 29, 2009).

CHAPTER 8

ARCHAEOLOGICAL INDICES OF ENVIRONMENTAL CHANGE AND COLONIAL ETHNOBOTANY IN SEVENTEENTH-CENTURY DUTCH NEW AMSTERDAM

Joel W. Grossman

ABSTRACT

This chapter analyzes the environmental implications of seventeenth-century ethnobotanical data from the initial shoreline block of the Dutch West India Company (WIC) in Lower Manhattan. In addition to the structural remains of the colony's early inhabitants, the excavation yielded a well-preserved sequence of colonial plant remains spanning the periods of Dutch and early English rule. This analysis of the archaeological chronology and plants: (1) provides new understandings of the continuity and shifts in the relative prevalence of European and indigenous plants between the seventeenth and the eighteenth centuries; (2) presents new archaeological insights about the introduction and nature of early Dutch cultigens in New Amsterdam; (3) suggests that many of the archaeologically recovered early-seventeenth-century plants may have been maintained or collected as foods, dyes, or medicines, from both European and Native American sources; and finally (4) building from new research in Dutch botanical history, suggests mechanisms and institutionalized protocols in the exchange of medicinal plant knowledge between Native American herbalists and Dutch botanists in the seventeenth century.

INTRODUCTION

The study of environmental history has two ways to go. As brought to my attention by my Dutch colleague Jaap Jacobs, in 2008 Geoffrey Parker—a British-trained military historian of sixteenth and seventeenth-century Europe—defined the dilemma as follows: "Either we 'fast-forward' the tape of history and predict what might happen on the basis of current trends; or we 'rewind the tape' and learn from what happened during global catastrophes in the past. . . . [Many] experts . . . have tried the former, few have systematically attempted the latter" (Parker 2008, 1078).

Parker's work supported the notion that much of contemporary environmental modeling is too shallow in time-depth to provide reliable bases for projecting into the future. He also cited the work of two Norwegian scientists, Nordås and Gleditsch, who summarized a recent military intelligence assessment entitled, "National Security and the Threat of Climate Change: Report from the Panel of Retired Senior US Military Officers" (Military Advisory Board 2007). This crossover report between the disciplines of military threat assessment and the study of climate change is relevant because it criticized the: "failure of the International Panel on Climate Change (IPCC) to undertake systematic analysis of historical evidence to show how climate change acts as a threat multiplier for instability in some of the most volatile regions of the world" (Nordås and Geditsch 2007, 627–38; in Parker 2008, 1078).

This inadvertent validation of the need for time-depth in environmental reconstruction is music to an archaeologists ears . . . and an opera to environmental historians working on issues of habitat change in colonial New York. We have the best of both worlds: an unmatched material record of early Dutch settlement, coupled with a trove of seventeenth-century archival sources, in Manhattan, Albany, and The Netherlands. Accordingly, while most of our regional environmental modeling has relied heavily on relatively recent nineteenth, and rarely eighteenth, century sources, I will use archaeological and ethnobotanical evidence from New Amsterdam to push the record back to the mid-seventeenth century.

Accordingly, consistent with the focus of this volume, *Environmental History of the Hudson Valley*, and the four hundred-year anniversary of the arrival of Henry Hudson, I will use the archaeological record of seventeenth-century New Amsterdam to characterize the environmental conditions and consequences of human interaction within the confines of the Dutch West India Company (WIC) property in Lower Manhattan, which fronted on the waterfront at Pearl Street, then also referred to as the Strand (Fig. 8.1). The 1984 NYC Landmarks Commission–mandated excavation, eight to twelve feet below the modern city (protected by the rising sea and the thick brick basement floors of early-nineteenth-century row houses), documented the survival of the four hundred-year-old structural remains of the colony's first inhabitants (Grossman et. al. 1985; Grossman 1985; 2000; 2003; 2008). In addition to the recovery of 43,000 well-preserved Dutch, English, and Native American artifacts, foundations, and cisterns, the deep urban dig disclosed a number of

FIG. 8.1. The Seventeenth-Century Environment of New Amsterdam. Extruded from Viele's 1865 topographic map of Manhattan, this 3D terrain model shows the environmental context of the seventeenth-century Dutch West India Company colony (red outline) and excavated western end of the block at Pearl Street and Whitehall (red rectangle). The initial settlement was bounded to the north by a two-pronged escarpment which stepped down from a higher plateau at City Hall Park, and to the east by a spring-fed marsh (Blommaerts Vly) which drained into the East River through a ditch (the "Graft") under modern Broad Street. The predominantly "open" vegetation illustrates not a "pristine" or "primeval" canopy of continuous tree cover at European contact, but instead an "anthropogenic" landscape representing centuries of Native American seasonal clearing, burning, cultivation, and selective tree harvesting. As put forth by Hammett (2000) and others (cf. Day 1953; Cronon 1983; Denevan 1992), these activities suggest a patchwork for Lower Manhattan of upward to thirteen humanly altered habitats. In addition to major thoroughfares (e.g., Broadway), these probably included fields and gardens, residential and defensive sites, food (fish and shellfish) processing stations, edge areas and meadows, parklands and orchards, hunting areas, old fields, and landing sites.

undisturbed features and deposits, each containing dated, but previously unanalyzed and unreported, samples of colonial seeds, and each important as an "environmental time capsule." This reanalysis of the artifact and botanical evidence documents a refined three-phase, century-long sequence dating back to the second quarter of the seventeenth century. It also revealed significant, and previously unreported, order-of-magnitude changes in plant diversity between the seventeenth and eighteenth centuries; shifts that help refine the onset of the "Historic Horizon" in the environmental history of the Hudson drainage (see Peteet, ch. 9 in this volume for long-term prehistoric change; see Vispo and Vispo, ch. 12, and Teale, ch. 13 in this volume for cases of eighteenth and nineteenth-century change).

Using this archaeological chronology (and its associated 3D computerized database—see Grossman 2003), the following ethnohistorical study: (1) defines the context and time frame of the ethnobotanical data, (2) uses quantified seed data to highlight continuities and changes in plant diversity between the 1630s to the 1730s (Fig 8.2; Tables 8.4, 8.5, 8.6), (3) incorporates new data to argue the presence of European vegetables, (4) evaluates each of the identified plants from the multiple perspectives of sixteenth and seventeenth-century European herbalism and botanical history, North American prehistoric archaeology, and contact-period ethnobotany to suggest that many of the archaeologically recovered seeds may in fact represent previously underappreciated indigenous foods, or Native American and European medicinal plants in seventeenth-century New Amsterdam, (5) uses new research into the training of Dutch botanists, doctors, and officials to suggest potential mechanisms of cross-cultural information exchange between the Dutch and Native Americans. Finally, parallels in the naming, qualities, and uses of the medicinal plants also, it is argued, may reflect the existence of interregional and often long-distance, networks (e.g., Interior-Coastal, Intercoastal, Upper Hudson-Lower Hudson) between Native American ethnic groups, as well as between Dutch, English, and possibly French colonial botanists and medical practitioners.

THE ARCHAEOLOGICAL SEQUENCE AND HISTORICAL CONTEXT

The age and timing of historic environmental impacts in the Lower Hudson have often been based on a limited number of radiocarbon determinations (generally with a standard deviation of +/- one hundred years, or more), localized historical accounts and assumptions, or estimates based on the interpolation between earlier and later dates to fill gaps in sediment core time scales, especially for the seventeenth century. In contrast, this study uses two lines of archaeological and historical evidence to date and define changing patterns of plant diversity between the seventeenth and eighteenth centuries in Lower Manhattan: (1) the reanalysis of the archaeological chronology of the Pearl Street excavation based on the availability of new artifact dates from excavations in Europe and the Americas, and (2) a reevaluation of historical land-use records based on when the first residents of the block arrived in New Amsterdam, in contrast to using the date of the earliest recorded (or surviving), and significantly later, land grants by the Dutch West India Company to parcels within the block. Both approaches need to be explicitly addressed because they define the earliest concrete evidence of plant use in New Amsterdam, and because they date significant shifts in the environmental record between the seventeenth and eighteenth centuries.

The excavation resulted in the documentation and reconstruction of three major phases, or periods of occupation: the second quarter of the seventeenth century, the late seventeenth century, and the early eighteenth century:

1. Early to mid-seventeenth century (ca. the early 1630s to ca. early 1650) deposits and features all belong to the Dutch period;
2. Late-seventeenth-century, post-1680, deposits pertain to the culturally Dutch, but politically "English" Period of occupation at the site (as per Goodfriend 1991);
3. Early-eighteenth-century, post-1710 to ca. post-1730, complex of features and structural remains deposited some forty to fifty years after the English takeover of New Amsterdam.

Of these three periods and primary units of comparison, only the time frame of the earliest group has been revised, from the mid- or late seventeenth century, back to the second quarter of the seventeenth century. This chronological shift also provides a three-phase framework for comparing continuity and change in the nature and diversity of colonial plant remains over a one hundred year time span between the early seventeenth and early eighteenth centuries at the site (Table 8.1).

Chronological Revisions

As is the case for the Dutch West India Company site, archaeological chronologies (dating schemes) are moving targets, subject to change with each new discovery. While the relative placement of individual features and deposits is fixed at excavation by the natural stratigraphic sequence of deposition, the absolute dates of the artifacts within them can change significantly as new data become available. Over the last decade, new artifact dates from early-seventeenth-century excavations at Jamestown, Virginia, new finds in The Netherlands, and the deep-sea discovery of tightly dated shipwrecks, have redefined many of the original time markers originally used to date the artifacts of the Pearl Street block. This transatlantic progress in historic archaeology is important because it underscores the significance of multinational collaboration and because it suggests that the initial occupation of the site is significantly earlier, by at least two decades, than initially thought.

When the Broad Street excavation took place in the early 1980s, the early-seventeenth-century excavations at Jamestown had not yet taken place, and local comparative material was limited to ongoing excavations in Manhattan and to preliminary results from the work of Paul Huey at Fort Orange in Albany, New York. Previous excavations in Manhattan east of Broad Street had demonstrated the survival of late-seventeenth-century remains (post-1670) under nineteenth-century basements in Lower Manhattan, but these were still being analyzed (Rothschild, Wall, and Boesch 1987; Cantwell and Wall 2001, 170). At the time, it was generally assumed that the earliest Pearl Street artifacts were roughly conterminous with two periods identified in Albany dating to 1640–47, and 1648–1657, and specifically post-1653 based on a land grant to one of the occupants in the New Amsterdam block (Huey 1988, 598). Others thought they were later. One ceramic specialist reassigned new letter designations to the excavated deposits, grouped/mixed the earliest and latest deposits from the Pearl Street excavations together into one assemblage, and assigned it to a single broad late-seventeenth-century period between 1653 to 1685 (Janowitz 1993, 13, Table I). New data now suggest that these treatments are no longer reliable.

Over the last decade, new research with a focus on the date of introduction (T.P.Q., or *Terminus Post Quem*—"date after which") of a number of "generic" pottery types (e.g., "Eng/Dutch Tin-glazed" earthenware and several kinds of stoneware), formerly thought to have been indicative of the mid-seventeenth century (post-1640 or 1650), are now dated in Jamestown to post-1600 (Kelso and Staube 1997, 14, Tables 2 and 3; Mallios 2000, 50, Fig. 58). Furthermore, new archaeological sequences from well-dated, single-component (unmixed with later periods or occupations) house and farmstead sites near Jamestown have shown that many of the pottery types recovered from the Pearl Street site, formerly thought to be statically "most popular" in the mid-seventeenth century, were subsequently recovered in the Chesapeake area between 1630 and 1650, or to at least a decade, if not two decades, earlier (Mallios and Fesler 1999, 3, Fig. 60; Mallios 2000, 50, Fig. 58).

Likewise, from Holland, new chronologies for Delft tiles, developed by Dutch scholars, show that particular design elements on tiles (specifically, "ox-head" and "spider's head" corner motifs) thought in the 1980s to postdate 1650 (Grossman et al. 1985, Plates V-4, V-17) may have actually been introduced in the second quarter of the seventeenth century, if not as early as the 1620s (Pluis 1998, 537; Van Lemmen 1997) (Table 8.1). In addition, two important time markers, large sherds of Wan-Li–decorated pottery, from two different features (Components 38 and 62) at Pearl Street, both with seeds, previously thought to postdate 1670, or even 1690 (Wilcoxen 1990, pers. com. to Diana Wall at the South Street Seaport Museum 1990, in Dallal 1996, 220), are now dated by Dutch scholars to sometime between 1650 and 1660; a shift that in turn suggests that the basket feature it was found in predates the mid-seventeenth century (Jan Baart, pers. com. Dec. 4, 2009). Furthermore, the possibility exists that they may be somewhat earlier; Wan-Li pottery has been repeatedly recovered from late-sixteenth to early-seventeenth-century shipwrecks (Table 8.1). This chronological revision suggests that the basket (which I interpret as a probable drain at the outside corner drip-line of two walls of an early-seventeenth-century shell-limestone foundation), and the seeds it contained, was in use before 1650. Although the presence of post-1676 leaded glass kept the barrel fill of Component 62 in the late seventeenth century, this adjustment in the age of Wan-Li pottery is also important because the a basket/cask with seeds (Component 38) can now be reassigned to the early to-mid-, versus the late, seventeenth century, as was previously thought (Table 8.1).

Finally, Dutch experts in the history of clay smoking pipes have now established that one former, and widely used, dating tool, the measurement of pipe stem bore diameters (based on the assumption that the wider the bore diameter, the earlier the stem fragment), which supported the initial interpretation that the earliest deposits at Pearl Street postdated the 1640s and 1650s, appear now to be no longer useful. Research by Don Duco, of the Pijpenkabinet Museum of Amsterdam, has invalidated the utility of this technique for seventeenth-century artifact dating by showing that pipe stem bore measurements from a single pipe can vary considerably in diameter (Duco 1987, 135–36). This elimination of pipe stem dating for seventeenth-century contexts effectively removed four mid-seventeenth-century age time-markers from six early deposits, and from four with seeds (Components 2, 6, 8, 12); a change that reassigned their probable age of deposition to sometime in the early seventeenth century, instead of the mid-seventeenth century (Table 8.1).

These multiple lines of revised chronological time markers—a post-1630 pipe bowl, decorative tile motifs now understood to have been introduced as early as the 1620s, the recovery of post-1620–30 raised glass "prunts," (adornos in the form of raised molded berries on the stems of goblets) from three early deposits (Components 8, 12, 13), and the elimination of previously presumed mid-seventeenth-century "pipe stem mean dates" for six features (Components 2, 6, 8, 9, 13, 38)—now suggest that the earliest archaeological features postdate the 1630s, and were probably deposited within the decade of 1630 to 1640 (Table 8.1).

Historical Evidence of Early Occupation

These corrections of the material record are paralleled by historical shifts in archival interpretation which suggest that the first inhabitants of the block arrived earlier than initially thought. At the time of the original study (1983–85), the consensus of a number of New York archaeologists and historians was that the earliest surviving land grants and deeds, dating to the mid-seventeenth century (1647–1653), referenced by Stokes in his *Iconography of Manhattan Island* (1915–1925), represented the initial dates of occupation for the waterfront along Pearl Street in Manhattan. This interpretation overlooked the fact that all land was originally owned and controlled by the Dutch West India Company, and that its workers and officials resided and worked on "company land"; none of which was "deeded," or transferred, to private ownership until later.

In addition, other historical sources, both primary and secondary, suggest that the initial occupation of the excavated block may have begun as early as the 1630s. In 1902, J. H. Innes suggested that "within a few years after 1633 [and following the completion of Fort Amsterdam between 1626 and 1635, (Innes1902, 5)] . . . they had extended easterly along the north side of Pearl Street (which here ran along the shore of the East River) almost as far as the present Broad Street, where at this time the tide ebbed and flowed through a small salt-water creek. . . . [to become] the seat of trade for the town and the focus of early shoreline commercial activities" (Van Laer 1974, I: 111; Innes 1902, 5, 45).

Innes explicitly noted that "[t]hough the deeds or ground briefs for most of the parcels of land at this locality [western end of the Pearl Street and the area of the excavation] were made from 1645 to 1647, it is difficult to believe that they had not been in several instances built upon at an earlier date" (Innes 1902, 45). Five stone workshops along the western end of the block may have been the first

TABLE 8.1. Table of Revised Artifact Dates
This revised chronology used the recent availability of new artifact dates from Jamestown, Va., Holland, and tightly dated shipwrecks to suggest that the earliest seventeenth-century features and structural remains (components) from the Pearl Street site (the Broad Financial Site), were significantly earlier than initially thought when first studied in the 1980s. As detailed above and in the text (see chapter 8, Section II), contemporary time markers from other subsequently excavated sites now strongly suggest that the initial occupation along Pearl Street took place within a twenty-year period between 1630 and 1650; with the earliest features probably constructed in the decade of 1630 to 1640, or at least 10–20 years earlier than previously estimated.

General Time Period	Component No.	Primary 17th Century Components (with seeds)	Revised Component TPQ Time Range (+/- 5 years)	Revised Ceramic TPQ	Original Component TPQ	Original Ceramic TPQ	Original Glass TPQ	Original Pipe TPQ	Original Pipe Mean Date (Note 8)
EARLY-MID 17th CENTURY									
Early-Mid. 17th.c.	8	BT-Lot8-N-Bar	1633–1650	1600	1640	1640	nd	nd	1635 (na)
Early-Mid. 17th.c.	12	Pit/BT-Lot 8-N-Bar	1633–1650	1600	1640	1640	1630	1630	nd
Early-Mid. 17th.c.	9	BT-Lot8-S-Bar	1633–1650	1630	1650	1650	nd	nd	1664 (na)
Early-Mid. 17th.c.	22	BT-Rect-Yel Brk Feat	1633–1650	1600	1640	1640	nd	1630	nd
Early-Mid. 17th.c.	10	BT-Oval Yel-Cistern	1633–1650	1600	1640	1640	nd	nd	nd
Early-Mid. 17th.c.	61	BT-Lot 14-Bar-BT	1633–1650	1600	1640	1640	nd	nd	nd
Early-Mid. 17th.c.	2	Below-Bld A-Floor	1633–1650	1600	1640	1650	nd	17th c.	1649 (na)
Early-Mid. 17th.c.	6	Bld. A Floor - Heerman's Warehouse	1633–1650	1620	1640 - St. Group IA	1762?	1676 ?	1645?	1665 (na)
Early-Mid. 17th.c.	13	Lot 8-N Barrel Fill	1633–1650	1630	1650	1650	1630	1630	1645 (na)
Early-Mid. 17th.c.	5	BT - Hermans Warehouse	1633–1650	1580	1640?	1580	nd	nd	nd
Early-Mid. 17th.c. /Mid-Late 17th c.	38	Rope Basket-/cask Drain-Fill	1650	1650–60	1670–80	1670–80	nd	nd	1616 (na)
LATE 17th CENTURY									
Late 17th c.	14	Lot 8-S Barrel Fill	Post-1680	1620	1680	1650	1678	1678	1697 (na)
Late 17th c.	16	1/2 cir YB Cist Fill	Post-1680						
Late 17th c.	62	Lot 14-Barrel-Fill	Post-1676	1630	1676	1680	1676	1664	1684 (na)

Diag. TPQ Data/Comments (See numbered footnotes 1–8)	17th Century Occupants— Innes 1902	Arrival Date (van der Donck 1656; Innes 1902)	Deed/Ground Briefs from Dutsh West India Company (Innes 1902; Stokes 1915–1928)
Delftware: Revised TPQ = 1600; Found in Post 1630 Contexts at Buck Site, Jamestown, Va. (2); Pipe MD from small sample and unreliable (Duco 1987)	Haie/ van Tienhoven?	1633	Ground brief July 16, 1645, pos. Jacob Haie (Stokes II, 266); van Tienhoven "Great House" or Warehouse, Post-1652 (Innes 1902, 57)
Delftware: Revised TPQ = 1600; Found in Post-1630 Contexts at Buck Site, Jamestown, Va.; Raspberry Glass Prunt TPQ = 1630 (1,2); Pipe date based on single stem frag 8/64"; not reliable (Note 8)	Haie/ van Tienhoven?	1633	Ground brief July 16, 1645, pos. Jacob Haie (Stokes II, 266); van Tienhoven "Great House" or Warehouse, Post-1652 (Innes 1902, 57)
Westerwald, Orig TPQ 1650 revised to Post 1630 based on Jamestown dates (4); Pipe MD from small sample and unreliable (Duco 1987); No Pb = pre-1676	Haie/ van Tienhoven?	1633	Ground brief July 16, 1645, pos. Jacob Haie (Stokes II, 266); van Tienhoven "Great House" or Warehouse, Post-1652 (Innes 1902, 57)
Delftware: Revised TPQ = 1600; Found in Post 1630 Contexts at Buck Site, Jamestown, Va.; "EB" Pipe bowl Post 1630 (1,2)	A. Heerman?	1633	Pre-1651 (Innes 1902); 1645 (Stokes 1915–1928)
Delftware: Revised TPQ = 1600; Found in Post 1630 Contexts at Buck Site, Jamestown, Va. (2)	A. Heerman?	1633	Pre-1651 (Innes 1902); 1645 (Stokes 1915–1928)
Delftware: Revised TPQ = 1600; Found in Post 1630 Contexts at Buck Site, Jamestown, Va. (2)	Kierstede/Steenwyck	1638	1646 (Innes 1902); 1647 (Stokes 1915–1928)
Delftware: Revised TPQ = 1600; Found in Post 1630 Contexts at Buck Site, Jamestown, Va. (Millios 1999); Pipe MD from small sample & unreliable (1,2,7) cf Duco 1987	A. Heerman?	1633	Pre-1651 (Innes 1902); 1645 (Stokes 1915–1928)
Component TPQ marked by Pb glass (pos intrusive); Two 18th century "post-1762 Creamware" sherds prob intrusive and excluded from sample; Ceramic TPQ = "Ox-head" dec on tile rev to 1620; 95 Pipe frags had MD of 1665, now rejected cf Duco 1987; "Rouletted" dec. bowl rims dated to ca 1645 cf (Hume 1976).	A. Heerman?	1633	Pre-1651 (Innes 1902); 1645 (Stokes 1915–1928)
Diag. Tile with "Ox-Head" motif Rev TPQ = 1620; Glass Raspberry Prunt, TPQ = 1630; Pipe MD = 1645 or pos. 1635 (1,5) invalid (Duco 1987); Pipe TPQ based on ca. EB = 1630–1683; bowl shape (ca 1645–1666) cf Duco 1981; Westerwald orig. dated to ca 1650, rev to 1630 cf. Jamestown dates	Haie/ van Tienhoven?	1633	Ground brief July 16, 1645, pos. Jacob Haie (Stokes II, 266); van Tienhoven "Great House" or Warehouse, Post-1652 (Innes 1902, 57)
Majolica: Ceramic TPQ = 1580 (1)	A. Heerman?	1633	Pre-1651 (Innes 1902); 1645 (Stokes 1915–1928)
WanLi design = Comp TPQ; Originally assigned TPQ of post-1670 in text & Plate III-C2 (Table I-A2 in Grossman et al. 1985 listing of "1664" is typo); Revised TPQ of 1650–1660 cf (Jan Baart-Pers. comm. 2009) TPQ for Comp38.; Note: General Wan-Li ca 1630–50 at site, Jamestown,Va. and post-1613 Shipwreck (Note 3); Tile w "spider's head" corner tile dec. Rev TPQ = 1640 (Note ,5), but common in second half of 17th century.	Haie/ van Tienhoven?	1633	Ground brief July 16, 1645, pos. Jacob Haie (Stokes II, 266); van Tienhoven "Great House" or Warehouse, Post-1652 (Innes 1902, 57)
Component TPQ set by Pipes and Pb Glass: Post-1678 "RT" mark on English Pipe; Bowl forms = 1680–1690; Lead Glass = TPQ of 1676; Tile w "Ox-head" corner Motif-post 1620; (1,5)	van Tienhoven?	1633	1652? (Innes 1902); Stokes II, 266
		1638	
Originally dated to post-1680 "buff bodied slipware" questionable def. type; Wan-Li-TPQ: Originally ident. as post-1670, Rev to 1620; Manganese Purple: orig. TPQ of 1670–75 cf Ft. Orange contexts, rev. to post-1630 cf. Jamestown, Va. data; Comp. TPQ: Pipe Bowls (HG Mark) Post 1664; Pipe MD of 1684 unreliable (1,3); Comp 62 TPQ from Pb glass—post 1676	Kierstede/Steenwyck?	1638	1646 (Innes 1902); 1647 (Stokes 1915–1928)

TABLE 8.1. continued

General Time Period	Component No.	Primary 17th Century Components (with seeds)	Revised Component TPQ Time Range (+/– 5 years)	Revised Ceramic TPQ	Original Component TPQ	Original Ceramic TPQ	Original Glass TPQ	Original Pipe TPQ	Original Pipe Mean Date (Note 8)
Late 17th c.	76	Pearl St.-Matrix	Post-1680	1620	1680	1670	nd	1680	1698 (na)
Late 17th c.	17	Build. E BT	Post-1680	1620	1800	1800	1680	1678	1688 (na)
EARLY 18TH CENTURY									
Early 18th c	63(Cx 102.02-04)	Lot 14 R-BrkCistern-02-04	Post-1720	na	1720	1720	1710	1720–1727	1725
Early 18th c	53	Bld. D Lower Fill	Post-1720	na	1720	1675	1705	1678	1706
Early 18th c	54	Bld. D Upper Fill	Post-1720	na	1720	1700	1705	1678	1716
Early 18th c	63(Cx 102.01)	Lot 14 R-BrkCistern-Late 01 Cx	Post-1734	1734	1734	1734	1710	1690–1720	1699
EARLY-MID 19TH CENTURY									
Early-Mid. 19th. c.	28	Pit - Stone Pier Fill	Post-1830	1795	1795	1726	1730	1711	
Early-Mid. 19th. c.	66	Pit Fill (N65 E25)	Post-1813	1813	1813	nd	1786	nd	
Early-Mid. 19th. c.	33	Brick Drain-Fill	Post-1850	1850	1780	1750			
Early-Mid. 19th. c.	15	Oval YL Brk Cis Fill (reused brick)	Post-1844	1844	1844	1800	1738	1730	
Early-Mid. 19th. c.	75	Interface w Floor??	Post-1857	1857	1780	1857			
EARLY 20th CENTURY									
Early-20th	35	StoneRubble-Bl Base	Post-1903	1903	1903	1903	1680		
Early-20th	68	Olive Silt-Lt 13-14?	Post-1903	1903	1903	1903	1832		

Footnotes: New TPQ Dates and Changes:

1. (Grossman et.al. 1985; Table I-A2); Pipe Mean and TPQ dates as per originally reported by D. Dallal (Chapter VII in Grossman et al., 1985); Glass Dates as originally reported by J. Diamond, Chap VI in Grossman et al., 1985); Revised Ceramic dates per Kelso and Stroub 2004 & Mallios 1999.
2. "Delftware": Originally dated to Post-1640 based on Mean Date at Fort Orange (Huey 1984 per. Com. in Grossman et al., 1985); Rev. to Post 1600 TPQ; Post-1630 at Buck Site, Va. Jamestown, Va.: (Mallios 1999, Fig. 60).
3. "Wan Li dec.": Orig TPQ 1670, Revised Date Range for Comp 14 and 62 TPQ examples = 1650–1660 (Pers. Comm. J. Baart Dec. 2009); Generic Wan-Li dec.; From 1630–1650 contexts at Buck Site, Va. (Mallios 1999, Fig 60, p. 48); Recovered from 1613 Shipwreck Witte Leeuw (van der Pijl-Ketel (ed) 1982; Sjostrand, 2007); See Dallal 1996 re earlier assessments that Comp 38 Wan Li charger post-dated 1670–1690 ; now disputed by Jan Baqart (pers. comm. 2009).
4. Westerwald post-1618 at Jamestown (Kelso & Straube 2004, 136); European Date Range of 1550–1775 (Mallios 1999, Fig 60, p 48; Hurst et al., 1986).
5. "Ox-Head" Tile Corner Motif: New TPQ 1620 vs. 1650 (Pluis 1998, 537; see Huey 1988, p. 436); "Spider's head corner motif on delft tile dated to post-1640 to ca. 1670 (Pluis 1998, 555)
6. "Buff bodied slipware" original dated to 1680 (cf.Huey pers. Com. in Grossman et al 1984, Pages, V-8, V-21) Revised to post 1588, cf. Huey 1988, 404); Assumed to be too generalize for site-specific TPQ.
7. Component 2 (Below cobble floor of building A) dateable only to early-mid 17th c.; Original Pipe TPQ of 1657 is typo (1659 rev down to 1649, cf McCashion, Dallal in Grossman et al.,1985). Orig. Glass TPQ of 1676 is error—no lead glass present (Grossman et al., 1985, vi–5, vii–14).
8. Pipe Bore Stem Mean Dates are now rejected as unreliable cf. (Duco 1987, 135–136).

Graphic: Joel Grossman, Ph.D. © 2010.

Diag. TPQ Data/Comments (See numbered footnotes 1–8)	17th Century Occupants— Innes 1902	Arrival Date (van der Donck 1656; Innes 1902)	Deed/Ground Briefs from Dutch West India Company (Innes 1902; Stokes 1915–1928)
"EB" Pipe Mark 1630–1683; 10 Pipe bowls post-1680 forms; Latest dateable ceramic was tile with "Ox-head" motif, Rev. to post-1620; Orig. Ceramic TPQ based on 3 late sherds—Westerwald, Pearlware and Whiteware, prob. Intrusive; 99% (292/298) were Early-Mid 17th c.; Earliest Ceramic type was a Weser red-slipware platter (1570–1630), [No seeds recovered due to mixture](1,5)	Street Matrix @ Pearl	"Laid Out" ca 1630; Paved-cobbles ca. 1680	Singleton 1909
Post-1725 Glass TPQ in is data entry error in Table I-A2, Actual Glass TPQ for Component was ca. Post-1680 Wine Bottle finish& Pb Glass = Post 1676-80; Orig. Ceramic TPQ of 1800 based on a probably intrusive sherd of embossed Pearlware; Profile suggests mixture from cap of 18th rubble over Late 17th c. Builders Trenches (Grossman 1985, Plate IV-12Top); Comp. TPQ revised from Post-1800 to Post-1680 based on pipes & glass 1) Grossman 1985 et. al V29;VI-21); Ceramic TPQ based on "Ox-Head" tile motif, Rev. to Post-1620	Post-Warehouse Building E Wall BT	Building Earlier than thought; Rev. from Early 19th c. to Late 17th; Wall suggests pos correlation in time and space with first Stat Hays.	na
Note: Top context (102.01) Stratigraphically more recent with Ceramic TPQ (1734 Soft Past-Porc.) than lower contexts.	Kierstede/Steenwyck	1638	1646 (Innes 1902); 1647 (Stokes 1915–1928)
Pipe Mean Date = 1706; Pipe bowl TPQ's = 1678; Kiersted property until 1710 (Stokes 1915–1928); Crossmend w Comp. 54 w MD of 1716; Original and 2008 Component TPQ of 1720 is approximate cf. Glass and Pipe TPQ's.	Kierstede Rear Shed/Cookhouse?	1638	1646 (Innes 1902); 1647 (Stokes 1915–1928)
Pipe stem Mean date = 1716; Pipe bowl TPQ =1678 (crossmnd: Cmp53)	Kierstede Rear Shed/Cookhouse?	1638	1646 (Innes 1902); 1647 (Stokes 1915–1928)
Note: Top context (102.01) Stratigraphically more recent with Ceramic TPQ (1734 Soft Past-Porc.) than lower contexts.	Kierstede/Steenwyck	1638	1646 (Innes 1902); 1647 (Stokes 1915–1928)
Cx-29 TPQ=1830; Early 19th C. Stone Pier Pits, Cmps. 27,28,29,49 = Early 19th C - Contemp. Structural Group		1830–1850	Early 19th c. Features
		1830–1850	Early 19th c. Features
Doc and Structural Evidence of mid-19th Cent.; Assoc w Comp. 47-Brick Drain assoc. w post 1820 Ceram & post 1850 ceramic (St.Grp I).(Grossman et. al. Table I-A2)		1830–1850	Early 19th c. Features
Flow Blue Transprint Whiteware (1844)		1830–1850	Early 19th c. Features
Comp. TPQ = Glass- Snap case base, Post 1857		1830–1850	Early 19th c. Features
ABM (Automatic bottle Machine) 1903			
ABM (Automatic bottle Machine) 1903			

buildings erected (Innes 1902, 5–6). A tavern and a brewery were in place, apparently across the street to the north, by 1631, and a church—"a mean barn"—was erected along the Strand (Pearl Street) by 1633 (Innes 1902, 3, 58; Stokes 1915–1925, 267). A surviving letter, referring to the decade before 1639, also documented that this early 1630s commercial activity, in the western end of the block near Whitehall, was matched by a zone of boat repair and construction facilities fronting the eastern end of the block at the outlet of the ditch or "Graft," later renamed as Broad Street (Van Laer 1974, I: 111). In 1934 Poole also described this it as a landing place for small "country shallops" (Poole 1934, 52).

This 1639 affidavit before Secretary van Tienhoven by a carpenter seeking compensation for work done during the administration of Van Twiller, the director of New Netherland between 1633 and 1638, provides a glimpse of the extent of building activity in the 1630s. Specifically referring to work outside the fort, he listed a bake house, a church with house and stable in the rear, a large shed in which boats and yachts were built, a goat house, a small house for the midwife (the mother of Sara Roelofs Kierstede?), a number of houses, the repair of sawmills and a gristmill, and the buildup of the fort bastion (Van Laer 1974, I: 108–109).

Additionally, the excavation exposed the rectangular stone foundation of a single large building that was originally interpreted as the warehouse belonging to Agustijn Heerman (variously spelled as Augustyn Heermans and/or Augustine Heerman), who arrived in New Amsterdam in 1633 (Jameson 1909, 289). Heerman, in actuality, administered the warehouse as an agent for the firm of Pieter Gabry and Sons; Pieter Gabry was the son of Charles or Carel Gabry, merchant of Amsterdam and director of the West India Company (Jameson 1909, 375; pers. com. Jaap Jacobs 2009). When excavated, it was thought that the warehouse postdated these surviving records of land transfer, interpreted by different historians to have taken place either in 1647 (Innes 1902, 18) or after 1645 (Stokes 1915–1935). But Innes suggested that the warehouse appears to have been rebuilt several times before 1647 (Innes 1902, 18).

Finally, the block included the early home, or compound, of one of the settlement's first doctors,

Dr. Hans Kierstede (built for him by the WIC), who arrived in 1638 and married Sara Roelofs in 1642 (Van Rensselaer 1898, 24). (As noted by Jaap Jacobs, Kierstede was in actuality a surgeon and would have been addressed and referred to as "meester," or Mister, instead of Doctor [pers. com. Jan. 24, 2010].) After a decade of service to the WIC, Dr. Kierstede was granted title to his parcel at the corner of Pearl and Whitehall streets in 1646 (Innes 1902, 18). But his home may have been built soon after his arrival in 1638 and possibly before his wedding. Innes described his company-built home as being "to the west of the Company's Warehouse on the Strand," which suggested (1) the absence of other residences in the intervening space, and (2) that both the warehouse and the Kierstede home may have been already built before 1642 (Fernow 1976; van Rensselaer 1898, 24; Innes 1902, 18).

Accordingly, when combined, the revised archaeological and archival evidence suggests that the earliest structural elements and ethnobotanical samples date to the second quarter of the seventeenth century. Given ambiguities over the date of introduction for different artifact types and the constantly evolving assessment of regional and international chronologies, it is safe to suggest that the early to mid-seventeenth-century components probably fall within a twenty-year time span between the 1630s and the 1650s. Of these, the earliest features appear, based on the historical references above, to have been constructed and deposited between 1633 and 1638, consistent with the above-referenced revised archaeological assignment to the decade of 1630. As detailed in Table 8.1, the earliest deposits and features were almost exclusively made up of the fill of builder's trenches for foundation walls, cisterns, and privies (denoted by "BT"). As such, these "BT" features, and their seeds, also reflect the earliest environmental conditions and plants when the site was initially occupied by the Dutch, prior to 1630 (Table 8.1).

ETHNOBOTANICAL CONTINUITY AND CHANGE

Using the artifact-based archaeological sequence for dating, this treatment will concentrate on the chronology and ethnobotanical significance of three primary topics: (1) Native American potherbs and starchy seed bearing plants, (2) identification of *Cruciferae/Brassica* or cabbage family vegetables, and (3) Native American and European medicinal plants, predominantly in the early seventeenth century.

While the archaeological record at Broad Street does not extend back to the decade of Henry Hudson's initial visit to the area in 1609, it documents several important trends in changing plant diversity that are otherwise not clearly in evidence from archival sources alone. As will be documented below, the stratigraphically sequenced, dated, and quantified plant remains suggest that the earliest Dutch settlers may have had access to, or actively exploited, a range of previously underappreciated indigenous plant foods and medicines; present new evidence for the appearance of European vegetables and fruits; and show a profound "dropoff" in plant diversity by the early eighteenth century.

The Botanical Flotation Samples

The basic units of ethnobotanical analysis were selected only from "hi-integrity," or unmixed, units of well-dated natural stratigraphic association and contemporaneity. These minimal units of association, the individual excavation "Contexts" (each distinguished by a unique computer number designation), were grouped in the stratigraphic reconstruction process into larger units of analysis and dating, called Components (each a discrete, and functionally distinct, feature—e.g., pit, cistern, builder's trench—with each comprised of one or more contexts). Botanical analysis was limited to only those reconstructed components that were both tightly dated and stratigraphically unmixed. Once the stratigraphic associations and relative age of each was defined based on the age of the most recent artifacts they contained, this subset of unmixed and dated deposits were subjected to archaeological "flotation," a technique that uses water suspension and jets of circular air streams to agitate and separate out fragile plant seeds, mostly charred, from their soil matrix.

With the exception of one large 8.5 liter sample from Component 38, and one two liter sample from Component 8, all other flotation samples were limited to one-liter volumes. A total of 32.5 liters from six components and fifteen stratigraphically distinct contexts were "floated," manually sorted, and prioritized for ethnobotanical identification. Additional specimens, especially the larger pumpkin and peach pits, were also recovered from the one-quarter-inch field screens during excavation from five additional components made up of eight contexts (Tables 8.2 and 8.3).

The resultant seed recovery was roughly comparable both in sample size and seed count for all three periods. Out of a total sample of 2,607 recovered seeds (1,148 unidentified), 1,458 seeds were identified to the genus level from twelve components and twenty-four contexts for all three periods. Out of twenty-four identified seed types, nineteen were identified from seven components and twelve contexts dating to the early-mid seventeenth century; thirteen varieties from three components and seven contexts were recorded for the late-seventeenth-century sample, and seven plant types were recovered from three components and five contexts for the early eighteenth century (Tables 8.2 and 8.3). See www.GeospatialArchaeology.com/BroadSeedData.html for a context-specific breakdown of seeds types by basic context-level units of association and contemporaneity.

These levels of recovery may be far from representative of the full range of plants once present. Samples from other historical sites have demonstrated that only between 8 to 32 percent of artificially introduced "control" seeds were recovered by flotation (Miller 1998, 65). Accordingly, the following analysis treats the recovered plants remains as gross, order of magnitude, indices of the changing diversity, and assumes that the actual range of variation may have been significantly broader for each period.

Continuity and Change in Plant Diversity

Despite these generic sampling issues, the quantified seed data suggests order of magnitude changes in plant diversity between the early seventeenth and early eighteenth centuries (Fig. 8.2).

In addition to these gross changes, the range of identified seed types were evaluated according to seven major ethnobotanical functional categories:

TABLE 8.2. Cross tabulation breakdown of seed counts per dated Components and Periods.

Seed Totals (Identified) per Dated Component and Period	Nicotiana sp. - tobacco	Galium sp. - bedstraw / "cheese rennet"	Polygonum sp. - knotweed / "Bistort"	Amaranthus sp. - amaranth	Chenopodium sp. - lambsquarters	Phytolacca sp. - pokeweed / pokeberry	Citrus sp. - citrus	Vitis sp. - (Lg. Seed 6.1 - 6.5mm)	Vitis sp. - (Sm. Seeds 4.5 - 5.5mm)	Portulaca sp. - purslane	Vaccinium sp. - blueberry	Cruciferae sp. - cabbage	Cucurbita sp. - pumpkin / squash	Fragaria sp. - strawberry	Rubus sp. - brambles (raspberry / blackberry)	Prunus persica - peach	Trifolium sp. - clover	Cyperus sp. - sedge / nut grass	Acalypha sp. - copperleaf	Mollugo sp. - carpetweed	Linum sp. - flax	Stachys sp. - "woundwort" / "betony" / "heal-all"	Prunus sp. - cherry	Acacia sp. - an acacia	GRAND TOTAL
EARLY-MID 17TH C.																									
Comp 13 - Lot 8 - North Barrel fill	1	1	1	1	1		1		23	1		1	6	6	2		1	1							47
Comp 10 -BT - Oval Yellow brick cistern													12												12
Comp 61 - BT - Lot 14 Barrel															2										2
Comp 9 -BT - Lot 8 South Barrel														6											6
Comp 8 -BT - Lot 8 North Barrel									5			1													6
Comp 6 -Warehouse floor - Bld. A															1										1
Comp 38 - Rope Basket / Cask - fill	1					5		4	4	7		1	1	10	9	1									43
EARLY-MID 17TH C. TOTAL	2	1	1	1	1	5	1	4	4	35	1	1	14	17	15	10	1	2	1						117
LATE 17TH C.																									
Comp 14 - Lot 8 - South Barrel fill									24	1	7		8	10	2		8	2	1	1	1				65
Comp 62 - Lot 14 - Barrel fill												2	2	2	1	1									8
LATE 17TH C. TOTAL									24	1	7	2	10	12	3	1	8	2	1	1	1				73
EARLY 18TH C.																									
Comp 53 - Building D - Lower fill								18				1	2	1173										1	1195
Comp 63 - Lot 14 - Red brick cistern / well -Cx.01													6												6
Comp 63 - Lot 14 - Red brick cistern / well -Cx.02															9										9
Comp 63 Lot 14 - Red brick cistern / well - Cx. 02-04													9												9
Comp 63 Lot 14 - Red brick cistern / well - Cx. 02-05													28												28
Comp 63 Lot 14 - Red brick cistern / well - Cx. 02-06														4											4
Comp 63 Lot 14 - Red brick cistern / well - Cx. 02-07														14											14
Comp 63 Lot 14 - Red brick cistern / well - Cx. 02-08																							2		2
EARLY 18TH C. TOTAL								18				1	44	2	1173	27							2	1	1267
GRAND TOTAL	2	1	1	1	1	5	1	4	22	59	2	8	60	29	1200	40	2	10	3	1	1	1	2	1	1457

Graphic: Joel Grossman, Ph.D. © 2010

(1) Native American potherbs, (2) indigenous starchy seed plants, (3) indigenous seasonably available fruits and berries, (4) Native American pumpkin or squash, (5) non-food plants (indigenous and European), (6) European fruits and vegetables, and (7) Native American and European medicinal plants. Each of these categories was further compared as a plot of continuity and change in three functional distribution tables designed to graphically show the shifting patterns of plant diversity between each period of the revised three-phase site chronology: early to mid-seventeenth century (Table 8.4), the late seventeenth century (Table 8.5), and early eighteenth century (Table 8.6). These comparisons also showed that some varieties continued to be represented in all three periods. Seeds of squash or pumpkins, strawberries, and brambles (as well as peaches and small-seeded

TABLE 8.3. Cross Tabulation Showing Number of Instances of Stratigraphically Distinct Deposits (Contexts) for Each Seed Type per Dated Component and Period.

Contexts per Component and Period	Nicotiana sp. - tobacco	Galium sp. - bedstraw / "cheese rennet"	Polygonum sp. - knotweed / "Bistort"	Amaranthus sp. - amaranth	Chenopodium sp. - lambsquarters	Phytolacca sp. - pokeweed / pokeberry	Citrus sp. - citrus	Vitis sp. - (Lg. Seed 6.1 - 6.5mm)	Vitis sp. - (Sm. Seeds 4.5 - 5.5mm)	Portulaca sp. - purslane	Vaccinium sp. - blueberry	Cruciferae sp. - cabbage	Cucurbita sp. - pumpkin / mustard fam. / squash	Fragaria sp. - strawberry	Rubus sp. - Brambles (raspberry / blackberry)	Prunus persica - peach	Trifolium sp. - clover	Cyperus sp. - sedge / nut grass	Acalypha sp. - copperleaf	Mollugo sp. - carpetweed	Linum sp. - flax	Stachys sp. - "woundwort" / "betony" / "heal-all"	Prunus sp. - cherry	Acacia sp. - an acacia	Grand Total
EARLY-MID 17TH C.																									
Comp 13 - Lot 8 - North Barrel fill	1	1	1	1	1		1			4	1		1	2	3	1		1	1						20
Comp 10 -BT - Oval Yellow brick cistern													2												2
Comp 61 - BT - Lot 14 Barrel																1									1
Comp 9 -BT - Lot 8 South Barrel																1									1
Comp 8 -BT - Lot 8 North Barrel								1						1											2
Comp 6 -Warehouse floor - Bld. A																		1							1
Comp 38 - Rope Basket / Cask - fill	1					1			1	1		1	1	1	2		1								11
EARLY-MID 17TH C. TOTAL	2	1	1	1	1	1	1	1	1	6	1	1	4	4	5	3	1	2	1						38
LATE 17TH C.																									
Comp 14 - Lot 8 - South Barrel fill								4	1	3			4	3	2			5	1	1	1	1			26
Comp 62 - Lot 14 - Barrel fill											2		1	1	1	1									6
LATE 17TH C. TOTAL								4	1	3	2		5	4	3	1		5	1	1	1	1			32
EARLY 18TH C.																									
Comp 53 - Building D - Lower fill								2				1	2	2									1		8
Comp 63 - Lot 14 - Red brick cistern / well -Cx.01												1													1
Comp 63 - Lot 14 - Red brick cistern / well -Cx.02														1											1
Comp 63 Lot 14 - Red brick cistern / well - Cx. 02-04												1													1
Comp 63 Lot 14 - Red brick cistern / well - Cx. 02-05												1													1
Comp 63 Lot 14 - Red brick cistern / well - Cx. 02-06																1									1
Comp 63 Lot 14 - Red brick cistern / well - Cx. 02-07																1									1
Comp 63 Lot 14 - Red brick cistern / well - Cx. 02-08																						1			1
EARLY 18TH C. TOTAL								2				4	2	2		3						1	1		15
GRAND TOTAL	2	1	1	1	1	1	1	1	3	10	2	4	10	11	11	9	2	7	2	1	1	1	1	1	85

Graphic: Joel Grossman, Ph.D. © 2010

grapes) were recovered from a variety of deposits from all three periods. Blueberries were restricted to only the early and late-seventeenth-century samples.

As graphically depicted in Table 8.4, twelve (12) or ca. 60 percent, of the nineteen different plant types identified from the early-seventeenth-century sample could be linked to Native American food sources (potherbs and seed bearing plants) and potential medicinal uses. However, the early-seventeenth-century sample also included two European orchard fruits, represented by multiple instances of peach pits and a single citrus seed. Like blueberries, strawberries, and brambles (raspberries/blackberries), peach pits were recovered from multiple deposits from all three seventeenth and eighteenth-century sample periods. While peach was of

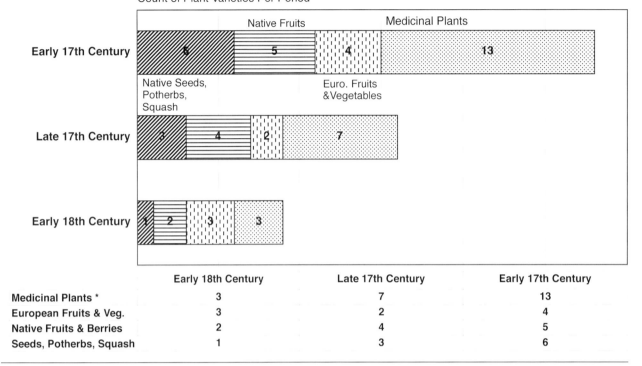

FIG. 8.2. Changes in Plant Diversity by Period.
This horizontal bar chart compares gross changes in the relative prevalence of major plant categories between each of the three main Periods. It illustrates an order of magnitude (ca. 50%) decrease in the number and diversity of plants between the early and late seventeenth centuries, with an even sharper reduction in plant diversity (ca. 80%) by the early eighteenth century (see Tables 8.4 to 8.6 for detailed plant-use breakdowns). These pronounced changes underscore the danger of relying on either contemporary or historical, eighteenth or nineteenth century, plant inventories to reconstruct conditions in the early seventeenth century.
*Medicinal plants include: tobacco, grapes, sedge, bedstraw, and squash as well as a number of indigenous and introduced potherbs and seed-bearing nutritional plants; exclude: citrus, clover, copperleaf, and cabbage family examples.
Graphic: Joel Grossman, Ph.D. © 2010

undisputed European origin, and peach orchards were documented in New Amsterdam by 1639 (Jacobs 2005, 107), the attribution of peaches to purely Dutch sources must be approached with caution. Peach orchards were cultivated by Cherokee farmers along the Gulf Coast, suggesting that peaches may have been introduced as early as the fifteenth century by Spanish conquistadors (Delcourt 2004, 107). Peach (as well as plum and cherry) pits were also among the seeds ordered by the Massachusetts Bay Colony to the north by 1629 (Hedrick 1919, 463).

The single citrus seed was recovered from the unmixed, single-component, interior fill of a double-barrel cistern that was abandoned in the second quarter of the seventeenth century (Component 13; Tables 8.2 and 8.3). The feature was undisturbed by later intrusions, and both its association and dating to the early seventeenth century appear reliable. No citrus seeds were recovered from later seventeenth and eighteenth-century deposits. However, its presence begs the question as to how it got into the site matrix. The native habitat of citrus is generally limited to tropical and subtropical environments; it does not tolerate temperatures below 47° to 57° F, and does not react well to frost or salty soils (Culture Sheet.org; www.culturesheet.org/rutaceae:citrus). Therefore, citrus trees probably could not have grown in New Amsterdam in the seventeenth century without the protection of a greenhouse-like structure against frost. One possibility is that the seed arrived in some form of preserve such as an early marmalade (a concentrate of boiled sugar and rinds) that was being made in Europe, origi-

TABLE 8.4. Early-Seventeenth-Century Plant Diversity

Archaeological Plants		Nicotiana sp. – tobacco	Galium sp. – bedstraw	Polygonum sp. – knotweed / "bistort"	Amaranthus sp. – amaranth	Chenopodium sp. – lambsquarters	Phytolacca sp. – pokeweed / pokeberry	Citrus sp. – citrus	Vitis sp. – (Lg Seed 6.1 – 6.5mm)	Vitis sp. – (Sm. Seeds 4.5 – 5.5mm)	Portulaca sp. – purslane	Vaccinium sp. – blueberry	Cruciferae fam. – cabbage fam.	Cucurbita sp. – pumpkin / squash	Fragaria sp. – strawberry	Rubus sp. – Brambles /raspberry	Prunus persica – peach	Trifolium sp. – clover	Cyperus sp. – sedge / nut grass	Acalypha sp. – copperleaf	Mollugo sp. – carpetweed	Linum sp. – toadflax	Stachys sp. – "betony" / "woundwort"	Prunus sp. – cherry	Acacia sp. – an acacia	Grand Total
Period	Early-Mid 17th C.	2	1	1	1	1	1	1	1	1	1	1	1	4	5	3	1		1							38
	Late 17th C.										4	1	3	2	4	4	1	2	5	1	1					33
	Early 18th C.							2						4	2	3			1			1	1	1	1	15
	Grand Total	2	1	1	1	1	1	3	1	1	6	2	4	10	11	10	2	2	7	1	1	1	1	1	1	86
Ethnohistorical and Ethnobotanical Plant Uses	Native American Potherbs			Members of Eastern Ag. Complex																						
	Native American Starchy Seed Foods																									
	Native American Pumpkin/Squash													✓												
	Native American Fruits and Berries								✓	✓		✓			✓	✓										
	Non-Food Plants (Indigenous & European)		✓																							
	European Fruits & Vegetables							?	?	?	?	?	?	?	?	?	?	?	?	?	?	?	?	?	?	
	Native American Medicines	+	+	+	+	+	+	+	+	+	+	+	+	+	+	+	+	+	+	+	+	+	+	+	Adopted by Indigenous healers	
	European Medicines	+	+	+	+	+	+	+	+	+	+	+	+	+	+	+	+	+	+	+	+	+	+	+	+	

Notes on non-food/dye column: "Native Americans Dye; Coffee substitute in N. Europe"

This cross-tabulation table shows the early to mid-seventeenth-century distribution of recovered plant remains in relation to eight ethnobotanical and functional use categories: (1) Native American Potherbs, (2) Native American Starchy Seed Foods, (3) Native American Pumpkin/Squash, (4) Native American Fruits and Berries, (5) Non-food Plants, (6) European Fruits and Vegetables, and (7 & 8) Native American and European Medicinal Plants. In addition to the presence of traditionally recognized Native American foods (Squash/pumpkin and berries), this breakdown by function and origin illustrates the early availability of citrus, peach, and possibly grapes, as well as a number of European vegetables. Furthermore, ethnobotanical and archaeological sources strongly suggest that the range of foods available to the earliest Dutch settlers may have been significantly broader (through the addition of five indigenous potherbs, of which three were also seed-bearing plants) than traditionally thought. They also indicate that at least thirteen out of nineteen (or ca. 70%) of the early-seventeenth-century plants may have been exploited as Native American and European medicines. Graphic: Joel Grossman, Ph.D. © 2010

TABLE 8.5. Late-Seventeenth-Century Plant Diversity

Archaeological Plants		Nicotiana sp. – tobacco	Galium sp. – bedstraw	Polygonum sp. – knotweed / "bistort"	Amaranthus sp. – amaranth	Chenopodium sp. – lambsquarters	Phytolacca sp. – pokeweed / pokeberry	Citrus sp. – citrus	Vitis sp. – (Lg Seed 6.1 – 6.5mm)	Vitis sp. – (Sm. Seeds 4.5 – 5.5mm)	Portulaca sp. – purslane	Vaccinium sp. – blueberry	Cruciferae sp. – cabbage fam.	Cucurbita sp. – pumpkin / squash	Fragaria sp. – strawberry	Rubus sp. – Brambles / raspberry	Prunus persica – peach	Trifolium sp. – clover	Cyperus sp. – sedge / nut grass	Acalypha sp. – copperleaf	Mollugo sp. – carpetweed	Linum sp. – toadflax	Stachys sp. – "betony" / "woundwort"	Prunus sp. – cherry	Acacia sp. – an acacia	Grand Total
Period	Early-Mid 17th C.	2	1	1	1	1	1	1	1	6	1	1	1	4	4	5	3	1	2	1						38
	Late 17th C.				1					4	1	3	2	5	4	4	1	5	1		1	1		1		33
	Early 18th C.							2					4	2	2	3			1						1	15
	Grand Total	2	1	1	1	1	1	3	1	10	2	4	10	11	11	10	2	7	2	1	1	1	1	1	1	86
Ethnohistorical and Ethnobotanical Plant Uses	Native American Potherbs																									
	Native American Starchy Seed Foods																									
	Native American Pumpkin / Squash																									
	Native American Fruits and Berries																									
	Non-Food Plants (Indigenous & European)																									
	European Fruits & Vegetables																									
	Native American Medicines																					Adopted by indigenous healers				
	European Medicines																									

This cross-tabulation table of late-seventeenth-century plant remains shows their distribution relative to the same eight ethnobotanical use categories as in the previous early-seventeenth-century breakdown (Table 8.4). In addition to an order of magnitude reduction in number and diversity of indigenous potherbs, nutritional seed sources, and potential indigenous medicinal plants (down from thirteen to five) by the late seventeenth century, it shows a continuity of indigenous squash/pumpkin, berries, and European fruits and vegetables. The late-seventeenth-century sample was also distinguished by the introduction of three (for a total of eight) plant types used in Europe as sixteenth and seventeenth-century medicines. Graphic: Joel Grossman, Ph.D. © 2010

TABLE 8.6. Early-Eighteenth-Century Plant Diversity

Archaeological Plants	Periods	Nicotiana sp. – tobacco	Galium sp. – bedstraw	Polygonum sp. – knotweed / "Bistort"	Amaranthus sp. – amaranth	Chenopodium sp. – lambsquarters	Phytolacca sp. – pokeweed / pokeberry	Citrus sp. – citrus	Vitis sp. – (lg Seed 6.1 – 6.5mm)	Vitis sp. – (Sm Seeds 4.5 – 5.5mm)	Portulaca sp. – purslane	Vaccinium sp. – blueberry	Cruciferae sp. – cabbage fam.	Cucurbita sp. – pumpkin / squash	Fragaria sp. – strawberry	Rubus sp. – Brambles / raspberry	Prunus persica – peach	Trifolium sp. – clover	Cyperus sp. – sedge / nut grass	Acalypha sp. – copperleaf	Mollugo sp. – carpetweed	Linum sp. – toadflax	Stachys sp. – "betony" / "woundwort"	Prunus sp. – cherry	Acacia sp. – an acacia	Grand Total
	Early-Mid 17th C.	2	1	1	1	1	1		1	6	1	1	4	4	5	3		1	1	1	1	1	1			38
	Late 17th C.							1		4	1	3	2	5	4	4	1	2	5	1						33
	Early 18th C.								2				4	2	2	3	1		1					1		15
	Grand Total	2	1	1	1	1	1	1	3	10	2	4	10	11	11	10	2	3	7	2	1	1	1	1	1	86

Ethnohistorical and Ethnobotanical Plant Uses
Native American Potherbs
Native American Starchy Seed Foods
Native American Pumpkin/Squash
Native American Fruits and Berries
Non-Food Plants (Indigenous & European)
European Fruits & Vegetables
Native American Medicines
European Medicines

This cross-tabulation shows the distribution of early-eighteenth-century plant remains relative to the same eight ethnobotanical use categories as in the previous early and late-seventeenth-century breakdowns (Tables 8.4 and 8.5). While it illustrates the continuity of indigenous berries, squash/pumpkin, and at last one European fruit (peach), it shows the near-disappearance of indigenous potherbs, nutritional seed-bearing and medicinal plants, as well as the lack of European vegetables by the first quarter of the eighteenth century. The recovery of a single cherry pit is inconsistent with archival references to its presence in the early seventeenth century. The disappearance of sedge (Cyperus sp.) suggests that the local wetlands may have been drained and filled or that it was no longer being harvested in the vicinity by the early 1720s–'30s.
Graphic: Joel Grossman, Ph.D. © 2010

nally as of the thirteenth century with quinces, but with oranges and limes by the seventeenth century (Davidson 2006, 483; Wilson 1999, 126). The other possibility is that that the seventeenth-century Dutch of New Netherland may have experimented with early examples of "orangery." An early heated, and apparently glassed-in, building had been built at the Hortus Botanicus of Leiden as of 1599, called the *Ambulacrum,* to house exotic collections and dormant plants, and to train students during the winter (Swan 1998, 11; Huxley 1978, 230; Cook 2007b, 120). Given the strong links between Dutch East and West India Company doctors, apothecaries, botanists, officials, and the University of Leiden (see below); it is plausible that similar protective structures may have been tried in New Amsterdam as well.

The recovery of sedge (*Cyperus* sp.) from the early and late-seventeenth-century deposits may reflect both environmental conditions and a combination of indigenous and European cultural patterns. Its "nut-like tubers" are edible, either raw or cooked, were known in the Rhine drainage as "German Sarsaparilla," and used there as a substitute for coffee (Fernand and Kinsey 1958, 107–10). Because of its pleasant odor, "sweet sedge" was used in Europe to cover the floors of churches and homes (Grieve 1931, 726–30). In addition to its Native American use as cordage and basket-making material, sedge was known in nineteenth-century America as a diuretic and "sudoric" treatment for profuse sweating (Ripley and Dana 1875, XIV, 748). Although present in the earlier deposits, no sedge was recovered from the early-eighteenth-century samples; a change that suggests either that the local wetlands may have been drained or filled, or that it was no longer being collected or growing in the vicinity by the early 1720s-'30s.

The transition from the early to late seventeenth century was characterized by three contrasting trends: (1) continuity of fruits and berries of both local and foreign origin; (2) the disappearance of most of the earlier indigenous potherbs and starchy-seed esculents (edible plants, either wild or cultivated); and (3) by the appearance of members of the *Brassica* or cabbage family. In addition to the dropping out of six plants belonging to Eastern Agricultural Complex, the transition to the late seventeenth century was demarcated by the appearance of toadflax (*Linum* sp.) and woundwort (*Stachys* sp.), both apparently alien introductions from Europe (Tables 8.2, 8.3, 8.5, 8.9). Potential medicinal plants dropped by one-half in the late seventeenth century, down from the early-seventeenth-century total of thirteen to six. Finally, although recognized as a member of the prehistoric Eastern Agricultural Complex, carpetweed *(Mollugo* sp.) was not identified until the late seventeenth century (Table 8.5).

The early-eighteenth-century sample was distinguished by a pronounced reduction in overall plant diversity. However, fruit and berry plants (strawberries and brambles), pumpkin/squash, as well as peaches, continued from the earlier seventeenth century into the first quarter of the eighteenth century (Table 8.6). The singular appearance of cherry pits only in the early-eighteenth-century deposits was late for the settlement's horticultural history; Van der Donck recorded the successful importation and cultivation of cherry trees at least by the first half of the seventeenth century (Goedhuys 2008, 25).

Sampling and recovery issues aside, the revised stratigraphic and artifact sequence, and quantified comparisons of shifting plant diversity between the early seventeenth and early eighteenth centuries suggests: (1) that what were potentially indigenous potherbs, starchy seed-bearing foods, and medicinal plants were concentrated only in the early-seventeenth-century phase of the sequence, but had disappeared from the archaeological record by the first quarter of the eighteenth century; (2) a sharp decline in indigenous plant diversity, both food-related, and of potential medicinal uses between the early and late seventeenth century; (3) a continuity of indigenous fruits, berries, and squash/pumpkin (as well as peach)—but no vegetables—into the early eighteenth century; and (4) the introduction of European vegetables and fruits in the early seventeenth century.

INDIGENOUS PLANTS OF THE "THE EASTERN AGRICULTURAL COMPLEX"

Pre-Contact Starchy Seed Plants and Potherbs

Although often dismissed as "introduced," "weedy," "alien," "emergent," "naturalized," "adventive [*sic*]," "invaders," "pioneering" species, or simply "pests" (Dudek et al. 1998, 66; Richardson etal. 2000, 93), and often interpreted as indicators of environmental trauma, nearly 30 percent of the nineteen seed varieties from the initial early to mid-seventeenth-century samples may have been derived from indigenous antecedents, that were exploited either as food sources, dyes, or as medicinal plants (Table 8.4). Archaeological findings from prehistoric sites throughout eastern North America, and historic ethnobotanical accounts, have underscored the important roles these formally underappreciated potherbs and high-carbohydrate seed-producing plants over the last two millennia in the Northeastern United States (see Smith 1989; 1992; Delcourt 2004).

Although no evidence for maize, beans, or sunflower cultivation was recovered from the historic seventeenth-century deposits at Broad Street, in addition to pumpkin/squash, fruits and berries, eight of the seventeenth-century seed types—amaranth (*Amaranthus* sp.), lambsquarters (*Chenopodium* sp.), knotweed (*Polygonum* sp.), purslane (*Portulaca* sp.), tobacco (*Nicotiana* sp.), bedstraw (*Galium* sp.), pokeweed (*Phytolacca* sp.), and carpetweed (*Mollugo* sp.) belong to what is now defined by North American archaeologists as prehistoric and contact-period potherbs and/or starchy seed-bearing components of the two thousand-year-old pre-maize "Eastern Agricultural Complex" (Smith 1989), or the "early Woodland garden complex" (Delcourt 2007, 42; Watson 1989). Five of the Broad Street plants have been identified in the archaeological and ethnobotanical literature as potherbs: pokeweed (*Phytolacca* sp.), purslane (*Portulaca* sp.), amaranth (*Amaranthus* sp.), lambsquarters (*Chenopodium* sp.), and carpetweed (*Mollugo* sp.) (Delcourt 2004, 42, 106). Three others from the early to mid-seventeenth-century contexts may have been exploited for their high-starch-yielding seeds, knotweed (*Polygonum* sp.), amaranth (*Amaranthus sp.*), and lambsquarters (*Chenopodium* sp.) (McAndrews and Boyko-Diakonow 1989; Byrne and Finlayson 1998; Delcourt 2004, 94) (Tables 8.2 and 8.3).

Both amaranth and chenopods have a long history in the archaeological and ethnohistorical record as significant Native American food plants, long recognized for Mexico and the Andes, but only recently for eastern North America (Safford 1917; Sauer 1950; Sauer 1967). "Amaranths are fast growing, cereal like plants that produce high protein grains in large, sorghum-like seed heads" (National Academy of Sciences 1975, 14). Both wild and domesticated South American and Mexican species have been recorded to produce yields of between eight hundred and one thousand pounds per acre; with nutritional qualities distinguished by high levels of protein (+15%), amino acids, especially lysine (6.2%), and fat (3–6%) (Cole 1979, 275–79). Chenopodium, like amaranth (as well as pokeweed and bedstraw), thrives in disturbed "anthropogenic" habitats "as an invasive plant . . . near barns, fields, and along roadsides" or "other humanly altered environments" (Martin et al. 1951, 389–90; Fernald and Kinsey 1958, 185; Delcourt 2004, 86). The Mohawk name for lambsquarters was "loves villages" (Fenton 1942, 525).

Over the last thirty years, archaeologists working in the eastern United States have argued that these indigenous potherbs and seed-producing plants began to be exploited, collected, or "quasi-cultivated," several thousand years before the appearance of maize (ca. AD 800 and 1100); and—after an initial period of transition as floodplain-adapted species between 2000–1500 BC—were under cultivation between 500 and 0 BC (Smith 1992, 12). Significantly, both knotweed *(Polygonum* sp.*)* and lambsquarters (*Chenopodium* sp.) were recovered from prehistoric storage pits or caches in caves outside their natural habitat range, with knotweed constituting upward of 30 percent of the "small seed assemblage" in some excavated prehistoric sites (Delcourt 2004, 42, 106). Likewise, roughly contemporary charred seeds of amaranth and *Chenopodium quinoa* were recovered from pre-Inca deposits dating to between 1000 and 1500 BC from a hilltop occupation site in the southern Andes of Peru (Grossman 1983, 86).

In North America, and building on early work by Jonathan Sauer (1952) on the floodplain adaptation of pokeweed, later scholars have argued that many of these plants (e.g., pumpkin/squash/gourds, amaranth, chenopods) were "tightly tethered," if not pre-adapted, to disturbed open floodplain environments created first by annual flooding and which later expanded into "open habitats created by human activities whenever opportunities arose" (Struever 1964, 102–103; Smith 1992, 29).

Likewise, Watson proposed that amaranth was exploited for its edible seeds as a member of a "panoply of tolerated, encouraged or quasi-cultivated plants" by at least 1000 BC (Watson 1989, 555–71). Subsequently, Watson and Kennedy (1991) also proposed that this process was both gradual and "gender-specific," and tightly linked to the role of women in planting (Smith 1992, 31). As such, their presence in the early Dutch deposits may no longer be easily dismissed as "emergent" "weeds"—indicative of environmental trauma—of "modern" origin, but instead as potential carryovers, or transplants, of long-established indigenous foods—perhaps extending back many centuries before the arrival of the Dutch.

However, given the fact that only single specimens of lambsquarters *(Chenopodium* sp.*)* and amaranth (*Amaranthus*) were recovered from early-seventeenth-century contexts, it is difficult to evaluate their presence based on either the contextual or morphological criteria of domestication (thickening of seed casing)—identifiable only with electron microscope scans not readily available at the time of the original analysis of the Dutch West India Company samples—set forth by Bruce Smith in his study of Midwestern prehistoric specimens of chenopods (Smith 1992, 110–23).

In addition, the recovery of pollen and seeds of purslane from cores near Iroquois sites in the Great Lakes region, dating to between the fourteenth and sixteenth centuries, suggests that purslane was also exploited as a North American potherb, and possibly as a source of nutritional seeds, both before and after European contact (Byrne and McAndrews 1975, 726–27). The persistent association of purslane with the better-known prehistoric Native American crops (corn, pumpkin/squash, and beans) has been interpreted by North American archaeologists as indicative of Iroquois agriculture beginning at least 650 years earlier than estimated—ca. AD 1350 (Delcourt 2004, 92–94; McAndrews and Boyko, Diakonio 1989, 528–30; Byrne and McAndrews 1975, 726–27). In addition, Delcourt and others classified purslane as a critical element of prehistoric sustenance, of equal import with other traditionally recognized indigenous cultigens: "Evidence of local cultivation of plants included pollen from maize and cucurbits [pumpkin/squash], pollen and seeds of sunflower, and *pollen and seeds of purslane*" (Delcourt and Delacourt 2004, 94; emphasis added; Byrne and Finlayson 1998, 94–107).

Two non-food plants, bedstraw (*Galium* sp.) and tobacco (*Nicotiana* sp.), both recognized as members of the Eastern Agricultural Complex, were recovered from seventeenth-century contexts at the site. A single seed of bedstraw (*Galium*) was recovered from an unambiguous early-seventeenth-century context (Component 13—see Tables 8.2 and 8.3). Although bedstraw has been included by prehistoric archaeologists in the Eastern Agricultural complex because of its utility as a late prehistoric Native American dye (Delcourt 2004, 42), its presence may also have been due to its importance as an indigenous and/or European medicinal plant. Finally, one tobacco seed was recovered from Component 38 (the "Tienhoven Basket/Cask"), which can now be dated to post-1650 to 1660, versus late in the seventeenth century (Jan Baart pers. com. Dec. 4, 2009; see Tables 8.1, 8.2, 8.3). A second possible example of a tobacco seed came from a 1630–1650 context (Component 13; Tables 8.1, 8.2, 8.3). However, a question mark in the original laboratory seed inventory notes puts its identity in question. Both only provided material evidence that tobacco was present in the mid-seventeenth century—a fact that was already well documented in the archival record (Jacobs 2005, 231, 261; 2009, 124–28).

Ethnohistorical Parallels and Analogues

Archaeological evidence is generally restricted to the recovery of either burned or waterlogged seeds; ethnohistorical archival sources may also include references to the use of soft tissue (leaves, roots, and stems) that are not generally preserved in the archaeological record. In addition to their long tenure in prehistoric archaeology, historic ethnobotanical accounts suggest that the recovery of pokeweed, amaranth, lambsquarters, and purslane in the seventeenth-century deposits in New Amsterdam may also reflect their continuity as "carryovers" or the residual byproducts of indigenous patterns of exploitation as esculents, or potherbs, and/or, as I suggest below, as medicinal plants. Members of the amaranth and *Chenopodium* families, pokeweed and purslane were exploited as potherbs both in Europe and by contact-period indigenous groups in the eastern United States (Delcourt 2004, 42; Hedrick 1919, 43–44; Foster and Duke 2000, 243). Pokeweed is native to eastern North America, and in addition to the use of its berries as a dye, its young leaves are edible and taste like asparagus (Peterson 1977, 46; Grieve 1931, 648). The Iroquois, the Mohegan, and the Ojibwa harvested lambsquarters (*Chenopodium* sp.) as a vegetable (Tantaquidgeon 1972, 83; Waugh 1916, 117; Arnason, Hebda, and Johns 1981, 2209; Regan 1928, 240). Both were documented as historic-era potherbs and "spinach" in North America (Hedrick 1919, 43,161).

Knotgrass or bistort (*Polygonum* sp.) was broadly recognized in Europe as a garden herb that was exploited both as a potherb and for its medicinal qualities. Its starchy root was eaten in eastern and northern Europe "in times of scarcity as a substitute for bread" (Hedrick 1919, 449). Where encountered, it was presumed to have been "an escape from cultivation" (Grieve 1931, 105) and was described by the sixteenth-century herbalist Fuchs as being "commonly found along paths" (Dressendorfer 2001, 901). Of potential relevance to its recovery in seventeenth-century contexts in New Amsterdam, Grieve advised that "when it has a corner in the Kitchen garden, it is well to pluck it now and then, even when it is not immediately required for culinary purposes" (Grieve 1931, 103).

Finally, purslane, or "pulsey", was used in seventeenth-century Europe as "a pleasant salad herb . . . with oil, salt and vinegar"; the younger shoots in salads and the older shoots as "potherbs . . . [and] . . . largely cultivated in Holland" (Grieve 1931, 660). The sixteenth-century German botanist Fuchs listed purslane as a vegetable and its buds as substitute for capers (Dressendorfer 2001, 903). It was also recorded in colonial-era Native American contexts in eastern Canada. In 1605 Champlain observed purslane in native gardens among the Maine coast and noted that it grew in "large quantities among the Indian corn" (Hedrick 1919, 451).

The presence of many of these exploited seed, potherb, and medicinal plants in the seventeenth-century contexts from the Pearl Street block both broadens the range of potential indigenous foods and resources available to the early Dutch inhabitants and supports the argument that they may have been more dependent on Native American foods and plants in the first half of the seventeenth century than previously recognized. Several scholars have pointed to poor crop yields in the first half of the seventeenth century, but not—other than corn—to the exploitation of other Native American food sources that may have been available (Jacobs 2005, 220; Jacobs 2009, 119: Folkerts 1996, 42–52). As alluded to in my introduction, recent research by students of climate history has suggested that the stressed agricultural production of the mid-seventeenth century (and specifically the decade of 1640) may be partially attributed to broader worldwide patterns of severe weather events, including spikes in volcanic activity, drought, and extreme cold during what has been called the Little Ice Age (Parker 2008, 1063–73; Gehring 2009, 78).

THE SEARCH FOR EUROPEAN VEGETABLES

In addition to underscoring the role of indigenous plants, this reanalysis has yielded new, and apparently the first, material evidence for the presence of European vegetables in the archaeological record of seventeenth-century New Netherland. This previous gap in the physical record was particularly perplexing because archival sources suggested that a broad range of Dutch garden vegetables, including members of the *Brassica* or cabbage family, should have been archaeologically visible in the seventeenth and eighteenth-century samples at Pearl Street.

The "discovery" of seeds of European-derived vegetables came about during the reanalysis of the original seed tabulations by the author, which led to the identification of a thirty-year-old data entry error in the laboratory and computer records. The 1984 laboratory inventory included entries of five seeds that were initially identified as members of the nightshade family (*Solanaceae*) (Grossman et al. 1985, Appendix II). However, comparison of the hard copy duplicates to the computer files revealed that although these entries had been corrected as "*Brassica*" in the original laboratory notes, they had not been transferred and corrected in the final database inventory as submitted with the official draft report (Grossman et al. 1985). Once rectified, the question became: (1) What vegetables were represented, and (2) how might they be distinguished?

Given the fact that most members of the *Brassica* or cabbage/mustard family are characterized by small (ca. 1.65 to 2.1 mm) round to oval seeds only distinguished from one another by small increments in size, two lines of evidence were used to define which vegetables may have been actually present: (1) a comparison of the seed sizes of the archaeologically recovered *Brassica* to control samples of modern garden varieties and to those of "wild" mustard seeds (introduced from Europe and adapted to the Northeast, and (2) a review of ethnohistorical literature to refine the range of potential *Brassica* cultigens in the archaeological record.

Metric Comparisons of Modern Brassica and Wild Mustard Seeds

Eight kinds of modern garden *Brassica* (cabbage, kale, Brussels sprouts, turnip, broccoli, cauliflower, and radish) were measured and averaged (from ten seeds per type) to yield a median diameter for each seed type. A "fudge factor" of 10 percent above and below the mean diameter was then plotted to show the size range of each seed type relative to the sizes of each of the five archaeologically recovered specimens. In addition, samples of seeds from beets and radishes were measured, but neither belonged to the *Brassica* family of vegetables, and both fell outside the size ranges of the "modern" seeds of that genus (Table 8.7).

The results showed that the five cases of archaeologically recovered *Brassica* or cabbage/mustard family seeds overlapped in size with six of the modern comparative samples: kale, turnip, broccoli, Brussels sprouts, cabbage, and cauliflower. In addition, the archaeologically recovered *Brassica* seeds were compared to size ranges of "Wild Mustard" seeds—from the published measurements in the Cornell University inventory *Weeds of the Northeast* (Uva, Neal, and diTomaso 1997). These overlapped in size with five introduced varieties: "Yellow rocket," Hedge mustard, Field pennycress, Virginia pepperweed, and "Wild Mustard" (Table 8.7).

When cross-referenced to *Sturtevant's Edible Plants of the World* (Hedrick 1919), (1) each of the metrically comparable "Wild Mustard" species was of European origin, naturalized in the northeast United States, and described as escaped "weeds" in the modern botanical literature, (2) most were also harvested in the wild or cultivated as garden herbs in both Europe and the eastern United States and Canada, and 3) all were classed as "esculents" or edible plants in eighteenth and nineteenth-century accounts. Three—"Yellow Rocket," "Hairy Bittercress" (also referred to as "Scurvy Grass"), and "Hedge Mustard"—were formerly used as "salads," potherbs, and/or "spinach" (Hedrick 1919, 82, 141, 536). "Wild Mustard"—also referred to as "Wild Radish" (Uva et al. 1997, 170–71)—was described by Sturtevant as a "troublesome weed of Europe naturalized in northeastern America," but its leaves were eaten as a salad and its pungent seeds used as a substitute for mustard (Hedrick 1919, 483–84).

Ethnobotanical and Historical Clues

Four sixteenth and seventeenth-century botanical accounts and plant catalogs (Table 8.8) were then surveyed to refine the range of the potential *Brassica* suggested by seed measurements (Table 8.7). Two, Van Tienhoven, secretary to the director of New Amsterdam, and Van der Donck, wrote about Dutch vegetables in New Netherland. The third and fourth sources came from lists of plants compiled by sixteenth and seventeenth-century Dutch botanists working in Holland. One of the latter came from the work of Van der Groen, the official gardener of William III, who published *The Dutch Gardener* (1669). The fourth continental source came from a recently published archive of 1,115 watercolor paintings of plants, known as the *Libri Picturati*, which was the first morphologically precise illustrated catalogue of native and exotic plants in the Low Countries (de Koning et al. 2008).

The *Libri Picturati* was apparently conceived, coordinated, technically defined, and annotated by Carolus Clusius (or Charles de L'Ecluse [1526–1609]), physician and botanist, who—as we will discuss below in the context of medicinal plants and the role of botanical training at Leiden University—later became professor of botany and designed the *Hortus Botanicus* at the University of Leiden (of which only 25%, versus 100%, as had been previously assumed by earlier scholars, was dedicated to

TABLE 8.7. Metric Comparison of *Brassica* Seeds

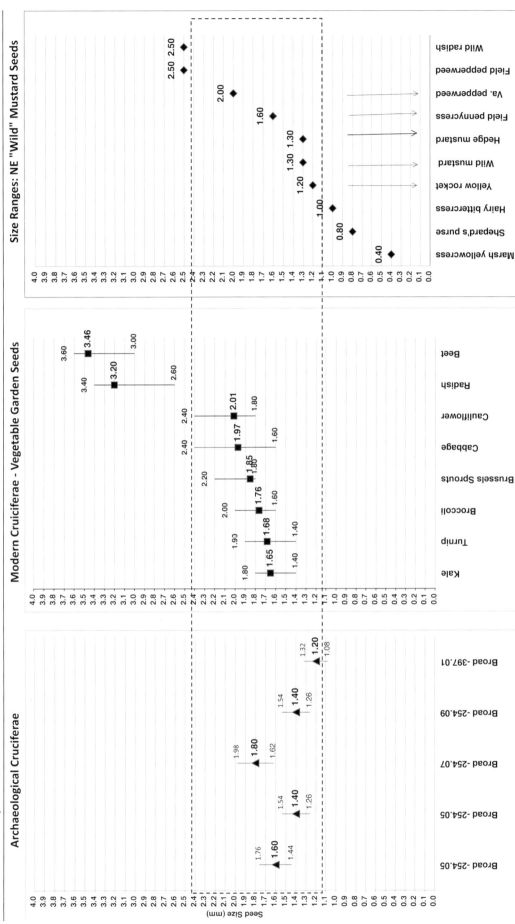

This three-part table uses measurements of seed size between (1) the archaeologically recovered specimens, (2) relative to modern (store-bought) varieties, and (3) published seed measurements for "wild" mustard varieties found in the Northeast United States, to help define which members of the cabbage/mustard family of vegetables actually may have been under cultivation in seventeenth-century New Amsterdam. As demarcated by the dotted rectangle outline, overlaps in size suggest that the historic/archaeological seeds may have corresponded to upward of six varieties of garden vegetables and/or five examples of European-introduced "wild" mustard (all of which served as seventeenth-century potherbs or condiments). When additionally evaluated relative to contemporary historical sources from New Amsterdam and Europe (see Table 8.8), the actual range of seventeenth-century vegetables appear to have been limited to cabbage, kale, turnip, and possibly cauliflower, beets, and spinach. Although mentioned in Holland at the time, neither the seed measurements or literature for New Amsterdam suggest the seventeenth-century presence of radishes in the range of garden vegetables. Graphic: Joel Grossman, Ph.D. © 2010

TABLE 8.8. Table of Sixteenth and Seventeenth-Century Archival References to *Brassica*

MODERN BRASSICA (Mustard/Cabbage Family)	HISTORIC 17th CENTURY LITERARY REEFERENCES				ARCHAEOLOGICAL SEEDS
	Van der Donck [New Amsterdam] (Goedhuys 2008, 28)	Van Tienhoven [New Amsterdam] (Singleton 1909, 14)	Van der Groen [Holland] (Oldenburger-Ebbers, 1990, 167)	Clusius/Saint Omer, [Holland] Libri Picturati - ca. 1564-59, (de Koning et. al, 2008; Uffelen 2008; Egmond 2008)	Projected Identifications— Seed Sizes & Historical References (See Table 13-8)
Cabbage	X	X	X	X (n=14)	§
Turnip	X			X (n=1)	§
Mustard				X (n=5)	§
Kale				X (n=3)	§
Broccoli					
Cauliflower				X (n=1)	?
Radish		X			
Brussels sprouts					
Kohlrabi *				X(n=1)	
Rape *				X(n=1)	

This comparison of contemporary New World and European ethnobotanical accounts and plant inventories, together with the metric (seed-size) identifications of archaeologically recovered seeds (see Table 8.7), suggests the possible presence of cabbage, kale, mustard, turnip, and possibly cauliflower, beets, and spinach, but no evidence, either physical or archival, for the seventeenth-century presence of broccoli, Brussels sprouts, or radish in New Amsterdam. (Note: * = not measured; n = number of varieties listed.)

medicinal plants) (Cook 2007b, 119; van Uffelen 2008a, 54–59; Egmond and Ramon-Laca 2008, 45; Hophouse 1977, 118). Clusius's detailed annotations included morphological attributes, information on the "ecological character" [*sic*] and geographic distribution, and advice on the best ways to grow plants in gardens (Egmond 2008, 20). This sixteenth-century source is important in this context (1) because of the caliber of scientific data it showed was available to seventeenth-century students of medicine and botany at Leiden, (2) because it reflected an "ecological" approach to the categorization of plants by habitat—i.e., plants growing in marshes, by the sea, in "rough, sandy, and sunny places," etc. (Savoiea et al. 2008, 91), and (3) because it included Cauliflower in the mid-sixteenth-century catalogue of Brassica, suggesting that it was probably present in seventeenth-century New Amsterdam as well.

The later-seventeenth-century work of Van der Groen, the official gardener of William III, included a formalized plant inventory and conceptual design templates for the layout for a typical Dutch "kitchen" garden and "fruit and berry" garden. His template for the ideal "kitchen" garden was divided into four functionally and spatially distinct quadrants: "Brassicas and roots," salad plants, medicinal herbs, and aromatic herbs. Van der Groen's list of "Brassicas and roots" included "Canadian Onion," asparagus, beet, cabbage, carrot, Spanish radish, and "others" (Oldenburger-Ebbers 1990, 167). Unfortunately, his grouping of "Brassicas and roots" into one category obscured the distinction between true cabbage and non–cabbage family vegetables.

The 1650 report by Van Tienhoven listed the contents of the first gardens in New Netherland as being "made and planted in season with all sorts of potherbs, particularly parsnips, carrots and cabbage, which bring great plenty husbandman's dwellings," including "whatever else is normally found in a cabbage and kitchen garden" (O'Callaghan 1856, 369; Jacobs, 2005, 28; 2009, 9). Likewise, Van der Donck also explicitly mentioned the cultivation of turnips: "Turnips are as good and firm as any sand turnip in this country [Holland] can be" (Goedhuys 2008, 31; Jacobs 2009, 9). Turnips were also under cultivation in seventeenth-century Canada, New England, and Jamestown (Hedrick 1919, 120).

Three of the historical sources—Van der Groen, Van der Donck, and Van Tienhoven—included cabbage in their lists of garden plants of the *Cru-*

ciferae/Brassica family (Table 8.8). Cabbage was recorded in Canada by 1540, observed in Haiti by 1556, in Brazil by 1647, and in Virginia by 1669 (Hedrick 1919, 114). Given that both Van der Donck and Van Tienhoven mentioned it, it is probable that cabbage was also one of the first *Brassicae* in New Amsterdam (Goedhuys 2008, 28; O'Callaghan 1856, 368). However, neither Van der Donck nor Van Tienhoven mentioned the wider range of vegetables in the modern inventory of *Brassica* or cabbage family produce, for example, mustard, broccoli, kale, or Brussels sprouts. Van der Groen mentioned the radish in Holland, but it was not mentioned by Van der Donck or Van Tienhoven in their lists of *Brassicae* in New Amsterdam. This omission may not have been an oversight (Table 8.8).

Although not explicitly mentioned by any of these archival references, kale—an "open" green without the closed head of cabbage or edible "flowers" of cauliflower or broccoli (Hedrick 1919, 107)—may have been present in New Amsterdam early on. Kale was observed in Haiti as early as 1565, and recorded in Virginia by 1669 (Hedrick 1919, 108–109). Not only is kale early in the New World historical record, but modern nutritional studies rank it highest (by 30 to 50%) among the vegetables for vitamin K and Lutein (a key source of carotenoids in the lens of the eye) relative to turnip greens, Swiss chard, and raw spinach (Liebman and Hurley 2009, 15). In his 1543 *New Herbal*, Fuchs discussed five kinds of *Brassica* as a group, but only explicitly mentioned cabbage and kale, an omission suggesting that the two were primary in the mindset of sixteenth-century herbalists (Dressendorfer 2001, 910).

The ethnobotanical record also suggests that two modern members of the *Brassica* or cabbage/mustard family, broccoli and Brussels sprouts, not mentioned by Van der Donck, Van Tienhoven, or Van der Groen, may not have been part of the seventeenth-century inventory of garden produce in New Amsterdam. Broccoli was not commonly mentioned or illustrated by European botanists until the early eighteenth century (Hedrick 1919, 110–11). Brussels sprouts were not documented in Belgium, France, or England until the early nineteenth century, and not in American gardens until 1806; and its seeds were not listed for sale here until 1828 (Hedrick 1919, 112). Finally, although not mentioned by the three other sources, a single variety of cauliflower was depicted and described, together with fourteen kinds of cabbage and four kinds of kale, in the mid-sixteenth-century *Libri Picturati* by Clusius (van Uffelen 2008b, 117, Fig. 4).

Thus, based on these multiple lines of archival and metric evidence, the comparison of the five archaeological samples to modern *Brassica* seeds, to seed sizes for "Wild Mustard," and, finally, to historic sixteenth and seventeenth-century botanical surveys, the actual diversity of garden vegetables may in fact have been quite limited for seventeenth-century Manhattan. Out of the range of potential candidates, the five archaeological seeds could have derived from either cabbage, kale, turnip, and possibly cauliflower, or from one of five varieties of introduced "wild" mustard, of which three were exploited as either potherbs or condiments (Tables 8.7 and 8.8). Finally, given the dynamic European and transatlantic trade in seeds from the sixteenth century onward, the inclusion of cabbage, cauliflower, turnip, beets, and spinach in a 1673 English catalogue of seeds suggests that these garden cultivars may also have been available in mid-seventeenth-century New Amsterdam (Thick 1990, 115–16).

Taken together, the archaeological and ethnobotanical evidence coalesce to suggest that the early-seventeenth-century plants represented an amalgam of both native and introduced varieties; some long recognized, such as indigenous fruits, berries, and squash/pumpkin; others only recently recognized as nutritional sources from indigenous potherbs and seed-bearing food sources; augmented by what can now be described as imported European members of the *Brassica/Cruciferae* family; and finally, as will be argued below, the recognition that at least half of the early-seventeenth-century archaeologically recovered seeds may have been present at the site due to their medicinal qualities.

INDIGENOUS AND EUROPEAN MEDICINAL PLANTS

The importance of medicinal plants among contact-period Native American groups is well established in the ethnobotanical and historical literature.

What is new here is the notion that many of the excavated seeds found in Lower Manhattan may be archaeological manifestations of these ethnobotanical patterns. As elaborated below, my idea that some of the archaeologically recovered seeds may have been used as medicines initially came from historical suggestions that a seventeenth-century medicinal garden may have been planted within the block by one of the Dutch West India Company surgeons, presumably Dr. Hans Kierstede, and from the fact that his wife played an important role with Native American women (Grossman 1985, 2000). This premise led me to incorporate the work of William A. Fenton, Native American ethnobotany, and that of European herbalists, to expand on the idea that some, if not most, of the identified plants may have been used both as foods and, perhaps more importantly, as medicines in seventeenth-century New Amsterdam.

Cross-Cultural and Interregional Patterns of Exchange

In his important 1942 study of indigenous medicinal plants and cross-cultural exchange, "Contacts between Iroquois Herbalism and Colonial Medicine," Fenton highlighted the fact that knowledge of medicinal plants was not restricted in function to specific ethnic groups or localized territories. His observations on intertribal networks of exchange throughout the northeast are important because they provide a basis for looking beyond the limits of Manhattan Island for Native American ethnobotanical analogues. In particular, his work with linguistic parallels documented that similar plant names cross-cut tribal and geographic boundaries, and that much of the Native American knowledge was interregional, with common linguistic cognates and uses shared between distinct indigenous groups throughout the northeast and mid-Atlantic regions. He explicitly wrote: "In comparing present Iroquois and Algonquian plant names we find some names that have similar meanings and yet we cannot be sure in which direction such ideas traveled [between different ethnic groups]" (Fenton 1942, 505). These interregional networks of medical and botanical knowledge were also at times long-distance. Fenton cited the example of an injured Mohawk warrior who traveled 2,100 miles to be treated by a tribal surgeon (Fenton 1942, 512).

Fenton furthermore noted that scholars and early botanists were "hard put to decide which plants a century after contact were native and whether Indians or colonists first used them medicinally" (Fenton 1942, 514). However, it is also clear that the exchange of medicinal information and plant knowledge was going on in both directions—clearly in the eighteenth century, and probably so in the seventeenth century as well. Fenton cited the observation of two eighteenth-century botanists working in North America to suggest that "the Indians were eager to learn the remedies of the white physicians" (Fenton 1942, 525).

His World War II–era ethnobotanical work on Native American medicinal plants also underscored the problems posed by the reticence of native herbalists to divulge traditional secrets, specifically concerning their uses and sources. He noted that native plant collectors and traders were aware of the financial gains possible and were thus reluctant to share their knowledge (Fenton 1942, 506). He also identified impediments to ethnobotanical interpretation caused by linguistic ambiguities and the issues of inconsistent transliterations between what the Dutch thought they heard and later botanical attributions, with many native names remaining unknown to European botanists until Peter Kalm and John Bartram began to apply the techniques of Linné (Carl Linnaeus) in the eighteenth century (Fenton 1942, 515).

Despite these constraints, similar networks of information exchange have been documented in the historical and archaeological record of Dutch, English, and French interregional trade. Not only do archaeologists and historians now recognize fluid interregional patterns of exchange between Dutch and English settlements along the eastern seaboard (Wilcoxen 1987, 23–37), but in addition, ceramics experts working in Jamestown have recognized the difficulty of distinguishing English from "Dutch" ceramics in the early seventeenth century. They concluded that "much of the material culture found in early 17th century sites in North America is the result of Dutch Traders who offered better rates . . . than the English" (Straub and Luccketti 1996, 20). Parallel historical research now also corroborates the existence of dynamic trade networks between

Jamestown and New Amsterdam in the first half of the seventeenth century (Matson 2009, 100).

These patterns of fluid trade of material goods and information between the English and the Dutch were paralleled by concurrent exchanges between the Dutch of New Netherland and the French Jesuits of Canada (Fenton 1942, 511). Jesuit missionaries were steeped in Dutch medical literature and maintained dynamic networks for the international exchange of drugs and medicinal knowledge through the publication of medical "handbooks" and broadly dispersed networks of pharmacies in Europe and the Americas (Anagnostou, 2007, 301–302). Similar to the writing of Clusius of Leiden, discussed below, "These [Jesuit] handbooks contain[ed] descriptions and drawings of many indigenous plants, information about the best period to collect them and optimal storage conditions, explanations about their medicinal effects, and advice for the preparation of different medications" (Anagnostou, 2007, 301). Their motives were similar to those of the seventeenth-century Dutch botanists, doctors, and apothecaries. For the Dutch, English, and Jesuits, European drugs were expensive, hard to come by, and often lost their effectiveness after long international voyages (Anagnostou 2007, 300). It is also probable that what the Dutch knew of indigenous medicinal plants was, like the material record, shared between the English settlers of Jamestown and, in all probability, with the French Jesuits of Canada.

Finally, Goedhuys's 2008 translation of botanical names and origins of plants listed in Van der Donck's 1655 *A Description of New Netherland* also suggests that the repertoire of medicinal plants known to the mid-seventeenth-century Dutch of New Amsterdam may have come from multiple ecological zones throughout the northeast, and in several cases from distant, and in one instance, international sources. Seven (7) or 20 percent of Van der Donck's inventory were introduced species. One (*Scholopentria*) came from Florida, and at least one other, "Dragon's Blood," was native to Indonesia (Goedhuys 2008, Appendix, 144).

These ethnohistoric observations are important for the following assessment of the plants found in the archaeological features of New Amsterdam because they imply that indigenous and European medicinal knowledge traveled in a fluid network of interregional exchange over considerable distances, and across tribal, and/or ethnic boundaries throughout the northeast and the mid-Atlantic regions of the eastern United States.

Ethnobotanical Evidence of Medicinal Plants

Despite the limits posed by the excavated seeds being defined to only the genus level, the difficulties of correlating pre-Linnaean plant descriptions by sixteenth and seventeenth-century herbalists to modern varieties, and ambiguities over the direction of information exchange, it is possible to identify multiple cases from North American and European ethnobotany to suggest that similar patterns of cross-cultural and interregional, if not international, botanical exchange were taking place in seventeenth-century New Amsterdam.

At the most general level, of the nineteen plants identified in the earliest deposits, at least ten were recognized, both in North America and Europe, for their medical qualities, and many as members of the household garden. Of these, no less than five (blueberry, knotweed/knotgrass, or "bistort," amaranth, raspberry, and lambsquarters) were recognized as *astringents*—substances that shrink tissue, dry up secretions, and restrict blood flow. At least three (including knotweed, or "bistort," and toadflax) served as *diuretics* that help in the elimination of liquids, especially urine. Two of the identified plants (pumpkin/squash and lambsquarters) were well-known *anthelmintic* cures for intestinal worms (Meyer 1972, 148–58).

At a more specific level, the following treatment of the recovered plants from the seventeenth-century Pearl Street site can be organized into three primary cross-cultural categories: (1) indigenous medicinal plants of probable local origin that were either analogues of recognized European plants or adopted by the Dutch, (2) medicinal plants that could have come from either indigenous North American or European sources, and (3) plants with documented indigenous and European medicinal uses of probable European origin. Together, they highlight multiple ethnobotanical parallels in indigenous and European medicinal uses between the properties and uses of potential medicinal plants by different, and often distant, indigenous groups in the northeast (Table 8.9).

TABLE 8.9. Table of Potential Plant Origins

		Linum sp. – toadflax (EU: diuretic, jaundice; NA: repertory)
		Stachys sp. – "Betony" (EU:17th c. "aspirin"; NA:VD, colic
		Citrus sp. – (scurvy, Late 17th c.)
		Trifolium sp. – clover (NA: colds, repertory & milk flow)
		Brassica sp. – cabbage/kale ? (EU: hair loss, cranial hematomas)
	Portulaca sp. Purslane (NA: intestinal & urinary tract, skin; EU: scurvy)	
	Polygonum sp.- knotweed, "Bistort" (astringent, skin wounds, bleeding)	
	Galium sp. - bedstraw (NA: orthopedic aid; EU: hysteria, epilepsy, skin)	
Chenopods sp. – amaranth & lambsquarters (astringent, worms)		
Phytolacca sp. – pokeberry (skin ailments, blood purifier)		
Curcurbita sp. – Squash/ Pumpkin (intestinal worms, bladder)		
Frageria sp. – strawberries (dysentery, bladder ailments)		
Nicotiana sp. – tobacco (diverse -cf. Monardes)		

Although limited to predominantly the genus level of botanical identification, and only suggestive, the identified plants can be grouped into three transatlantic categories: (1) those of probably indigenous or of Native American origin, (2) those that could be from either continent, herein defined as Bilateral or Analogous (origin undetermined, or parallel uses at the genus level), and (3) those of probable European origin. This "best guess" depiction of potential origins suggests a balanced mixture of both indigenous and introduced species—many with parallel transatlantic ethnobotanical functions—most visibly during the early to mid-seventeenth century, but incrementally less so by the early eighteenth century. Graphic: Joel Grossman, Ph.D. © 2010

Indigenous Medicinal Plants

In addition to their nutritional value, at least five of the early plants (amaranth, lambsquarters, pokeweed, pumpkin/squash, and strawberries) may have been present because of their medicinal qualities.

CHENOPODS (*AMARANTHUS* SP. AND *CHENOPODIUM* SP.)—"LAMBSQUARTERS," "GOOSEFOOTS," AND "WORMSEED"

Chenopods were recognized by both Native American and European herbalists for their medicinal qualities. The seventeenth-century English herbalist Culpeper lauded Amaranth for stopping blood flow in both men and women, and bleeding, either from the nose or a wound, and specifically recommended it as a "most gallant anti-venereal and a singular remedy for the French Pox" (Potterton 1983, 15). The herbalist Grieve pointed to its use to treat chronic diarrhea, dysentery, fevers, and malaria ... and commented that it was superior to quinine (Grieve 1931, 30). Additionally, a synonym for lambsquarters (several varieties of chenopods) in seventeenth-century Europe was "pilewort" (Meyer 1972, 95), or "smearwart," reflecting its use as an ointment to clean and heal chronic skin sores, which the English herbalist Gerard said "they do scour and mundify" (Grieve 1931, 365). In tandem with their European medicinal analogues, the leaves of both amaranth and lambsquarters were used by Native Americans as astringents to reduce swelling, to treat dysentery, diarrhea, and ulcers, and to stop intestinal bleeding (Foster and Duke 2000, 243). The Mohegan used an infusion of amaranth leaves for sore throats (Tantaquidgeon 1972, 70, 128). The Iroquois used lambsquarters to treat diarrhea, as a salve for burns, and to aid with milk flow (Herrick 1977, 315–16).

But perhaps the most striking parallels in medicinal qualities were manifested by the use of several species of chenopods to treat intestinal worms. In both continents, varieties of *Chenopodium* were seen as effective *anthelminic* treatments for the removal of round worms and hookworms, "especially in children" (Grieve 1931, 885; Chevalier 1996, 186). One species of *Chenopodium*, native to the northeast, was commonly referred to as "wormseed" in Europe and as "American wormseed" in the United States (Grieve 1931, 189, 854–55). In 1895, the active ingredient from *Chenopodium* seeds was distilled to yield "Wormseed or Chenopodium Oil," which was used extensively in World War I as a preferred prescription capable of removing 95 percent of a patient's worms with three treatments (Grieve 1931, 856).

Significantly, three scholars, working on the ethnobotany and indigenous medicines of three different eastern Native American groups, documented parallel medicinal uses of *Chenopodium* seeds for the treatment of worms. The Natchez, derived from the Mississippian Moundbuilders, gave the plant as a pediatric treatment for worms in children (Taylor 1940, 22). The Rappahannock, who in the seventeenth century lived near the English settlement of Jamestown, Virginia, gave children a concoction of stewed *Chenopodium* seeds for worms (Speck 1942, 30), and the Seminole administered a decoction of the whole plant for "worm sickness" (Sturtevant 1954, 241).

Squash/Pumpkin (*Curcurbita* sp.)

Like *Chenopodium* seeds, those of pumpkin and squash have long been seen as important cures for both intestinal worms and urinary tract ailments. The Iroquois used an infusion of pumpkin seeds to treat children with reduced urination (Rousseau 1945, 66). As an early introduction to Europe, the sixteenth-century German herbalist Fuchs recommended pumpkin seeds, which he lumped together with cucumbers, melons, and cantaloupe—an ambiguity perhaps reflecting its recent arrival from America—"when the bladder is being difficult," a prescription that coincided with the modern use of extract of pumpkin seeds for urinary and prostate problems (Dressendorfer 2001, 918, 928). Pumpkin seeds have also been long recognized as a Native American cure that was adopted by American doctors in the early nineteenth century as "among the most valued anthelminics for the removal of tapeworm" (Ripley and Dana 1875, Vol. XIV, 87–88). They are still used by modern herbalists as a nontoxic treatment to excise tapeworms in pregnant women and children (Chevallier 1996, 194).

Pokeweed (*Phytolacca* sp.)— "American Nightshade," "cancer root," "American spinach"

The twentieth-century herbalist Grieve described pokeweed as "one of the most important of indigenous American plants" (Grieve 1931, 648). It was widely viewed as a dermatological cure for skin diseases by both Europeans and Native American healers (Grieve 1931, 648; Speck et al. 1942, 29). The Delaware Indians used it as a stimulant to treat rheumatism, as a blood purifier, for chronic sores, and to treat glandular swelling (Tantaquidgeon 1972, 27, 32, 78). Illustrating parallel medicinal uses between often distant groups, the Rappahannock of Virginia used pokeweed as a dermatological aid to treat poison ivy, rheumatism, warts, and piles (Speck et al. 1942, 29). Likewise, the Iroquois in New York also used it as a dermatological treatment for sprains, rheumatism, bruises, swollen joints, bunions, "skin lumps," as an expectorant to treat liver sickness, as a blood purifier, and as a love medicine (Parker 1910, 93; Herrick 1977, 316–17). The Mohegan used a salve from its leaves to treat sore breasts and as an antidote against poison (Tantaquidgeon 1972: 74, 83; Parker 1910, 93).

Strawberries (*Fragaria* sp.)

Strawberries were important to both Native Americans and Europeans for their medicinal qualities. The University of Michigan Database of Ethnobotany currently documents fourteen specific medicinal uses for strawberries by the Iroquois alone, many pertaining to stomach ailments (Moerman 2004; http://herb.umd.umich.edu/). These included its use as a blood remedy, as a treatment for stomach bleeding, for the regulation of menstrual flow, for bloody diarrhea, for sties, for babies with colic, for gonorrhea, strokes, as a wash for chancre sores, to soothe teething babies, as a general blood remedy, and as an antidote for snakebite (Herrick 1977, 352; Moerman 2004). The Ojibwa used strawberries for stomach aches, especially with children (Smith 1932, 384). Similarly, the Chippewa used the berries to treat "cholera infantism," or children's dysentery (Densmore 1928, 346). Although farther away, the Cherokee also used strawberries to treat dysentery, urinary and bladder problems, kidney disease, jaundice, scurvy, and nerves (Hamel and Chiltoskey 1975, 57). Referring specifically of its use by native peoples in New Amsterdam, Van Rensselaer wrote: "They would brew cat-nip for the sick or strengthen an invalid with a decoction of strawberry leaves" (Van Rensselaer 1898, 74). The European herbalist Grieve described strawberries as a common medicinal component in seventeenth-century "pharmacopoeias" and cited the seventeenth-century herbalist Culpeper who saw them as "singularly good for the healing of many ills" (Grieve 1931, 777).

Transatlantic or Bidirectional Analogues

Three of the identified seed types (purslane, bedstraw, and knotgrass) occurred in both European and North American contexts and could have come from either source (Table 8.9).

PURSLANE (*PORTULACA* SP.)

In addition to now being recognized as a potherb on both continents, purslane was also used in the seventeenth century as a medicinal herb by both North American indigenous peoples and European herbalists. As Fenton warned, it also represents a good example of the difficulty of establishing the direction of these transatlantic parallels in its use as a medical plant. It was appreciated as an important medicinal plant at least by the sixteenth century and was listed in Fuchs's *New Herbal* of 1543 as a cure for many ailments (Grieve 1931, 661). Also known as "pulsey," it was prescribed by the mid-seventeenth-century herbalist Culpepper as a treatment for gout (Grieve 1931, 660). In the 1650s, Gerard recommended the raw leaves to ease teeth "that are set on edge with eating of sharpe [*sic*] and soure [*sic*]

things" (Grieve 1931, 660). Purslane was also seen by sixteenth and seventeenth-century Dutch explorers as a cure for scurvy, perhaps due to its high vitamin content. In his 1593 voyage to the South Sea, near Cape Saint Thomas off Brazil, Sir Richard Hawkins found a "great store of the herbe [*sic*] purslane . . . which he used to treat his scurvy-suffering crew" (Hedrick 1919, 451). Its medicinal qualities may have to do with the fact that it contains several neurohormones, reported to reduce tissue hemorrhage, its high levels of vitamins A, C, and E, riboflavin, calcium, phosphorus, magnesium, and iron; and like fish oils, it is one of the richest natural sources of omega-3 fatty acids (Shimer 2004, 98; Peterson 1977, 72; Foster and Duke 2000, 110).

The ethnobotanist Shimer wrote that "Native American people ate purslane, but were more interested in its medicinal applications" (Shimer 2004, 100). Although it was cooked and seasoned by the Iroquois as a potherb (Waugh 1916, 118), they also used the juice from its leaves as a dermatological treatment for burns, insect bites, and bruises (Herrick 1977, 318). Tea from its leaves was used for diarrhea, stomach aches, and urinary tract infections (Shimer 2004, 100). Likewise, the Rappahannock of Virginia used the leaves to make a topical salve to treat "footage trouble," or sore feet (Speck 1942, 28).

BEDSTRAW (*GALIUM* SP.)

Galium is both a European and American genus and as such could have derived from either continent. It was recognized by European herbalists and North American indigenous peoples as a dye, for its medicinal qualities and for the shared chemical characteristics among different species of the genus (Grieves 1931, 92; Delcourt 2004, 42; de Koning 2008, 121). As reflected by the sixteenth-century reference to bedstraw as "cheese rennet," for its ability to curdle milk, these organic characteristics contributed to its medicinal qualities as well. Used mainly as a diuretic and for skin problems, in 1735 the Irish herbalist K'oeh wrote of bedstraw that "when applied to burns, the crushed flowers alleviate inflammation, and when applied to wounds, they heal them" (Chevallier 1996, 212). Among European herbalists it was also formerly "highly esteemed as a remedy for epilepsy and hysteria and externally for cutaneous eruption, and is currently recognized as a popular remedy for gravel, stone and urinary tract diseases" (Grieves 1931, 91). The seventeenth-century English herbalist Gerard described it as "good for weary traveler" and his contemporary Culpeper recommended it for interior bleeding (Grieve 1931, 91). In his *New Herbal* of 1543, Fuchs listed it as a protection against the bite of poisonous animals as well as a treatment for earaches and goiters (Dressendorfer 2001, 901).

In addition to its contact-period use as a dye, the genus *Galium* has also been documented by ethnobotanists as a widely used medicinal plant among a number of northeastern Native American groups. The Iroquois used it to treat swollen testicles and ruptured skin, as an eye medicine, to treat babies with backaches, as a treatment for venereal disease (presumably of European origin), and as a "love medicine" (Herrick 1977, 440). The Ojibwa also prescribed it as a dermatological drug, for kidney and urinary tract ailments, and, following European contact, to treat tuberculosis (Smith 1932, 387). The Penobscot of Rhode Island used it to treat gonorrhea, as well as for kidney ailments (Speck 1917, 331).

Knotweed (*Polygonum* sp.)

Known in Europe as bistort, or "bistorta" in seventeenth-century contexts, knotweed/knotgrass is a worldwide genus that, like bedstraw, shares common chemical and medicinal properties between diverse European and American species (Grieve 1931, 105, 205). In the 1930s, the herbalist Grieve recognized knotweed as "one of the strongest astringent medicines in the vegetable kingdom for internal and external bleeding" and "of proved excellence in diarrhea, dysentery, cholera and all bowel complaints and in hemorrhages" as well as for the treatment of infant diarrhea, hemorrhoids, piles, ulcerated tonsils, and discharges of the nose, vagina, urethra, and ears (Grieve 1931, 106–107).

In his *New Herbal* of 1543, Fuchs prescribed bistort or knotgrass to stop bleeding, evacuate the bladder, and sink fevers and lauded the plant for its utility in the treatment of "wounds, diarrhea, menstrual problems" (Dressendorfer 2001, 916). In 1652, Culpeper described its "Diverse Medical Uses" and recommended it for stings or bites; its root "hinders abortion or miscarriage," its leaves kill worms in children, stop inflammation of mouth and throat, and with plantain, form an external salve for gonorrhea (Potterton 1983, 29). In 1682, the herbalist Salmon specifically recognized its astringent properties and prescribed knotgrass to treat the "spilling of blood," kidney infections, inflammation, and because it "cleanses and heals old filthy wounds" (Grieve 1931, 458).

This European recognition of the medicinal qualities of the various species of *Polygonum* was matched by equally diverse, and often parallel, medicinal uses by Native American herbalists in the northeast United States. The University of Michigan ethnobotanical database (Moerman 2004; http://herb.umd.umich.edu/) listed twenty-one medicinal uses of *Polygonum* among the Iroquois and the Ojibwa. The Algonquin of Quebec used the astringent qualities of its leaves to stop bleeding (Black 1980, 188).

A related species, Pennsylvania Smartweed *(Polygonum pennsylvanicum)*, was used by unspecified groups of American Indians as a tea to treat diarrhea, bleeding of the mouth, and epilepsy (Foster and Duke 2000, 180). The Iroquois adopted the introduced variety of knotgrass, *Polygonum persicaria,* to treat rheumatism in the feet and legs, and as a heart medicine (Herrick 1977, 315). They used *Polygonum hydropiper* (marshpepper) to treat chills "when cold," as a gastrointestinal aid for indigestion, and to treat children with swollen stomachs (Herrick 1977, 314). The Iroquois used a third variety of knotweed *(P. arenastrum)* to treat injuries from miscarriages, as a love medicine, and to heal sore backs (Herrick 1977, 314). They used "prostrate knotweed" *(P. aviculare)* for children's diarrhea and bleeding from cuts and wounds (Herrick 1977, 313).

Knotweed or "bistorta" may have also been one of the earliest medicinal plants imported from New Netherland to the University of Leiden Medical Garden in the early seventeenth century. Writing from Leiden in 1633, Johan de Laet noted that "there are a great variety of herbaceous plants, some of which bear splendid flowers and others are considered valuable for their medicinal properties. I cannot avoid describing here two of this class, although their use is not yet known" (Jameson 1909, 55). He continued to describe how "[t]wo plants were sent to me from New Netherland that grew finely last year (1632) in the medical garden of this city [Leiden]" (Jameson 1909, 55, footnote 1; 56, footnote 1). Johan de Laet, both a director of the Dutch West India Company and an accomplished seventeenth-century botanist who maintained a herbarium in Leiden, included a drawing and description of the two plants, which Jameson identified as *Polygonum artifolium*, or "heart-leaved tear-thumb," and *Polygonum sagittatum*, or "arrow-leaved tear-thumb" (Jameson 1909, 56).

While it is not possible to link the genus-level seed identifications from the Pearl Street flotation samples with either of these two "species-specific" identifications by Jameson, the presence of *Polygonum* in the early-seventeenth-century contexts in New Amsterdam suggests that, given De Laet's treatment of these plants as important medicinal specimens, worthy of import to the Hortus Botanicus of Leiden, their transport to Holland may have reflected parallel Transatlantic uses, and/or the possibility that they were "recognized" as similar to known European varieties.

European Medicinal Plants

In addition to the *Brassica*, three other plants were probably introduced and possibly utilized for their medicinal qualities by both Dutch and native herbalists: betony or *Stachys* sp., clover, and toadflax (Table 8.5). Although not specifically discussed, some of the excavated European cultivars may also have been used in the sixteenth and seventeenth centuries as medicines. In addition to their value as foods, cabbage and kale were listed by Fuchs for the treatment for hair loss and cranial hematomas (Dressendorfer 2001, 910).

EUROPEAN BETONY (*STACHYS* SP.)— "WOUNDWORT," "HEAL-ALL"

The potential presence of betony was represented by one seed from a single late-seventeenth-century context (Component 14; Tables 8.2, 8.3, 8.5). One of more than three hundred worldwide species, *Stachys* sp. is alien to North America and was probably introduced from Europe (USDA 2009). Its species diversity and wide distribution was matched by an equally broad range of medicinal uses and applications. Variously known as "Woundwort," "Heal-all," "Self Heal," and betony, it was viewed as the aspirin of the seventeenth century. Pavord described it as "one of the most important cure-alls in the medieval canon" (2005, 18). It was valued as a treatment for headaches, facial pain, "frayed nerves," premenstrual cramps, poor memory, tension, and as an astringent for headaches and congestion; during the first century AD, the physician to the Emperor Augustus "claimed that betony would cure 47 different illnesses" (Chevallier 1996, 270). Of possible relevance to its recovery at Pearl Street, the herbalist Grieves noted, "It was largely cultivated in the physic gardens, both of the apothecaries and the monasteries, and may still be found growing in the sites of these ancient buildings" (Grieve 1931, 97). Its recovery in the late-seventeenth-century deposits—in association with other seventeenth-century cultivars (toadflax, clover, and *Brassica*)—suggests that it may have been introduced as a medicinal plant.

Documented for three indigenous groups in the northeast (the Chippewa, Ojibwa, and the Delaware), multiple ethnobotanical references suggest that *Stachys* sp. may also have been adapted by them as a medicinal plant after its presumed introduction from Europe. Two references to its use by the Delaware tell of its use with nightshade and snakeroot to treat venereal disease—presumably of European origin (Tantaquidgeon 1942, 29, 35, 80). The Chippewa of the northern United States and southern Canada employed an infusion of its leaves to treat abdominal pain described as "sudden colic" (Densmore 1928, 344).

CLOVER (*TRIFOLIUM* SP.)

Clover was found in both early- and late-seventeenth-century, but not eighteenth-century, samples (Tables 8.2, 8.3, 8.4, 8.5, 8.6). The presence of clover is pertinent to this discussion of medicinal plants because it was of unambiguous European origin and because it appears to have been widely adopted by a diverse number of Native American groups in the northeast after its introduction in the seventeenth century. Three northeastern indigenous groups, the Iroquois and Mohegan of New York, and the Algonquin of Quebec, adopted clover as a cold remedy and for whooping cough (Black 1988, 188; Hamel et al. 1975, 29).

TOADFLAX (*LINUM* SP.)

Although only detected in one late-seventeenth-century context (Tables 8.2, 8.3, 8.5), archival sources suggest that toadflax was probably in place by the early seventeenth century. It was reported to have been under cultivation in New Netherland and New England by the 1620s and 1640s (Ripley and Dana 1875, 292; Hedrick 1977, 338). Van der Donck listed it as an herb of the mid-seventeenth-century gardens of New Amsterdam (Goedhuys 2008, 28). Toadflax was perceived to have had medicinal benefits in sixteenth-century Europe, at least a century before the arrival of the Dutch in New Netherland. In 1543, Fuchs mentioned its benefits as a diuretic and as a remedy for jaundice (Dressendorfer 2001, 914), as did the seventeenth-century English herbalist Gerard (Grieve 1931, 816). Its seeds were lauded by Culpeper for multiple remedies, including for "pains of the breast [and to] softens [*sic*] hard swellings" (Grieve 1931, 230). It also represents a clear example of a European medicinal plant that was both naturalized early on and adopted by Native American healers. Although no local ethnobotanical references are documented for New York, the Cherokee adopted toadflax to treat "violent colds, coughs and diseases of the lungs, fevers, and to relieve "gravel or burning during urination" (Hamel and Chiltoskey 1975, 34; Taylor 1940, 34).

CROSS-CULTURAL VECTORS OF ETHNOBOTANICAL EXCHANGE

Archaeological and Ethnohistorical Evidence

> Fresh wounds and dangerous injuries they know how to heal wonderfully with virtually nothing. They also have a cure for lingering sores and ulcers. They can treat gonorrhea and other venereal diseases so easily as to put many an Italian physician to shame. They do all this with herbs, roots, and leaves from the land, having medicinal properties known to them and not made into compounds.
> —Adriaen Van der Donck, ca. 1655

There are no proofs in archaeology, only parallels and patterns. However, multiple lines of evidence, archival, archaeological, and ethnobotanical, converge to suggest that the waterfront block of the Dutch West India Company, and the focus of the excavation on the Strand (Pearl Street) in lower Manhattan was a center of Native American and Dutch interaction in the early seventeenth century. This long-standing locus of Native American occupation may also have contributed to its being a focal point for information exchange, especially concerning medicinal plants, between, as I will argue, the women of both cultures. This suggestion is based on six lines of evidence: (1) the recovery of late prehistoric Woodland or contact-period Native American artifacts at the site; (2) historical references to the long-term use of the excavated site at and near Pearl Street by Native Americans; (3) historical sources pointing to the presence in this waterfront block at Pearl and Whitehall of one of the colony's first doctors, Dr. Hans Kierstede, who worked for the Dutch West India Company sometime between 1638 and the middle of the seventeenth century; (4) the close association of his wife, Sara Kierstede, with native traders and women as a multilingual speaker of indigenous dialects; (5) historical references to the presence of a medicinal garden maintained by an unnamed doctor (presumably Dr. Kierstede) in the first half of the seventeenth century; and finally, (6) the identification of indigenous and European medicinal plants among the recovered seeds from the site.

Archaeological evidence suggests that the shoreline Pearl Street block, between modern Broad and Whitehall (also generally referred to as the Strand), was a locus of indigenous activities, before, during, and after the arrival of the Dutch. The site was bounded on the east by a key marine landing spot, referred to as "Canoe Place" (Van Rensselaer 1898, 32), which also served as a hub linking marine and terrestrial transport to the southernmost end of a major Native American roadway (modern Broadway) up the spine of the Island (Bolton 1922). It also corresponds with the location of the subsequent Dutch boat repair area at the mouth of the tidal marsh outlet at what would become Broad Street (Van Laer 1974, I: 111; Innes 1902, 5, 45).

The excavation at Pearl Street also documented a number of Native American artifacts from both mixed and unmixed historic-era deposits. A total of eleven indigenous ceramic shards, including a broken pipe stem, and thirty-one indigenous chipped stone tools (flakes, cores) were recovered. These contact-period or Late Woodland artifacts, dating to between the thirteenth and seventeenth centuries, appear to have been either utilized or deposited during or shortly before the seventeenth-century Dutch occupation at this site. In addition, the excavation encountered five shell wampum beads in the wooden-bottomed basket or cask (Component 38; see Table 8.1) which was abandoned and filled in sometime after 1650, and initially cut into the surface, sometime before that date (Grossman et al. 1985, Plate VIII6). These finds were important, because they dovetail with multiple historical references to Native American interaction and trade with the Dutch along Pearl Street during the seventeenth century.

The association between the Pearl Street block and the growing of medicinal plants can be dated to the seventeenth-century tenure of Dr. and Mrs. Kierstede at the site. Although not named directly, an intriguing reference in Van der Donck's mid-seventeenth-century *A Description of New Netherland* suggests that some of the potential medical plants excavated from within the block may have reflected the efforts of Dr. Kierstede, one of the settlement's first Company doctors: "A certain surgeon once laid out a fine garden and, *as he was a botanist as well*, planted many medicinal species he found growing wild, but with his departure this came to an end" (Goedhuys 2008, 32; emphasis added).

Hans Kierstede arrived as a prominent officer of the Company with Director Kieft in 1638, was given the parcel of land and a dwelling in what was the westernmost lot of the excavated block at the corner of Pearl Street and Whitehall, and worked as the West India Company doctor from his arrival to his departure from company employment to take up private practice in 1648 (Fernow 1976). This time frame suggests that it may have been his "fine garden" of medical plants, and that it existed sometime in the late 1630s to the early 1640s, or about the time he married Sara Kierstede in 1642 (Van Rensselaer 1898, 24).

Despite the lack of any in-depth references to the medical background of Dr. Kierstede, or his botanical training or studies in New Amsterdam, knowledge of local Native American medicinal plants was particularly well documented for his wife, Sara Kierstede. While many of the surviving historical references to her come from secondary turn of the century sources, primarily from the work of Singleton and Van Rensselaer, and are often dismissed by historical scholars as what might be called "oral traditions" or even multigenerational folklore, these accounts begin to take on a new and more credible stature in light of the archaeological and ethnobotanical data. According to these traditions, Sara Kierstede and her three sisters were multilingual, Dutch and English speakers, daughters of Anneke Jans, the first midwife in the settlement, and, "having been born and brought up among the 'Wilden', they had learned the Algonquin language, which they understood and spoke with fluency" (Van Rensselaer 1898, 22). One historian described Sara Kierstede as "being probably more learned in the native Indian tongues than anyone in the province" (Singleton 1909, 172). As the daughter of a native-speaking midwife, it is probable that her interest in, and knowledge of, local medical plants may have come as much from her mother, one of the first Dutch West India Company midwives, as it did from her later union with Dr. Hans Kierstede.

Her ability to communicate with the native women also appears to have contributed to making her new compound at the corner of Pearl and Whitehall a "safe haven" for local Dutch–Native American interaction, at least between the women of both cultures in the mid-seventeenth century:

The Dutch women had become well acquainted with the wild people who surrounded them and were on friendly terms with them. Madame Kierstede was particularly kind to them, and as she spoke their language fluently, she was a great favorite among them; and it was owing to her encouragement that the savages ventured within the city walls to barter their wares. . . . For their better accommodation and protection Madame Kierstede had a large shed erected in her backyard, and under its shelter there was always a number of squaws who came and went as if in their own village, and plied their industry of basket and broom-making, stringing wampum and sewing, and spinning after their primitive mode; and on market days they were able to dispose of their products protected by their benefactress, Madame Kierstede. (Van Rensselaer 1898, 26)

Furthermore, in addition to her linguistic skills, one reference clearly links the women of the Kierstede family to a tradition of knowledge concerning locally derived medicines and medicinal plants. A century later, Van Rensselaer lauded Mrs. Alexander's (the granddaughter of Dr. and Mrs. Hans Kierstede) medical skills and the fact that she was held in high esteem by the native ladies, "as a great 'medicine woman', and with her salve for burns, which her grandmother [Sara Kierstede] had been taught to prepare by the great Dr. Kierstede, and which is to-day [ca. 1740] sold under his name" (Van Rensselaer 1898, 355).

Other, predominantly secondary, references also hint that (1) the waterfront block at Pearl Street was both, given the Native American artifacts recovered, a pre-contact landing site, and (2) in tandem with the above quote, it may have continued as a safe place for the native women well into the mid-eighteenth century. Speaking of the annual permission granted to New Jersey Indians to visit Manhattan, Mrs. Van Rensselaer noted that after landing, the native women "proceeded in procession to the open space provided for them behind Mr. Phillipse's house, which had been kindly set apart for their use by that gentleman, when the ancient camping ground on the Strand, by Dr. Kierstede's house, had been required by the builder" (Van Rensselaer 1898, 352–53).

The historical references to the interplay between Sara Kierstede and early-seventeenth-century Native American women coming to Manhattan are important because they are consistent with regional patterns of native women being tied to botanical knowledge in general, and to knowledge of medicinal plants in particular. Speaking of the native women of New Netherland in the mid-seventeenth century, Van der Donck wrote that "[t]he women do all the farming and planting," and thus by extension had firsthand knowledge of medicinal plants (Goedhuys 2008, 97). Similarly, in 1644 the Reverend Johannes Megapolensis observed that "the women are obliged to prepare the land, to mow, to plant and to do everything [involved with plants and agriculture]" (Jameson 1909, 174; Jacobs 2005, 25).

Augmenting these historical references to indigenous women as botanical experts, one account by Fenton underscores their role as keepers of medicinal knowledge in both the Hudson River drainage and the Great Lakes region. Fenton told of the early-eighteenth-century explorer Lafitau, who "made field trips and questioned Mohawk herbalists" (Fenton 1942, 519). After an unsuccessful search for an American species of ginseng, he returned in three months only to "unexpectedly encounter the mature plant growing within striking distance of a [native woman's] house; to his dismay, a Mohawk woman, whom he had employed to search for it on her own, recognized it as one of their ordinary remedies" (Fenton 1942, 518–19).

The association of medicinal plants with the contact-period Native American and Dutch-era occupation site in Lower Manhattan is also consistent with the suggestion, put forward by Gordon Day nearly sixty years ago, that "[p]lants used by Indians for medicinal purposes may owe their existence in many localities to the transplanting hand of an Indian herbalist" (Day 1953, 340). These ethnohistorical examples may also help explain both the ethnobotanical role of Mrs. Kierstede in the transference of Native American medicinal knowledge and the presence of so large an assortment of indigenous and European medicinal plants in the seventeenth-century deposits at Pearl Street site.

Institutionalized Protocols of Plant Collecting and the Role of Women Informants

The role of multilingual Dutch women and indigenous informants as culture brokers for the systematic collection and exchange of medicinal plants and knowledge can be traced to long-standing policies of the Dutch East and West India Companies, to the teaching of company officials, doctors, medical students, and botanists at the University of Leiden, and to strong corporate links between the University of Leiden and the *Hortus Botanicus* of Leiden under Clusius—all of which came together to play a central role in the development of scientific and administrative protocols for the collection of botanical specimens and medicinal information for

Dutch expeditions. These antecedents in turn derived from two specific traditions in European medicine and botany. The first reflected official corporate practices mandating the collection of exotica and plants in search of profit and new medicines in newly discovered territories. The second stemmed from the long-standing tradition of using local and foreign, multiethnic and multilingual, women informants to garner information on local medicinal plants and cures.

While close corporate-university ties were previously documented for the late seventeenth century (Stern 1989, 181; Oldenburger-Ebbers 1990, 166), new research by Dutch scholars at the *Hortus Botanicus* archives at Leiden (*The Clusius Project*), has established that these links were firmly in place by the early seventeenth century; before and during the initial settlement of New Amsterdam. As early as 1601, Clusius of the *Hortus Botanicus* and Professor Pauw of the School of Medicine of Leiden wrote a formal memorandum to officials of the Dutch East India Company with the aim of implementing rigorous procedures for plant collecting entitled *Instructions to Apothecaries and Surgeons who will Board the Fleet to the East Indies in the Year 1602* (van Uffelen 2008a, 57). Their instructions were precise and exacting. Like their Jesuit counterparts mentioned above, they stipulated what to collect and how specimens were to be collected, listing "branches bearing leaves, fruits, and flowers . . . pressed between paper . . . together with sketches of . . . how they grow, whether they are large or small, deciduous or not, the names of trees and how they are used (Swan 2007, 235–36). They also gave guidance and mandates "to question and learn from people of all stations and sexes—from statesmen, scholars, and artists as well as from craftsmen, sailors, merchants, peasants and '*wise women*'" (Schiebinger 2007, 131).

The importance of women informants was not new to seventeenth-century European doctors and students of *Materia Medica* (Egmond 2007, 28–31; Barona 2007, 102; Cook 2007b, 204; Schiebinger 2007, 132). Various fourteenth to eighteenth-century herbalists credited their insights and sources to "highly expert old women" as the chief repositories of multigenerational folk knowledge on the herbs and medicinal plants of Europe (Arber 1986, 319–20). One sixteenth-century herbalist confided that he was "not ashamed to be the pupil of an old peasant woman" (Arber 1986, 321). Even the eighteenth-century Swedish botanist, Linnaeus (who also spent time at the *Hortus Botanicus* of Leiden) wrote: "It is the folk whom we must thank for the most effacious medicines, which they keep [*sic*] secrete" (Schiebinger 2007, 130–31). A modern scholar writing about one of Dr. Kierstede's contemporary medical counterparts, Dr. Bontius, who studied at Leiden and was serving at the Dutch East India Company outpost of Batavia (modern Jakarta), has argued that this hurdle and veil of indigenous secrecy was only breached, as may have been the case for Mrs. Kierstede of New Amsterdam, by a "growing population of mixed heritage and multilingual abilities, many of whom became crucial information brokers" (Cook 2007a, 115).

In this context, Clusius of the *Hortus Botanicus* of Leiden is important for his role in disseminating awareness of cross-cultural methods of information gathering to Dutch and European students of botany and medicine. These graduates often subsequently served as officials in company expeditions and settlements. Clusius did this through his own research and through his wildly disseminated translations—into many languages—of, and commentaries on, the botanical studies of the sixteenth-century Spanish physician and botanist Nicolás Monardes and other Iberian scholars. In particular, these studies detailed Monardes's botanical and medical experiments with the "newe Medicines and newe Remedies" coming from the New World (Barona 2007, 101; Pavord 2005, 303), and served as the principal reference texts for seventeenth-century students and practitioners of medicine (Thomás 2007, 176). Of particular relevance, Monardes wrote of the importance of indigenous knowledge and the value of what we today call "oral history" from local informants, especially indigenous women informants. As his medical disciples would later reiterate, Monardes specifically lauded native women as seasoned practitioners and for the quality of their cures, which he described as being "very good and in accordance with good medicine" (Bleichmar 2007, 96). He also taught his students the importance of women herbalists as information brokers. Quoting an informant who had written him, Monardes specifically noted: "If we know anything of the matters I have treated . . . we learned it from the female Indians" (Bleichmar 2007, 95).

These antecedents involving the role of women informants in Europe, Asia, and the New World and institutionalized protocols for organized plant collecting and experimentation suggest intriguing parallels with the multilingual and cross-cultural links to indigenous women suggested for Mrs. Kierstede on the Strand of New Amsterdam. It is furthermore probable that these parallels are not happenstance, idiosyncratic or unique to any one region, but instead suggest broader historical patterns that in all probability influenced, if not prescribed, Dutch and Native American mechanisms of information exchange in New Netherland in the third and fourth decades of the seventeenth century. For a historical analogue, one can turn to the Old Testament (Exodus 3:22) and to what Robert Alter has described as the "social phenomenon" of the "sojourner," a Biblical female noun which recognized women as "the porous boundary between adjacent ethnic communities: borrowers of the proverbial cup of sugar, sharers of gossip and women's lore" (Alter 2004, 324).

SUMMARY OF RESULTS

This reanalysis of the archaeological sequence, ethnobotanical records, and historic plant remains suggests:

1. The archaeologically dated sequence of early-seventeenth, late-seventeenth, and early-eighteenth-century samples provides new quantified evidence documenting major temporally specific patterns and shifts in the relative prevalence and diversity of European and indigenous plant types between the early seventeenth and eighteenth centuries. These trends showed specifically that: (a) Indigenous potherbs and starchy seed food sources (and the potential medicinal plants) were restricted to the early seventeenth century and dropped out of the sequence by the late seventeenth century; (b) The European garden vegetables of the cabbage/mustard family were restricted to early- to mid- and late-seventeenth-century contexts, but were not identified in any early-eighteenth-century deposits; (c) The late-seventeenth-century sample was distinguished from earlier deposits by a ca. 50 percent reduction in plant diversity and by the introduction of carpetweed, toadflax, and woundwort (Fig. 8.2); (d) The early-eighteenth-century sample of edible plant foods was characterized by a sharp reduction of ca. 80 percent in the number and diversity of all varieties relative to the early-seventeenth-century sample and was limited to four types, three of local origin (pumpkin/squash, strawberries, and brambles) and one, peach, of European origin (Fig 8.2).

2. Many, if not most, of the identified early-seventeenth-century plants were characterized by the 1630s by "emergent" species adopted to disturbed open habitats that were anthropogenic in origin (influenced by human intervention), which may have been intentionally selected, collected, protected, transplanted, and/or cultivated. The mere presence of many of these "emergent" species suggests that (a) they were probably humanly introduced or symbiotic, and (b) that Lower Manhattan was heavily disturbed by the second quarter of the seventeenth century, if not earlier by Native American land-use patterns.

3. Insights from prehistoric North American archaeology and ethnobotany suggest that what had been commonly dismissed as invasive weeds, may have served as both Native American and colonial-era starchy seed sources and potherbs. With the exception of the ubiquitous peach pit, most of the plants from the first half of the seventeenth century were dominated by indigenous squash, collectable fruits and berries, and an assortment of what are suggested to be both prehistoric and contact-period Amerindian foods, medicinal and craft plants. This diversity exclusively in the early-seventeenth-century deposits significantly broadens the range of locally available edible food sources of indigenous origin. It also supports the idea that the early Dutch inhabitants may have been more dependent on a broader range of Native American food sources than previously recognized (Table 8.4).

4. The sharp transformations in the diversity of indigenous and potentially introduced plants also dovetails with modern studies of later New York City historical habitats to suggest a long record of disturbance and change in even our supposedly most pristine, or "primeval" habitats in the metropolitan area (Horenstein 2007; Brash 2007; see Peteet, ch. 9 in this volume). By the early to mid-seventeenth century, the local urban setting had

already undergone profound environmental transformations. The magnitude of these changes, between the early seventeenth and early eighteenth centuries, also underscores the danger of relying on either contemporary or recent historical inventories of supposedly pristine ecological "type" sites for environmental reconstruction. While some recent reconstructions have attempted to describe conditions as far back as 1609, no archaeological evidence exists to establish the identity or changing diversity of colonial plants in Manhattan prior to a ca. twenty year period, plus or minus five years, between the 1630s and 1650s.

5. The reevaluation and correction of the original 1980s laboratory and database records, together with the metric analysis of the colonial seeds relative to modern varieties, suggested the probable presence of several varieties of European garden vegetables belonging to the cabbage family (*Brassica*), in the seventeenth-century deposits, clearly by the late seventeenth century, and possibly as early as the second quarter of the century (Tables 8.4, 8.5, and 8.6).

6. The archival and historical folk references to Doctor Kierstede and his multilingual wife Sara's practice of providing shelter to Native American women, recent insights into the importance of women informants to Dutch East and West India Company doctors and botanists, the wide range of potential medicinal plants, the breadth of ethnobotanical references to their use, come together to underscore the import of women, both Native American and Dutch, as primary information brokers in the exchange of botanical and medicinal knowledge in seventeenth-century New Amsterdam.

7. Coupled with new insights into Dutch traditions of plant collecting and the transatlantic exchange of new medicinal plants and knowledge, the role of women and Dutch/Native American informants in New Amsterdam can now be partially attributed to the nexus of influences. These included early traditions of relying on "old wise women" for "folk" knowledge of medicinal plants. Through the translations of Clusius, they integrated the training of Dutch doctors and botanists in techniques of oral history and the use of informants, especially women informants, much based on, or influenced by, the sixteenth-century work of the Iberian doctor Monardes. Finally, they incorporated formalized methods and protocols, taught at seventeenth-century university-based botanical gardens and aimed at sensitizing students to the economic and scientific potential for cross-cultural transfer of knowledge from native women herbalists and practitioners.

8. Finally, this revised chronology and historic ethnobotanical sequence illustrates the potential for archaeology to provide independent "proxies," and/or "ecological benchmarks," to help refine otherwise ill-defined episodes of environmental change in the Hudson River drainage, in general, and the onset of the "Historic Horizon," in particular. Estimates—based primarily on historical assumptions, geomorphological and pollen core data, often interpolated, or by a radiocarbon dates blurred by a large +/- one hundred-year sigma, or standard deviation—for the timing of this transition have spanned from the early seventeenth to the mid-eighteenth centuries (Pederson et al. 2004, 246; Koster and Pienitz 2006, 521, Fig. 5; Russell et al. 1993, Fig. 2, 654–58; Hilgartner and Brush 2006, 482; Gehrels et al. 2006, 954, 958; Maenza-Gmelch 1997, 27, Table 2, 33). The dominance of "emergent" species in the earliest samples and the identification of order of magnitude shifts in the diversity of colonial plant remains within the Dutch West India block suggest that the advent of the "Historic Horizon" in the Lower Hudson is visible in the archaeological record by the second quarter of the seventeenth century in general, and probably by the 1630s in particular.

ACKNOWLEDGMENTS

The findings and conclusions, as well as any factual errors, of this reanalysis of the chronology and plant remains are entirely my own. It is critical that I acknowledge the team members who worked with me at Greenhouse Consultants in 1984 and 1985 to create the quantified 3D computer database that made this study possible, some thirty years after the excavation. Special thanks go to the original field and laboratory specialists: Melba J. Meyers and Nancy Stehling (ceramics and conservation); Joseph Diamond (glass); Diane Dallal (pipes); William Roberts (stratigraphy); Haskell Greenfield (faunal); Michael Davenport and George Meyers (computer mapping and photogrammetry), Lisa Panet (floral), Peter Namuth and Karen Bluth (field and laboratory photography), Leo Hershkowits (historic research), and Mindy Washington, Sara Stone, and Bonnie Bogumil (laboratory and computer database). The initial flotation work, seed sorting and preliminary identification were done by Lisa Panet with the greatly appreciated assistance of staff of the New York Botanical Garden and the New York State Museum; with particular thanks to Chuck Fisher (d) and Nancy Davis for their assistance over the last decade. I want to express appreciation to Paul Huey for his unstinting generosity and sharing of his research and findings throughout. I am also indebted to Sidney Horenstein for his encouragement, sage advice, and assistance in the identification of critical ethnobotanical sources and for his help in the refinement of the plant classifications so as not to exceed the limits of the data.

Heartfelt appreciation is extended to the many persons who painstakingly read and commented on the numerous drafts of this manuscript, including, in alphabetical order: Eric Axelson, Jonathan and Benjamin Gell, Jesse Grossman, Mary Grossman, Hollee Haswell, Robert Henshaw, Sidney Horenstein, Jaap Jacobs, Lucille Johnson, Rande Mas, Dorothy Peteet, Peter Rose, and Bernice Zentner for their editorial advice and improvements. My sincere appreciation also goes to Robert Henshaw, Steve Wilson, Dorothy Peteet, Lucile Johnson, and Roger Panetta of the Hudson River Environmental Society for facilitating my participation in the conference and the publication of its proceedings.

The critical role of my colleagues in The Netherlands I heartedly acknowledge: Jaap Jacobs for suggesting me, Dirk Tang of the National Library of the Netherlands, Hans Krabbendam of the Roosevelt Study Center, Astrid Weij, Barbara Consolini, and Marjolein Cremer (responsible for the International Visitors Programme) of the Netherlands Institute of Heritage for the gracious invitation and coordination of my participation in the Dutch Quadricentennial celebration of Henry Hudson. Grateful appreciation is also extended to Gerda van Uffelen of the Hortus Botanicus of Leiden; Lena Euwens and Erik Zevenhuizen of the Hortus Botanicus of Amsterdam for taking the time to provide access to their most current research and publications concerning seventeenth-century Dutch botanical history. Finally, I am especially indebted to my archaeological colleagues: Benedict Goes, Don Duco, Fleur Cools, Anouk Fienieg, Oscar Hefting, Hans van Westing, and Jan Baart for their collaboration and patient assistance in facilitating entrée to the most current data—both published and not—on seventeenth-century artifact dates.

REFERENCES CITED

Alter, Robert. 2004. *The Five Books of Moses: A translation with commentary.* New York: W. W. Norton.

Anagnostou, S. 2007. The international transfer of medicinal drugs by the Society of Jesus (sixteenth to eighteenth centuries) and connections with the work of Carolus Clusius. In *Carolus Clusius: Towards a cultural history of the Renaissance naturalis,* ed. Florike Egmond, Paul Hoftijzer, and Robert P. W. Visser, 293–312. Amsterdam: Royal Netherlands Academy of Arts and Sciences.

Arber, A. 1986. *Herbals: Their origin and evolution: A chapter in the history of botany 1470–1670.* Third Edition. Cambridge: Cambridge University Press.

Arnason, T., R. J. Hebda, and T. 1981. Use of plants for food and medicine by Native Peoples of Eastern Canada. *Canadian Journal of Botany* 59(1): 2189–2325.

Barona, J. L. 2007. Clusius' exchange of botanical information with Spanish scholars. In *Carolus Clusius: Towards a cultural history of the Renaissance naturalist,* ed. Florike Egmond, Paul Hoftijzer, and Robert P. W. Visser, 99–116. Amsterdam: Royal Netherlands Academy of Arts and Sciences.

Black, M. J. 1980. *Algonquin ethnobotany: An interpretation of Aboriginal adaptation in South Western Quebec.* Ottawa: National Museums of Canada, Mercury Series Number 65.

Bleichmar, D. 2007. Books bodies and fields: Sixteenth-century transatlantic encounters with the New World Materia Medica. In *Colonial botany: Science, commerce, and politics in the early modern world,* ed. Londa Schiebinger and Claudia Swan, 83–99. Philadelphia: University of Pennsylvania Press.

Bolton, R. P. 1922. *Indian paths in the great metropolis. Indian notes and monographs, miscellaneous series 23.* New York: Museum of the American Indian, Heye Foundation.

Brash, A. R. 2007. New York City's primeval forest: A review characterizing the type ecosystem. In *Natural history of New York City's parks and Great Gull Island,* ed. Alice Deutsch, Vol. X, 55–79. New York: Transactions of the Linnaean Society of New York.

Byrne, R., and J. H. McAndrews. 1975. Pre-Columbian purslane (*Portulaca oleracea L*) in the New World. *Nature* 253: 726–27.

Byrne, R., and W. D. Finlayson. 1998. Iroquoian agriculture and forest clearance at Crawford Lake, Ontario. In *Iroquoian peoples of the land of rock and water, AD 1000–1650: A study in settlement archaeology,* ed. W. D. Finlayson, Vol. 1, 94–107. London: London Museum of Archaeology, University of Western Ontario.

Cantwell, A.-M., and D. deZerega Wall. 2001. *Unearthing Gotham: The archaeology of New York City.* New Haven: Yale University Press.

Chevallier, A. 1996. *The encyclopedia of medical plants.* New York: D. K. Publishing.

Cole, J. N. 1979. *Amaranth: From the past for the future.* Emmaus: Rodale Press.

Cook, H. J. 2007a. Global economics and local knowledge in the East Indies: Jacobus Bontius learns the facts of nature. In *Colonial botany: Science, commerce, and politics in the early modern world,* ed. Londa Schiebinger and Claudia Swan, 100–18. Philadelphia: University of Pennsylvania Press.

———. 2007b. *Matters of exchange: Commerce, medicine, and science in the Dutch golden Age.* New Haven and London: Yale University Press.

Cronon, W. 1983. *Changes in the land: Indians, colonists, and the ecology of New England.* New York: Hill and Wang.

Dallal, D. 1996. Van Tienhoven's basket: Treasure or trash? In *One man's trash is another man's treasure,* ed. A. G. A. van Dongen, 215–24. Rotterdam: Museum Boymans-van Beuningen.

Davidson, A. 2006. *The Oxford companion to food,* 2d ed. Oxford and New York: Oxford University Press.

Day, G. M. 1953. The Indian as an ecological factor in the northeastern forest. *Ecology* 34(2): 329–46.

De Koning, J., G. van Uffelen, A. Zemanek, and B. Zamanek. 2008. *Drawn after nature: The complete botanical watercolours of the 16th-century Libri Picturati.* Zeist: KNNV Publishing.

Delcourt, P. A., and H. R. Delcourt. 2004. *Prehistoric Native American ecological change: Human ecosystems in eastern North America since the Pleistocene.* New York: Cambridge University Press.

Denevan, W. M. 1992. The pristine myth: The landscape of the Americas in 1492. *Annals of the Association of American Geographers* 82(3): 369–85.

Densmore, F. 1928. Uses of plants by the Chippewa Indians. *SI-BAE Annual Report* 44: 273–379.

Dressendorfer, W. 2001. Medical plants in modern herbal medicine. In L. Fuchs, *The new herbal of 1543,* 925–28. Paris: Tachen.

Duco, D. H. 1987. *De Nederlandse kleipijp, handboek voor dateren en determineren [The Dutch clay tobacco pipe, manual for dating and determination].* Leiden: Pijpenkabinet Foundation.

Dudek, M. G., L. Kaplan and M. M. King. 1998. Botanical remains from a seventeenth-century privy at the Cross Street site. *Historical Archaeology* 32(3): 63–71.

Egmond, F. 2007. Clusius and friends: Cultures of exchange in circles of European naturalists. In *Carolus Clusius: Towards a cultural history of the Renaissance naturalist,* ed. Florike Egmond, Paul Hoftijzer, and Robert P. W. Visser, 9–48. Amsterdam: Royal Netherlands Academy of Arts and Sciences.

———. 2008. The making of the Libri Picturati A16–30. In *Drawn after nature: The complete botanical watercolours of the 16th-century Libri Picturati,* ed. J. De Koning, G. van Uffelen, A. Zemanek, and B. Zamanek, 12–21. Zeist: KNNV Publishing.

———, P. Hoftijzer, and R. P. W. Visser. 2007. *Carolus Clusius: Towards a cultural history of the Renaissance naturalist.* Amsterdam: Royal Netherlands Academy of Arts and Sciences.

Egmond, F., and L. Ramón-Laca. 2008. Dramatis personae. In *Drawn after Nature: The complete botanical watercolours of the 16th-century Libri*

Picturati, ed. J. De Koning, G. van Uffelen, A. Zemanek, and B. Zamanek, 44–48. Zeist: KNNV Publishing.

Fenton, W. N. 1942. Contacts between Iroquois herbalism and colonial medicine. *Smithsonian Report for 1941*: 503–26.

Fernald, M. L., and A. C. Kinsey. 1965. *Edible wild plants of eastern North America*. Reprint of 1958 edition. New York: Dover.

Fernow, B., ed. 1976. *Records of New Amsterdam 1653–1674*. Baltimore: Genealogical Publishing.

Folkerts, J. 1996. The failure of West India Company farming on the island of Manhattan. In *de Halve Maen. Magazine of the Dutch colonial period in America* 69, 42–52.

Foster, S., and J. A. Duke. 2000. *Medicinal plants and herbs—eastern/central*. Peterson Field Guides. Boston, New York: Houghton Mifflin.

Fuchs, L. 2001. *The new herbal of 1543*. New Kreuterback. [A Complete Colored Edition]. Cologne: Tachen.

Gehrels, R.W., W. A. Marshall, M. J. Gehrels, G. Larsen, J. R. Kirby, J. Eiriksson, J. Heinemeier, and T. Shimmield. 2006. Rapid sea-level rise in the North Atlantic Ocean since the first half of the nineteenth century. *The Holocene* 16(7): 949–65.

Gehring, C. 2009. New Netherland: The formative years, 1609–1632. In *Four centuries of Dutch American relations,* ed. Hans Krabbendam, Cornelis A. van Millen, and Giles Scott-Smith, 74–84. Amsterdam: Uitgeverij Boom.

Goedhuys, D. W., trans. 2008. *Adriaen Van der Donck: A description of New Netherland,* ed. Charles T. Gehring and William A. Shorto. Lincoln and London: University of Nebraska Press.

Goodfriend, J. D. 1992. Before the melting pot: Society and culture in colonial New York, 1664–1730. Princeton: Princeton University Press.

Grieve, Mrs. M. 1981. *A modern herbal, Vols. I and II*. Reprint of 1931 edition. New York: Dover.

Grossman, J. W. 1983. Demographic changes and economic transformations in the south central highlands of pre-Wari Perú. *Nawpa Pacha* 21: 45–126. Berkeley: Institute of Andean Studies.

———. 1985. Environmental shifts and changing food patterns at the 17th century Dutch West India site in Lower Manhattan. Paper presented at the annual meeting of the Council of Northeast Historical Archaeology (CNHEA), Ottawa, Canada, October 27.

———. 2000. Mrs. Kierstede's rear yard: The archaeological discovery and ethno-botanical, cartographic, and archival reanalysis of the 17th century Dutch West India Company remains in Lower Manhattan, New York. Paper prepared for the Regia Civitas and Institute of Archaeology of the Hungarian Academy of Sciences Conference: Medieval Towns and its Citizens, Budapest, Hungary, June 1–4.

———. 2003. From Raritan Landing to Albany's riverfront: The path toward total 3D archaeological site recording. In *People, places, and material things: Historical archaeology of Albany, New York. Chapter 15: Battles and breakthroughs,* ed. Charles Fisher. New York State Museum Bulletin 499 (2003): 167–86.

———. 2008. Future of archaeology in 21st century: Human-landscape interactions. In *2008 Encyclopedia of archaeology,* ed. Deborah Pearsall, Vol. 2: 1458–76. Oxford: Elsevier/Academic Press.

———, M. Meyers, N. Stehling, J. Diamond, D. Dallal, W. Roberts, H. Greenfield, M. Davenport, G. Meyers, and M. Washington. 1985. *The excavation of Augustine Heerman's warehouse and associated 17th century Dutch West India Company deposits: The Broad Financial Center mitigation final report, 1985.* Report on file at the New York City Landmarks Preservation Commission, and the New York State Museum, Albany.

Hamel, P. B. and M. U. Chiltoskey. 1975. *Cherokee plants and their uses—A 400 year history*. Sylva: Herald Publishing.

Hammett, J. E. 2000. Ethnohistory of aboriginal landscapes in the southeastern United States. In *Biodiversity and Native America*, ed. P. E. Minnis and W. J. Elisens, 248–99. Norman: University of Oklahoma Press.

Hedrick, U. P. 1977. *Sturtevant's edible plants of the world*. Reprint of 1919 edition. New York: Dover.

Herrick, J. W. 1977. Iroquois medical botany. PhD diss., State University of New York, Albany.

Hilgartner, W. B., and G. S. Brush. 2006. Prehistoric habitat stability and post-settlement habitat change in a Chesapeake Bay freshwater tidal wetland, USA. *The Holocene* 16(4): 479–94.

Horenstein, S. 2007. Inwood Hill and Isham Parks: Geology, geography, and history. In *Natural history of New York City's parks and Great Gull Island*, ed. Alice Deutsch, 1–54. New York: *Transactions of the Linnaean Society of New York*, Vol. X.

Huey, P. 1988. Aspects of continuity and change in colonial Dutch material culture at Fort Orange, 1624–1664—A dissertation in American civilization. Philadelphia: University of Pennsylvania.

Hunt, J. D., ed. 1990. *The Dutch garden in the seventeenth century.* Washington: Dumbarton Oaks Research Library and Collection.

Huxley, A. 1998. *An illustrated history of gardening.* Reprint of 1978 edition. New York: Paddington Press.

Innes, J. H. 1922. New Amsterdam and its people: Studies, social and topographical, of the town under Dutch and early English rule. New York: Charles Scribner's Sons.

Jacobs, J. 2005. *New Netherland: A Dutch colony in seventeenth century America.* Leiden-Boston: Brill.

———. 2009. *The colony of New Netherland: A Dutch settlement in seventeenth-century.* Ithaca and London: Cornell University Press.

Jameson, F., ed. 1909. Description of the Towne of Mannadens, 1661. In *Narratives of New Netherland, 1609–1664. Digital archives: Original narratives of early American history.* Reproduced under the Auspices of the American Historical Association. New York: Charles Scribner's and Sons.

Janowitz, M. F. 1993. Indian corn and Dutch pots: Seventeenth-century floodways in New Amsterdam/New York. *Historical Archaeology* 27(2): 6–24.

Kelso, W. M., with B. Straub. 1997. *1996 interim report on the APVA excavations at Jamestown, Virginia.* Richmond: Association for the Preservation of Virginia Antiquities.

———. 2004. *Jamestown rediscovery 1994–2004.* Richmond: Association for the Preservation of Virginia Antiquities.

Köster, D., and R. Pienitz. 2006. Late-Holocene environmental history of two New England ponds: natural dynamics versus human impacts. *The Holocene* 16(4): 519–32.

Krabbendam, H., C. A. van Millen, and G. Scott-Smith, eds. 2009. *Four centuries of Dutch American relations: 1609–2009.* Amsterdam: Uitgeverij Boom.

Liebman, B., and J. Hurley. 2009. Rating rutabagas: Not all vegetables are equal. *Nutrition Action Newsletter* 36(1) (January/February): 9–15. Washington: Center for Science in the Public Interest (www.cspinet.org).

Maenza-Gmelch, T. 1997. Holocene vegetation, climate, and fire history of the Highlands, southeastern New York, USA. *The Holocene* 7(1): 25–37.

Mallios, S. (with contributions by G. Fesler). 1999. *The Reverend Richard Buck site: Archaeological excavations at 44JC568.* Richmond: Association for the Preservation of Virginia Antiquities.

———. 2000. *At the edge of the precipice: Frontier ventures, Jamestown's hinterland, and the archaeology of 44JC802.* Richmond: Association for the Preservation of Virginia Antiquities.

Martin, A. C., H. S. Zim, and A. L. Nelson. 1951. American wildlife and plants: A guide to wildlife food habits—The use of trees, shrubs, weeds, and herbs by the birds and mammals of the United States. New York: Dover.

Matson, C. Economic networks of Dutch traders and the British colonial empire. In *Four centuries of Dutch American relations,* ed. Hans Krabbendam, Cornelis A. van Millen, and Giles Scott-Smith, 97–107. Amsterdam: Uitgeverij Boom.

McAndrews, J. H., and M. Boyko-Diakonow. 1998. Pollen analysis of varved sediment at Crawford Lake, Ontario: Evidence of Indian and European farming. In *Quarterly Geology of Canada and Greenland,* ed. R. J. Fulton, 528–30. Ottawa: Geological Survey of Canada.

Meyer, J. E. 1972. *The herbalist.* Ninth ed. of 1918 printing. New York: Self-published.

Military Advisory Board. 2007. National security and the threat of climate change: Report from a panel of retired senior US military officers. CNA Corporation, 3. Virginia.

Miller, N. 1989. What mean these seeds: A comparative approach to archaeological seed analysis. *Historical Archaeology* 23(2): 50–59.

Moerman, D. E. 2004. *Native American ethnobotany*. Portland, OR, and Cambridge, MA: Timber Press. University of Michigan Database of Ethnobotany: http://herb.umd.umich.edu/.

National Academy of Sciences. 1975. *Unexploited tropical plants with promising economic value*. Washington: National Academy of Sciences.

Nordås, R., and N. P. Gleditsch. 2007. Climate change and conflict. *Political Geography* 26: 627–38.

O'Callaghan, E. B., and B. Fernow, trans. 1856. *Documents relative to the colonial history of the State of New York*. 15 vols. Albany: Weed, Parsons.

Oldenburger-Ebbers, C. 1990. Notes on plants used in Dutch gardens in the second half of the seventeenth century. In *The Dutch garden in the seventeenth century,* ed. J. D. Hunt, 159–74. Washington, DC: Dumbarton Oaks Research Library and Collection.

Parker, A. C. 1910. *Iroquois uses of maize and other food plants*. Albany: University of the State of New York.

Parker, G. 2008. Crisis and catastrophe: The global crisis of the seventeenth century reconsidered. *American Historical Review* (October): 1053–79.

Pavord, A. 2005. *The naming of names: The search for order in the world of plants*. New York: Bloomsbury.

Pederson, D. C., D. M. Peteet, D. Kurdyla, and T. Guilderson. 2005. Medieval warming, little ice age, and European impact on the environment during the last millennium in the lower Hudson Valley, New York, USA. *Quaternary Research* 63: 238–49.

Peterson, L. A. 1977. *Edible wild plants: Eastern/central North America*. Peterson field Guides. New York: Houghton Mifflin.

Pluis, J. 1998. *De Nederlandse tegel: Designs and names 1570–1930 [The Dutch tile: Designs and names]*. Leiden: Nederlands Tegelmuseum/ Vrienden van het Netherlands Tegelmuseum, in samenwerking met Primavera Pers.

Poole, Sidman P. 1934. Geographic interpretation of New Amsterdam. *Bulletin of the Geological Society of Philadelphia* 32: 42–64.

Potterton, D., ed. 1983. *Culpepper's color herbal*. New York: Sterling.

Ramon-Laca, L. 2008. The additions by the Count of Arenberg. In *Drawn after nature: The complete botanical watercolours of the 16th-century Libri Picturati,* ed. Jan De Koning, Greta van Uffelen, Alicja Zemanek, and Bogden Zamanek, 54–69. Zeist*:* KNNV Publishing.

Reagan, A. B. 1928. Plants used by the Bois Fort Chippewa (Ojibwa) Indians of Minnesota. *Wisconsin Archeologist* 7(4): 230–48.

Richardson, D. M., P. Pysek, M. Rejmanek, M. G. Barbour, F. D. Panetta, and C. J. West. 2000. Naturalization and invasion of alien plants: Concepts and definitions. *Diversity and Distributions* 6: 93–107.

Ripley, G., and C. A. Dana, eds. 1873–76. *The American cyclopedia: A popular dictionary of general knowledge*, XIV (1875). New York: D. Appleton, 88–89.

Rothchild, N. A., D. deZ. Wall, and E. Boesch. 1987. *The archaeological investigation of the Stadt Huys block: A final report*. Report on file with the New York City Landmarks Preservation Commission.

Rousseau, J. 1945. Le folklore botanique de Caughnawaga. Montreal, Canada. *Contributions de l'Institut botanique l'Universite de Montreal* 55: 7–72. Montreal: University of Montreal.

Russell, E. W. B., R. B. Davis, R.S. Anderson, T. E. Rhodes, and D. S. Anderson. 1993. Recent centuries of vegetational change in the glaciated north-eastern United States. *Journal of Ecology* 81: 647–64.

Safford, W. E. *A forgotten cereal of ancient America. 1917. Proceedings of the 19th International Congress of the Americanists, Washington, D.C.* 1915: 286–97.

Sauer, J. D. 1950. The grain amaranths: A survey of their history and classification. *Annals of the Missouri Botanical Garden* 37: 561–632.

———. 1952. A geography of pokeweed. *Annals of the Missouri Botanical Garden* 39: 113–25.

———. 1967. The grain amaranths and their relatives; A revised taxonomy and geographic survey. *Annals of the Missouri Botanical Garden* 54: 103–37.

Savoiea, A. U., A. Zemanek, and B. Zemanek. 2008. The beginnings of ecological thought. In

Drawn after nature: The complete botanical watercolours of the 16th-century Libri Picturati, ed. J. De Koning, G. van Uffelen, A. Zemanek, and B. Zemanek. Zeist: KNNV Publishing.

Schiebinger, L. 2007. Prospecting for drugs: European naturalists in the West Indies. In *Colonial botany: Science, commerce, and politics in the early modern world,* ed. Londa Schiebinger, and Claudia Swan, 119–33. Philadelphia: University of Pennsylvania Press.

———, and C. Swan, eds. 2007. *Colonial botany: Science, commerce, and politics in the early modern world.* Philadelphia: University of Pennsylvania Press.

Shimer, P. 2004. *Healing secrets of the Native Americans: Herbs, remedies, and practices that restore the body, refresh the mind, and rebuild the spirit.* New York: Black Dog and Leventhal.

Singleton, E. 1968. *Dutch New York.* Reprint of 1909 edition. New York: Ayer.

Smith, B. D. 1989. The independent domestication of indigenous seed-bearing plants in eastern North America. In *Emergent horticultural economies of the eastern woodlands, Occasional Papers No. 7,* ed. W. Keegan, 3–48. Carbondale: Center for Archaeological Investigations, Southern Illinois University.

———, with contributions by C. W. Cowen and M. P. Hoffman. 1992. *Rivers of change: Essays on early agriculture in eastern North America.* 1st. ed. Washington, DC, and London: Smithsonian Institution.

Smith, H. H. 1932. Ethnobotany of the Ojibwa Indians. *Bulletin of the Public Museum of Milwaukee* 4: 327–525.

Sjostrand, S., and Sharipah Lok bt. Sted Idrus. 2007. *The Wanli shipwreck and its ceramic cargo.* Nanhai Marine Archaeology Sdn. Bhd. in collaboration with the Ministry of Culture, Arts and Heritage Malaysia and the Department of Museums Malaysia.

Speck, F. G., R. B. Hassrick, and E. S. Carpenter. 1942. Rappahannock herbals, folk-lore, and science of cures. *Proceedings of the Delaware County Institute of Science* 10: 7–55.

Stern W. T. 1989. Horticulture and botany. In *The age of William III & Mary II: Power politics and patronage 1688–1702,* ed. Robert P. Maccubbin and Martha Phillips, 179–85. Williamsburg: The College of William and Mary in Virginia.

Stokes, I. N. P. 1967. *The iconography of Manhattan Island, Vol. I–IV.* Reprint of 1915–1925 edition (New York: Robert H. Dodd). New York: Arno Press.

Straub, B., and N. Luccketti. 1996. *1995 interim report.* Richmond: APVA Jamestown Rediscovery.

Struever, S. 1964. The Hopewill interaction sphere in riverine–western Great Lakes culture history. In *Hopewellian studies,* ed. J. R. Caldwell and R. L. Hall, 87–106. Springfield: Illinois State Museum, Scientific Papers No. 12.

Sturtevant, W. 1954. The Mikasuki Seminole: Medical beliefs and practices. PhD diss., Yale University.

Swan, C. 1998. *The Clutius botanical water colors: Plants and flowers of the renaissance.* New York: Harry N. Abrams.

———. 2007. Collecting naturalia in the shadow of early modern Dutch trade. In *Colonial botany: Science, commerce, and politics in the early modern world,* ed. Londa Schiebinger and Claudia Swan, 223–36. Philadelphia: University of Pennsylvania Press.

Tantaquidgeon, G. 1972. *Folk medicine of the Delaware and related Algonquian Indians.* Harrisburg: Pennsylvania Historical Commission Anthropological Papers no. 3.

Taylor, L. A. 1940. *Plants used as curatives by certain southeastern tribes.* Cambridge: Botanical Museum of Harvard University.

Thick, M. 1990. Garden seeds in England before the late eighteenth century—II, The trade in seeds to 1760. *Agricultural History Review* 38(II): 105–16.

Thomás, J. P. 2007. Two glimpses of America from a distance: Carolus Clusius and Nicolás Monardes. In *Carolus Clusius: Towards a cultural history of the renaissance naturalist,* ed. Florike Egmond, Paul Hoftijzer, and Robert P. W. Visser, 173–93. Amsterdam: Royal Netherlands Academy of Arts and Sciences.

USDA, ARS, National Genetic Resources Program. 2009. Germplasm Resources Information Network (GRIN). National Germplasm Resources Laboratory, Beltsville, MD. http://www.ars-grin.gov/cgi-bin/npgs/html/taxon.pl?312174.

Uva, R. H., J. C. Neal, and J. M. DiTomaso. 1997. *Weeds of the Northeast.* Ithaca and London: Comstock Publishing Associated, a Division of Cornell University Press.

Van der Groen, Jan. 1670. *Den Nederlandtsen Hovenier.* Amsterdam: Marcus Doornick.

Van Laer, A. J. F. 1974. *New York historical manuscripts: Dutch. Vol. 1, Register of the provincial secretary, 1638–1642.* Baltimore: Genealogical Publishing Co.

Van Lemmen, H. 1997. *Delftware tiles.* Woodstock: The Overlook Press.

Van Rensselaer, Mrs. J. K. 1898. *The Goede Vrouw of Mana-ha-ta: At home and in society 1609–1760.* New York: Charles Scribner's Sons.

Van Uffelen, G. 2008a. The *Libri Picturati* and the early history of the Hortus Botanicus Leiden. In *Drawn after nature: The complete botanical watercolours of the 16th-century Libri Picturati,* ed. J. De Koning, G. van Uffelen, A. Zemanek, and B. Zamanek, 54–59. Zeist: KNNV Publishing.

———. 2008b. Food plants. In *Drawn after nature: The complete botanical watercolours of the 16th-century Libri Picturati,* ed. J. De Koning, G. van Uffelen, A. Zemanek, and B. Zamanek, 116–17. Zeist: KNNV Publishing.

Viele, Egbert L. 1865. *Sanitary and topographical map of the city and island of New York prepared for the Council of Hygiene and Public Health of Citizens, Association of New York upon the sanitary condition of the city.* New York: D. Appleton.

Watson, P. J. 1989. Early plant cultivation in the eastern woodlands of North America. In *Foraging and farming: The evolution of plant exploitation,* ed. D. R. Harris and G. C. Hillman, 555–71. London: Unwin Hyman.

Watson, P. J., and M. Kennedy. 1991. The development of horticulture in eastern woodlands of North America: Women's role. In *Engineering archaeology,* ed. J. Gero and M. Conkey, 255–75. Oxford: Basil Blackwell.

Waugh, F. W. 1916. *Iroquois foods and food preparation.* Ottawa: Canada Department of Mines.

Wilcoxen, C. 1987. *Dutch trade and ceramics in America in the seventeenth century.* Albany: Albany Institute of History and Art.

Wilson, A. C. 1999. *The book of marmalade.* Philadelphia: University of Pennsylvania Press.

CHAPTER 9

LINKING UPLANDS TO THE HUDSON RIVER

Lake to Marsh Records of Climate Change and Human Impact over Millennia

Dorothy M. Peteet, Elizabeth Markgraf, Dee C. Pederson, and Sanpisa Sritrairat

ABSTRACT

The wetlands of the Hudson Valley provide valuable archives for understanding how climate change and anthropogenic impact have influenced the region. Recent pollen, macrofossil, charcoal, and loss-on-ignition (LOI) data from a suite of cores on a downriver transect allow comparison of uplands and river marshes. In the mid-Hudson Region, wetland sediments from Rhododendron Swamp, Shawangunk Mountains, and Black Rock Forest pond/fen are compared to Hudson National Estuarine Research Reserve (HNERR) Tivoli Bay, Iona Marsh, and Piermont marsh records. All of these sites indicate forest decline about four hundred years ago, coeval with a rise in invasive species such as ragweed. While upland lakes record the regional forest history, small wetlands such as fens, bogs, and swamps with a rich macrofossil component enhance our ecological and climatic reconstructions. In the river marsh environments where sedimentation rates are high, the Medieval Warm Period (MWP), with elevated charcoal from the watershed, is conspicuous.

In the lower Hudson Valley, records from Piermont Marsh and the Hackensack Marsh are compared to an Alpine Swamp record from atop the Palisades in Alpine, New Jersey. Upland farming/lumbering/industry is recorded by a decline of trees concurrent with a rise of weedy species such as ragweed and plantain and paralleled by the same markers in the river wetlands. At the same time, the marshes document dramatic disappearances of a diverse local marsh flora with the encroachment of cattail and the highly invasive common reed. Both inorganic sediment input and nutrient supply were altered by human activity. Sensitive to climate changes and droughts in particular, Hudson marshes record the shifts in upland vegetation along with watershed increases in inorganic input, charcoal, and foraminifera. Learning from the historical perspective enhances our ability to preserve, restore, and manage both upland wetlands and marshes as the region prepares for rising sea level, warmer temperatures, and further invasions.

INTRODUCTION

Lake sediments are the traditional archives of Quaternary paleoclimate history throughout North America (Wright, 1983). However, wetlands such as bogs, fens, swamps, and marshes also serve as archives with valuable complementary information. We select several of a variety of wetland archives and compare these records to lake/bog stratigraphy to assess the strengths of each environmental repository. Included is a transect of sites from 125 km upriver in the Shawangunk Mountains near New Paltz, New York, south to the mouth of the Hudson at the Hackensack Meadowlands, New Jersey, and Jamaica Bay, New York. Each of the marshes is located at the mouth of a tributary of the Hudson, which enhances the signal of human disturbance in the marshes as derived from the uplands.

GEOLOGY, CLIMATE, AND MODERN VEGETATION

The Hudson Valley uplands contain glacially derived small lakes and wetlands such as Rhododendron Swamp (Fig. 9.1), which is adjacent to Mohonk Lake atop the Shawangunk Ridge.

This site lies along parallel ridges of Silurian-age quartz pebble conglomerates and quartz sandstones (Shawangunk Formation) that are part of the southeasternmost ridge of the Appalachian Mountains (Bernet et al. 2007). Slightly to the south, Sutherland Fen is located adjacent to Sutherland Pond (Maenza-Gmelch 1997) near West Point in Black Rock Forest (Fig. 9.1). Farther south near the New York/New Jersey state line, the Palisades sill rises 150 m, where Alpine Swamp is located. A focused examination of the age of deglaciation in this part of the southeastern margin of the late Pleistocene Laurentide ice sheet has resulted in a major controversy over the timing of ice retreat (Peteet et al. 2006b). In contrast to a relatively early age for deglaciation around twenty thousand years ago from bulk dates, more recent AMS dating of basal sediments from lake/bog sites throughout the area suggest a substantially younger age for deglaciation at fifteen thousand years ago (12.5 ^{14}C) (Peteet et al. 1990; 2006b). Marsh cores retrieved from low elevations along the Hudson, including Tivoli Bay, Iona, Piermont, Staten Island, Jamaica Bay and Hackensack (Fig. 9.1), are much younger than the lakes, most forming as the river became an estuary (Peteet et al. 2006a).

The modern climate of southeastern New York is temperate, with warm summers and cool winters. According to the Mohonk Lake Cooperative Weather Station, the average temperature over the last 105 years (1896–2000) on the Shawangunk ridge was 8.9°C with a mean January temperature of -4.0°C and a mean July temperature of 21.4°C. Annual rainfall has averaged 120 cm/year with an annual snowfall average of 149 cm. The temperature is 2°C cooler in winter and 2.3°C cooler in summer than nearby West Point, the meteorological records used for 50 km south of the site and very close to Sutherland Pond/Fen in Black Rock Forest. Southern Hudson Valley climate, including Jamaica Bay, has a mean July temperature over 22°C, with about the same amount of rainfall as along the Shawangunk ridge.

Vegetation of the Hudson Valley is classified as oak-chestnut (Braun l950) with the Hudson Highlands dominated by chestnut oak (*Quercus montana*), red oak (*Quercus rubra*), and some white oak (*Quercus alba*) (Thompson l996; Thompson and Huth, ch. 10 this volume). Higher elevations of the Shawangunks grade from pitch pine (*Pinus rigida*)—oak-heath communities to mesic sugar

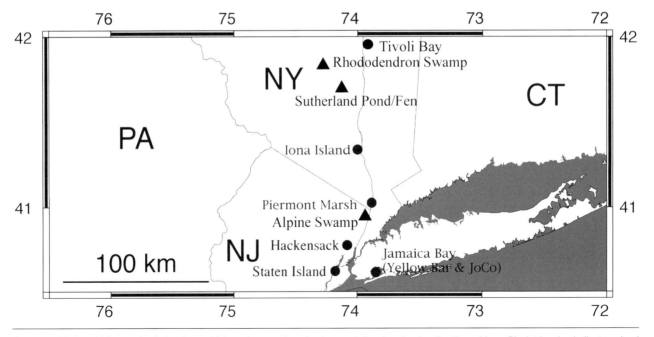

FIG. 9.1. Hudson Valley wetland sites from which we have retrieved paleovegetational and paleoclimatic archives. Black triangles indicate upland swamps/fens while circles indicate Hudson tidal marshes.

maple (*Acer saccharum*), beech (*Fagus grandifolia*), yellow birch (*Betula lutea*), and hemlock (*Tsuga canadensis*) communities. Hickory (*Carya* sp.) and ash (*Fraxinus* sp.) are also regionally present, particularly in more southern parts of the valley. The fresh and brackish Hudson marshes today are dominated by common reed (*Phragmites*) and cattail (*Typha angustifolia*), while higher salinity marshes on the coast retain saltwater grass (*Spartina alterniflora*) and cordgrass (*S. patens*) dominance.

METHODS

Field Research

Sediment cores were taken in a N-S transect throughout the last decade in a focused effort to understand the timing and environment of deglaciation, as well as the vegetational history, the climate history, and the human impact throughout the region. Several types of corers were utilized, including a Dachnowski sampler, a Hiller peat sampler, and a modified Livingstone piston corer (Wright et al. 1984). Cores were extruded, wrapped in saran and foil, and stored in the LDEO cold room repository for analysis.

Pollen, Spore, LOI, and Macrofossil Laboratory Analysis

Lake, fen, and marsh samples were analyzed for loss-on-ignition (LOI) using procedures outlined in Dean (1974). Samples for pollen analysis were processed every four or five centimeters using standard procedures (Faegri and Iverson 1989). A minimum of three hundred terrestrial pollen grains was counted at 400x magnification. Pollen sums were calculated to include all terrestrial pollen types, and spore percentage calculations were based on the sum of pollen and spores. Charcoal pieces greater than fifty microns were counted on the pollen slides. These percentages were plotted in Tilia-Graph (Grimm 1992).

Macrofossils were analyzed from 20 cm^3 of sediment according to methods of Watts and Winter (1966), identifying seeds, fruits, needles, stems, bryophytes, foraminifera, and charcoal. Macrofossils were picked and counted using a binocular microscope and plotted based on numbers of a particular type per volume. Selected macrofossils were dated using AMS radiocarbon dating techniques at Lawrence Livermore Labs, California. Radiocarbon ages were calibrated to calendar ages using CALIB (version 5.0) (Stuiver and Reimer 1993).

RESULTS

Mid-Hudson Valley

Rhododendron Swamp (Fig. 9.2) is a small (2.4 hectare) wetland that lies southwest of Lake Mohonk at an altitude of 275 m.

A sediment core taken in 2002 provides an 8,800-year vegetational record thus far that records the dominance of oak (30%) along with the major role of birch (20–30%) at the site. Maximum pine percentages at the base of the diagram indicate a drier climate in the early Holocene, followed by a wetter zone (RH-2) with hemlock and shining clubmoss (*Lycopodium lucidulum*) well represented. The mid-Holocene hemlock decline at 75 cm (beginning of RH-3) is a regional event, previously linked to a pathogen and possibly drought (Foster et al. 2006). Additional downcore pollen and macrofossil research in progress will define the earliest plant communities that colonized the region after deglaciation, prior to the Holocene oak dominance.

The advent of European impact (Russell 2001) is shown in the top 25 cm of the pollen diagram (RH-5) dated by ^{14}C to 1700 CE with a decline in trees such as birch and chestnut concurrent with an increase in weedy species such as ragweed (*Ambrosia*), grass (*Gramineae*), plantain (*Plantago*), and dock (*Rumex*). Interestingly, the top 5 cm of the core records a resurgence in some of the trees (pine, hemlock, but not chestnut) and stabilization of oak and birch, probably due to forest regrowth of the twentieth century as agricultural activity declined. The human impact is similar to that recorded in Sutherland Pond (Maenza-Gmelch 1997) in Black Rock Forest, where the top 50 cm is marked by a decline in tree pollen replaced by the same invasive ragweed, grass, and plantain followed by a slight reforestation signal.

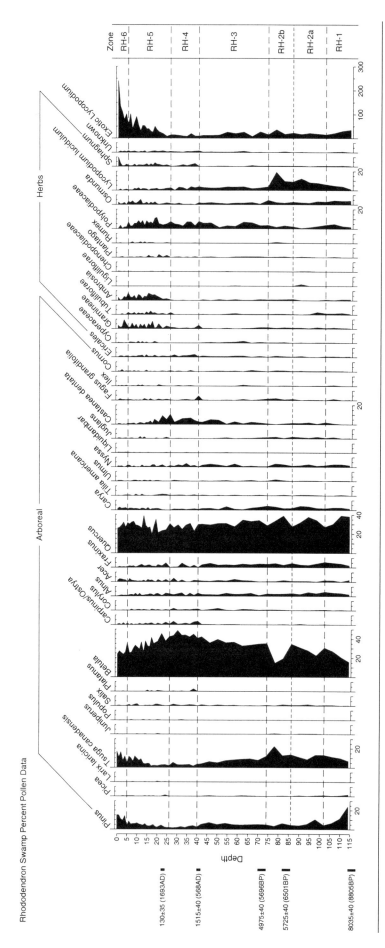

FIG. 9.2. Rhododendron swamp pollen and spore stratigraphy, Mohonk Ridge, New York.

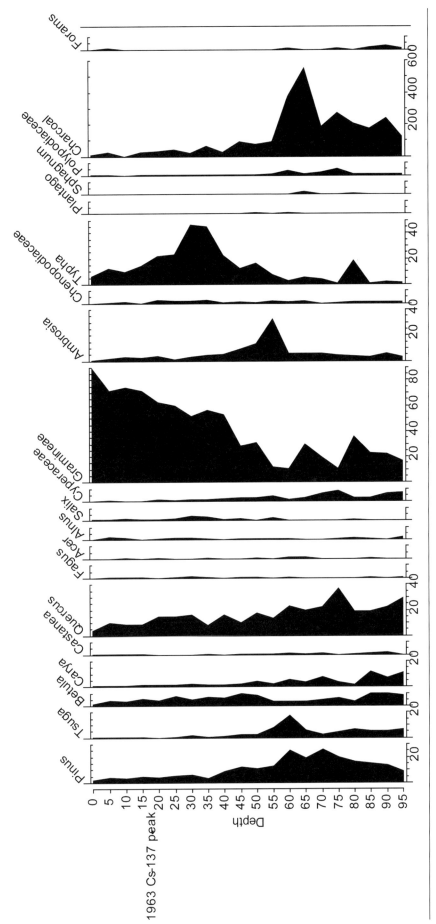

FIG. 9.3. Iona Marsh, Hudson Valley pollen and spore record of the human impact beginning about 60 cm with the rise in *Ambrosia* and Gramineae.

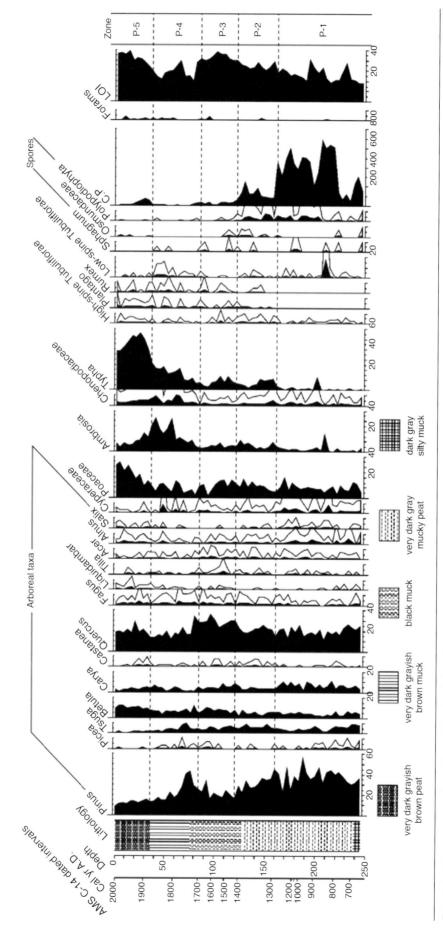

FIG. 9.4. Pollen and spores from Piermont Marsh, Hudson River, along with charcoal/pollen (C/P) ratios, foraminifera (Forams), and loss-on-ignition (LOI). (Pederson et al., 2005).

Marsh investigation from Tivoli Bay (Sritrairat et al. 2008) records a shift in marsh vegetation from a more aquatic/marsh environment to a shrubbier, weedier regime with European impact. The shift to a drier wetland was caused by an increase in inorganic flux based on loss-on-ignition measurements probably due to erosion from farming/industry, which raised the elevation of the marsh. Invasives such as pigweed (*Chenopodiaceae*), jewelweed (*Impatiens*), and loosestrife (*Lythrum*) become more abundant at this time.

Lower Hudson Valley

Analysis of Alpine Swamp, a nine-hectare freshwater swamp atop the Palisades Sill (Peteet et al. 1990) records the same four invasive taxa (ragweed, grass, plantain, dock) in the top 30 cm demonstrating the widespread signature of these weedy plants as regional deforestation took place.

A pollen and spore record from Iona Island State Park (Fig. 9.3) demonstrates the same decline in trees (pine, hemlock, oak) regionally beginning above 60 cm depth in the core that accompanied the rise of grasses, ragweed, and cattails in the marsh and adjacent upland and which dates to European impact around 1700 CE.

The date of 1963 using the Cs-137 bomb peak indicates the timing of an aggressive common reed expansion, resulting in this plant replacing cattails as the dominant wetland species in many of the Hudson marshes over the last half-century.

This pattern of invasives is also evident in the pollen and spore record from Piermont Marsh (Fig. 9.4), which reveals the similar declines in trees accompanied by increases in weedy taxa beginning about 350 years ago.

In both Iona and Piermont Marshes, charcoal is much higher prior to European impact, inferring drought-induced fires during the MWP from 800 to 1350 CE. (Pederson et al. 2005).

The Hackensack River Meadowlands, spanning 34 km², lie adjacent to the Hudson watershed. A pollen and spore record from this marsh reveals the same invasive species (Carmichael 1980), but the accompanying detailed macrofossil record reveals the various taxa that comprised the sedge (*Cyperaceae*) community in the Meadowlands prior to the modern dominance of common reed, pigweed, and *Compositae* (Fig. 9.5).

A remarkable assemblage of various sedges (*Cladium, Scirpus, Cyperus, Carex, Eleocharis*) prior to European impact demonstrate the wetland biodiversity now lost at the Meadowlands. This diversity depletion is a hallmark result of human impact, and has multiple causes, including habitat destruction, urbanization, and nitrogen fertilization.

DISCUSSION AND CONCLUSIONS

While previous lake pollen/macrofossil stratigraphy has set the stage for our understanding of large-scale climate and human impacts since the Laurentide ice retreat, the records from wetlands and marshes in the Hudson Valley augment our understanding of the landscape in several new ways. The following table gives a comparison of strengths from each archive type:

TABLE 9.1. Comparative Strengths of Different Hudson Valley Archive Wetland Types

Hudson Valley Lakes/ Ponds/Fens/Swamps	Hudson Valley Tidal Marshes
Long-term (present–15,000 years)	Shorter term (present–6,000–11,000 years)
Sedimentation rate lower	Sedimentation rate higher
Regional forest composition	Some regional forest composition; local marsh flora
Deforestation and invasive history	Detailed invasive history
Twentieth-century reforestation	Continued detailed invasive history
Regional fire history	Local and watershed fire history
Hydrological shifts	Salinity shifts Sediment decline from dams Detailed watershed pollution history

Most lake and pond records were formed by glacial excavation and retreat fifteen thousand years ago (Peteet et al. 2006b), and give a continuous representation of the history of the upland region from ice age to modern environment. In contrast, the marshes appear to owe their origin to the timing of the formation of the Hudson River as an estuarine environment around 7000 BCE), with the exception of the Staten Island record which began as a freshwater pond (Peteet et al. 2006a). However, the

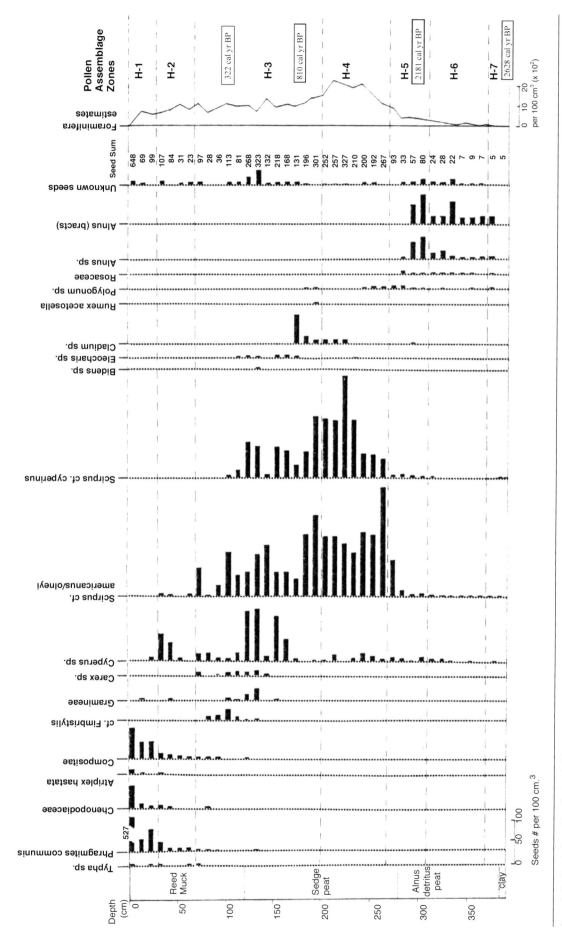

FIG. 9.5. Seeds from Meadowlands Marsh showing dramatic shift from a very diverse sedge assemblage to the dominance of the present weedy assemblage (adapted from Carmichael, 1980).

sedimentation rates in marshes are usually much higher (0.1–0.33 cm/yr and up to 0.7 cm/yr) than in lakes/ponds (0.1 cm/yr), enabling much higher resolution studies from marsh sediment, which is particularly useful in the centuries since European impact.

Prior to European impact, forests were composed of more pine, oak, hemlock, and hickory than today, as documented by pollen records from both lakes and marshes. Climate changes suggested by pollen and charcoal include droughts, fire, and cooler intervals, including the Little Ice Age and the MWP. While the lake records provide regional patterns of species decline (i.e., hemlock decline about five thousand years ago), the small wetlands and marshes reveal significant details about the ecosystem, enabling refinement of the climatic regime. For example, Rhododendron Swamp shining clubmoss spore (*Lycopodium lucidulum*) stratigraphy parallels the hemlock moisture curve, reinforcing the interpretation of a decline in moisture around 5000 BCE. The Hackensack Marsh macrofossil stratigraphy enables us to envision a local plant community that today does not exist in the Hudson Valley region, and provides a very detailed understanding of the invasive history. All of the marshes indicate extensive community disturbance of the landscape from the time of European settlement up to present, with large losses of forest. They also provide a remarkable documentation of the sequential invasive history, beginning with cattail dominance in the l960s and the subsequent takeover by common reed in the last decades. However, lacking in the marshes is the history of twentieth-century reforestation in the region, primarily because the regional pollen record has been overshadowed by local marsh invasives.

Charcoal presence in Piermont, Tivoli Bay, and Iona marshes reveal dramatic shifts throughout the last millennium. These shifts are much larger than those recorded in lakes/ponds, and reflect a watershed signal of drought as they span the MWP. Foraminifera increases in nearby Hackensack Marsh may be correlative with the charcoal increase, suggesting a salinity increase at that site. As precipitation and river flow was reduced, the salt wedge would have moved northward. However, further dating is needed in the Hackensack watershed.

Upland lake/pond cores often reveal an inorganic increase in cores concurrent with the ragweed rise resulting from increased sedimentation due to soil destabilization as trees were cut and farms replaced forests. Hudson marshes also record sedimentation increases (Tivoli Bay, Piermont). However, in Jamaica Bay the opposite is true. There, declines in sediment supply, up to a factor of 5, are attributed to the numerous dams built by the Dutch colonists for mills and the subsequent destruction of the watersheds due to urban pavement of the watershed (Peteet et al. 2006a). This decline may be a contributing factor in the loss of Jamaica Bay marshes today, along with a rise in nitrogen.

IMPLICATIONS FOR MANAGEMENT

Upland forests in the Hudson Valley are facing serious challenges due to deer browse, insect outbreaks, and fire suppression (Thompson and Huth, ch. 10 this volume). Sediment archives from lakes provide a historical framework for understanding these shifts in the long term, and various wetlands with their abundant plant macrofossil records give us additional clues to historical biodiversity and past droughts, wet intervals, and temperature shifts. Identifying past losses and the reasons for them can enhance our ability to make the best choices for managing these resources.

Providing historical context for Hudson River marshes is vital to our definition of what is "natural" and in developing plans for restoration. River marsh archives document a tremendous loss in biodiversity due to the intensive human impact of dams, farming, industry, urbanization, and land use (see Vispo and Knab-Vispo, ch. 12 this volume). Major marsh habitat for algae, invertebrates, vascular plants, fish, and birds has drastically changed since European impact (Peteet et al. 2006a). Undoubtedly the plant species loss has affected the estuarine function. Restoration efforts can be undertaken with the knowledge of what communities have been lost. Charcoal and inorganic input into the watershed can be understood in the context of past climate and anthropogenic forcing, and future estimates/models of how the watershed will be affected as global warming and development continue are enhanced by historical data. Understanding past sedimentation rates in the context of past land use and climate shifts can aid us as we prepare for sea level rise and strive for marsh preservation.

ACKNOWLEDGMENTS

We thank Paul Huth and John Thompson of Mohonk Preserve as well as the Smiley family of Mohonk Mountain House for access to the Rhododendron Swamp site and for field assistance in retrieving the Rhododendron Swamp core. Thanks are also due the NYS Dept. of Environmental Conservation (DEC) and the Hudson River National Estuarine Research Reserve (HRNERR) staff for permission to core the estuarine marshes, including Betsy Blair and Sarah Fowell. The National Park Service granted us permission to core in Jamaica Bay, and we appreciate the assistance of George Frame and his staff. Many Columbia University students aided in our coring efforts in the upland wetlands and marshes. We are grateful to Robert Henshaw and Kirsten Menking for helpful comments on the manuscript. LDEO Climate Center and NASA/GISS provided financial support, and sample material used in this project was stored in the LDEO Sample Repository, supported by the National Science Foundation (Grant OCE06-475574).

REFERENCES CITED

Bernet, M., D. Kapoutsos, and K. Basset. 2007. Diagenesis and provenance of Silurian quartz arenites in south-eastern New York state. *Sedimentary Geology* 201: 43–55.

Braun, E. L. 1950. *Deciduous forests of eastern North America.* New York: Hafner.

Carmichael, D. (Peteet). 1980. A record of environmental change during recent millennia in the Hackensack tidal marsh, New Jersey. *Bulletin of the Torrey Botanical Club* 107(4): 514–24.

Dean, W. E. 1974. Determination of carbonate and organic matter in calcareous sediments and sedimentary rocks by loss-on-ignition: Comparison with other methods. *Journal of Sedimentary Petrology* 44: 851–61.

Faegri, K., and J. Iversen. 1975. *Textbook of pollen analysis.* New York: Hafner.

Foster, D., W. Oswald, E. Faison, E. Doughty, and B. Hanson. 2006. A climatic driver for abrupt mid-Holocene vegetation dynamics and the hemlock decline in New England. *Ecology* 87: 2959–66.

Grimm, E. C. 1992. TILIA and Tilia-Graph software, version 2. Illinois State University.

Maenza-Gmelch, T. 1997. Holocene vegetation, climate, and fire history of the Hudson Highlands, southeastern New York, USA. *The Holocene* 7: 25–37.

Pederson, D., D. Peteet, D. Kurdyla, and T. Guilderson. 2005. Medieval warming, Little Ice Age, and European impact on the environment during the last millennium in the Lower Hudson Valley, New York, USA. *Quaternary Research* 63: 238–49.

Peteet, D., D. Pederson, D. Kurdyla, and T. Guilderson 2006a. Hudson River paleoecology from marshes. In *Hudson River fishes and their environment,* ed. John R. Waldman, Karin Limburg, and David Strayer, 113–28. American Fisheries Society Monograph.

Peteet, D., J. Schaefer, and M. Stute. 2006b. Enigmatic Eastern Laurentide ice sheet deglaciation. *EOS* 87(15): 151.

Peteet, D., J. Vogel, D. Nelson, J. Southon, R. Nickman, and L. Heusser. 1990. Younger Dryas climatic reversal in northeastern USA? AMS ages for an old problem. *Quaternary Research* 33: 219–30.

Peteet, D., J. Rayburn, K. Menking, G. Robinson, and B. Stone. 2009. Deglaciation in the Southeastern Laurentide Sector and the Hudson Valley—15,000 years of vegetational and climate history. *New York State Geological Association 2009 Field Trip Guide* 4, 4.1–4.18.

Russell, E. 2001. Three centuries of vegetational change in the Shawangunk Mountains. Report to The Nature Conservancy. Newark: Rutgers University.

Sritrairat, S., T. Kenna, and D. Peteet. 2008. Multiproxy analysis of droughts, landscape changes, sediment dynamics, and human disturbances in Hudson River marsh peat, using pollen, spores, macrofossils, LOI, and X-ray fluorescence spectroscopy, *EOS Trans. AGU* 89(53), Fall Meet. Suppl., Abstract B013B-0444.

Stuiver, M., and P. Reimer. 1993. Extended ^{14}C database and revised CALIB radiocarbon calibration program. *Radiocarbon* 35: 215–30.

Thompson, J. 1996. Vegetation survey of the Northern Shawangunk Mountains, Ulster County, New York. The Nature Conservancy, Troy, NY.

Watts, W., and T. Winter. 1966. Plant macrofossils from Kirchner Marsh, Minnesota; A paleoecological study. Geol. Soc. Am. Bulletin 77, 1339–59.

Wright, H. Jr., ed. 1983. *Late-quaternary environments of the United States.* Minneapolis: University of Minnesota Press.

———, D. Mann, and P. Glaser. 1984. Piston corers for peat and lake sediments. *Ecology* 65: 657–59.

CHAPTER 10

VEGETATION DYNAMICS IN THE NORTHERN SHAWANGUNK MOUNTAINS

The Last Three Hundred Years

John E. Thompson and Paul C. Huth

ABSTRACT

The northern Shawangunk Mountains support more than thirty-five natural communities, including three that are globally rare and eight that are rare in New York State, and forty-two state rare species. The unique combination of climate, bedrock geology, soils, and physiography of the Shawangunk landscape give rise to a remarkable array of species adapted to these conditions. Higher elevations are dominated by ridgetop pine barrens and oak forest, ravines by eastern hemlock. European settlement of the area in the eighteenth century led to farming of deeper soils for cultivation and pasture, logging in scattered woodlots, and cutting trails to access other natural resources of the area. From the mid-nineteenth to the early twentieth century, nearly all land was cleared except for inaccessible talus slopes, cliffs, and remote swamps. The debris left from tree harvesting provided fuel for intense fires that burned over the land. The low yield of this mountainous land was barely able to support the needs of the local people.

The decline of industries and agriculture concurrent with the rise of the resort industry led to less intensive land use. Resorts provided the first regular seasonal employment for locals, lessening people's reliance on cottage industries. Property values rose as well, and some local people sold out and moved on. These factors and a slackening demand for local products eventually led to expanded tree cover during the latter half of the twentieth century.

At the end of the twentieth century, two main ecological forces began driving succession at higher elevations: fire suppression and overbrowsing by white-tailed deer. The spatial arrangement of twentieth-century ecological communities is similar to the pattern of ecological communities that existed before people logged, burned, and farmed the land. Today, the "Gunks" are a world-renowned recreational resource for rock climbing and hiking and are an important economic driver for neighboring communities. Changes in forest structure and composition are affected by both past and current processes. Understanding the dynamics of this system will help land managers to make informed decisions about stewardship of this biologically rich landscape.

INTRODUCTION

The last three hundred years of vegetation change in the northern Shawangunk Mountains may be divided into four periods: European settlement, intensive extractive forest use, mountain resorts, and land conservation. European settlement in the 1700s was followed by a period of intensive use by people living in the mountains trying to scratch out a subsistence living. Coincident with the exhaustion of forest resources and changing demands outpaced by economic need, land was bought up by large resorts at Mohonk and Minnewaska Lakes. A movement of land conservation in the late twentieth century resulted in the protection of more than 12,000 ha (30,000 ac.) of summit land by New York State and private land trusts.

The Shawangunk Ridge begins in Rosendale, Ulster County, New York, strikes southwest through Sullivan and Orange Counties into New Jersey as Kittattiny Mountain, forms the Delaware Water Gap, and continues into Pennsylvania as Blue Mountain (Fig. 10.1). Northwest of the northern Shawangunk Mountains is Rondout Creek, to the east is the Shawangunk Kill, which flows into the Wallkill River, which in turn joins with Rondout Creek. The highest elevation of the northern Shawangunks is about 700 m (2,289 ft.). Minnewaska State Park Preserve, which manages 7,000 ha (18,000 ac.) of the northern Shawangunks, is bordered by the 2,700 ha Mohonk Preserve, located near the northeastern end of the ridge, and by the 2,000 ha (5,000 ac.) Sam's Point Preserve to the southwest, which is managed by The Nature Conservancy.

The northern Shawangunk Mountains have been the subject of numerous field investigations due to the unique ecological communities and many rare and endangered species found on the ridge (e.g., Britton 1883, 105; McIntosh 1959; Olsvig 1980; Selender 1980; Snyder 1981; Dirig 1994; Town et al. 1994; Abrams and Orwig 1995; Seischab and Bernard 1996; Evans et al. 2003).

Due to their rich biodiversity, the Shawangunk Mountains in 1993 were designated by The Nature Conservancy as one of the "Last Great Places" on Earth. The northern Shawangunk Mountains support forty-two New York State rare species, three globally rare ecological communities, and eight state rare communities. Forests of the Shawangunks are

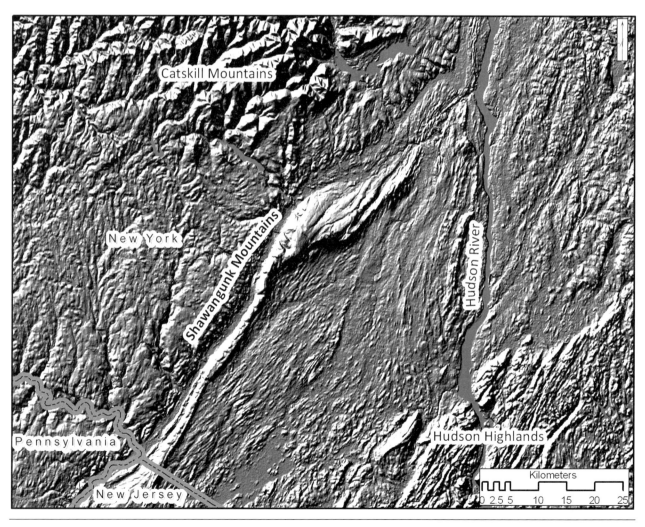

FIG. 10.1. The The Shawangunk Mountains, New York, are found between the Catskills and the Hudson River. The single ridge of the southern Shawangunks broadens to form the northern Shawangunks.

relatively unfragmented, supporting the second largest contiguous chestnut oak forest in New York State (Evans et al. 2003).

The Shawangunk Mountains are capped by the highly resistant quartz pebble conglomerate of the Shawangunk Formation (Rutstein 1976; Leeds 1989). Below the Shawangunk conglomerate, and underlying much of the southeastern slope of the ridge, is the predominantly shale Martinsburg Formation (Rutstein 1976; Leeds 1989). The surrounding valleys are formed in shales, siltstones, and limestones mantled with glacial outwash. The complex geology underlying the current plant communities is significant because it influences soil development, hydrology, and human land use, all factors that affect plant growth.

NORTHERN SHAWANGUNKS ECOLOGICAL COMMUNITIES

The Shawangunk Mountains support more than thirty-five natural communities (Fig. 10.2) (Thompson 1996). The Shawangunks boast a nearly continuous 15,400 ha (38,000 acre) patch of chestnut oak forest, the predominant forest of higher Shawangunk elevations (Evans et al. 2003). The chestnut oak forest occurs on well-drained, thin soils on both the Shawangunk conglomerate and the Martinsburg shale bedrock Pitch pine-oak-heath rocky summit and other pitch pine–dominated communities occur at the highest summits, generally surrounded by chestnut oak forest. Appalachian oak-hickory forest, beech-maple mesic forest, and successional old fields are dominant at lower elevations.

The northern Shawangunks' chestnut oak forest is dominated by chestnut oak (*Quercus montana*) and red oak (*Q. rubra*), with red maple (*Acer rubrum*) and some white oak (*Q. alba*) (Thompson 1996). American chestnut (*Castanea dentata*) sprouts may appear in the understory. Common shrubs include mountain laurel (*Kalmia latifolia*), black huckleberry (*Gaylussacia baccata*), and lowbush blueberries (*Vaccinium* spp.). Characteristic groundlayer species include Pennsylvania sedge

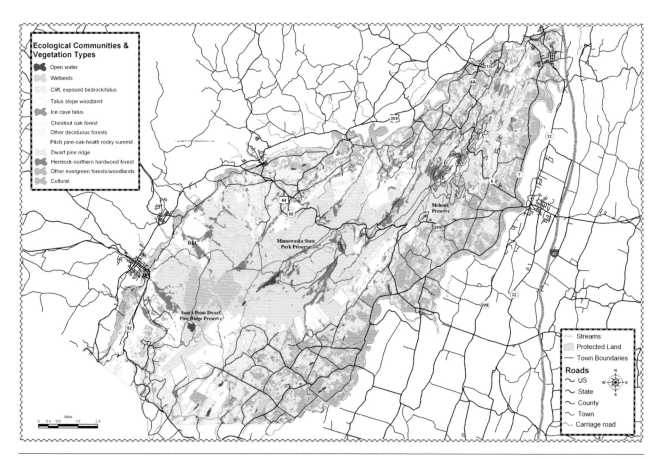

FIG. 10.2. Northern Shawangunk Mountains Ecological Communities (Thompson 1996).

(*Carex pensylvanica*), wild sarsaparilla (*Aralia nudicaulis*), wintergreen (*Gaultheria procumbens*), and pincushion moss (*Leucobryum glaucum*).

Ridgetop pine barrens communities, such as pitch pine-oak-heath rocky summit and the globally rare dwarf pine ridge, cover 2,850 ha (7,040 ac.) in the northern Shawangunks (Thompson 1996). Pitch pine-oak-heath rocky summit occurs on shallow soils over Shawangunk Conglomerate. Pitch pine (*Pinus rigida*) is the dominant tree, with black huckleberry, scrub oak (*Quercus ilicifolia*), low-bush blueberry (*Vaccinium angustifolium*), tufted hairgrass (*Deschampsia flexuosa*), sheep laurel (*Kalmia angustifolia*), mountain laurel (*Kalmia latifolia*), cow-wheat (*Melampyrum lineare*), and wintergreen occurring in the understory.

The hemlock-northern hardwood forest, covering 2,200 ha (5,400 ac.), is dominated by hemlock (*Tsuga canadensis*), with white pine (*Pinus strobus*), sugar maple (*Acer saccharum*), red maple, yellow birch (*Betula alleghaniensis*), and black birch (*B. lenta*) sharing the forest canopy.

Appalachian oak-hickory forest covers 4,400 ha (10,900 ac.) in the northern Shawangunks. The dominant trees are white oak and red oak. Associated species are hickories (*Carya ovata* and *C. glabra*), American elm (*Ulmus americana*), white ash (*Fraxinus americana*), red maple, eastern hop-hornbeam (*Ostrya virginiana*), flowering dogwood (*Cornus florida*), witch hazel (*Hamamelis viginiana*), and shadbush (*Amelanchier arborea*). Low shrubs include maple-leaf viburnum (*Viburnum acerifolium*), lowbush blueberry (*Vaccinium angustifolium*), and red raspberry (*Rubus ideus*). Groundcover may include speedwell (*Veronica officinalis*), tufted sedge (*Carex platyphylla*), and hogpeanut (*Amphicarpaea bracteata*).

Beech-maple mesic forest dominated by sugar maple, covering 2,700 ha (6,700 ac.), generally occurs on sites that were farmed in the nineteenth century.

Cliffs are vertical exposures, up to 100 m (350 ft.), of Shawangunk conglomerate. Talus is the broken rock below the cliff that was plucked by the Wisconsin Glaciation or fallen to the base from the weathering of the cliff face. Globally rare ice cave talus communities, which occur in crevices in the talus slopes below the cliffs, retain ice and snow into the late spring and early summer months.

FIG. 10.3. 1626 White Pine cored by Ed Cook of Lamont Doherty Tree Ring Laboratory. Photo by John Thompson.

Some trees of the cliff and talus areas may be described as "old growth." Inaccessible trees were less likely to be harvested. The white pine in Fig. 10.3 dates to 1626; the tree was cored by Dr. Ed Cook of the Lamont-Doherty Earth Observatory Tree Ring Lab in 1974.

EUROPEAN SETTLEMENT

The Dutch first settled fertile valleys surrounding the Shawangunks, followed by the English. One of the first land patents that included part of the Shawangunk Mountains was granted to Captain John Evans in 1694 (Russell 2001).

European settlers first moved into the Shawangunk Mountains from the Rondout Valley in the eighteenth century. Unlike in the adjacent valleys, few early mountain homes were built of stone, but were log structures (Haynes 1999). One of the first cabins was constructed of hand-hewn logs in 1771 by Adam Yaple, son of German immigrants, (L. G. Yeaple 1998).

FIG. 10.4. Enderly Sawmill, circa 1860. DSRC archives.

FIG. 10.5. Settlers lived a subsistence lifestyle. Photo c. 1920. DSRC archives.

Descendants of the Europeans settled a landscape characterized by pitch pine barrens at the highest elevations, surrounded by an oak-dominated deciduous forest, with hemlock-dominated forests in the ravines. Witness trees found in the earliest land surveys (eighteenth and early nineteenth centuries) at higher elevations show a forest of white oak, chestnut oak, American chestnut, black oak, and red oak (Russell 2001). These trees were an attractor for the locals, as more people traveled out of the valleys to explore the mountains, the value of the vast amount of standing timber became apparent.

Forests were cleared for small farms, and sawmills were quickly erected to provide lumber for building (Fig. 10.4). In 1799, when "up and down" water-powered sawmills were first established along swift flowing mountain streams, such as the Coxingkill, "plank" houses began to be constructed. This type of construction, which uses thick, vertical-sawn planks as upright supports, has been designated as "Shawangunk batten-plank frame" (Haynes 1999). The Enderly Mill was operated on the Coxingkill at least until 1920 (Haynes 1999).

Each settlement family cleared enough mostly rocky, poor land for their subsistence lifestyle, raising vegetables, including corn, potatoes, root crops (carrots, beets, and turnips), and buckwheat (Fig. 10.5). Enough land was cleared for hayfields and pasture to keep a few domestic livestock, horses, and chickens. Fuelwood was cut from nearby woodlots for home use and for sale. At the peak of the Trapps, a mountain community that had grown up, there were forty-five homes and about 150 residents. The hamlet no longer exists, although a few

FIG. 10.6. Hiram Van Leuven and his family lived in the Trapps, c. 1890. At that time, there were 50 homes and about 200 residents in a village that is now largely gone. Courtesy of Roger Van Leuven.

houses remain (Fig. 10.6) (Robert A. Larsen, personal communication).

INTENSIVE PRODUCT EXTRACTION

In 1828, the Delaware and Hudson Canal opened. The canal ran from the Hudson at Rondout to Pennsylvania and the Delaware River, paralleling the Shawangunk Ridge (LeRoy 1980). The New York Ontario and Western Railroad extended to Ellenville by 1871 (Daniel Smiley Research Center Archives). The Ellenville to Kingston spur of the O&W was constructed in 1902. Whereas the canal provided seasonal transportation, the railroad could provide year-round transport. The Wallkill Valley Railroad, built in 1870–72 from Montgomery to Kingston, paralleled the Shawangunks on the east (Mabee 1995). Both the canal and the railroads, immediately adjacent to the Shawangunks, provided an outlet for Shawangunk forest products and other raw materials.

Much of what we know about the lifestyle of the people living in the Trapps and the generations of people that inhabited the Shawangunks in the nineteenth century is due to a scientist of the twentieth century. Daniel Smiley (1907–1989) of Mohonk Lake is today regarded as one of the preeminent American naturalists of the twentieth century (Burgess 1996). Smiley saw people as living "within" and "from" the landscape and using the natural resources surrounding them for their survival needs and income. He was exposed to and documented examples of this lifestyle through interacting with local people employed at the Mohonk Mountain House, people who had, for a number of generations, lived locally off the land on mostly small subsistence farmsteads. What is clear from the records of land use made by Smiley, and from other sources presented in this chapter, the use of Shawangunk natural products by early first-generation European settlers and their descendents for home use and in the harvesting and sale of forest commodities for commercial consumption waxed and waned as demand, industries, and economies changed. This makes for a less than concise and, from today's perspective, logical progression of dates for any one type of forest product extraction. Some harvests were going on at the same times that others would be started or stopped, as influenced by price and demand. All mostly faded due to resource exhaustion or as new technologies and cheaper products (e.g., manufacture of steel and use of anthracite coal) became available.

One of the earliest forest occupations centered around the supply of hemlock and oak bark for the tanning industry, which started regionally as early as 1822, and in the Shawangunks by 1849 (Russell 2001). Straight, mature hemlocks were harvested near Plateau Path (Huth and Smiley 1985), Mossy Glen, and Palmaghatt Ravine (Pike 1892). A number of local tanneries, including the McKinstry Tannery, were in operation in Gardiner along the Shawangunk Kill; the DeGarmo Tannery in Butterville (Lefevre 1909) and in the Napanoch and Ellenville area on the west side of the mountain. Hundreds of thousands of marketable hemlock and oak trees were cut just for their tannin-rich bark, their trunks and limbs left in the forest to decay or burn. In July 1864, John Stokes lost $1,000 worth of tanbark to a 120 ha forest fire that burned up the slope west of Mohonk Lake (Partington 1970). This harvest lasted in the Mohonk area until the mid-1860s. Only the hemlocks in the most inaccessible swamps and talus slopes were spared.

One of the most intensive clearings of hardwood forest centered around the production of charcoal, locally called "chark." Thousands of cubic meters of "chark" were produced from open pit burning, which occurred extensively on the ridge. Each seven-to-ten-day burn consumed 36–54 m^3 (10–15 cords) of wood resulting in 14–18 m^3 (400–500 bushels) of charcoal (Smiley 1986).

Multiple generations of the Van Leuven family lived in the Trapps, and the family was known for their "skill in burning charcoal" (Smiley 1986). In 1943, Irv Van Leuven demonstrated charcoal production for Dan Smiley. Smiley photographed and described the stages of the process. Forty-seven cubic meters (13 cords) of gray birch and red maple were stacked to form a central chimney with vent holes (Fig. 10.7). The stack was covered with leaves and dirt and then fired. The burn yielded 18.4 m^3 of charcoal (Fig. 10.8).

At its peak, charcoal was bagged in burlap sacks and used locally for heating and cooking, some sold to the Mohonk Mountain House and some shipped via the D&H Canal to Poughkeepsie and New York City and beyond for multiple uses, including industrial (Russell 2001). Iron furnaces created a ravenous demand for charcoal. It took 240 ha (600 ac.) of wood to produce enough charcoal to keep a blast furnace fueled for one year (Polhemus and Polhemus 2005).

Many pit sites were used repeatedly into the early twentieth century. Charcoal was loaded into wagons and taken away and new wood was brought to the same site to be stacked, covered, and burned, with the resulting charcoal hauled away, then the process was repeated in the same spot. Today we find some of these pits still with bits of charcoal on the surface. Pits are round, level areas, about 10 m in diameter, with low encircling dirt berms. About fifty pits can still be found on the Mohonk Preserve in mature forest.

FIG. 10.7. Charcoal burning was a slow, meticulous process. On May 27, 1943, Irv Van Leuven stacked 13 cords of gray birch (*Betula populifolia*) and red maple to be burned. Photo taken by Daniel Smiley, DSRC archives.

FIG. 10.8. The resulting burn yielded 522 bushels of charcoal, 40.1 bushels/cord. Photo taken by Daniel Smiley, DSRC archives.

As early as 1830, hoop poles were cut and shaved to supply the extensive cooperage industry. The need for hoop poles was driven by the large demand for barrels, kegs, tubs, and firkins, and the demand in the Shawangunks lasted into the twentieth century. The cement production region, north of the Shawangunks, produced 2.25 million barrels of cement in 1886, each barrel requiring twelve hoops (Forest Commission of the State of New York 1887). American chestnut, red oak, red maple, birch, and hickory were some of the species used — each species having favored uses and prices dependent on quality and use (Forest Commission of the State of New York 1887). Stump resprouts and saplings were cut down and stripped of their branches and cut into 1.4 m (4.5 ft.) or 2.4 m (8 ft.) lengths, transported via wagon to hoop pole sheds to be split and shaved to 3.8 cm (1.5 in.) diameter in the fall and winter when other work was in short supply and the wood was easiest to work (Fig. 10.9) (Forest Commission of the State of New York 1887; Anonymous 1897a; Smiley 1986). Many farms had long sheds (hoop shops) and work was accomplished by the shaver sitting on a bench (called a "horse") and splitting each pole into two to five strips (Anonymous 1893). Shaved hoops were sorted with the hickory providing the most valuable hoops used for the finest barrels and kegs; oak hoops were used to "strap" packing boxes, mountain ash was used to bind firkins, butter tubs and sugar barrels; American chestnut was used for gunpowder kegs (Anonymous 1893). Seconds, "shorts," were used in cement barrels (Anonymous 1893). Once poles were cut on a property resprouts could be re-harvested in three to four years, (Forest Commission of the State of New York 1887).

FIG. 10.9. Irv Van Leuven displays a hoop he shaved. DSRC archives.

Mohonk Mountain House allowed, and profited from, hoop pole cutting on the area that had burned in the 1864 fire. A December 2, 1882, note from Daniel Smiley (Sr.) states, "Contracted with Case Elmendorf to cut hoop poles on land below Pine Hill this winter . . . [Elmendorf] to shave and market them handing over to us one half the receipts" (DSRC archives).

In 1887, an estimated thirty million hoops were produced from the Shawangunks in Ulster County alone, generating more income than did grain production in the county (Forest Commission of the State of New York 1887), which was remarkable, considering 80 percent of the land in 1900 was in farms (DSRC archives). In 1908, Theodore Wiklow's barrel hoop manufacturing plant in Ellenville was making more than seventy-five million hoops per year, the largest output of hoops in the country at that time (Smiley 1986).

George Davis operated a large sawmill along the Peterskill, from ca 1860 to 1879. As reported in the 1865 census, Davis owned 770 ha (1900 ac.) of land, of which 750 ha (1,850 ac.) was "unimproved" woodlot. In 1877, he contracted with the Wallkill Valley Railroad to provide 2.4 m (8 ft.) railroad ties (Anonymous 1877).

By the late 1880s, 300 ha of forest was cleared each year in town of Wawarsing alone, 3,800 m^3 (1.6 million board feet) of timber was sawed, forty-three million hoops were produced, and two hundred thousand railroad ties (Forest Commission of the State of New York 1887). Wawarsing and Ellenville provided a multitude of wood products including "chair stock," veneer, excelsior, and chestnut posts that were shipped by rail (Forest Commission of the State of New York 1887). Some of the 44,190 m^3 (12,275 cords) of cordwood cut in Wawarsing were used to supply the 3,600 to 7,200 m^3 (1,000–2,000 cords) of wood needed each year by the Ellenville glassworks (Forest Commission of the State of New York 1887).

Barrel headers were produced in Rochester and Wawarsing for use in the cement and lime industries (Forest Commission of the State of New York 1887). Local wood was used in Napanoch as raw material in axe, tobacco-knife, and rake factories and in paper mills (Forest Commission of the State of New York 1887).

ECONOMIC AND ECOLOGICAL TRANSFORMATION

A feedback developed between humans and the landscape with humans changing the structure and composition of the forest and the changed ecosystem providing products that were more quickly cropped and sold. Humans changed the output of the landscape from mature hemlocks, oaks, and chestnuts, used for tannins and timber, to short rotation coppice used for hoops and charcoal. Repeated cutting favored vigorously sprouting trees such as chestnut oak, red oak, and American chestnut. In turn, harvesters selected for trees that could resprout and grow quickly, to shorten the period between harvests. Increasing industry provided increasing demands on raw materials, including the transportation industry, which required raw materials in its construction (e.g., railroad ties) and an ever-increasing market outside the local area. Forest harvest continued until cheaper materials could be provided from other localities, or technology replaced the raw materials produced in the Shawangunks.

By 1893, iron was beginning to replace wooden straps for packing boxes (Anonymous 1893), and eventually displaced wooden barrel hoops. As Rosendale cement was displaced by Portland cement the local hoop-making industry declined and adapted. People of the Trapps made straps from white birch that were shipped to areas of citrus production to be used in binding orange boxes (Knickerbocker 1937). Though the industry slowly faded, it is not forgotten. One area on Mohonk Preserve land, where trees today demonstrate multiple trunks from the base, is still known as the Hoop Pole Lot.

The ecological communities of the Shawangunks produced much more than wood products. From the mid-nineteenth century to the 1940s "huckleberry" (blueberry) picking was a major Shawangunk industry, with as many as 350 people employed at one time in picking (Smiley 1986), some living in camps on the mountain (Fig. 10.10). Wild blueberry crops "filled dozens of wagons, each carrying 40–50 half-bushel boxes [1.4–1.6 m^3] in 1878" (Russell 2001). Pickers produced $35,000 worth of berries in 1887 and $50,000 of produce in

FIG. 10.10. During the berry season, pickers loaded dozens of wagons each carrying 40–50 of these half-bushel boxes filled with blueberries. Arthur Van Lueven and others at camp on Smiley Road c. 1915. DSRC archives.

1895 (Anonymous 1887; Anonymous 1895). Berries were shipped to Kingston and to Newburgh, and as many as three railroad cars per day were filled with berries going to New York City (Forest Commission of the State of New York 1887; Smiley 1986).

Huckleberry pickers likely set fires intentionally to promote the berry crop (Anonymous 1897b; Russell 2001). Light surface fires activated sprouting and increased productivity of lowbush blueberry plants, at the same time killing taller plants that outcompete blueberries for light. Blueberry plants are most productive in full sunlight and become less productive or fail to set fruit entirely when shaded by other plants (Chandler and Mason 1946; Hall 1955). Heaths are able to withstand fire due to belowground stems and roots that are protected from the heat of fire by mineral soil; these shrubs are able to vegetatively reproduce, also a factor in their survival of fires. Frequent burns also prevented fuel loads from building up. Berry producers would try to avoid severe fires, which can burn or kill blueberry rhizomes, causing a decline in their population and productivity (Smith 1968). Frequent, low severity fires not only increased berry production but favored other fire-adapted plants, such as mountain laurel, scrub oak, and pitch pine.

It is clear that at least by the 1890s all of the marketable timber in the Shawangunks had been exhausted. Much of the smaller, younger forest growth was intensively consumed for cordwood, hoops, and charcoal. The resulting "forest" was a shrubland of resprouts with ample sunlight on the forest floor, enough to support oak regeneration (Fig. 10.11). Though a tremendous amount of forage was available for white-tailed deer to browse, few to no deer were present to eat these sprouts.

Overharvesting, land clearing, and heavy winters had removed deer from this area by 1875. Deer virtually disappeared from the Catskill area at about

Fig. 10.11. Smoke from "chark" burning filled the valley, as oaks and other trees resprout, eventually to grow to the chestnut oak forest we have on the Mohonk Preserve today. Photo taken by geologist N. H. Darton, October 4, 1892. DSRC archives.

this time due to deep snow covered by a thick crust and the subsequent killing of large numbers of deer (Forest Commission of the State of New York 1887).

MOUNTAIN RESORT ERA

Nestled in the northern Shawangunks are high elevation "sky" lakes. Mohonk Lake, Minnewaska Lake, Awosting Lake, and Maratanza Lake each came to support either resorts or camps (Russell 2001). Spectacular natural settings of these lakes have, for more than a century, provided an attraction to the population of the New York City metropolitan area and beyond (see Flad, ch. 20 this volume). Though the surrounding mountain lands had succumbed to a combination of ax, saw, and fire, there still existed beautiful niches surrounded by cliffs and panoramic views. Mohonk Mountain House was founded by twins Albert and Alfred Smiley in 1869, when Albert purchased the Stokes Tavern on Mohonk Lake (Fig. 10.12). Alfred Smiley first saw at Mohonk Lake "the dark pines and the glittering water—and beyond it the wonderful cliffs rising from the western side of the lake. It had for him all the sensation of a discovery" (Partington 1911). Alfred later discovered the beauty of Minnewaska lake and left Mohonk to start Cliff House at Minnewaska Lake in 1879, and in 1887 opened Wildmere.

The construction and maintenance of these mountain houses along with the development of an extensive network of carriage roads and other infrastructure provided regular employment for many local families as either a supplement to their current income, or as an alternative to seasonal work.

With the soil of the small substance farms eroding and deteriorating, many woodlots had been cleared, and the profit from hoop poles, charcoal, and cordwood declining with the rise of iron and coal, it became more and more difficult for the in-

FIG. 10.12. "Lake House at Lake Mohunk," Stokes Tavern c. 1865. DSRC archives.

FIG. 10.13. "Harvesting at 10 Below" photo taken by Daniel Smiley of cordwood removal on Oakwood Drive. DSRC archives.

FIG. 10.14A. American chestnut, photo by Daniel Smiley, 1954. DSRC archives.

FIG. 10.14B. Chestnuts provided an important food source for wildlife and humans. Photo by Daniel Smiley, 1954. DSRC archives.

habitants of these mountains to eke out a living. Surrounded by willing sellers, the mountain resorts bought up land to supply the resources they needed to sustain their operations and also to protect their viewsheds (Partington 1911). Local forests supplied the mountain houses with fuelwood. Mohonk alone was cutting some 7,200 m^3/year in the early twentieth century (Fig. 10.13) (DSRC archives).

The resulting landscape of the late nineteenth century was recognized as being overharvested and awakened a conservation ethic for forest lands and wildlife. Predators such as wolves and mountain lions were extirpated along with deer and other wildlife. To protect deer, the deer hunting season was restricted in duration in 1886; the number harvested was limited to two in 1892; hounding and using jack-lights were outlawed in 1897 (Kelsey 1978). At Mohonk in 1896, due to the lack of local deer, the Mohonk Mountain House acquired nine white-tailed deer to put in a fenced paddock near the gardens so that guests could see deer for their entertainment (Anonymous 1896). In 1908–09 only a few areas were open to hunting in New York State and in 1919 the deer harvest was limited to one buck (Kelsey 1978).

One of the last major assaults on the forest was not by axe, saw, or fire, but by an imported

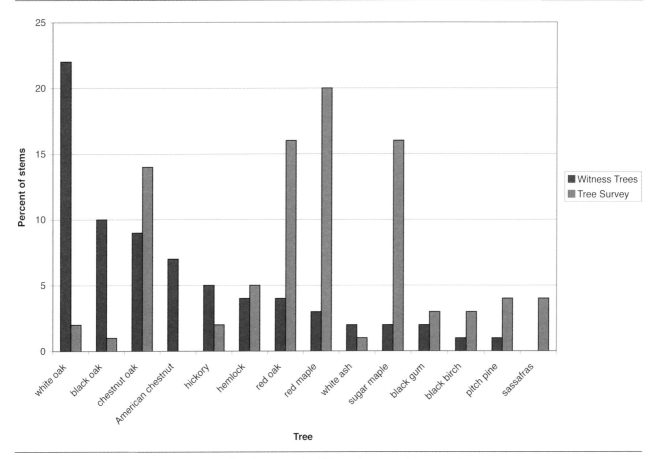

FIG. 10.15. Tree species composition in 2001 field study compared to Eighteenth century witness tree data.

disease—chestnut blight (*Cryphonectria parasitica*). The disease was first noted at Mohonk in 1912, and by 1915 had killed off all of the American chestnut trees and pole wood. American chestnut, the fourth most common precolonial forest tree in the Shawangunks, a majestic tree, providing one of the most important rot resistant hardwoods and a rich source of food for people and wildlife alike, survived merely as scattered clumps of stump sprouts and small, lanky trees (Fig. 10.14a, 10.14b).

A NEW FOREST

By the late 1920s and early 1930s, cheaper, alternative products and energy sources saw the broad-scale, intense harvesting of wood products come to an end in the Shawangunks. The resilience of harvested forest lands and large areas of abandoned farm fields was remarkable. Abandoned lands soon began overgrowing intensely and relatively quickly, for the most part, with the same species that had made up the presettlement forest, but with different species concentrations reflecting land use.

Wide-scale improvement in habitat due to forest regrowth and agricultural abandonment was a significant factor that led to an increasing deer population. Unchecked by overhunting or natural predators and slowed only by heavy winters, the deer population grew quickly. By 1951, Dan Smiley counted 150 deer at one time at Brook Farm on the Mohonk Mountain House land.

On February 26, 1963, Dan Smiley joined with other Smiley family members and friends to create the Mohonk Trust. The first land purchase of the Trust was the 200 ha (500 ac.) Trapps parcel in 1966. The Trust became the Mohonk Preserve in 1978 and today owns 2,700 ha (6,700 ac.) of land contiguous to Minnewaska State Park Preserve and Sam's Point Preserve. The "Gunks" are a world-renowned rock climbing area and a recreational destination.

The forest that recovered from the repeated cutting and fires of the late nineteenth and early

FIG. 10.16A. The Rondout Valley c. 1895, "View from Prospect Hill." Most land is in agriculture with woodlots on small ridges and trees along riparian corridors. DSRC archives.

FIG. 10.16B. The Rondout Valley in 2005, is largely forested, with few, scattered fields. Photo taken by John Thompson.

twentieth century had a different composition than the presettlement forest (Fig. 10.15). White oak, black oak, and American chestnut did not recover in abundance, while red maple, sugar maple, red oak, and chestnut oak increased (Russell 2001).

From 1948 to 1994, ridgetop pine barrens declined by 30 percent, from 4,050 ha (10,000 ac.) in 1948 to 2,800 ha in 1994 (Russell 2001). Many of the smaller barrens patches were invaded by hardwood forests and slowly shaded out. The most resistant patches are those on thinner soils near cliff lines or ledges. Fields that were not maintained succeeded to beech-maple mesic and Appalachian oak hickory forest (Fig. 10.16a, 10.16b).

Composition of the chestnut oak forest is changing over time. Lack of regeneration of the dominant canopy oaks is of concern for land managers in the Shawangunks. Red maple has become dominant in the tree and shrub layers of some areas of chestnut oak forests (Batcher 2000; Russell 2001). Gypsy moth (*Lymantria dispar*) defoliation combined with drought has killed tree oaks in some of this forest. Thousands of chestnut oak trees were observed to die after nearly 100 percent gypsy moth defoliation in the summers of 1986 and 1987, and in the spring of 1988 coincident with below normal rainfall in the 1987 and 1988 growing seasons (Huth 1989). As the Shawangunk deer population increased during the twentieth century, coincident with an era of aggressive fire suppression, forest canopy gaps were filled by red maple and associated species, such as sassafras (*Sassafras albidum*) and black gum (*Nyssa sylvatica*). As aging chestnut oak and other oaks succumb to mortality, other species will begin to dominate in these forests.

Vegetation plots placed on Mohonk Trust land in 1977 were prescribed burned in 1978 and 1979 (Thompson and Sarro 2008). Chestnut oak declined in importance from 1980 to 2003, while red maple, pitch pine, hemlock, black birch, and red oak increased. Despite its decline, chestnut oak continued to have the highest importance value in 2003. Chestnut oak saplings declined from 42 percent of the saplings in plots surveyed in 1980 to zero when resurveyed in 2003.

Tree canopy cover and competition with shrubs on the forest floor are limiting tree seedling growth by decreasing available light on the forest floor. Periodic droughts limit the number of seedlings that survive, and oak seedlings that do establish are browsed by deer and rarely grow to sapling size.

While the Shawangunk Mountains of today support large, unbroken habitat, two ecological drivers are determining the trajectory of ecological succession: preferential browsing of palatable woody and herbaceous plants by white-tailed deer and fire suppression. Deer herbivory has greatly altered and simplified the composition and structure of the ground and shrub vegetation layers and eliminated regeneration in much of the typical chestnut oak forest type. Overabundant deer degrade the vertical structural diversity of forest habitats by decreasing species diversity or by completely eliminating the shrub and sapling layer and decreasing the diversity of the ground layer. Deer-attributed changes to forest structure and composition translate directly to acute, negative effects on the diversity and abundance of forest fauna (Casey and Hein 1983; deCalesta 1994; McShea and Rappole 2000). Decline in diversity results from both the loss of ecological niches due to habitat simplification and increased exposure as habitat complexity decreases (McShea and Rappole 1997). Due to a combination of deer and lack of light on the forest floor, oak regeneration is severely limited in the Shawangunks.

To ensure a viable chestnut oak forest, it is important that chestnut oak be a major part of the canopy, the subcanopy, and the sapling layers. The decline of oaks due to gypsy moth and drought and the subsequent invasion by mesophytic hardwood species threaten the key ecological attributes that maintain the chestnut oak forest (Nowacki and Abrams 2008). An increase in red maple within the chestnut oak forest changes forest dynamics and ecological processes to a degree that will become increasingly difficult to reverse. Gypsy moth defoliation and drought are shifting dominance toward red and sugar maple. Maples produce leaf litter that decays more readily than oak litter and increases the minerals available in the soil, which may make it easier for other hardwoods to invade. The shrub layer has become dominated by mountain laurel, heaths, and shrubs that are not preferred forage for deer.

Oak forests have been the principal forest over much of eastern North America for more than six thousand years, and are documented as being important in the Shawangunks for the past six thousand years (Margraf 2003). Currently, oak forests appear to be experiencing a range-wide decline. The decline of oak dominance within oak forests may have catastrophic impacts on wildlife. McShea and others (2007) state that ninety-six bird and mammal species consume oak mast, with many of those animals dependent on acorns to sustain them through the winter. Since no other tree fills the functional niche of oaks, a large decline in oaks will have disastrous impacts on wildlife of eastern forests.

Hemlock–northern hardwood forests are declining since 1991 due to hemlock woolly adelgid (*Adelges tsugae*). The adelgid feeds on hemlock twigs at the base of the needle, causing needle drop, bud mortality, branch dieback, and eventual tree death (Orwig and Foster 1998; McClure 2001; McClure et al. 2001). The adelgid, originally from Asia, is present in most hemlock stands and many trees are declining or already dead. As the hemlock canopy opens up, more light will reach the forest floor; impacting species diversity, vegetation structure, environmental conditions, and ecosystem processes.

On April 18, 2008, a discarded cigarette started the Overlooks Fire at Minnewaska State Park Preserve. The fire burned approximately 3,100 acres (1,250 ha) and was declared contained on April 22, 2008 (Fig. 10.17). This was the largest fire in the Shawangunks since 1947 and only the second largest fire documented since 1842 (Smiley and Huth 1982). With this wildland fire, researchers have the opportunity to investigate the effects of a large-scale disturbance on an oak forest in the northern Shawangunks, which will augment ongoing prescribed burn research.

Ecological communities are dynamic, continually changing abiotic and biotic conditions. The pre-European Shawangunk landscape pattern of pine-dominated communities at the higher elevations grading into oak-dominated forests is largely the arrangement that we see today. The general recovery of these forests from the nineteenth to the twentieth century is a testament to the resilience of these natural communities, but the forests, woodlands and wetlands of today have not recovered equally. The twenty-first-century Shawangunk pitch pine barrens communities appear to be less changed than the oak forest communities (Russell 2001). Oak forests are much different in species assemblage

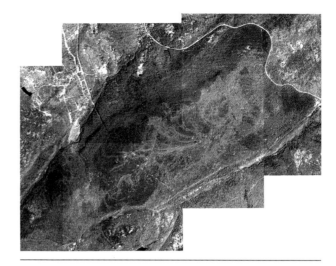

FIG. 10.17. The April 2008 Overlooks Wildfire burned mostly chestnut oak forest, pitch pine-oak-heath rocky summit and red maple-hardwood-heath forest. This was the largest fire in the Shawangunks since 1947. Natural color aerial photo taken on May 5, 2008, courtesy of New York State.

and structure than the forest at European settlement. To sustain oak forests and our biodiversity into the future, land managers in the Shawangunk Mountains, the lower Hudson Valley, and throughout the region need to implement management actions aimed at improving forest condition through thinning the canopy, prescribed burning, and reducing deer browsing. There is a great conservation challenge here with great opportunity.

REFERENCES CITED

Abrams, M. D., and D. A. Orwig. 1995. Structure, radial growth dynamics, and recent climatic variations of a 320-year-old *Pinus rigida* rock outcrop community. *Oecologia* 101: 353–60.

Anonymous. 1877. *New Paltz Independent*, June 28.

Anonymous. 1893. The hoop-pole industry. Talk with a man who has made money following it. *New York Times,* July 30.

Anonymous. 1895. Huckleberries. *Ellenville Journal,* September 20.

Anonymous. 1896. *New Paltz Independent*, July 31.

Anonymous. 1897a. A barrel hoop magnate some revelations by an old farmer about the queer business he followed successfully. A little advice thrown in. Interesting story of the way in which a shrewd farmer boy found the way to a snug fortune. *New York Times,* April 11.

Anonymous. 1897b. Huckleberry days here where New York's supply of the fruit comes from and how it is gathered. Work for 10,000 pickers interesting statistics regarding the crop that is sent to this city—battles with snakes always accompany a berrypicking campaign. *New York Times*, July 7.

Batcher, M. B. 2000. Ecological processes and natural communities of the northern Shawangunk Mountains. Research Report Prepared for The Shawangunk Ridge Biodiversity Partnership.

Britton, N. L. 1883. Notes on a Botanical Excursion to Sam's Point, Ulster Co., N. Y. Bulletin of the Torrey Botanical Club 10: 105–106.

Burgess, L. E. 1996. *Daniel Smiley of Mohonk: A naturalist's life*. Fleischmanns, NY: Purple Mountain Press.

Casey, D., and D. Hein. 1983. Effects of heavy browsing on a bird community in deciduous forest. *Journal of Wildlife Management* 47: 829–36.

Chandler, F. B., and I. C. Mason. 1946. *Blueberry weeds in Maine and their control*. Orono, ME: Maine Agricultural Experimant Station Bulletin 443.

Cook, E. R. 1976. A tree ring analysis of four tree species growing in southeastern New York State. MS thesis. University of Arizona.

deCalesta, D. S. 1994. Effect of white-tailed deer on songbirds within managed forests in Pennsylvania. *Journal of Wildlife Management* 58: 711–18.

Dirig, R. 1994. Lichens of pine barrens, dwarf pine plains, and "Ice Cave" habitats in the Shawangunk Mountains New York. *Mycotaxon* 52: 523–58.

Evans, D. J., J. W. Jaycox, and T. W. Weldy. 2003. *Rare species and ecological communities of Minnewaska State Park Preserve*. Albany: New York Natural Heritage Program.

Fagan, J. 1996. *Time and the mountain*. New Paltz, NY: Mohonk Preserve.

Forest Commission of the State of New York. 1887. Second Annual Report of the Forest Commission. Albany: Argus Company.

Hall, I. V. 1955. Floristic changes following the cutting and burning of a woodlot for blueberry production. *Canadian Journal of Agricultural Science* 35: 143–52.

Haynes, W. 1999. *Trapps Mountain Hamlet Ulster County, New York cultural resources survey*. Report to the Mohonk Preserve.

Huth, P. C. 1989. Letter to Shirley A. Briggs. Daniel Smiley Research Center Archives. Mohonk Preserve, New Paltz, NY.

Kelsey, P. M. 1978. The history of deer in New York State. *N.Y.S. Conservation Council Comments*. Daniel Smiley Research Center Archives, December: 4–5.

Knickerbocker, N. L. 1937. Minnewaska, "Frozen Waters," the intimate story of the origin and growth of the Lake Minnewaska Mountain Houses. New York: The author.

Leeds, T. 1989. Structural geology of the northern termination of the Shawangunk Mountains, Ulster County, New York. MS thesis. State University of New York-New Paltz.

LeFevre, R. 1909. *History of New Paltz, New York, and its old families (From 1678 to 1820)*. 2nd ed. Albany: Fort Orange Press.

LeRoy, E. D. 1980. *The Delaware and Hudson Canal and it's gravity railroads*. Honesdale, PA: Wayne County Historical Society.

Mabee, C. 1995. *Listen to the whistle: An anecdotal history of the Wallkill Valley Railroad in Ulster and Orange Counties, New York*. Fleischmanns, NY: Purple Mountain Press.

Margraf, E. S. 2003. Effect of climate variations and land use changes over the last 8,000 years on the vegetation at Rhododendron Swamp, Mohonk Preserve, New York. MA thesis. Columbia University.

McClure, M. S. 2001. Biological control of hemlock woolly adelgid in the eastern United States. Morgantown, West Virginia: USDA Forest Service, Forest Health Technology Enterprise Team.

———, S. M. Salom, and K. S. Shields. 2001. Hemlock woolly adelgid. Morgantown, West Virginia: USDA Forest Service, Forest Health Technology Enterprise Team.

McIntosh, R. P. 1959. Presence and cover in pitch pine-oak stands of the Shawangunk Mountains, New York. *Ecology* 40: 482–85.

McShea, W. J., W. M. Healy, P. Devers, T. Fearer, F. H. Koch, D. Stauffer, and J. Waldon. 2007. Forestry matters: Decline of oaks will impact wildlife in hardwood forests. *Journal of Wildlife Management* 71: 1717–28.

McShea, W. J., and J. H. Rappole. 1997. Herbivores and the ecology of forest understory birds. In *The science of overabundance: Deer ecology and population management*, ed. W. J. McShea, H. B. Underwood, and J. H. Rappole, 298–309. Washington, DC: Smithsonian Institution Press.

———. 2000. Managing the abundance and diversity of breeding bird populations through manipulation of deer populations. *Conservation Biology* 14: 161–70.

Nowacki, G. J., and M. D. Abrams. The demise of fire and "mesophication" of forests in the Eastern United States. *BioScience* 58: 123–38.

Olsvig, L. S. 1980. A comparative study of northeastern pine barrens vegetation. PhD thesis. Cornell University, Ithaca, NY.

Orwig, D. A., and D. R. Foster. 1998. Forest response to the introduced hemlock woolly adelgid in southern New England, USA. *Journal of the Torrey Botanical Society* 125: 60–73.

Partington, F. E. 1911. *The story of Mohonk,* 1st ed. Fulton, NY. The Morrill Press.

———. 1970. *The story of Mohonk,* 4th ed. Annadale, VA: Turnpike Press.

Pike, M. H. 1892. Shongum—III. *Garden and Forest* V: 483–84.

Polhemus, J., and R. Polhemus. 2005. *Up on Preston Mountain, the story of an American ghost town.* Fleischmanns, NY: Purple Mountain Press.

Rees, C. 1997. *Fire and Pinus rigida* rock outcrop communities of the northern Shawangunk Mountains. MS thesis. Bard College, Annandale-on-Hudson, NY.

Russell, E. W. B. 2001. *Three centuries of vegetational change in the Shawangunk Mountains.* Report to The Nature Conservancy. Newark, NJ: Rutgers University.

Rutstein, M. S. 1976. *A guide to the geologic evolution of the New Paltz area of the mid-Hudson Valley.* Unpublished manuscript. State University of New York-New Paltz.

Seischab, F. K., and J. M. Bernard. 1996. Pitch pine (Pinus rigida Mill.) communities in the Hudson Valley Region of New York. *American Midland Naturalist* 136: 42–56.

Selender, M. D. 1980. Increment borings of pitch pine (*Pinus rigida Mill.*, Pinacea) from sites on the Shawangunk Ridge and the Ramapo Mountains of southeastern New York state: Age and growth dynamics. *Skenectada* 2:1–9.

Smiley, D. 1986. *Resource industries of the Shawangunks Historical/Cultural Note No. 16.* New Paltz, NY: Mohonk Preserve.

———, and P. C. Huth. 1982. *Shawangunk Forest Fires 1842 to 1982.* New Paltz, NY: Mohonk Preserve.

Smith, D. W. 1968. Surface fires in northern Ontario. *Proceedings of the Tall Timbers Fire Ecology Conference* 8: 41–54.

Snyder, B. 1981. *The Shawangunk Mountains: A history of nature and man.* Mohonk Preserve, Inc., New Paltz, NY.

Thompson, J. 1996. Vegetation survey of the northern Shawangunk Mountains, Ulster County, New York. Troy, NY: The Nature Conservancy.

Thompson, J. E., and T. J. Sarro. 2008. *Forest change in the Mohonk Preserve: A resurvey of two vegetation studies.* Research Report. New Paltz, NY: Mohonk Preserve.

Town, W. R., M. Corey, and M. Pudiak. 1994. A preliminary report of the moss flora of the northern Shawangunk Mountains of Ulster County, New York. *Evansia* 11: 22–27.

Yeaple, L. G. 1998. *The Adam Yaple Cabin and the Clove community,* 2nd Edition. Daniel Smiley Research Center Archives.

CHAPTER 11

AGRICULTURE IN THE HUDSON BASIN SINCE 1609

Simon Litten

ABSTRACT

Since 1609 agriculture has played a central role in Hudson Basin land use. The technology and economics of farming have changed as other aspects of society evolved. Here we examine some of the changes that have fed a growing population.

SETTING THE STAGE

Agriculture, like every other aspect of human culture, is continually evolving, affected by technology, social and political forces, and economics. The story of agriculture in the Hudson Basin begins with an accelerating use of cultivated foods by the Native Americans as they sequentially adopted the agronomically and nutritionally successful "three sisters" of squash (*Cucurbita pepe*), corn (*Zea mays*), and beans (*Phaseolus vulgaris*). The system of polycropping extends back millennia in Central and South America but the complete package had been in place in New York for only three hundred years at the initiation of European contact.

Indian gardens in the Mohawk and Hudson valleys were often situated on flood plains where the soils were refreshed by springtime floods while villages were defensively located on higher ground. Woods around the villages supplied material for construction, fuel, and tools. Underbrush was burned off to reduce insects and hiding places for enemies. Villages were abandoned after five, ten, or twenty years when the people relocated, usually not a great distance away. The habit of moving villages resulted in a checkerboard landscape of cleared land and lands reverting back to forest. The Indians also modified the landscape with fires that may have encouraged nut-bearing trees—chestnut, hickory, hazelnut, butternut, walnut, and oak—which formed a significant part of their diet (Engelbrecht 2003; Hart 2008). While the Indians foraged and gardened in the Hudson Basin for thousands of years, agriculture as we know it came with European settlement. The earliest Dutch and associated people came to trade manufactured goods for pelts, but required at least subsistence agriculture to supply the foods they were accustomed to eating. Europeans quickly adopted corn, beans, and tobacco from the Native Americans. Intentionally and inadvertently imported European grasses such as bluegrass and red clover spread rapidly and turned out to be nutritious for native American animals such as elk and deer. European red clover was seen at Claverack in the Hudson Estuary by 1615 (Bidwell and Falconer 1941). Daniel Boone found great fields of bluegrass when he entered Kentucky in 1765 (Brown 2008). Honeybees were introduced to Virginia in 1622, to New England in 1638, and were seen in New York by 1670 (Crane 1999).

MALTHUS

Watershed population statistics show how the Hudson Basin has developed (Table 11.1). The Lower Hudson is the region drained by waters entering the

Hudson north of New York City and south of the Mohawk River. The Mohawk drainage includes the Schoharie system, and the Upper Hudson is drained by waters entering the Hudson north of the Troy Dam. Since 1790 the population of the Hudson Basin has grown at 3 percent per year from 270,000 to 5.6 million. At the end of the Revolution, New York City was depopulated and shattered, but it grew quickly after the 1825 opening of the Erie Canal. Other factors promoting growth included the importation of coal through the Delaware and Hudson Canal (1828) which provided energy for developing industries, water supplied by the Croton Aqueduct (1842), and influxes of Northern European immigrants in the 1840s, and Southern and Eastern Europeans around the turn of the twentieth century. After World War II New York City's population declined as people fled to the suburbs but rebounded after the crises of the 1970s (2008a; 2008b; 2008c; 2009b; 2009d). The Lower Hudson had the greatest population at the close of the Revolution but lost relatively to New York City until after World War II when suburbanization shifted people out of cities. These statistics demonstrate how little the Mohawk River region benefited from the Erie Canal.

Thomas Malthus, concerned that an arithmetically increasing food supply could not keep pace with a geometrically growing population, published his dismal observations on food and population in 1798. The French Revolution showed that widespread hunger could destabilize an ancient regime. Outside of a few instances of scarcity induced high prices, for example the 1837 bread riot in New York City, food supply has been sufficient for the residents of the Hudson Basin. The Malthusian disaster has been dodged, but how?

CLEARING THE LAND

Land clearing was a prerequisite to agriculture. When the pioneers came into the wilderness of New York they found a patchwork of clearings created by Native Americans for gardens, particularly in river bottoms and around current and former dwelling places. They also found forests so dense that the ground was in perpetual twilight. The pioneer's task of deforestation remains stunning. U. P. Hedrick writes that a "corneous-handed American son of toil," with the assistance of a "keen-edged, shining, trenchant American axe housed on its helve of American hickory, and efficiently swung," could chop, log, burn, plow, and sow ten acres of forest land in a year (Hedrick 1933). Without transportation that could bring the timber to a market, the only way to remove the wood was by burning. From a very early date, however, there was a market for potash.

Manufacture of woolen goods required that the fleeces be washed in soap to remove the animal oils. Alkali for soap had been supplied by burning kelp, from mining sesquicarbonate in the Egyptian desert, from the Mediterranean salt marsh plant barilla, or saltwort, but it could also be made from wood ashes. By 1760 England needed several thousand tons a year of potash for its wool and glass industries. Russia was its chief supplier but mer-

TABLE 11.1. Population in Millions

	1790	1800	1810	1820	1830	1840	1850	1860	1870	1880	1890	
Lower Hudson	0.18	0.19	0.21	0.24	0.28	0.33	0.39	0.47	0.54	0.57	0.61	
Mohawk	0.05	0.06	0.11	0.13	0.16	0.18	0.19	0.20	0.21	0.23	0.25	
Upper Hudson	0.01	0.05	0.07	0.08	0.10	0.11	0.13	0.15	0.16	0.17	0.18	
NYC (Manhattan)		0.03	0.06	0.10	0.12	0.20	0.31	0.52	0.81	0.94	1.21	1.52

	1900	1910	1920	1930	1940	1950	1960	1970	1980	1990	2000	
Lower Hudson	0.65	0.75	0.78	0.97	1.04	1.14	1.42	1.65	1.73	1.82	1.92	
Mohawk	0.27	0.33	0.36	0.39	0.39	0.42	0.46	0.48	0.47	0.47	0.47	
Upper Hudson	0.19	0.19	0.18	0.19	0.20	0.21	0.24	0.28	0.31	0.34	0.36	
NYC (Manhattan)		2.05	2.76	3.02	3.13	3.28	3.41	3.12	3.01	2.60	2.69	2.87

Census data from counties were multiplied by the proportion of the counties' areas in each drainage region. ArcGIS software was used to determine the proportion of counties in each of the watersheds. The proportions and county populations were multiplied and summed to obtain watershed populations. For the purposes of this chapter, county statistics were assumed to be evenly distributed geographically. Bronx County devolved in stages from Westchester County between 1874 and 1914. This detail was ignored in making the table.

cantilist authorities saw the American colonies as a potentially cheaper source and sent experts here to provide technical advice on potash production.

Homesteaders poured fresh ashes from elm, ash, sugar maple, hickory, beech, and basswood into a V-shaped trough slightly open at the bottom and lined with grass and straw to filter the ash. Water leached out the readily soluble potassium and sodium carbonates. These "black salts" were then boiled to dryness in big iron pots. Local asheries purified the black salts made by the farmers into a higher grade product suitable for export called pearl-ash (Roberts 1972). Sale of the black salt became an essential component of homesteading economy. Black salt from a single tree could pay for two acres of land. The light ash was easily transported. In 1810 pioneer developer Judge William Cooper observed, "A man who is careful of his ashes and profits by the advantage which new cleared lands afford, that of raising his forest crop without the experience of either plowing or weeding, is rather a gainer by the wood which he has cut down. . . . [A] man's profits are never greater than at the time of clearing his lands" (Hedrick 1933). Potash was an important cargo on the new Erie Canal. By 1791 other technologies began to replace the need for wood ash and after the 1870s mined potassium salts eliminated the market. The principal use of potash now is for agricultural fertilizer. Removal of potassium and other nutrients in harvested vegetable products has a deleterious effect on soil fertility, which is made up through replenishment by mined fertilizers.

Logging, mining, primitive agriculture, charcoal burning, tar and potash making, and harvesting bark for tanning increased rates of erosion in the Hudson Basin (Fox 1901); (McMartin 1994). Forest fires, particularly caused by farmers clearing land, hunters, and sparks from woodburning steam locomotives, destroyed huge swaths of timberland. Soils unprotected by forest or agriculturally tilled are far more prone to erosion. By the turn of the twentieth century sediment loads in Hudson River reached their zenith and were estimated to be twenty times background (Ayres et al. 1985; Ayres and Rod 1986) This loss of forest and soil precipitated the forest conservation movement. Replacement of charcoal with coal, of untreated wood with chemically treated wood, of lumber with concrete and steel, and firewood with fossil fuels greatly reduced the demands on forests (MacCleary 1994).

HUDSON BASIN CROPS AND THEIR EVOLUTION

In the mid-nineteenth century good agricultural statistics for the nation and New York began to be assembled. The 1865 New York Census provides a detailed look at New York society and economy at the close of the Civil War. Table 11.2 shows the proportion of land area, plowed land, and agricultural production in the three major regions of the Hudson Basin.

TABLE 11.2. Proportion of NYS Agricultural Production from the Three Major Hudson Basin Watersheds in 1864

	Lower Hudson	Mohawk	Upper Hudson
area of state	10.28%	7.13%	8.29%
land plowed, of all NYS plowed lands	10.16%	6.25%	4.42%
land plowed, of subbasin area	11.18%	9.90%	6.03%
cheese, pounds	2.51%	20.45%	1.69%
cider, barrels	18.50%	5.91%	5.19%
flax, pounds of lint	4.10%	6.27%	8.99%
liquid milk, gallons	63.60%	1.00%	1.20%
oats, bushels	9.99%	7.04%	3.95%
potatoes, bushels	9.81%	5.71%	9.07%
rye, bushels	38.54%	5.97%	7.62%
wheat, spring and winter, bushels	2.70%	1.23%	0.42%
wool, pounds shorn	5.72%	3.06%	5.01%

County-level data were partitioned among the watersheds using ArcGIS statistics as described in Table 11.1.
Source: New York Secretary of State 1867.

Significant changes in the production and consumption of important crops in the Hudson Basin have occurred. Four key Hudson Basin crops, potatoes, grain, dairy, and apples are examined in more detail. Per capita potato consumption in the United States has been relatively flat during the last century. Potatoes are now almost exclusively consumed as human food, but in the nineteenth century potato starch was used for coating yarns to increase durability in mechanical weaving. Some potatoes were fed to pigs. The Russet Burbank, a disease-resistant strain of the Burbank potato, thrives in volcanic soil in mountainous areas with high day and low night temperatures. It also needs specific soil moisture levels that can be achieved by irrigation. This potato is particularly good for freezing and frying (due to its high sugar content), and is extensively cultivated in Idaho, Washington, and Oregon (Davis 1992). The prime potato region in Idaho, the Snake River Valley, is a high desert plain that receives about eleven inches of rain per year. Large-scale irrigation projects enable intense agricultural production on desert lands where insect and fungal losses are much less than those in more humid New York.

In 1959–60 only 4 percent of the U.S. potato crop was processed into frozen products. By 1989–90, 32 percent of the crop went to making French fries. In 2006 70 percent of potatoes were consumed as "processed." Modern farming methods have increased potato yields in New York fivefold from 4,800 kg/ha in the 1880s to 25,000 kg/ha in the 2000s but in Washington State, almost eightfold (from 6,900 to 54,000 kg/ha) in the same period (2009a).

New York state led the nation in potato production until the 1940s. Refrigerated railcars and large-scale irrigation allowed raising foodstuffs thousands of miles from consumers, so western states could compete in eastern markets. Increases in potato yields on New York farms since World War II are dwarfed by gains elsewhere. New York state acreage in potatoes has been decreasing since the beginning of the twentieth century (Lucier et al. 1991). After peaking at the beginning of the twentieth century, potatoes have become a minor New York crop.

In Colonial and Revolutionary days, the Lower Hudson and the Mohawk Valley were very important wheat-growing districts. The Revolution was particularly devastating to wheat production in the Mohawk Valley where about one-third of the prewar white population was killed and half the buildings were burned. In addition, the "Hessian fly" an invasive species of gall midge (*Mayetiola destructor*) was introduced, probably in animal bedding brought from Europe by British or mercenary forces. It was first seen in New York City but rapidly infested the northeast, especially the already devastated Mohawk Valley (Pauly 2002). The Hessian fly is still a highly destructive pest to wheat farmers in the midwest. Mohawk valley farmers substituted rye, which was not a target of the Hessian fly, for wheat. By the mid-nineteenth century, wheat was a minor crop in the Hudson Basin. Now New Yorkers primarily get their bread from wheat grown in places better suited to cultivation with modern machinery.

New York corn acreage (currently around 3 percent of the state's area) is less than the national average of about 5 percent. Nationally, nonfood uses for corn, chiefly for animal feed, predominate. These also include industrial starch and ethanol. Since 1927 only 2 to 11 percent of the US domestic corn supply has gone to human food. In 2007 Americans consumed 13.4 kg per capita of corn as cereal and 50 kg of corn converted to sweeteners, mostly high fructose corn syrup.

Corn yields remained virtually flat from the beginning of organized USDA statistics in 1866 until 1900. Between 1900 and 1950 low but positive increases (5.2 to 10 kg/ha/year) in corn yield occurred. After World War II continuous improvement in fertilizers, crop strains, and pest management have produced a steady rise in yield (kg/ha/year) of 126 (California) and 68 (New York) over the fifty-seven years between 1951 and 2008.

Corn is principally raised in the lower Hudson and Mohawk regions of the Hudson Basin. Of this, 84 percent is for grain and 16 percent is for silage. Modern corn production rates in the Hudson Basin are about 127,000 tonnes/year. A spike in 1998 (21 million kg) was in anticipation of ethanol production (2005a).

Rye was formerly used to make bread, beer, and tough rye straw used for animal bedding. Nonfood uses have often dominated rye consumption. Until the late nineteenth century rye straw was still grown in Manhattan for use in the city's stables. The decrease in rye production mirrors the replacement of horses with motor vehicles and, perhaps, Prohibition.

At the beginning of the nineteenth century paper makers experimented with an array of materials ranging from mummies to manure as replacements for rags. Success came in 1854 when straw, principally from rye, could be economically used as fiber for paper manufacture. Columbia, Rensselaer, and Ulster counties were early centers for rye straw paper making. This industry provided an important market for local farmers, particularly those with marginal land less suitable for other crops (Munsell 1870; Weeks 1916; Smith 1997). After 1859 improvements in the process, straw paper became common. The silicious character of the straw gave straw paper a hard, brittle surface, which wore out type about twice as fast as the older softer surfaced paper, and it was less durable than rag paper, yet nearly all newspapers were printed on it until after the Civil War when wood pulp began to be used (Munsell 1870). A peculiar application of straw paper was the manufacture, in Hudson, New York, of paper-cored Pullman railroad car wheels. The paper core damped vibration and noise (Wright 1992). From the 1860s until the end of the nineteenth century, E. Waters & Sons in Troy, New York, manufactured canoes and architectural elements made from paper. In the last decades of the nineteenth century, New York produced around 76,000 tonnes of rye per year but in the 1920s production fell to 14,000 tonnes; it is now around 7,600 tonnes.

Dairy is the largest agricultural activity in the Hudson Basin. New York is the nation's third-largest dairy producer, falling behind California and Wisconsin. However, dairy production in New York has been relatively flat in comparison with California, Idaho, and New Mexico. A USDA report shows that northeastern dairy farms are smaller, less efficient, and operated by people with less education than western farms. Northeastern operators are less likely to engage in forward purchases of feed to lock in favorable prices and more interested in leaving the industry (Blayney 2002). As with other commodities, milk production has shown phenomenal and continuous gain since World War II (Table 11.3). This has been achieved through a variety of changes in genetics, nutrition, antibiotics, and other means. Monsanto's recombinant bovine somatotropin, also called bovine growth hormone, first became available in 1994 but milk yields had begun increasing before then and continue to do so.

TABLE 11.3. Milk Production per Cow in California, New Mexico, New York, and Wisconsin

kg/year/cow increase	California	New Mexico	New York	Wisconsin
1924–1949	21	16	15	18
1950–1975	95	120	78	62
1976–2008	102	118	101	93
kg/year/cow				
1924	2,190	1,067	1,902	1,969
2008	8,334	8,679	7,407	7,291

In the early 1920s New York had 1.4 million cows producing 3.2 million tonnes of milk; by the late 2000s New York's 0.64 million cows produced 5.5 million tonnes of milk. Dairying is undergoing concentration: in 1940 the average dairy herd was five animals but by 2000 the number of dairies had shrunk and the number of cows was up to eighty-eight per farm (Blayney 2002).

Increased operating efficiency of distant farms would not necessarily be significant if milk required rapid transport from farm to consumer, but the long-term trends are away from fresh milk and toward cheeses. In 1909 the annual U.S. per capita consumption of fresh milk was 129 liters but it has fallen to less than 87 liters in 2001. Cheese consumption, 1.4 kg per capita per year in 1909, has risen to 11 kg per capita per year in 2001. Milk is being replaced in people's diets by soda and bottled water. More than half of cheese now comes in mass-produced foods such as fast food sandwiches and packaged snack foods (Putnam and Allshouse 2003).

Per capita butter consumption has decreased due in large part to substitutes (margarine) available to the home cook as well as to substitutes used in industrial and commercial food preparation. Cheese consumption increased rapidly after World War II. Cheese production was encouraged by the Agriculture Act of 1949, which permitted the government to buy milk (as cheese, nonfat dry milk, or butter) to support prices.

Apples were nearly ubiquitous on New York farms by 1774. Apples, used primarily in hard cider, were a principal Erie Canal cargo (Cohen 1992). By the end of the nineteenth century New York apples were outcompeting English apples in the London market (Large 1962). New York state has 17,000 ha of apple orchards. Second in the nation (behind Washington—producing 57% of the U.S. crop), New York produces 10 percent of the country's apples. Apple yields in New York lag behind those from Washington, Oregon, and Idaho (2005c). Since 1959 the slope of yield growth has been 1.7 times greater in Washington state than in New York.

Consumption of fresh apples has been relatively constant (about 6.7 kg per year per capita) but apple juice and highly processed apple products have increased at the rate of 0.21 kg per year since 1970.

TABLE 11.4. Apple Yields in Metric Tons/ha, New York and Washington

	1880s	1890s	1900s	1910s	1920s	1930s	1940s
New York	4.8	5.1	5.1	5.6	6.3	7.0	10.6
Washington	6.9	7.3	7.1	6.8	9.0	9.7	15.3
	1950s	1960s	1970s	1980s	1990s	2000s	
New York	18.0	22.3	23.5	23.2	25.0	25.0	
Washington	23.5	31.3	40.1	47.6	51.7	53.9	

Per capita consumption of juice rose from 2.2 kg per year in 1970 to 10 in 2007 (2009c). This trend again encourages competition from distant sites where production is cheaper. Chinese apples comprise almost half the world's crop and are exported as juice concentrate halfway around the word (2004). U.S. apple exports are rapidly decreasing in response to Chinese and European competition. New York apple farms are decreasing; since 1959 almost one-quarter of the apple-bearing acreage was released to other uses.

NUTRIENTS

Under subsistence conditions, when production and consumption of agricultural foodstuffs was concentrated in space, nutrients remained relatively close to the site of growing. However, as labor-saving machinery became more prevalent and efficient, fewer people and, after the adoption of tractors, fewer animals were required on the farm. Labor was released to work at sites remote from farms. Table 11.5 illustrates this.

TABLE 11.5. Proportion of Regional Population Engaged in Agriculture

Census year	Lower Hudson	Mohawk	Upper Hudson
1820	10.99%	18.45%	18.53%
1840	11.10%	21.18%	21.59%
2007	1.57%	0.49%	0.53%

(2009b; USDA NASS New York Field Office 2009)

Nutrients are lost as agricultural products are moved off farms and as soil is washed away by erosion. Soil exhaustion and the benefits of manure have long been known. In the first half of the nineteenth century, a scientific basis for this began to emerge through the work of Nicolas-Théodore de Saussure in Switzerland (1804), Phillipp Carl Sprengel (1831 and 1832), and Justus von Liebig (1840 and 1855) (van der Ploeg, Bohm, and M. J. Kirkham 1999). Plant growth could be enhanced and "worn out" soils replenished through the addition of mineral salts. Liebig popularized the "law of the minimum," which stated that plant growth is always limited by one nutrient. If that nutrient were supplied in greater amounts, some other nutrient would then become limiting. Progressive farmers used animal manure, street sweepings containing horse manure, ground-up bison bones, fish and seaweed, "poudrette" (night soil mixed with charcoal or gypsum), leached ashes, or swamp muck. By the 1840s South American guano was being sold worldwide. In the 1860s mineral potash was discovered in Germany and in 1911 the Harber-Bosch process made anhydrous ammonia from natural gas and air with catalysts under high temperatures and pressures. The process uses 938 cubic meters of natural gas to make a tonne of anhydrous ammonia. Today 20 percent of the energy used in agriculture goes to the manufacture of fertilizer—mostly nitrogen (U.S. Environmental Protection Agency 2008). Intensification of fertilizer application after World War II coincided with dramatic increases in crop yields. It also led to increased eutrophication of water bodies from farm and suburban yard runoff. Nutrients originating on the farm are a source of pollution when flushed into urban waters. The natural cycle of consumption and reuse has been broken and urban sewage must be managed at public expense.

TRACTORS

The nineteenth century was a great era of agricultural gadgets. Hopeful inventors filed patents for improved plow blades, corn shuckers, apple corers, manure spreaders, scythes, reapers, and the like. Most of these devices suffered from limitations in the amount and form of available power. Animals of various sizes were placed on treadmills or walked in a circle, either attached to a lever or threshing grain with their feet. Power for almost any function in the field, beyond simply moving forward, required some kind of gearing off a wheel in contact with the ground. Horses or oxen pulled almost all large farming equipment. This is a distinct disadvantage in harvesting standing crops such as wheat or corn. Modern farm tractors have power take-offs (PTOs) that can deliver hundreds of rotary horsepower to operate mobile machinery.

The introduction of tractors in the twentieth century changed the necessary skills of farmers and increased the area of land that a single farmer could manage. In 1915, 38 million hectares of U.S. cropland, roughly 22 percent of the total, went to feeding horses and mules; 85 percent of that area went to feed work animals on farms. The first practical tractor, the Bull, appeared in 1913; the 1917 Ford Fordson was the first mass-produced tractor; and the 1924 McCormick-Deering Farmall was the first tractor capable of operating between growing crops and among the first to have a PTO. The amount of power these tractors brought into the field enabled an explosion of associated farm machinery (Olmstead and Rhode 2001). Tractors reduced the labor necessary to farm an acre of ground. Since one man could now do the labor of several, the size of the farm that a family could manage increased.

Table 11.6. Power Available to Mid-Atlantic Farmers (million joules/sec), 1910–1960

	1910	1920	1930	1940	1950	1960
animal	859	823	518	498	238	96
tractor		74	656	1,265	2,810	4,019
total	859	897	1,174	1,762	3,048	4,114

SOIL

The agriculture typical of colonial and modern farming in the Hudson Basin is machine-based monoculture. The basic machine is the plow, pulled by animals or tractors. The plow assists in pulverizing the soil to a size suitable for the seed, eases new root growth by loosening the soil, buries weeds and crop residue to prepare the seed bed and retard weed growth, and creates furrows or ridges desired for specific crops such as potatoes. By stripping off protective vegetation, plowing hastens wind and water erosion. Exposing more soil to the atmosphere, oxidizes organic matter in the soil and releases significant amounts of carbon dioxide. It has been estimated that U.S. soils have, since Contact, lost 30–50 percent of soil organic carbon (Lal 2004).

Soil management improvements have numerous benefits including enhanced biodiversity, reduced pesticide use and topsoil runoff, and more sustainable agriculture. "Conservation tillage" encompasses a variety of techniques that reduce soil disruption by plowing, disking, and cultivation. No-till is a conservation tillage practice that dispenses with the plow. Instead of mechanically turning over the soil and cutting or burying weeds, no-till uses specialized seed drills that penetrate stubble and a variety of pesticides such as 2,4-D, Atrazine, and Paraquat to control weeds. The stubble left on the ground helps retain moisture and reduces the impact of rain on the soil. The structure of the soil is less disrupted encouraging earthworms (another invasive species) and decreasing soil oxidation. The nutrients and carbon in the stubble return to the earth on site. Elimination of plowing, disking, and cultivation significantly reduces labor, fuel, and machinery costs. Despite these advantages, no-till has the drawbacks of retarding soil warming, relying on high use of pesticides, and enhancing fungal pests due to higher moisture content.

FARMS

Loss of agricultural land in the Hudson Basin was rapid from 1940 until 1970 when it slowed. The ways farmers use their lands have also changed. Area devoted to crops increased at the expense of woodland and permanent pasture, reducing habitat and diversity on farms. Between 1940 and 2003 the Hudson Basin has lost 850,000 hectares of farmland, releasing land from farming that could revert to natural succession or be a site for development and sprawl. In 1940, the average Hudson Basin farm was 43.7 hectares; by 2003 it had grown to 72.5 hectares.

Between 1910 and 2008, the number of New York farms decreased almost sixfold; nationally the number of farms decreased almost threefold. Farm size has grown since 1950 but the rate of growth has declined. New York farms are smaller than the national average.

The total amount of land farmed in New York has almost halved since the early 1950s (26.9 million hectares to 14.7 million hectares). This may reflect competing uses for land where agricultural economics are weak more than technological drivers such as tractorization, which was largely accomplished by the early 1950s . The average age of New York's farmers has risen from 50.1 in 1978 to 56.2

in 2007. Since more than half of the New York state dairy farmers have indicated that they would like to leave the business in the next ten years, the loss of farmland may soon accelerate.

TABLE 11.7. Changes in Hudson Basin Farm Area (2005b)

	Lower Hudson	Mohawk	Upper Hudson
1940, farm hectares	621,980	460,180	311,289
2003, farm hectares	212,410	207,573	102,533
hectares lost each year			
cropland	3,435	1,878	1,336
pasture	1,111	1,260	527
woodland	1,617	748	1,016
TOTAL	6,286	3,981	2,942

Statistics from the 2007 Agricultural Census show that in five of the twenty-three Hudson Basin counties (Dutchess, Essex, Schenectady, Westchester, and Orange) farmers spend more to produce crops than they make from selling them. Land held in uneconomic farms is less likely to be sold as farms, and farming it is less attractive to the next generation. Table 11.8 shows return on investment (ROI—value of farm products sold minus the cost of producing them divided by the capital invested in land, building, and machinery). The Lower Hudson has 78 percent of the state's agricultural capital and 78 percent of its agricultural profits (USDA NASS New York Field Office 2009).

TABLE 11.8. New York Agricultural Economic Statistics (Dollars in Millions)

	NYS	Lower Hudson	Mohawk	Upper Hudson	Outside Basin
profit	$83.7	$62.0	$3.1	$2.2	$18.6
capital	$1,928.1	$1,496.9	$92.0	$138.9	$339.2
ROI	4.34%	4.14%	3.34%	1.55%	5.49%

PESTICIDES

Pesticides are used to decrease losses from attack by insects, fungi, and nematodes, and from competition with other plants. They change the way farming is practiced by allowing denser plantings or easier cultivation and harvesting. They enhance cosmetic qualities and improve storage. Heavy metal–based fungicides and insecticides appeared in the mid-nineteenth century in response to the *Phytophthora infestans* epidemic that destroyed the Irish potato crop and a million Irish human lives. Lead arsenate (LA) was first extensively used in 1892 to control gypsy moths. LA was superior to the pigment/pesticide Paris green (copper acetoarsenate) introduced in 1867. Paris green was a potent insecticide but it is also toxic to plants. Lead arsenate lacked Paris green's plant toxicity and adhered very well to vegetation, making it a long-lasting agent. Lead arsenate was used extensively on fruits, particularly apples, but also on garden crops, turf grasses, and elsewhere. It accumulated in soils to such an extent that significant lead and arsenic contamination occurs in some apple orchards. LA was also incompletely removed from fruits by washing. DDT made LA obsolete in the late 1940s. New York banned DDT in 1970 (the EPA followed, banning DDT in 1972) and EPA banned LA in 1988 (Peryea 1998).

DDT, famously effective in halting the spread of typhus in Naples in late 1943, was extensively used in the Hudson Basin. Newspaper reports in the mid 1940s, soon after DDT became available to civilians, describe extensive spraying for nuisance insect control in the New York City area and in the Hackensack Meadowlands. High DDT concentrations persist in sediments in the Wallkill River and in the Arthur Kill near Staten Island. The largest use of DDT in New York state occurred in 1957 during an massive but unsuccessful USDA aerial spraying campaign to eradicate Eurasian gypsy moths (Purdue Extension 2005). Target insect populations quickly became resistant to DDT, which is also a potent hormone disruptor to many nontarget species.

The largest amounts of pesticides applied to potatoes are "other pesticides." This class includes chlorpropham and thiabendazole (sprout inhibitors), dichloropropene (to control nematodes), and metam-sodium (a fungicide). Sulfuric acid is heavily used to kill off potato shoots for ease of harvesting (Murphy 1997). Historically, corn was planted in hills of three or more plants or in check rows, which allowed farmers to cultivate the corn in two directions for weed control. Effective herbicides let farmers switch from hill planting to drilled, narrow-row planting. The plant density increased from 25,000–30,000 plants per ha to 62,000–74,000 plants per ha. High-yield hybrids tolerate high population densities. Herbicides also allow corn to be planted earlier in the growing season resulting in a higher yield potential for the crop. Before herbicides, corn had to be planted later so that the first flushes of weeds could be killed with tillage. The development of soil-applied insecticides also allows farmers to grow corn for multiple years and increases productivity on an area-wide basis (Committee on the Future Role of Pesticides in U.S. Agriculture 2000).

Pesticide use for apples, which is very intensive, has been decreasing through an increased use of integrated pest management or IPM (Cornell Cooperative Extension 2009). IPM practices include a wide range of agronomic techniques such as site, rootstock, and planting systems, soil management and irrigation, tree training, pest monitoring and forecasting, weed management, disease management, vertebrate management, and proper harvest and storage methods. Consistent application of sound agronomic principles reduces the need for expensive and dangerous chemicals.

Pesticides play an important role in modern agriculture, but their use was excoriated by Rachael Carson in her landmark 1962 book *Silent Spring*. She reserved her harshest criticism for massive and arrogant campaigns that were often ineffective against relatively inoffensive species. She describes in excruciating detail the destruction of wildlife in unsuccessful attempts to control Japanese beetles and the role that pesticide manufacturers played in elevating fire ants to being seen as a significant threat to human and animal well-being. Damaging campaigns to stop the spread of Dutch elm disease were unsuccessful. Carson wrote at a time of increasing public anger over atomic testing (Kopp 1979; Lutts 1985). The first substance mentioned in *Silent Spring* was the atomic bomb fallout product strontium-90. She was part of a wave of growing skepticism of authority that became a hallmark of the later 1960s. In the Hudson Basin this was epitomized by reaction to Consolidated Edison's 1962 plan to convert the Storm King scenic landmark into a pumped storage facility.

yields and reductions in operating expenses relative to competitors. For many foodstuffs it is almost irrelevant where in the world production occurs. Consumption patterns are changing with more food manufactured instead of being prepared at home, thus Americans are eating more frozen French fries and fewer fresh potatoes, more cheese and less milk, more high fructose corn syrup and less corn as grain, and more reconstituted apple juice concentrate. Manufacturers of snacks and operators of chain restaurants seek out minimum costs for high calorie foods with long shelf lives.

The long range history of U.S. agriculture shows that food demand was met from Colonial times until the beginning of the twentieth century primarily through the farming of more land. Incremental improvements in techniques and in machinery led to farming requiring fewer people. As transportation improved, food could be moved longer distances. The introduction of gasoline-powered tractors had the effect of greatly reducing labor and also freeing up very large amounts of land previously needed to support draft animals. Following World War II, science started a remarkable rise in yields per unit area. These yields continue to rise for many crops. The effect of this rise in efficiency has meant that more food can be raised on less land. The methods behind this continuous growth in yield include heavy applications of fertilizer, particularly energy-rich nitrogen, use of a wide variety of insecticides, herbicides, and fungicides, and land-use practices that reduce diversity and beauty. Soils have been eroded, stripped of nutrients, and oxidized. Consumer preferences are moving away from consumption of fresh locally grown food toward mass-produced prepackaged foods made from inexpensively produced components where the true costs are often hidden. Ultimately, these trends have given us generations that have not seen hunger. They have profoundly altered the economics of farming such that agriculture is declining in the Hudson Basin.

FEEDBACK

A combination of factors reduces the competitiveness of New York farms. Success in agricultural commodities requires continual improvements in

REFERENCES CITED

2004. Apple update. Available from http://www.fas.usda.gov/htp/horticulture/Apples/Apple%20Update%20-%20December%202004.pdf.

2005a. New York NASS. Available from http://www.nass.usda.gov/ny/.
2005b New York NASS. Available from http://www.nass.usda.gov/ny/.
2005c. Noncitrus fruits and nuts. Available from http://usda.mannlib.cornell.edu/reports/nassr/NoncFruiNu/2000s/2005/NoncFruiNu-07-06-2005.pdf.
2008a. Census '90. Available from http://www.census.gov/main/www/cen1990.html.
2008b. Census of population and housing: 1970 census. Available from http://www.census.gov/prod/www/abs/decennial/1970.htm.
2008c. Census of population and housing: 1980 census. Available from http://www.census.gov/prod/www/abs/decennial/1980.htm.
2009a. Economic Research Service: Data sets. Available from http://www.ers.usda.gov/Data/FoodConsumption/FoodAvailSpreadsheets.htm.
2009b. Historical census browser; University of Virginia Library. Available from http://fisher.lib.virginia.edu/collections/stats/histcensus/php/state.php.
2009c. National Agricultural Statistics Service. Available from http://www.nass.usda.gov/Data_and_Statistics/Quick_Stats/index.asp.
2009d. United States Census 2000. Available from http://www.census.gov/main/www/cen2000.html.
Ayres, R. U., L. W. Ayres, J. McCurley, M. Small, J. A. Tarr, and R. C. Widgery. 1985. *An historical reconstruction of major pollutant levels in the Hudson-Raritan Basin 1880–1980*. Pittsburgh: Variflex Corp.
Ayres, R. U., and S. R. Rod. 1986. Patterns of pollution in the Hudson-Raritan basin. *Environment*, 28: 14–43.
Bidwell, P. W., and J. I. Falconer. 1941. *History of agriculture in the northern United States: 1620–1860*. New York: Peter Smith.
Blayney, D. P. 2002. The changing landscape of U.S. milk production. Available from http://www.ers.usda.gov/Publications/sb978/.
Brown, M. M. 2008. *Frontiersman: Daniel Boone and the making of America*. Southern Biography Series. Baton Rouge: Louisiana State University Press.
Cohen, D. W. 1992. *The Dutch-American farm*. New York: New York University Press.
Committee on the Future Role of Pesticides in U.S. Agriculture. 2000. *The future role of pesticides in U.S. agriculture*. Washington, DC: National Academies Press.
Cornell Cooperative Extension. 2009. Elements of IPM for apples in New York State.
Crane, E. 1999. *The world history of beekeeping and honey hunting*. New York: Routledge.
Davis, J. W. 1992. *Aristocrat in Burlap: A History of the Potato in Idaho*. Eagle, ID: Idaho Potato Commission.
Engelbrecht, W. 2003. *Iroquoia*. Syracuse: Syracuse University Press.
Fox, W. F. 1901. History of the lumber industry in New York. In *Annual Report of New York State Forest, Fish and Game Commission for 1900*. Albany: James B. Lyon, State Printer.
Hart, J. 2008. Evolving the three sisters: the changing histories of maize, bean, and squash in New York and the greater northeast. In *Current Northeast Paeloethnobotany II*, ed. John P. Hart. New York State Museum Bulletin 512. Albany: The University of the State of New York, The State Education Department.
Hedrick, U. P. 1933. *Agriculture in the state of New York*. Albany: J. B. Lyon.
Kopp, C. 1979. The origins of the American scientific debate over fallout hazards. *Social Studies of Science* 9(4): 403–22.
Lal, R. 2004. Soil carbon sequestration impacts on global climate change and food security. *Science* 304(5677): 1623–27.
Large, E. C. 1962. *The advance of the fungi*. New York: Dover.
Lucier, G, A. Budge, C. Plummer, and C. Spurgeon. 1991. U.S. Potato statistics 1949–89. Statistical Bulletin No. 829. U.S. Department of Agriculture. Commodity Economics Division.
Lutts, R. H. 1985. Chemical fallout: Rachel Carson's *Silent Spring,* radioactive fallout, and the environmental movement. *Environmental Review* 9(3): 210–25.
MacCleary, D. W. 1994. *American forests: A history of resiliency and recovery*. Durham, NC: Forest History Society.
McMartin, B. 1994. *The great forest of the Adirondacks*. Utica: North Country Books.
Munsell, Joel. 1870. *A chronology of paper and paper-making*. Albany: J. Munsell.

Murphy, K. 1997. Sustainable potato production: innovative cropping systems can replace hazardous pesticides. *Journal of Pesticide Reform* 17(4): 2–7.

New York Secretary of State. 1867. *Census of the State of New York for 1865.* Albany: C. Van Benthuysen and Sons.

Olmstead, A. L., and P. W. Rhode. 2001. Reshaping the landscape: The impact and diffusion of the tractor in American agriculture, 1910–1960. *The Journal of Economic History* 61(3): 663–98.

Pauly, P. J. 2002. Fighting the Hessian Fly: American and British responses to insect invasion; 1776–1789. *Environmental History* 7(3): 485–507.

Peryea, F. J. 1998. Historical use of lead arsenate insecticides, resulting soil contamination, and implications for soil remediation. Available from http://soils.tfrec.wsu.edu/leadhistory.htm.

Purdue Extension. 2005. A brief history of gypsy moth in North America. Available from http://www.entm.purdue.edu/GM/history/HistOfGypsyDoc.htm.

Putnam, J., and J. Allshouse. 2003. Trends in U.S. per capita consumption of dairy products, 1909 to 2001. Available from http://www.ers.usda.gov/Amberwaves/June03/DataFeature.

Roberts, W. I. 1972. American potash manufacture before the American Revolution. *Proceedings of the American Philosophical Society* 116(5): 383–95.

Smith, M. 1997. *The U.S. paper industry and sustainable production: An argument for restructuring urban and industrial environments.* Cambridge: The MIT Press.

U.S. Environmental Protection Agency. 2008. Report on the environment: Fertilizer applied for agricultural purposes. Available from http://cfpub.epa.gov/eroe/index.cfm?fuseaction=detail.viewMidImg&ch=48&lShowInd=0&subtop=228&lv=list.listByChapter&r=201565#10309.

USDA NASS New York Field Office. 2009. 2007 census publications. Available from http://www.agcensus.usda.gov/Publications/2007/Full_Report/index.asp.

van der Ploeg, W. Bohm, and M. J. Kirkham. 1999. On the origin of the theory of mineral Nutrition of plants and the law of the minimum. *Soil Science Society of America Journal* 63: 1055–62.

Weeks, L. H. 1916. *A history of paper-manufacturing in the United States, 1690–1916.* New York: The Lockwood Trade Journal Company.

Wright, H. E. 1992. George Pullman and the Allen paper car wheel. *Technology and Culture* 33(4): 757–68.

CHAPTER 12

Ecology in the Field of Time

Two Centuries of Interaction between Agriculture and Native Species in Columbia County, New York

Conrad Vispo and Claudia Knab-Vispo

ABSTRACT

This chapter summarizes the history of agriculture's influence on the inland habitats of native plants and animals in an east bank Hudson Valley county. We follow agriculture in Columbia County, New York, since the early nineteenth century, highlighting the land covers created by farming and describing how these evolved as agriculture changed. Certain native species gained or lost habitat as land cover changed, and we use geospatial analysis of historical census data together with historical natural history accounts and our own recent fieldwork to depict the county's changing natural history. Understanding this interaction of agriculture with habitats for native species will be important if future efforts to meld agriculture and nature conservation are to be successful.

INTRODUCTION

This chapter traces the last two hundred years of agricultural history in Columbia County, New York, from the perspective of its implications for the ecology of native species. Farmland (both "improved" and "unimproved"; improved referred to land that was opened and actively used for farming) once covered nearly 90 percent of Columbia County's surface area; today, it accounts for less than 30 percent. The influence of that land cover change on native species, and on Hudson River sedimentation (Peteet et al., ch. 9 in this volume; Pederson et al. 2005), has been dramatic. Using geographically specific historical research and present-day observation, we describe the county's agricultural changes and their ecological consequences.

We use the concept of "ecological analogy" as a tool in our description. For our purposes, an ecological analogy occurs when human activities create habitat for a given species not by completely restoring that species' original habitat but by creating a new habitat that is sufficiently similar (i.e., offers enough analogies) so as to function. An example would be a mature, northeastern hayfield that, while sharing almost no plants with a Midwestern prairie, offers enough structural similarities so as to provide nesting habitat for certain birds whose demographic heartland was (and in some cases still is) the prairies. In this case, we would say such hayfields are "analogous" to prairies *from the perspective of these birds*. It is important to note that such analogies are almost never complete and that, while these new habitats may serve some native species, they likely exclude others.

After a brief introduction to the county, we explore three overlapping stages in the county's agricultural history and the associated consequences for its nature. We conclude with a brief consideration of the net effects and of the current forces influencing the ecology of native species. This chapter is not exhaustive. It focuses on changes in terrestrial cover types; additional factors such as exotic species (Teale, ch. 13 in this volume) and agrochemicals have also had pronounced ecological influences but are not considered here.

BACKGROUND AND CONTEXT

Columbia County borders the Hudson River to the west and Massachusetts and Connecticut to the east. Its elevation varies from less than 10 m along the Hudson to nearly 700 m in the Taconic Hills along its eastern edge (Fig. 12.1). The length of the growing season varies by about three weeks from the southwest corner of the county to the northeast corner. The county's 166,700 hectares can be roughly halved into a western Hudson Valley region and an eastern hill region. Limestones and dolomite formations that underlie parts of Ancram, Copake, Hillsdale, New Lebanon, Canaan, and Greenport substantially influence wild and cultivated plants. Biogeographically, Columbia County is in a "tension zone" (*sensu* Curtis 1959; Cogbill et al. 2002). This means that it harbors a mix of more boreal and more southerly species.

During the time frame covered here (ca. 1820–2009), Columbia County agricultural production evolved (Fig. 12.2), and its population grew from around thirty thousand to around sixty thousand, albeit with a dip in numbers between about 1870 and 1920. As has occurred throughout much of the Northeast, the extent of the county's farming has declined precipitously since the late 1800s. Currently, there is about one-quarter the farmland and one-sixth the number of farms of peak nineteenth-century levels. Ellis (1878) and Stotts (2007) are the classic references on the county's history, although these sources provide little information on overall landscape patterns or ecology. Litten (ch. 11 in this volume) provides an overview of agricultural history in the Hudson watershed.

Unless otherwise noted, population and agricultural statistics come from New York State and Federal censuses, the vast majority of these are available on line (New York State Library; U.S. Census Bureau; USDA-NASS). The Federal government began to collect agricultural data in 1820; New York State conducted its decennial censuses of agricultural production from 1845 to 1875. While these various censuses are imperfect and their methods

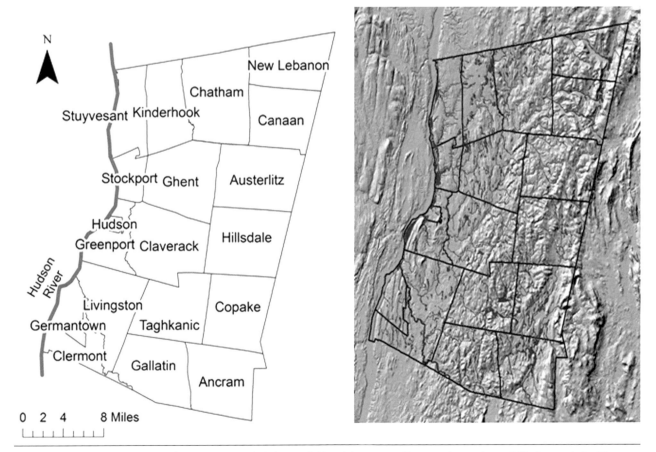

FIG. 12.1. The outline of Columbia County and the included towns (left) and the same outline superimposed on satellite image–derived topography. This study focuses on land-cover change associated with agriculture in Columbia County. Topography has had a major influence on the County's agriculture.

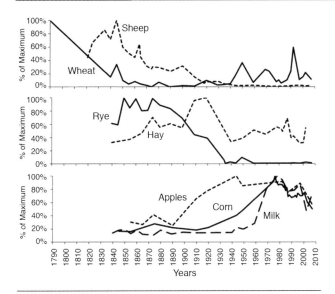

FIG. 12.2. The evolution of some important Columbia County agricultural commodities over the past 200 years. Values are represented as percent of observed maximum production (or density in the case of sheep). Data are from state and federal censuses, except for earliest wheat production estimate which is extrapolated from censused population size and estimated per capita needs (Bruegel 2002). During the past two centuries, the county has transitioned through periods of wheat, wool, rye, hay, fruit, and milk production, along with some additional products not illustrated here (e.g., potatoes, beef).

changed over time, they are probably suitable for outlining the general patterns discussed here. We use town-level statistics to describe the mode and spatial distribution of agriculture; we use regional historical literature together with our own fieldwork to hypothesize ecological effects.

Indigenous activity prior to European settlement (Lindner, ch. 7 in this volume) and European-spurred activity prior to 1820 (e.g., Henshaw, ch. 1 in this volume) no doubt affected the ecology of Columbia County. We selected our time period because of its immediate relevance to the current state of the land and the relative abundance of local information.

THE STARTING POINT: CREATING ANALOGIES

The typical farm of the 1820s was probably fairly diversified, providing many of the familial needs, but also creating some surplus for market (Bruegel 2002). By 1820, about 60 percent of the county was already in "improved acreage"; during subsequent years, that percentage did not exceed 75 percent (Fig. 12.3a). Thus, understanding our starting point helps explain much of what followed.

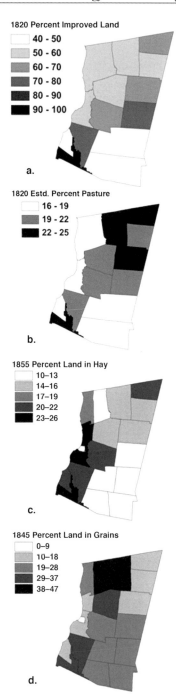

FIG. 12.3. The distribution of early Columbia County agriculture as derived from census data. Land uses were calculated as percent of a given town's total surface area. In 1820, improved land (a.) was defined as all land opened for agriculture. Estimated pasture (b.) is derived from census information on livestock multiplied by per animal land requirements from Lemon (1972); it is unlikely that these requirements derived from Lemon's Pennsylvania research are exactly true for us, but relative values might be more accurate. Grains (d.) included corn, wheat, oats, and rye. The earliest available agricultural census information is from 1820, however, hay and grain production were first censused in the later years indicated here. The number and extent of the towns within the county evolved between 1820 and 1845. Pasture, hay meadow, and grain land had distinct distributions, probably due in part to distinct soil types and climates within the county. These spatial differences helped lead to distinct ecological consequences for these land uses.

In this section, we will ask two questions: First, where did a given type of agriculture occur and hence which natural habitats were probably replaced or greatly modified? And, second, which organisms benefited from ecological analogies created by the new agricultural cover types?

During the first half of the nineteenth century, Columbia County could be described as having three forms of farmland: (1) *early cropland,* largely for grain growing and centered in the mid-county flats of present-day Claverack, Ghent, and Kinderhook (Fig. 12.3d); (2) *early pastures* in the northeast and southwest portions of the county (Fig. 12.3b); and (3) *early hay meadows* with hay production occurring mainly in the southwest corner (Fig. 12.3c). Below, we consider these three agricultural cover types in terms of their use, the natural habitats they may have impacted, and the new ecological analogies they may have created.

EARLY CROPLAND

Definition and Location

Early cropland was largely used for grain production. This was primarily intended for the consumption of the farm families and their livestock (Bruegel 2002 and census-based estimates of early yields). However, in good years, grain was also an important cash crop. By 1680, wheat was being shipped south on the Hudson from Columbia County landings (Danckaerts 1680 [1913]). The Hudson River was the county's major agricultural thoroughfare into the late nineteenth century; indeed, the City of Hudson was founded largely as a safe hub for such commerce (Schram 2004). While grain was grown in all towns of the county by the time it was first censused in 1845, it was most common on the county's "prime agricultural soils" (USDA 1989), a north/south band of relatively flat and well-drained soils lying some 3–13 km inland from the Hudson.

Habitats Lost

Although Native American clearings formed the core of some early settlements in the County (Ellis 1878), European settlers felled substantial forest. On much of the best flatland soils, there are few if any pockets of old forest remaining, and much of the land is still being farmed. Knowing the original forest composition is thus difficult. Our reconstruction of early forests in the county based on witness trees (Vispo, unpublished data) suggests that oak (mostly white oak) and hickory dominated on many of these flatland soils (Fig. 12.4). White oak is now much less common in the county than previously (USDA Forest Service; personal observation). At least part of this decline can be ascribed to widespread removal of the white oak–dominated forests on rich farmlands (other factors likely include browsing by white-tailed deer [Thompson and Huth, ch. 10 in this volume] and the preferential use of white oak for construction).

We have no account of the native herbaceous plants that grew in these forests before clearing. However, Braun (1950) suggests that Hudson Valley white oak forests were similar to forests on the Harrisburg Peneplain in Pennsylvania. Her list of herbaceous plants in a white oak forest remnant on the Peneplain is our best approximation of the native plants that might have occurred in our white oak forests: wild geranium, perfoliated bellwort, false Solomon's seal, hogpeanut, blue-stem goldenrod, asters, and tick-trefoil. These species do not thrive on the dry, acidic soils typical of modern second-growth, oak-hickory forests. They are still found on some richer forest soils, but their numbers are probably significantly lower than during precolonial times.

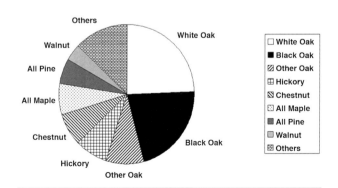

FIG. 12.4. The composition of early forests on Columbia County's central and western flatlands. White oak dominated these forests but is much rarer today, perhaps due in part to its occurrence on what proved to be agriculturally rich soils. These data (Vispo unpublished) were derived from witness tree information in late-eighteenth and earliest-nineteenth-century land deeds available in the county clerk's office in Hudson.

Analogies Created

Most cropland offers relatively few ecological analogies for animals because it is so heavily managed. Killdeer, a shorebird that strays inland, may have found some open cropland to be analogous to the beaches it had favored. More important than its role in providing structural habitat, however, was the fact that cropland provided food for wild animals. Woodchucks quickly arrived (Godman 1831), and a few butterflies, such as the black swallowtail (whose caterpillars feed on parsley and carrots) and our native whites (who feed on the brassicas) relished some crops (Fitch 1869; Harris 1862). A handful of native plants became cropland weeds. These included common ragweed in wheat, bur-cucumber and devil's beggar-ticks in gardens and corn fields, milk purslane and witch-grass in corn fields, and Pennsylvania smartweed in barnyards (Torrey 1843; Darlington 1859).

EARLY PASTURES

Definition and Location

Livestock played an early role in shaping the Columbia County landscape. Based on Lemon's (1972) estimates of the pasture requirements of colonial livestock (and roughly corroborated by correlational analysis of livestock and land use from later, more detailed Columbia County censuses), about one-third of all improved acreage in 1820 could be accounted for by the pasture needs of local sheep, horses, and cattle. Livestock (and hence estimated pastureland) were located primarily in the northeastern and, to a lesser degree, southwestern portions of the county (Fig. 12.3b). Early farmers used not only open pasture but also woodland pasture; however, we will not consider the latter.

In 1855 (when pasture was first tallied specifically), more than one-third the area of some eastern towns, but less than one-tenth that of some western towns, was in pasture. The nature of a pasture depends in part on who grazes it. The majority of pasture was probably accounted for by the needs of bovine cattle. At the peak of sheep populations around 1845, about one-fifth to one-quarter of the county's pastures were probably used by sheep; although this value averaged closer to one-third in some towns of the eastern hills (calculations based on livestock census information and estimates of per head land requirements from Lemon 1972).

Habitats Lost

Witness tree information (Vispo unpublished data) indicates that the county's forests on the steeper land where hill pasturing may have occurred were composed of chestnut and pine (mostly white pine) with lesser amounts of hemlock, beech, and maple, and interspersed oaks and hickories. Because these eastern forests were the main habitat for the county's more boreal organisms, such species probably decreased as forests were cleared for pasture (and for the production of tannins and charcoal, see Thompson and Huth, ch. 10 in this volume). Birds such as blackburnian, pine, black-throated-blue, and Canada warblers breed in the higher hills today (personal observation; McGowan and Corwin 2008) and were likely more common before forest clearing. More boreal plant species that may have declined include hobble bush, mountain maple, beaked hazel, yellow birch, paper birch, wood lily, painted trillium, bead-lily, trailing arbutus, poke milkweed, bunchberry, fly-honeysuckle, red-berried elderberry, and whorled aster, as well as some ferns, clubmosses, shade-tolerant grasses, and sedges (McVaugh 1958; personal observation).

Analogies Created

Scruffier pastures may have provided some analogies to prairie grasslands and savannahs. However, closely cropped pastures are too clean to provide many analogies to natural grasslands. Eaton (1910) provides a damning description of sheep pasture as bird habitat:

> [T]he principal harm of pasturing, to bird life, is found in the destruction of ground cover which inevitably results in woods and thickets. This is especially noticeable in sheep pastures where all the vegetation is destroyed to a height of three or four feet above the ground. In such pasture land the

thickets and undergrowth, which usually support an abundant bird life, are eliminated and the birds must seek other coverts.

The lack of bushes and potential close-cropping of the pastures left room for few birds, although certain species (such as savannah and field sparrows, and kingbirds) may have used these lands, especially when there was scattered brush. Where vegetation crept in along fence rows, species such as bobwhite quail, yellow warbler, song sparrow, and catbird probably entered.

Intensively grazed pastures did not harbor many native plants. Few eastern North American plants tolerate intensive grazing. In eastern soils and climate, even prairie plants that had coexisted with grazing Buffalo did not compete well with the pasture grasses and forbs introduced from Europe. During the initial period of relatively good soil fertility (and sufficient topsoil), native plants likely composed very little of the pasture vegetation. As we'll describe below, this changed as some pasture soils became depleted.

EARLY HAY MEADOWS

Definition and Location

In the Northeast, early hay meadows were primarily wet meadows. They produced reliable hay crops due to the regular input of nutrients from flooding (indeed, in some places early efforts were made to reroute floodwaters through fields in order to "fertilize them by flooding" (Donahue 2004). While we have found little direct evidence for such lowland hay meadows in Columbia County, mapping of adjacent Berkshire County, Massachusetts, done in the 1830s indicated that all hay meadows were in lowlands (Hall et al. 2002). Inspection of a 1762 property map from the Kinderhook area (in the collections of the Columbia County Historical Society) shows a long lot plot configuration with lots extending out from creeks; a configuration perhaps associated with assuring farmer access to a diversity of soils, including streamside meadows (Chelsea Teale, unpublished manuscript). There are reports of lowland haying by New Lebanon Shakers at least for the decade or so after 1790 (Anderson 1950).

Much of the early haying apparently occurred in the southwest corner of the County (Figure 12.3c). Some of this hay may have been cut from the Hudson River tidal floodplain, but some probably also came from inland swales. The topography of Germantown and Clermont is dominated by a series of north/south ridges with small wetland valleys in between. As Spafford (1824) put it, "The surface is but gently undulated, and the soil is good for grass." Haying also was common in New Lebanon in the northeastern corner of the county. Spafford (1824) describes that town as "good farming lands, dry and warm or *wet and grassy*" (emphasis added).

Habitats Lost

Some wet grasslands may have initially been floodplain or swamp forests. The clearing of *floodplain forest* would have removed habitat of plants such as silver maple, sycamore, cottonwood, bitternut, green ash, leatherwood, marsh pea, false mermaid weed, ostrich fern, green dragon, wild rye species, Canada brome, and certain sedges (e.g., *Carex davisii* and *C. spengelii*) and of animals such as wood turtles and select ground beetles, dragonflies, and damselflies (Knab-Vispo and Vispo 2009; Knab-Vispo and Vispo 2010; Thompson 1842). Through an examination of the early aerial photos (1940s), we estimate that, at the most, around 16 percent of the floodplain area maintained its forest cover over the last two hundred years. *Swamp forests* may have harbored red maple, winterberry, swamp white oak, buttonbush, black ash, and poison sumac (personal observation); we have no assessment of swamp forest extent in the county.

Analogies Created

Some of the new wet meadows were partial analogies for a habitat that humans had removed from the landscape some two hundred years prior: beaver meadows. Beavers were already "exceedingly scarce" in the Hudson Valley by the end of the seventeenth century (DeKay 1842; see too Müller-Schwarze and Sun 2003 and Henshaw, ch. 1 in this volume), and the ecosystems that they had created were largely missing prior to the beaver's partial return in the late twentieth century. Today, beaver densities are probably 20–50 percent of precolonial levels (based on current beaver density estimates for Massachusetts [Massachusetts Division of Fisheries and Wildlife] and Connecticut [Wilson 2001] together with estimated maximum beaver densities in areas/eras with little or no harvesting of beaver [Wright et al. 2002; Seton 1929, see also Hill 1982 and references therein]). Numerous native species are found in wet meadows that were created or are maintained by agriculture. These include rare species such as bog and spotted turtles, ribbon snakes, leopard frogs, and harriers (personal observation; Kiviat and Stevens 2001). We have found wetland butterflies including bronze copper, eyed and Appalachian browns, black dash, mulberry wing, and Baltimore checkerspot around wet meadows on farms (Vispo 2011). Plants of historical (Torrey 1843) and modern wet meadows (personal observation) include iris, blue-eyed grass, common monkeyflower, common vervain, sweetflag, golden ragwort, green-headed coneflower, yellow avens, and meadowsweet, as well as native sedges and grasses. However, some plants previously associated with wet meadows now rarely occur there. These include Canada lily, ragged-fringed orchid, purple-fringed orchid, nodding lady's tresses, blood milkwort, swamp saxifrage, and the adder's tongue fern (Torrey 1843; personal observation). Their consumption by increased white-tailed deer populations may partially account for the modern rarity of these species (McVaugh personal communication; personal observation).

In sum, as continues to be the case with farming, the ecological ramifications of early-nineteenth-century agriculture were likely mixed. By removing forests, farming caused certain organisms to lose habitat; by creating new cover types, it provided certain organisms with new space. In the section that follows, we move on from this starting point and explore the evolving analogies associated with two agricultural cover types: pasture and hay meadow.

EVOLVING ECOLOGICAL ANALOGIES ON ACTIVE FARMLAND

The ecology of farmlands after 1820 evolved in at least two ways: first, modes of production changed as markets rose and fell. These changes produced major variation in the proportions of different agricultural cover types. Second, technological developments meant that the ecology of a cover type and the analogies that it offered changed as techniques and practices evolved.

We will focus on pastures and hayfields in this section. At their peaks, these lands together covered more than eighty thousand hectares or around half of the county. Our central question is: "How did the agricultural techniques associated with each mode of production vary over time, and how did these developments influence the value of these lands as ecological analogies?"

Pastures

"Pastures were New England's stepchild," states Whitney (1994), implying that they got only the attention and manure that was left for them after croplands and hayfields. The result, in New England at least, was a decline in pasture quality. Cooper et al. (1929) depict the plant succession on pastures undergoing progressive soil depletion. Most of the introduced agronomic grasses outcompete native ones when nutrients are high, but are then unable to maintain themselves as nutrients decline. Thus, as soil quality declines, native plants become *more* common.

Many of the native plants that came into these exhausted pastures found analogies to their original, thin-soiled habitats on ridge tops, steep hillsides, sand barrens, etc. Examples of native plants that were common on "dry hillsides" or "sterile fields" and which still occur on such lands today are: pussytoe, gray goldenrod, mountain-mint, sweet fern, poverty oatgrass, little bluestem, pasture rose, dewberry, and arrowhead violet (Torrey 1843; personal observation). The native grasses such as little bluestem are, in turn, followed by a set of grassland skipper butterflies, specifically Leonard's skipper, cobweb skipper and Indian skipper (Cech and Tudor 2007; Vispo 2011). Again, there is a group of native plants described as "not rare," "frequent," or "common" in these habitats by Torrey (1843), but which are now quite rare. These include whorled milkweed, upland boneset, Venus looking-glass, American pennyroyal, clammy cuphea, yellow wild indigo, wild sensitive plant, rattlebox, downy trailing lespedeza, Virginia yellow flax, and little sundrops. It is not clear why these species are now rare; some of these may have always been relatively less common in our area (e.g., clammy cuphea) and so may have now become actually rare as the availability of their habitat declined.

Sheep pasture and cattle pasture are distinct ecological habitats. We have already quoted Eaton's damning description of sheep pastures as bird habitat. Sheep and bovines differ in their grazing behavior. Specifically, most cattle browse less intensively than most sheep. The result is that cattle pastures are more apt to fill-in with unpalatable shrubs. The net effect of both degradation of soil quality and of increases in cattle was the "shrubby pasture" that is still familiar to us today. We will discuss "shrublands" and their ecological analogies in greater detail in our section on abandonment; the point here is that pastures probably provided ecological analogies for the most native species when those pastures were agriculturally marginal—it was these conditions that allowed both native plant species and native shrubland birds to find homes.

Pastures imply fencing and hedgerows, and so we consider these technologies briefly here. The first fences in our county were likely of wood and were probably relatively rare. They fenced free-roaming livestock out of crops (Cronin 1983). Eventually, the containment of livestock became the main role of fencing. Around that time, rock walls sprang up as freeze-thaw cycles pushed more rocks to the surface and as timber scarcity led to moderation in wood use (Thorson 2002; Allport 1990). By the late 1800s, wire fencing was appearing. Strands of barbed wire and woven sheep fence still border many fields, even if their job has now been taken by high-tensile wire or other substitutes.

Different field margins provided different habitats. A variety of wild animals inhabited rock walls (e.g., snakes and rodents). Squirrels and chipmunks, in turn, helped disperse the nuts and acorns that have now grown up into towering oaks and hickories. Wire fences have proved excellent bird perches, and so tend to become neighbored by bird-dispersed plants such as cherries, viburnum, and shadbush (Whitney 1994; personal observation). Some suggest that the "cleaning up" of the fence line that followed the widespread acceptance of wire fencing was partially responsible for the sharp decline of bobwhite quail (Forbush 1912).

Unlike hedgerows in some other, more deforested regions, hedgerows in the forested Northeast do not currently appear to be important sanctuaries for woodland plants or animals, although they provide habitat for some and conduits for others (personal observation; Freemark et al. 2002). Their ecological role in our county may have been greater during the height of agriculture.

Hayfields

Hayfields reached their commercial zenith in the late 1800s when their area topped 42,000 hectares in the County. Much of this hay was sent via river to fuel New York City horse power.

Upland hay increased during the nineteenth century. We have no statistics for Columbia County, but Whitney (1994) describes the situation in Worcester, Massachusetts, where upland hayfields accounted for 49 percent of all hay meadows in 1780 and for around 75 percent by 1850. Between 1850 and 1875, all Columbia County towns reported increased hayfield area. The increase averaged over 800 hectares per town. At the same time, pasture and cropland *decreased* by an average of about 87 and 250 hectares respectively (presumably in part due to conversion to hayfield), and total improved acreage increased by roughly 230 hectares per town. Thus, although the pattern varied across towns, increased upland hayfields apparently came from a combination of pasture and cropland conversion and the opening up of new land. Additional hayfield apparently came from the shifting use of extant improved acreage (not all open agricultural land, i.e., "improved acreage," was apparently used in a given year).

To the degree that upland hayfields replaced former sheep pasture or cropland, the increase in hayfields may have signaled an overall increase in the ecological analogies provided by the agricultural landscape, at least for the grassland birds who found prairie-like structure in such fields. Early naturalists were quick to link farming, with its extensive hayfields, to the increased abundance of these avian species (Wilson 1829).

Upland hayfields provide ecological analogies for certain prairie species, especially some Tall Grass Prairie organisms. Many of the birds and some of the plants currently and/or historically found in hayfields originally had their demographic heartlands in the prairies of the Midwest. Bobolinks, meadowlarks, dicksissels, upland plovers, vesper sparrows, and grasshopper sparrows, for example, are all birds that occupy, or at least occupied, eastern hayfields but which probably had their largest precolonial populations on the prairies (Wells and Rosenberg 1999; note however that these species do not all necessarily co-occur in the same types of hayfield or prairie).

All else being equal, nesting habitat structure and extent seem to be the key parameters determining the occurrence of grassland birds (e.g., Swanson 1996). Such an emphasis on structure means that the structural analogies between hayfield and prairie are sufficient for these birds, even if the plants in the hayfields are nearly 100 percent European. Vegetation height, density, and herbaceous versus woody nature are among the parameters used to describe the habitats of grassland birds. This is in contrast to butterflies who, given their caterpillars' close links to food plants, are probably more strongly affected by the botanical composition of a field than by its structure. Hayfields—unless on thin soils—harbor few unique native butterflies (personal observation; Vispo and Knab-Vispo 2006).

The wet meadow hayfields probably favored the red-winged blackbird who seeks just such wet, grassy, reedy, or sedgy areas. Early accounts of bobolink also refer to them as being birds of wetter meadows (Macauley 1829; Thompson 1842). However, as ground nesters, they were probably most common not in true wet meadows but on moister upland fields where good watering made for a thick thatch. As drier upland hayfields expanded, so too did these species. In the mid 1800s, bobolinks were very common. Kent (1933, cited in DeOrsey and Butler 2006), for example, describes their "great flocks in migration" along the Hudson. However, as noted below, this boom was soon dampened by the changing calendar of mowing.

Upland hayfields may have changed little in plant composition throughout most of the nineteenth and twentieth centuries. They have been predominantly composed of nonnative grasses, especially timothy. Timothy or timothy and clover accounted for some 50–60 percent of haylands in the county at the beginning of the twentieth century. Earlier accounts suggest that the use of "English Grasses" was well established by the end of the eighteenth century. By the second half of the twentieth century, alfalfa hay was becoming more common. According to the Census of Agriculture, in 2007, it accounted for nearly one-quarter of all hayland, although "other tame hay," including timothy, still made up 55 percent. Torrey (1843) lists a number of plant species that occurred in "meadows" and which we still find in "wild" hayfields today. These include fleabanes, black-eyed Susan, spiked lobelia, evening primrose, and small-flowered crowfoot. However, he also lists slender lady's tresses (an orchid) as "common," common lousewort as "very common," and blue toadflax as "not rare" in meadows. None of these later species are now easily found in Columbia County (personal observation).

In the first half of the nineteenth century, hay cutting was with a scythe. It was slow and laborious. In Columbia County, it generally began in early or mid-July (Emmons 1846; Anderson 1950) and may have extended for several weeks. A practical, horse-drawn hay cutter was introduced before the Civil War. Prior to the end of the 1800s, mechanization and new ideas of progressive agriculture favored a cut in June, possibly followed by a second, later cut. Mechanization, for a variety of farm activities, continued apace in the twentieth century. Around 1944, the number of horses and mules on U.S. farms was surpassed by the number of tractors (White 2008). By 1950, for example, of 1,517 Columbia County farms censused, 19 percent used only horses, 32 percent used both tractors and horses, and 49 percent used only tractors. Early (i.e., May) haying became even more intense late in the twentieth century as the concept and technology for haylage stored in those blue "Harveststore" silos or, more recently, for plastic-wrapped baleage spread in the county. ("Haylage" and "baleage" are hays that are allowed to ferment, a process that, as with silage, results in greater nutrient availability in the feed.)

The results of these changes in harvesting techniques were momentous for birds. A key consideration for grassland birds is the timing of the hay cut relative to when the young leave the nest. If the hay cut occurs before fledging, then the hayfields become "ecological traps" that entice birds to nest but then foil reproduction. In Columbia County, bobolinks fledge around the first week of July (personal observation). When haying *began* at this time, most bobolink nestlings may have survived. When haying moved back into June and became more rapid, fewer clutches could survive to fledging. By the end of the nineteenth century, birders in the Northeast were noting steep declines of bobolink and meadowlark, and attributing this to changing farming methods (Eaton 1910). Bagg and Elliott (1937) put the beginning of this decline in the Connecticut Valley at as early as 1875.

The effect on the few native plants that were able to grow in hayfields may also have been substantial. In the 1930s, ragged-fringed orchid still was a common native plant in hay meadows, and McVaugh (personal communication) attributed its subsequent drastic decline at least in part to the change in haying schedule.

Butterflies also are affected by the timing of the hay cut. For those grassland species whose eggs and caterpillars are deposited in the fields, a cut that is made prior to when the adults take wing can destroy many individuals (Massachusetts Butterfly Club). While there is concern about the effects of early hay mowing (Massachusetts Butterfly Club), data from North America are sparse. Our observations suggest that intensively managed hayfields have host plants for the caterpillars of relatively few butterfly species (Vispo and Knab-Vispo 2006).

The trend toward early hay cuts in the county has been slowed somewhat by the modern spread of "estate" hayfields—hayfields cut once per year, often late, by contracted farmers who invest little in improvement and are thus sometimes satisfied by a late cut of relatively poor quality hay (personal observation). Landowners receive a property tax break for this "agricultural use" of their land. In 1910, around 1 percent of hay was "wild"; by 2007, nearly 20 percent was "wild" ("wild," in this context, refers to hay from a field that has not recently been seeded and thus tends to contain a higher diversity of plants). Yield has also begun to drop from 2.6 tons per acre in 1987 to less than 2 tons per acre in 2007.

Mechanization was also associated with drainage, because it both facilitated (through digging and tile-laying equipment) and required (wet ground could not support the heavier machinery) that practice. Large-scale drainage with clay tiling began in New York after 1850 (for example, New York State Agricultural Society 1858). Farmers could create cropland from areas that had been too wet to support more than occasional hay cuts. Once drained, many soils were rich in organic matter and offered high yields, at least initially. With the spread of subsurface drainage, the wet hay meadows were divided into those drained and used for crops and those left in hay and which, with the decline in the hay markets, eventually began to revert to floodplain or swamp forest.

This history of drainage has interacted closely with natural habitats in the county. Standing water or regular floods impose particular demands on native organisms and unique habitats result: red maple swamp forests, buttonbush swamps, sycamore floodplain forests, and sedge meadows are the names for unique communities that can occur on these lands. We estimate that some 40–60 percent of the plants, birds, and butterflies found in Columbia County wetlands are rare and/or experiencing declines (Vispo and Knab-Vispo 2006). For example, the New England cottontail, a species whose listing as an endangered species is pending, may have favored the shrubby cover associated with damper sites (Arbuthnot 2008). Statewide, wetlands are estimated to have decreased by 60 percent since 1790 (Dahl 1990). The 1923 soil survey of the county (Lewis and Kinsman 1929) lists 19,328 acres as being in muck and wetland while 1993 remote sensing by the IRIS program of Cornell put wetland area at 5,620 acres. A study of land change in the Hudson Valley (Amielle DeWan, unpublished data) estimated a 27 percent decline in Columbia County wetlands between 1986 and 2002. The techniques used in these studies differed but substantial wetland decrease is suggested.

Farms can be described as both the bane and the blessing of wetlands in Columbia County. Farmland drainage and clearing has resulted in significant loss or modification of wetlands, yet at the same time, because of the agricultural desirability of valley soils, the majority of wetlands do occur on farms. When managed in a compatible way, farms can help maintain important wetland habitats, such as the wet meadows that provide some ecological analogies to beaver meadows (Vispo and Knab-

Vispo 2007). Commercial and residential development is often less kind—few rules govern the use of small wetlands and their manipulation is frequent (personal observation).

The ecological changes we described above were all caused by farmers' efforts to improve their agriculture. In contrast, during the twentieth century, farmland abandonment and subsequent "rewilding" was one of the main causes of ecological change in Columbia County. In the section that follows, we describe the timing and distribution of abandonment, and sketch some of its ecological consequences.

FARMLAND ABANDONMENT: MAKING TRANSIENT, WILD-CRAFTED ANALOGIES

"Improved" farm acreage in the county began a steep decline around 1900 (Fig. 12.5). This drop probably reflected various, interacting factors including the spread of alternative, nonagricultural employment; the unprofitability of certain farms in the face of expanding Midwestern agriculture; and the shifts in regional styles of farming and, hence, changes in land requirements (see for example, Whitney 1901; Jones 1912; Vaughan 1929). This abandonment had two general ecological consequences: first, certain lands began to revert to con-

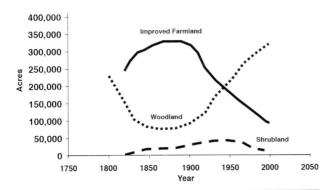

FIG. 12.5. Census data and extrapolations indicating the course of land use in Columbia County. Improved acreage is from census data; wooded acreage is partially extrapolation from land not in other uses and partially from forest cover estimates done by the state and federal agencies; shrubland extent is estimated based on change in forest extent and its recognition as a transitional state. The definition of "improved acreage" varied somewhat over time (in later years, emphasis was placed on ploughed lands); however, its general trend was probably more or less as indicated and can be corroborated by trends in total farmland.

ditions somewhat similar to pre-clearing and, second, in the process, large stretches of somewhat novel, highly transient shrubland and old field cover types were created. These habitats had not been completely absent from the county, however they now encompassed wider extents, and provided new opportunities for native organisms.

Figure 12.5 shows the general pattern of rapid reforestation and the shrubland "peak" that occurred in the county during the first half of the twentieth century. A botanical glimpse of this period comes from pollen core data from nearby Stockbridge Bowl in Berkshire County, Massachusetts (Patterson 2000). From the late 1700s through the early 1900s, those data show that forest trees declined and grasses, native field weeds, and native wetland plants increased. Sharp changes occurred between 1900 and 1950, when most field plants dropped precipitously, and pioneer forest trees and then mature forest trees began to increase, along with a slight increase in shrubland vegetation. Somewhat similar patterns are reported in cores taken from Hudson marshes (Peteet et al., ch. 9 in this volume).

To understand the resulting ecological consequences, one needs to understand the patterns of abandonment. Table 12.1 shows how, at least in one eastern town, the steeper, higher terrain was the first abandoned. Abandoned tracts also had poorer soils and were more likely to have a northerly exposure. Flinn et al. (2005) found similar patterns in central New York. The 1923 soil survey of the County (Lewis and Kinsman 1929) noted abandoned farms on the eastern hills; this is confirmed by census data (Fig. 12.6). The hilltops and ridgelines, rarely used for agriculture or abandoned much earlier, were soon covered by extensive forests where forest-interior animals experienced a relatively sheltered existence until fashion, affluence, and engineering combined to now make the hills and ridgetops favored housing locations (personal observation).

The ecological succession that followed abandonment meant that many of the open pastures and hay meadows of the early and middle nineteenth century initially grew into old fields filled with weeds and native herbaceous plants. These were largely new cover types: in 1843 Torrey reported goldenrod as a weed of roadsides and edges; he did not mention our now-common, goldenrod-dominated "old fields."

TABLE 12.1. A comparison of certain landscape characteristics on abandoned and active farmland in the town of Hillsdale.

Timing of Agriculture	Soil Quality Rank (from 0 to 6 with 6 being best)	Elevation (meters)	Incline (ratio of drop to run)	Exposure (ratio of southern to northern)
None Evident	1.6 ± .7	338 ± 28	0.24 ± .06	1.36
Pre-1940s, not 2006	1.0 ± .4	314 ± 20	0.13 ± .04	1.09
1940s, but not 2006	2.7 ± .8	269 ± 23	0.13 ± .04	4
1940 & 2006	3.9 ± .8	262 ± 25	0.13 ± .04	2

Soil quality rank is based on USDA agricultural production data (USDA 1989). Data are based on a GIS analysis of randomly placed points. Each category was represented by 25 points, and the values represent the averages (± 2SE) for each set of 25 points. Modern and historical (1942) aerial photographs were used in these calculations; pre-1940s agriculture was deduced from evidence (e.g., obvious traces of field margins) in the early photographs. Pre-1940s abandonment tended to occur on lands that were of poorer soil quality, higher elevation, and more southerly exposure. Lands with no evidence of agriculture tended to be steeper.

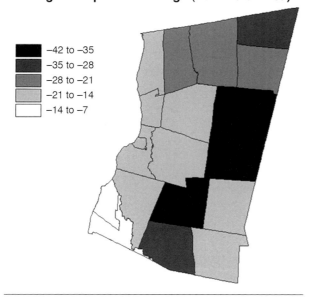

FIG. 12.6. Change in improved acreage between nineteenth-century maximum and 1930, expressed as percentage of total surface area of each town, as derived from census data. Eastern hill towns experienced noticeably higher rates of abandonment. While all towns experienced substantial abandonment during the remainder of the twentieth century (not shown), the hill towns still have the highest total farmland loss since nineteenth-century maxima.

Later in succession, shrubs (for example, dogwood) arrived and, finally, came forest (with "pioneer trees" such as birch, white pine, or ash often leading the way). However, succession is not a deterministic process, but rather a tendency colored by history, local particularities, and chance (see Wessels 1999). For example, gradually abandoned pastures often pass through a thorny shrub stage with hawthorn, raspberries, buckthorn, multiflora rose (after its mid-twentieth-century introduction), and red cedar. In contrast, suddenly abandoned plowlands might transition rapidly to a forest of whatever wind-dispersed tree species happens to be nearby and having a good masting year. White pine is a common colonizer. Alternatively, hayfield succession might be retarded somewhat as tall herbaceous growth delays the advance of woody plants.

From the perspective of animal ecology, shrubland, not old field, is perhaps the most interesting early stage of succession. The grassland birds of maintained but mature hayfields (e.g., bobolinks and meadowlarks) are not particularly common in old fields with rougher native "weeds" like goldenrod and ragweed and the beginnings of shrubby vegetation (personal observation). The ecological analogies to these birds' native prairies seem to break down as grasses become less common and broad-leaved plants become dominant. Even the butterfly community of old fields seems unspecialized and dominated by species typical of field/forest edges and hayfields (personal observation). Perhaps natural upland fields were rare in the original landscape, and few species are "pre-adapted" to them.

The arrival of shrublands, on the other hand, ushered in new plants and animals (see Litvaitis 2003 and accompanying articles). Some of these were species that had previously been found around beaver ponds and other wetlands. In the Northeast, rufous-sided towhees, chestnut-sided warblers, and catbirds, for example, may have originally occurred in such habitats (Birds of North America). Mockingbirds, field sparrows, brown thrashers, prairie warblers (a misnomer), and yellow warblers also settled into the shrublands.

The shrublands that develop from our old fields do not contain unique plants that are important for butterflies and moths; the ecologically important shrubs for butterflies and moths are those that are or simulate blueberry and scrub-oak dominated barrens (Wagner et al. 2003). Nonetheless, as noted in relationship to pastures, when old fields are on dry, poor soil then they support native plants and, in turn, native butterflies.

Wet old fields and shrubland host a variety of plant species, which may find these areas analogous to the beaver meadows and stream edges where they existed prior to the expansion of agriculture. In addition to the wet meadow herbs listed earlier, native shrubs such as dogwood species, arrow-wood, nannyberry, willow species, swamp rose, meadowsweet, and steeplebush colonize wet old fields.

Abandonment of farmland not only meant direct changes in surface cover but also abandonment of the maintenance of drainage. Many fields that were wet meadows in the 1820s were likely drained by the end of that century. They stayed relatively dry until lack of drainage maintenance or intentional release from management saw them return to wetland in the last quarter of the twentieth century. Many of the organisms already mentioned in relationship to wet meadows benefited from such reversion.

Old field and shrubland were succeeded by forest. The reforestation of the Northeast has had a huge effect on its wildlife (Foster et al. 2002). Many native animal species that had disappeared prior to 1800 have returned; moose, fisher, bobcat, black bear, and wild turkey have become substantially more common in Columbia County during the past thirty years (personal observation). White-tailed deer were among the first to return, in part because the old fields and shrublands that followed agriculture provided ideal habitat (Mattfield 1984). Historically, northeastern deer probably had survived in large part by utilizing openings created by fire, wind-throw, flooding, ice-scouring, or other disturbance (McCabe and McCabe 1984). The shrublands that followed farming, while probably somewhat different from the original shrublands, provided functional analogies, at least in terms of the food plants deer favored. The result of this increased habitat and of decreased predation/hunting has been a swelling of deer numbers to the point where forest succession is likely being affected today (e.g., Thompson and Huth, ch. 10 in this volume; Rooney and Waller 2003; personal observation).

These secondary forests are not botanical restorations of pre-European settlement forests (e.g., Singleton et al. 2001). Disease, logging, deer, and natural succession have all contributed to this change. Furthermore, McVaugh (1958) described the soils of the secondary forests as usually thinner, drier, and poorer than those of pre-settlement forests. We have few data on the ground flora of our pre-settlement forests, but poor soil species such as Pennsylvania sedge, wild sarsaparilla, Canada mayflower, and starflower may have now increased at the expense of rich soil species such as blue cohosh, bloodroot, wild ginger, Jack-in-the-pulpit, red trillium, and wild leek.

CONCLUSIONS

As a way of summary, we can take the changing landscape described above for which we have more or less firm statistics and hypothesize, based on some of the relationships we have mentioned, the resulting demographic chronology for select groups of wild plants and animals (Figure 12.7). Modern trends, which may only be beginning to show themselves, are especially hard to identify and so are particularly speculative.

Today, the county (and probably much of the Hudson Valley) is at a stage when relatively little habitat for native species is being created and existing habitat is being eroded. Most lands that will revert to forest have reverted and forest area, at least regionally, has begun declining; land in agriculture is shrinking and the farming on existing farmland is intensifying; and, compared to earlier levels, relatively little open land is succeeding to brush.

Specifically, for probably the first time since the mid-nineteenth century, forest area has begun to decline in the Northeast including the Hudson Valley (Tyrrell et al. 2004; DeWan and Zucker unpublished data; Foster et al. 2010). This "second

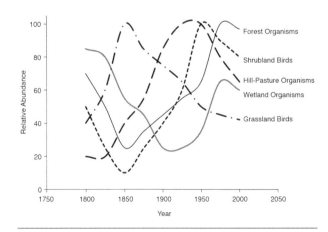

FIG. 12.7. An approximation of the county-wide relative abundance of select ecological groupings of organisms. Inspired by and compare with Foster et al.'s (2002) work in Massachusetts. The patterns seen here are hypothetical chronologies derived from the ecological relationships and changing agricultural landscape described in the text.

clearing" is due to development pressure, rather than agricultural expansion. The ecological effects reflect both absolute loss of forest area and the effective fragmentation of the remaining forest (e.g., Glennon and Kretser 2005). Much of this impact is due to regional changes in human population distribution and lifestyle, rather than absolute increases in population (Pendall 2003).

At the same time, land in farms continues to decline in the county. Part of this reflects a decline in the former "staples" of Columbia County agriculture: dairy farming and fruit production. It also reflects the growth of "niche farms." While some of these specialize in grass-fed livestock and dairy, most are intensive vegetable operations that use relatively small amounts of land compared to former modes of production. Average farm size has declined from nearly one hundred hectares in 2002 to fewer than eighty in 2007. From a nature conservation perspective, these patterns are especially troubling for those organisms that depend upon grasslands, shrublands, or open wetlands. We have estimated (Vispo and Knab-Vispo 2006) that 60 percent of the county's grasslands and perhaps 70 percent of its wetlands and shrublands (these categories are combined in agricultural statistics) occurred on farms in 1993.

Major landscape change in Columbia County is incipient. Rates of urbanization in Columbia County itself have been relatively modest, and forest loss, if any has occurred in the county, has been low. However, increase in developed lands and forest loss have been more marked in adjacent areas such as the Capital District, more southerly Hudson Valley counties, and the lands of Massachusetts and Connecticut (Tyrrell et al. 2004; DeWan and Zucker unpublished data; Foster et al. 2010; Loveland and Acevedo 2010). The current status can perhaps be best described as a slow or impending increase in the human domination of habitats. If trends evident in adjacent areas spread into the county, then this conversion would be expected to increase. Some of the habitats that largely avoided agricultural influences (such as ridgelines) are now being impacted (personal observation), and, due to fragmentation and other influences, the spread of human ecological impacts is advancing faster than the absolute rate of clearing (e.g., Glennon and Kretzer 2005).

While this chapter has not expressly linked upland land changes to impacts on the Hudson River itself, it is well documented that such land use changes within a watershed can result in direct impacts on the main waterways (e.g., Limburg et al. 2005; Cunningham et al. 2009). Many of the chapters in this volume have focused on the Hudson River itself, but this chapter has followed some of its tributaries upward and explored historical land use patterns which, while likely having water quality impacts, also had immediate, on-the-ground consequences for terrestrial ecology.

The changes in land cover and consequent ecological effects described in this chapter have been incidental insofar as they were driven by a variety of forces other than conscious conservation considerations. Despite increased public discussion of nature conservation, most nature conservation or habitat destruction continues to be "accidental." As human impacts in the Hudson Valley increase and as global stresses mount, a more conscious approach to landscape-level conservation will be needed if the trends highlighted by future historians are to reflect enhanced ecological analogies or actual habitat restoration.

ACKNOWLEDGMENTS

Much of our early fieldwork was funded by the DEC's Hudson River Estuary Program. Substantial additional funding came from private donors. Numerous volunteers, interns, and landowners were crucial in this work. Our colleague, Anna Duhon, helped with some analyses and stimulated others. Joel Grossman and Bob Henshaw provided helpful suggestions on drafts of this paper. David Foster and colleagues at the Harvard Forest provided useful critique of some of these ideas. We thank them all. Any errors of fact and concept here are ours alone.

REFERENCES CITED

Allport, S. 1990. *Sermons in stone: The stone walls of New England and New York.* New York: W. W. Norton.

Anderson, R. H. 1950. Agriculture among the Shakers, chiefly at Mount Lebanon. *Agricultural History* 24: 113–20.

Arbuthnot, M. 2008. *A landowner's guide to New England cottontail habitat management.* New York: Environmental Defense Fund.

Bagg, A. C., and S. A. Eliot Jr. 1937. *Birds of the Connecticut Valley in Massachusetts.* Northampton, MA: The Hampshire Bookshop.

Birds of North America. The Birds of North America, online. bna.birds.cornell.edu/bna/; accessed Aug. 15, 2010.

Braun, E. L. 1950. *Deciduous forests of eastern North America.* New York: Hafner Press.

Bruegel, M. 2002. *Farm, shop, landing.* Durham: Duke University Press.

Cech, R., and G. Tudor. 2007. *Butterflies of the East Coast: An observer's guide.* Princeton: Princeton University Press.

Cogbill, C. V., J. Burk, and G. Motzkin. 2002. The forests of presettlement New England, USA: Spatial and compositional patterns based on town proprietor surveys. *Journal of Biogeography* 29: 1279–1304.

Cooper, H. P., J. K. Wilson, and J. H. Barron. 1929. Ecological factors determining the pasture flora in the northeastern United States. *Journal of the American Society of Agronomy* 21: 607–27.

Cronin, W. 1983. *Changes in the land: Indians, colonists, and the ecology of New England.* New York: Hill and Wang.

Cunningham, M. A., C. M. O'Reilly, K.M. Menking, D. P. Gillikin, K. C. Smith, C. M. Foley, S. L. Belli, A. M. Pregnall, M. A. Schlessman, and P. Batur. 2009. The suburban stream syndrome: Evaluating land use and stream impairments in the suburbs. *Physical Geography* 30: 269–84.

Curtis, J. T. 1959. *The vegetation of Wisconsin.* Madison: University of Wisconsin Press.

Dahl, T. E . 1990. *Wetlands losses in the United States 1780's to 1980's.* Washington, DC: U.S. Department of the Interior, Fish and Wildlife Service.

Danckaerts, J. 1680 [1913]. *Journal of Jasper Danckaerts.* Ed. Bartlett James and J. Franklin Jameson. New York: Charles Scribner's Sons.

Darlington, W. 1859. *American weeds and useful plants.* New York: A. O. Moore.

DeKay, J. 1842. *The natural history of New York. Part 1.* New York: D. Appleton.

DeOrsey, S., and B. Butler. 2006. *The birds of Dutchess County, New York today and yesterday : A survey of current status with historical changes since 1870.* Millbrook, NY: Grinnell and Lawton.

Donahue, B. 2004. *The great meadow: Farmers and the land in colonial Concord.* New Haven: Yale University Press.

Eaton, E. 1910. *Birds of New York.* Albany: The University of the State of New York.

Ellis, F. 1878. *History of Columbia County, New York.* Philadelphia: Everts and Ensign.

Emmons, E. 1846. *Agriculture of New York.* Volume 1. Albany: C. Van Benthuysen.

Fitch, A. 1869. Thirteenth report on the noxious, beneficial, and other insects of the State of New York. *Transactions of the New York State Agricultural Society* 29: 495–566.

Flinn, K. M., M. Vellend, and P. L. Marks. 2005. Environmental causes and consequences of forest clearance and agricultural abandonment in central New York, USA. *Journal of Biogeography* 32: 439–52.

Forbush, E. H. 1912. *A history of the game birds, wild-fowl, and shore birds of Massachusetts and adjacent states.* Boston: Massachusetts State Board of Agriculture.

Foster, D. R., G. Motzkin, D. Bernardos, and J. Cardoza. 2002. Wildlife dynamics in the changing New England landscape. *Journal of Biogeography* 29: 1337–57.

Foster, D. R., B. M. Donahue, D. B. Kittredge, K. F. Lambert, M. L. Hunter, B. R. Hall, L. C. Irland, R. J. Lilieholm, D. A. Orwig, A. W. D'Amato, E. A. Colburn, J. R. Thompson, J. N. Levitt, A. M. Ellison, W. S. Keeton, J. D. Aber, C. V. Cogbill, C. T. Driscoll, T. J. Fahey, and C. M. Hart. 2010. *Wildlands and woodlands: A vision for the New England landscape.* Petersham, MA: Harvard Forest, Harvard University.

Freemark, K., C. Boutin, and C. J. Keddy. 2002. Importance of farmland habitats for the conservation of plant species. *Conservation Biology* 16: 399–412.

Glennon, M., and H. Kretser. 2005. *Impacts to wildlife from low density, exurban development: Information and considerations for the Adirondack Park*. Adirondack communities and conservation program technical paper no. 3, New York: Wildlife Conservation Society.

Godman, J. 1831. *American mastology*. Philadelphia: Stoddart and Atherton.

Hall, B., G. Motzkin, D. R. Foster, M. Syfert, and J. Burk. 2002. Three hundred years of forest and land-use change in Massachusetts, USA. *Journal of Biogeography* 29: 1319–35.

Harris, T. W. 1862. *A treatise on some of the insects injurious to vegetation*. Boston: Crosby and Nichols.

Hill, E. P. 1982. Beaver. In *Wild Mammals of North America*, ed. Joseph A. Chapman and George A. Feldhamer, 256–81. Baltimore: John Hopkins University Press.

Jones, G. J., ed. 1912. *Review of agricultural and natural resources of New York State*. Albany: J. B. Lyon Company.

Kiviat, E., and G. Stevens. 2001. *Biodiversity assessment manual for the Hudson River Estuary corridor*. Albany: New York State Department of Conservation.

Knab-Vispo, C., and C. Vispo. 2009. *The plant and animal diversity of Columbia County, NY floodplain forests: Composition and patterns*. Harlemville, NY: Hawthorne Valley Farmscape Ecology Program. www.hawthornevalleyfarm.org/fep/floodplains.html; accessed Aug. 15, 2010.

Knab-Vispo, C., and C. Vispo. 2010. *Floodplain forests of Columbia and Dutchess counties, NY: Distribution, biodiversity, classification, and conservation*. Harlemville, NY: Hawthorne Valley Farmscape Ecology Program. www.hawthornevalleyfarm.org/fep/floodplains.html; accessed Mar. 14, 2011.

Lemon, J. T. 1972. *The best poor man's country*. New York: W. W. Norton.

Lewis, H., and D. Kinsman. 1929. *Soil survey of Columbia County, New York*. Washington, DC: U. S. Government Printing Office.

Limburg, K. E., K. M. Stainbrook, J. D. Erickson, and J. M. Gowdy. 2005. Urbanization consequences: Case studies in the Hudson River watershed. *American Fisheries Society Symposium* 47: 23–37.

Litvaitis, J. 2003. Shrublands and early-successional forests: Critical habitats dependent on disturbance in the northeastern United States. *Forest Ecology and Management* 185: 1–4.

Loveland, T. R., and W. Acevedo. 2010. Land cover change in the eastern United States. USGS Land Cover Trends Project. landcovertrends.usgs.gov/east/regionalSummary.html; accessed Aug. 15, 2010.

Macauley, J. 1829. *The natural, statistical, and civil history of the State of New-York*. New York: Gould and Banks.

Massachusetts Butterfly Club. Butterfly Conservation. www.naba.org/chapters/nabambc/butterfly-conservation.asp; accessed Aug. 15, 2010.

Massachusetts Division of Fishery and Wildlife. Department of Fish and Game. www.mass.gov/dfwele/dfw/wildlife/facts/mammals/beaver/beaver_management.htm; accessed Aug. 15, 2010.

Mattfield, G. F. 1984. Northeastern hardwood and spruce/fir forests. In *White-tailed deer: Ecology and management*, ed. L. K. Halls, 305–30. Harrisburg, PA.: Stackpole Books.

McCabe, R. E., and T. R. McCabe. 1984. Of slings and arrows: A historical retrospection. In *White-tailed deer: Ecology and management*, ed. L. K. Halls, 19–72. Harrisburg, PA.: Stackpole Books.

McGowan, K., and K. Corwin. 2008. *The second atlas of breeding birds of New York State*. Ithaca: Comstock Publishing Associates.

McVaugh, R. 1958. *Flora of Columbia County area, New York*. Albany: The University of the State of New York.

Müller-Schwarze, D., and L. Sun. 2003. *The beaver: Natural history of a wetland engineer*. Ithaca: Cornell University Press.

New York State Agricultural Society. 1858. *Transactions of the New York State Agricultural Society*. Volume 18. Albany: Charles van Benthuysen.

New York State Library. Selected digital historical documents. www.nysl.nysed.gov/scandocs/historical.htm; accessed Aug. 15, 2010.

Patterson, W., III. 2000. Stockbridge Bowl pollen core data. North American Pollen Database. ftp://ftp.ncdc.noaa.gov/pub/data/paleo/pollen/asciifiles/fossil/ascfiles/gpd/stockbri.txt; accessed Aug. 15, 2010.

Pederson, D., D. Peteet, D. Kurdyla, and T. Guilderson, T. 2005. Medieval warming, Little Ice Age, and European impact on the environment during the last millennium in the Lower Hudson Valley, New York, USA. *Quaternary Research* 63: 238–49.

Pendall, R. 2003. *Sprawl without growth: The upstate paradox.* The Brookings Institution. Survey Series. Washington, D.C.: The Brookings Institution. www.brookings.edu/~/media/Files/rc/reports/2003/10demographics_pendall/200310_Pendall.pdf; accessed Aug. 15, 2010.

Rooney, T. P., and D. M. Waller. 2003. Direct and indirect effects of white-tailed deer in forest ecosystems. *Forest Ecology and Management* 181: 165–76.

Schram, M. B. 2004. *Hudson's merchants and whalers.* Hensonville, NY: Black Dome Press.

Seton, E. T. 1929. *Lives of game animals.* Vol. 4, Part 2. Garden City, NY: Doubleday, Doran.

Singelton, R., S. Gardescu, P. L. Marks, and M. A. Geber. 2001. Forest herb colonization of postagricultural forests in central New York State, USA. *Journal of Ecology* 89: 325–38.

Spafford, H. G. 1824. *A gazetteer of the State of New York.* Albany: B. D. Packard.

Stotts, P. 2007. *Looking for work.* Kinderhook: Columbia County Historical Society.

Swanson, D. A. 1996. *Nesting ecology and nesting habitat requirements of Ohio's grassland-nesting birds: A literature review.* Ohio Department of Natural Resources, Division of Wildlife, Ohio Fish and Wildlife Report 13. Jamestown, ND: Northern Prairie Wildlife Research Center. www.npwrc.usgs.gov/resource/birds/ohionest; accessed Aug. 15, 2010.

Thompson, Z. 1842. *History of Vermont natural, civil, and statistical.* Burlington: Chauncey Goodrich.

Thorson, R. M. 2002. *Stone by stone: The magnificent history in New England's stone walls.* New York: Walker.

Torrey, J. 1843. *A flora of the State of New York.* Albany: Carroll and Cook.

Tyrell, M. L., M. H. P. Hall, and R. N. Sampson. 2004. *Dynamic models of land use change in northeastern USA.* Global Institute of Sustainable Forestry. Yale University. GISF research paper 003. gisf.research.yale.edu/assets/pdf/ppf/Dynamic%20Model%20of%20Land%20Use%20Change%20Report.pdf; accessed Aug. 15, 2010.

U.S. Census Bureau. Decennials: Census of population and housing. www.census.gov/prod/www/abs/decennial/index.htm; accessed Aug. 15, 2010.

USDA. 1989. Soil survey of Columbia County, New York. (no additional information).

USDA Forest Service. Northeastern forest inventory and analysis—USDA Forest Service. www.fs.fed.us/ne/fia/; accessed Aug. 15, 2010.

USDA-NASS. USDA-NASS-Census of agriculture. www.agcensus.usda.gov/; accessed Aug. 15, 2010.

Vaughan, L. M. 1929. Abandoned farm areas in New York. *Journal of Farm Economics* 11: 436–44.

Vispo, C. 2011. Columbia county openland butterflies: A draft ecological classification. Harlemville, NY: Hawthorne Valley Farmscape Ecology Program. www.hawthornevalleyfarm.org/fep/Openland butterflies.pdf; accessed Mar. 14, 2011.

Vispo, C., and C. Knab-Vispo. 2006. *The flora and fauna of some Columbia County farms: Their diversity, history, and management.* Harlemville, NY: Hawthorne Valley Farmscape Ecology Program. www.hawthorncvalleyfarm.org/fep/onfarmbio.html; accessed Aug. 15, 2010.

———. 2007. *Ponds of Columbia County: Patterns in their biodiversity, thoughts on their management.* Harlemville, NY: Hawthorne Valley Farmscape Ecology Program. www.hawthornevalleyfarm.org/fep/ponds.html; accessed Aug. 15, 2010.

Wagner, D. L., M. W. Nelson, and D. F. Schweitzer. 2003. Shrubland Lepidoptera of southern New England and southeastern New York: Ecology, conservation, and management. *Forest Ecology and Management* 185: 95–112.

Wells, J. V., and K. V. Rosenberg. 1999. Grassland bird conservation in northeastern North America. *Avian Biology* 19: 72–80.

Wessels, T. 1999. *Reading the forested landscape : A natural history of New England.* Woodstock, VT: Countryman Press.

White, W. 2008. Economic history of tractors in the United States. In EH.Net Encyclopedia, ed.

Robert Whaples. www.eh.net/encyclopedia/article/white.tractors.history.us; accessed Aug. 15, 2010.

Whitney, G. C. 1994. *From coastal wilderness to fruited plain: A history of environmental change in temperate North America, 1500 to the present.* Cambridge: Cambridge University Press.

Whitney, M. 1901. *Exhaustion and abandonment of soils.* U.S. Department of Agriculture, Report No. 70. Washington, D.C.: Government Printing Office.

Wilson, A. 1829. *American ornithology; Or the natural history of the birds of the United States.* New York: Collins.

Wilson, J. M. 2001. *Beavers in Connecticut: Their natural history and management.* Hartford: Connecticut Department of Environmental Protection, Wildlife Division.

Wright, J., C. Jones, and A. Flecker. 2002. An ecosystem engineer, the beaver, increases species richness at the landscape scale. *Oecologia* 132: 96–101.

CHAPTER 13

THE INTRODUCTION AND NATURALIZATION OF EXOTIC ORNAMENTAL PLANTS IN NEW YORK'S HUDSON RIVER VALLEY

Chelsea Teale

ABSTRACT

The Hudson River Valley became the vanguard of American landscape interpretation and design in the nineteenth century, inspired by European Romanticism and facilitated in part by New York City fortunes. As the home of the Hudson River School of painting and the first generation of nationally renowned architects and gardeners, the region has since become recognized by the National Park Service as a "landscape that defined America." This chapter is concerned with ornamental plants that were imported to the region to create fashionable, artistic landscapes but have subsequently become naturalized. These plants are often termed "exotic" because they are not native to the northeastern United States and were initially imported to be unique. Despite actual and potential damage to Hudson River Valley ecosystems by the species described here, risks are often unknown or not communicated and many remain gardening favorites. Popular appreciation for their aesthetic and historic qualities, for example, has resulted in ongoing sales and efforts to maintain heritage gardens. Although many exotic ornamental plants continue to be accepted additions to the Hudson River Valley flora, landscape gardeners today are slowly beginning to favor native species as resident alternatives are promoted and legislation restricts the use of exotic ornamentals.

DESIGNING LANDSCAPES

Plants and animals have overcome natural geographic barriers for millennia with human assistance, typically for food, fiber, and labor. Although many introductions were haphazard or accidental, improvements in shipping beginning in the sixteenth century allowed for more successful and purposeful movement of organisms for an increasing variety of purposes. Historically, however, most intentional plant introductions have been driven by horticulture—first as crops and medicinals and later as ornamentals (Mack and Erneberg 2002; Todd 2001). By the eighteenth century aesthetics had become an especially important driver as British landscapers popularized a new landscaping style that favored imported ornamental plants, a trend one botanist referred to as "painting with living pencils" (Chambers 1991, 66). Within a century, British landscape gardeners were actively importing and planting trees from southern Europe, Asia, and North America (Jarvis 1973). This landscaping trend spread to continental Europe and the United States, making the interest in exotic ornamental plants an Atlantic phenomenon.

This chapter explores the nineteenth-century introduction and establishment of exotic ornamental plants to New York's Hudson River Valley, the birthplace of American landscape gardening.

Although today the region's biota is not unlike elsewhere in the Northeast, it received a large number of exotic ornamental plants at an early date and landscaping practices in New York frequently became national trends. Within the Hudson River Valley the emergence of a popular aesthetic motivated the introduction of exotic plants to improve domestic landscapes and many species have become established and accepted as permanent additions to the region's flora. As unintended negative consequences are recognized, action is increasingly taken by conservationists and policymakers—including initiatives to promote use of native plants and control the sale of known pest plants.

LANDSCAPE GARDENING IN THE HUDSON RIVER VALLEY

When landscape gardening took root in North America during the first decades of the nineteenth century, infrastructure was already in place to meet the growing demand for exotic ornamental plants: the first commercial nursery in the United States was opened in Flushing, New York, in 1737 and seed importers and catalog-based businesses were also established (Mack and Erneberg 2002; Reichard and White 2001). By the 1830s widespread introduction and naturalization of exotic plants was underway in the northeastern United States as American landscapers increasingly sought "rare and foreign species" (Doell 1986; Downing 1865, 76; Mack 2003). This trend was encouraged and facilitated by horticultural and botanical societies and gardens that formed in Boston, Philadelphia, and New York City (Leighton 1987; Mickulas 2002). Philadelphia and Brooklyn were also made early landscaping centers by the "how-to" advice published by nurserymen Bernard McMahon and André Parmentier, but neither widely applied his skills.

Alexander Jackson Downing of Newburgh, New York, was perhaps the nation's first true practicing, publishing, and popular landscape gardener. Together with the founder of the New York Horticultural Society, Dr. David Hosack of Hyde Park, Downing advocated the introduction of exotic ornamental trees, shrubs, and vines to American gardens in order to give an area "a highly elegant or polished air" (Downing 1865, 76; Mickalus 2002).

Both men were influenced by the British Romantic landscaping traditions of the Beautiful and Picturesque, themes also reinforced in paintings by the Hudson River School. An example of the relationship of landscape gardening to landscape painting is the planting of trees to structure views of the Hudson River and Catskill Mountains, reminiscent of the "Claude glass" (Bunce 1994) or "verdant frame" (Doell 1986) employed by earlier European Romantic artists.

The species and designs promoted by Hosack and Downing are epitomized in the dozens of estates clustered in the Hudson River Historic District of Columbia and Dutchess counties. Indeed, Downing believed that "[t]here is no part of the Union where the taste in Landscape Gardening is so far advanced, as on the middle portion of the Hudson" (1865, 28). These country seats of wealthy businessmen, industrialists, politicians, writers, and artists exemplify the Romantic ideals of a high society with the financial ability to produce them in landscape. Landscape gardening with exotic plants was not limited to estates in the Historic District, however; Jarvis (1973), for example, described the carefully constructed Picturesque landscapes of many smaller lots in middle and upper-class neighborhoods in the Hudson River Valley where a pastoral ideal is created through strategic placement of features such as ponds, trees, paths, and stone walls.

Downing's influence in suburban areas is not surprising, as he sought to improve the architecture and landscaping of all types of residences. He promoted landscape gardening as an everyman's pursuit and in 1842 published a book focused on suburban properties entitled *Cottage Residences: A Series of Designs for Rural Cottages and Cottage Villas, and their Gardens and Grounds; Adapted to North America*. Seven years later he published the first of many editions of the best-selling *A Treatise on the Theory and Practice of Landscape Gardening, Adapted to North America; With a View to the Improvement of Country Residences*. These publications were available throughout the country, along with the *Horticulturist* magazine of which he was editor, and each recommended exotic ornamental plants to homeowners based on personal observations as well as experience at his Newburgh nursery and elsewhere in the Hudson River Valley. Downing effectively transformed entire regions by introducing landscape gardening to both large and small landowners.

NATURALIZED AND INVASIVE SPECIES

Throughout the Northeast, exotic ornamental plants have spread widely through catalogs, nurseries, and by friends and family sharing seeds and clippings; many species ultimately escaped cultivation to become established in natural habitats (Mack and Erneberg 2002; Mack 2003). When a species is able to sustain populations over time without human assistance it is said to be naturalized, which is often attributed to organism-specific characteristics including high reproduction rates, widespread dispersal, tolerance of a range of environmental conditions, and lack of predators in the new area (Mitchell and Power 2003; Starfinger et al. 2003; Thuiller et al. 2006). Environmental traits such as similarity of climate and habitat between native and introduced ranges also contribute to successful introductions, so it is no surprise that many of the naturalized exotic plants in the Hudson River Valley are native to temperate Eurasia (Stohlgren et al. 2006). Cheap plants are also more likely to become naturalized because large quantities are purchased by many people (Dehnen-Schmutz et al. 2007).

Within the United States, nonnative species are most commonly found in coastal regions, where human population is greatest, and where local economies are strongest (McKinney 2004; Stohlgren et al. 2006; Taylor and Irwin 2004). Nationwide there is also a positive correlation between nonnative plant distribution and real estate gross state product, an indicator of economic strength that includes the value of the landscaping industry (Taylor and Irwin 2004). The introduction and dispersal of exotic plants in the Northeast has particularly been encouraged by a long settlement history, high concentration of ports, dense transportation networks, and extensive commercial and landscaping activities—the Hudson River Valley is therefore well positioned to receive and support naturalized species (Mack 2003; Pauchard and Shea 2006). Whitney (1994) observed that urban centers are especially characterized by exotic flora and surrounded by a ring of suburban landscapes hosting exotic species brought by residents moving out from the city. Proximity to New York City is indeed a likely factor in the establishment of exotic ornamentals in the lower Hudson River Valley's higher-income suburban areas (Duncan 1973).

A comparison of data from the New York State Flora Atlas (Weldy and Werier 2009) with the 2000 United States Census reveals that a higher number of nonnative species indeed tends to correspond to higher population density, housing density, and number of native plants in the Hudson River Valley. This finding follows those of other studies observing that nonnative species often cluster around human habitation and that prime habitats with high native biodiversity are likely to have more naturalized species (Lonsdale 1999; Stohlgren et al. 1999; Stohlgren et al. 2005; Stohlgren et al. 2006). Today there are approximately 1,400 nonnative plant species in New York State, representing nearly 40 percent of the total flora; within the Hudson River Valley the proportion is similar (Weldy and Werier 2009).

In recent decades there has been an increasing interest in naturalized species because many become invasive, meaning they establish large populations with harmful impacts in their new environments. In 1999 the National Invasive Species Council defined "invasive species" as any nonnative species likely to cause economic and/or environmental harm in its introduced range. Invasive plants are known to change landscapes by causing extinction of endemic species, altering ecosystem processes, disrupting relationships between native species, and impacting fire frequency and pest outbreaks (Bergman et al. 2000; Clavero and Garcia-Berthou 2005; Ehrenfeld 2003; Manchester and Bullock 2000; Mooney and Cleland 2001). From data given by the Invasive Plant Council of New York State in 2007, it is estimated that 18 percent of naturalized plants within the Hudson River Valley are invasive (172 species). Of 129 species assessed by the Brooklyn Botanic Garden in 2009, the vast majority of invasive species were introduced through intentional cultivation—especially as ornamentals (Cornell University 2008). Because their environmental impacts are of greatest concern, this chapter uses invasive species to illustrate how exotic ornamental plants contribute to the Hudson River Valley landscape.

LANDSCAPING TRENDS

What follows is a description of landscaping trends in America that first became popular in the Northeast, primarily in the nineteenth century. Lists of invasive plants in the Hudson River Valley were then evaluated to determine which species were introduced for use as ornamentals, with emphasis on species recommended by Downing in his 1865 *Treatise*. It is believed that these species were widely available in the Hudson River Valley and elsewhere in the Northeast, an assumption verified by garden historian Denise Adams in her 2004 book *Restoring American Gardens: An Encyclopedia of Heirloom Ornamental Plants 1640–1940*, which lists plants available to the public by nurseries and catalog. Similarly, Ann Leighton's 1987 *American Gardens of the Nineteenth Century "For Comfort and Affluence"* lists plants available before 1900. These and other exotic ornamentals have since been found to alter landscapes in the Hudson River Valley, but after nearly two centuries of establishment and ongoing marketing they have generally become accepted additions to native flora and many of their impacts remain unknown.

Trees

Downing believed that "[a]mong all the materials at our disposal for the embellishment of country residences, none are at once so highly ornamental, so indispensable, and so easily managed, as trees, or wood" (1865, 69). Trees frame views, hide landscape blemishes, and separate fields; evergreens and broadleaf trees can be combined to ensure year-round greenery, and a variety of tree shapes lends itself to smooth transitions between woodlots, houselots, roadways, and other landscape features. Advertised for their blossoms or foliage, the invasive trees European sycamore maple (*Acer pseudoplatanus*), Japanese princess tree (*Pawlonia tomentosa*), and Chinese White mulberry (*Morus alba*), were likely established by 1859 and definitely by 1870 (Cornell University 2008; Leighton 1987). Among the species recommended by Downing that are currently invasive in the Hudson River Valley are the Eurasian Norway maple (*Acer platanoides*) and Chinese tree of heaven (*Ailanthus altissima*).

Downing specified the Norway maple as an exotic alternative to native maples for its wider crown and faster growth; different varieties have since been developed to have colored or variegated leaves. It creates dense stands that shade out native species, and because it is unpalatable to deer, shade-tolerant, has a high reproductive rate, and seeds tend to fall near the parent tree, it is able to maintain these monocultures (Cornell University 2008; Martin 1999; Swearingen et al. 2002). Furthermore, Norway maples experience earlier leaf-out and later leaf-drop than native species, giving it an advantage in colonizing open or disturbed areas. Norway maples are the preferred host of the Asian long-horned beetle (*Anoplophora glabripennis*), another invasive species.

Downing recommended the tree of heaven because "its fine long foliage catches the light well, and contrasts strikingly with that of the round-leaved trees" (1865, 203). The tree of heaven was introduced to the United States by 1784 and sold in New York State by 1826 from a Flushing nursery; it was also sold in the New York metropolitan area by at least two other nurseries in the 1840s (Adams 2004). By the mid-nineteenth century Downing noted that it was already "one of the commonest ornamental trees sold in the nurseries" (1865, 204), desirable as a tolerant shade and street tree that readily naturalized and was able to survive in the most forbidding of places (Leighton 1987; Mack 2003). The species also has allelopathic properties, which prevent other species from germinating or surviving near it (Heisey 1990). Its offensive odor was considered a serious disincentive for planting (Leighton 1987) but it continues to be successful in a range of soils, habitats, and moisture and disturbance regimes; it is a favorite city tree for its tolerance to pollution (Cornell University 2008; Hu 1979). Today the tree of heaven is highly visible in cities and along roads and railroads in the Hudson River Valley.

Hedges and Shrubs

Hedges have been used for centuries to mark property boundaries and divide pastures and lawns, and beginning in the 1850s an increased variety of exotic plant species were introduced to North America for these purposes (Adams 2004). Downing advocated their use as an aesthetically pleasing alternative to fences or walls, and shrubs were also used in lawns, beside walkways, and, beginning in the 1920s, along house foundations. Unfortunately, many of these domestic plantings have escaped and become established in forests and fields where they may form impenetrable thickets. Many invasive shrubs including the Asian autumn olive (*Eleagnus umbellata*) and burning bush (*Euonymus alatus*) create new canopy layers in forests in the northeast and inhibit growth of native species below them.

Many invasive shrubs are successful because they are unpalatable to deer, including the thorned Japanese barberry (*Berberis thunbergii*) which was valued for its white flowers and red berries (Fig. 13.1). Recommended for use by 1839 (Leighton 1987), it is known to have been brought to Boston's and New York's botanical gardens as an ornamental

FIG. 13.1. Japanese barberry in downtown Albany, Albany County (photo by author, 2009).

FIG. 13.2. Young Japanese barberry beginning to form the understory of a forest on the Wilderstein estate in Rhinebeck, Columbia County (photo by author, 2009).

shrub around 1859 (Adams 2004; Silander and Klepeis 1999; Taylor 1965). Japanese barberry was commercially available by 1910 and commonly found in New England and mid-Atlantic nursery catalogues within fifteen years (Adams 2004). By the 1920s it was established in areas where residents from eastern cities vacationed, and at mid-century it was naturalized in forests throughout the Hudson River Valley (Ehrenfeld 1997; Silander and Klepeis 1999). It creates a dense, spiny understory in northeastern and midwestern forests, including much of the Hudson River Valley, and changes soil chemistry and microbiology (Fig. 13.2). Japanese barberry can serve as an alternate host for a type of cereal rust and is outlawed in some states (Taylor 1965).

The honeysuckles (*Lonicera* spp.) are other Asian imports intentionally introduced to North America in the 1860s and now present throughout the northeast and Hudson River Valley (Adams 2004; Belote and Weltzin 2006; Schierenbeck 2004). The bush honeysuckles (Tartarian [*L. tatarica*], Morrow's [*L. morrowii*], and Amur [*L. mackii*]) have high reproductive rates and can form thickets that create deep shade, preventing native species from germinating. They are also unpalatable to deer and therefore unaffected by browsing. Tartarian honeysuckle was praised as the best among ornamental shrubs in 1870 and others were advertised for their blossoms and as hedges after 1839 (Leighton 1987). Additional hedge species that have recently become invasive in the Hudson River Valley are privets (*Ligustrum* spp.), which are common in urban environments and likely to become established in forests throughout the Northeast. Privets are native to Eurasia, North Africa, and Australasia and in the nineteenth century at least one species was advertised for its flowers (1859) and clipping tolerance (1870) (Leighton 1987). Privet is also unpalatable to deer.

Among the species not yet identified as invasive but with the potential to become harmful is the paper mulberry (*Broussonettia papyrifera*), native to Japan and some southeast Asian islands. In the Hudson River Valley it is found mainly in Rockland and Westchester Counties, but elsewhere is common in gardens for the "exotic look" of its foliage, red berries, and rapid growth (Downing 1865, 187). The species tolerates pollution and is a vigorous sprouter, and although known to be a nuisance it was still pushed for cautious planting in 1965: "So easy as to be dangerous, as its ability to become a pest is notorious" (Taylor 1965, 115).

VINES AND CLIMBING SHRUBS

Vines were promoted for landscaping and house adornment in America as early as the mid-eighteenth century (Adams 2004). They were used to obscure landscape and house defects, to make homes appear picturesque, and created a harmonious landscape by providing a living link between the house and grounds (Doell 1986; Downing 1865). The uses of vines were so varied that Adams declared, "If we could designate any specific plant type as symbolic of historic American garden style, it would have to be the vine or climbing shrub" (2004, 133). Some vines like the Asian chocolate vine (*Akebia quinata*—"already quite a favorite" by 1870) appear to be restricted to house sites but have the potential to spread into nearby forests (Leighton 1987, 374).

Among the vines found to be invasive in forests is the Japanese honeysuckle (*Lonicera japonica*), related to the bush honeysuckles and found as a dominant understory plant in temperate deciduous forests in much of the United States (Belote and Weltzin 2006). It was valued for its fast climbing ability, drought- and shade-tolerance, and semi-evergreen nature; it was used as an ornamental screen but can break branches with its weight, girdle trees by tightly winding around them, and effectively smother small plants (Adams 2004; Cornell University 2008). At least one white-flowered variety, "Halliana," was listed as early as 1823 in a seed catalog (Mack and Erneberg 2002) and was widely available by 1870 (Adams 2004).

First mentioned in the commercial literature around 1860, oriental bittersweet (*Celastrus orbiculatus*) is also extremely invasive (Adams 2004) (Fig. 13.3). Its high growth rate and flashy berries made it an attractive ornamental, and with the aid of berry-eating birds this vine has spread throughout Hudson River Valley forests and fields (Silveri et al. 2001) (Fig. 13.4). It has been shown to inhibit forest regeneration, leaving some areas dominated by the vine and herbaceous species rather than shrubs and trees (Fike and Niering 1999). It also competes

FIG. 13.3. Oriental bittersweet planted along a fence in Castleton, Rensselaer County (photo by author, 2009).

FIG. 13.4. Oriental bittersweet strangles a young tree near Staatsburg, Columbia County (photo by author, 2009).

and hybridizes with the rare native bittersweet (*C. scandens*) and there is tentative evidence for changes in soil chemistry and microbiology (Swearingen et al. 2002).

The Mediterranean black swallow-wort (*Cynanchum louiseae*) and Eurasian swallow-wort (*C. rossicum*) are vines with small black or pale star-shaped flowers. The earliest records date from 1854 near Boston and 1871 in Brooklyn; within a decade they were reported naturalized in New Rochelle, New York, and found to have escaped from gardens from New England to Pennsylvania and west to Ohio (Sheeley and Raynal 1996). As a member of the milkweed family, swallow-wort is problematic because monarch butterfly (*Danaus plexippus*) eggs laid on its leaves typically do not survive (Mattila and Otis 2003). Swallow-wort infestations are linked to the presence of insect and fungal pests, decreased grassland bird nesting, and reduced invertebrate and vertebrate diversity (DiTommaso et al.

2005). Today these vines have population centers in the Hudson River Valley and along the southern shore of Lake Ontario. This two-part distribution may be due to both regions' horticultural history; the Mt. Hope Botanical and Pomological Gardens in Rochester was the largest in the world in the 1880s after being founded by a previous employee of the Flushing nursery (Parks 1983).

Porcelain berry (*Ampelopsis brevipedunculata*) is another potentially invasive Asian ornamental vine, mainly naturalized in disturbed areas and observed in portions of the southern Hudson River Valley and the City of Albany. The vine is valued for its blue berries and changeable foliage (young leaves of the commercial cultivar "Elegans" are white and pink, later turning to green). It can tolerate a range of environmental conditions and is a rapid grower that reproduces both by seed and vegetatively, allowing it to shade out or smother native vegetation; porcelain berry is also a preferred food of the invasive Japanese beetle (*Popillia japonica*) (Swearingen et al. 2002).

WATER GARDENS

Terrestrial landscaping has long been complemented by natural and constructed aquatic features. Plumptre (1993) stated that water bodies were typically architectural and without plantings in Western landscaping, but in eighteenth-century Britain water gardening began to embrace a more "natural" look that moved away from geometry and toward integration with local topography and vegetation (Currie 1989). The installation of plants in American ornamental ponds was inspired by British as well as Asian traditions and became popular in the 1850s (Adams 2004). The purpose of aquatic vegetation is multifold: to provide additional ornament to grounds, oxygenate water, and provide food and shelter for fish (Plumptre 1993; Sawyer and Perkins 1928). Fish themselves were often introduced as exotic ornamentals, and Asian goldfish (*Carassius auratus*) are now naturalized in the Hudson River. Similarly, the invasive Eurasian mute swan (*Cygnus olor*) was present by 1900 as an ornamental on coastal New York and New Jersey estates and park grounds. Like landscape gardening, water gardening was popular among all types of property owners—at least one guide to creating an ornamental pond was not only for estate owners, but directed toward the "everyday man or woman" (Sawyer and Perkins 1928, 8).

One American technique used to create the impression of a wild scene in and around ornamental ponds was random planting and scattering of seeds (Doell 1986). Among the plants included in these efforts was the Eurasian and North African yellow flag (*Iris pseudacorus*), a yellow-flowered, rapidly growing wetland plant that forms single-species stands. The rhizomes in these stands may become so dense as to interfere with the establishment of other aquatic plants, including native *Iris* species, and alters hydrology by increasing sedimentation and elevating topography (Cornell University 2008; Stone 2009). Yellow flag is found in American horticultural literature by 1859 (Leighton 1987) and historic gardens may continue using this plant to recreate period water gardens (Fig. 13.5). It is widely found in the tidal wetlands of Columbia, Dutchess, and Putnam Counties (Fig. 13.6).

Another species that may have become naturalized through the process of scattering seeds in wetlands is the Eurasian purple loosestrife (*Lythrum salicaria*). Like yellow flag, purple loosestrife can rapidly colonize an area and outcompete native aquatic vegetation such as cattail (*Typha* spp.) due to its high seed production and fast growth as well as unpalatability to grazers (Thompson et al. 1987). It grows in a variety of soils and can adapt to most site conditions. Listed in commercial literature by 1835, purple loosestrife was commonly found in New England nursery catalogues by the following decade

FIG. 13.5. Yellow flag at Clermont State Historic Site, Germantown, Columbia County (photo by author, 2009).

FIG. 13.6. Yellow flag naturalized in Stockport Flats on the Hudson River, Columbia County (photo by author, 2009).

and widely used in garden plans at the turn of the century as a "Handsome border flower" (Adams 2004; Leighton 1987, 338). Purple loosestrife was deemed problematic in the mid-twentieth century but is still available in nurseries (Shadel and Molofsky 2002). Today it is found in nearly every state and in every Hudson River Valley county, including the four intertidal wetlands of the Hudson River National Estuarine Research Reserve (Laba et al. 2008).

LANDSCAPE CHANGE IN THE HUDSON RIVER VALLEY

The most significant similarity among the plants listed here is the initial care given to them during introduction, which protected them from environmental instability. They also share a few additional characteristics that have helped them become naturalized, including unpalatability to deer and tendency to outcompete native biota. Because deer avoid many naturalized exotic plants, browsing is directed toward edible native species. The disturbance and open space created by heavy deer browsing also creates ideal habitats for the further establishment of some exotic species. Exotic plants may therefore increase in number because they are not consumed and because additional habitats are opened.

The rate at which many exotic species can colonize an open area also has consequences for native plants that require well-lit habitats. Some species of birch (*Betula*), aspen (*Populus*), American ash (*Fraxinus americana*), and other sun-loving trees may not germinate underneath the dense canopy that can form by some of the exotics described here. Even shade-tolerant native species may have difficulty establishing under the canopy of some exotics; sugar maple (*Acer saccharum*), for example, has difficulty competing with the exotic Norway maple.

Within wetland habitats, the vigorous growth of yellow flag and purple loosestrife is known to inhibit the establishment of native species like cattail. The extensive stands of purple loosestrife in Hudson River Valley marshes reduce available habitat for bog turtles (*Clemmys muhlenbergi*), forage for waterfowl and muskrats (*Ondatra zibethicus*), and nesting sites for birds such as long-billed marsh wrens (*Cistothorus palustris*) that prefer cattail stands (Thompson et al. 1987). Diverse native northeastern riparian areas may be replaced by simplified stands of yellow or purple flowers that do not support as many native species.

Not all native species will decrease in number or size, however; among the native species that could become more prevalent in forests invaded by exotic ornamentals is the black birch (*Betula lenta*) because it rapidly reseeds in disturbed areas and deer avoid it (Ward et al. 2006). Beech (*Fagus grandifolia*) can persist for more than a century in a forest understory and may likewise become more common. In wetlands, red-winged blackbirds (*Agelaius phoeniceus*) have been shown to prefer nesting in purple loosestrife stands and may become more visible (Thompson et al. 1987).

While the exact impact of exotic ornamental species on Hudson River Valley forests and riparian areas is unknown and unlikely to be fully recognized until more areas are invaded and research progresses, regional landscape change is unavoidable. Fortunately, awareness of the problem has increased and possible remedies suggested—including regulations guiding what plants may be sold and publications suggesting native alternatives. A new era favoring native flora may indeed be on the horizon.

REGULATIONS AND RECOMMENDATIONS

Despite the tremendous cost incurred by the United States in managing nonnative species, estimated by Pimental et al. (2005) as approaching $120 billion per year, landscape gardening remains a major industry demanding attention by conservationists (Niemeira and von Holle 2009). In 1997 alone, the United States Department of Agriculture estimated that the floriculture/horticulture sector earned $11.2 billion (Reichard and White 2001), placing economic concerns at the center of regulatory initiatives. Accordingly, the New York State Department of Environmental Conservation is developing an assessment tool that would account for the financial impacts of control. A phaseout period may be instated, for example, before a particular plant is regulated so that nurseries may sell their remaining stock without penalty. Nursery professionals are also working with scientists, municipal land managers, and Cornell Cooperative Extension agents to approve invasive species assessments done by Brooklyn Botanic Garden. In 2003 the New York State Nursery and Landscape Association also joined the New York State Invasive Species Task Force. Reichard and White (2001) have shown that people would avoid purchasing or planting exotic species known to have negative environmental impacts, making such involvement by professional landscapers and nursery owners essential.

Because gardening has proven to be a trend-driven practice influenced by popular taste, strong avocation for the use of native plants in landscaping will also be effective at controlling invasive species. Efforts to inform the public on native species that may be used instead of exotic ornamentals have ranged from brief suggestions on the New York State Department of Transportation Web site (www.nysdot.gov) to entire volumes of instruction on how to grow native North American plants (e.g. Cullina 2002). Some municipalities, including Westchester County, have passed legislation mandating use of only native plants in county-owned properties; Westchester Community College also runs The Native Plant Center, whose mission is "Educating people about the environmental necessity, economic value and natural beauty of native plants in the Northeast" (www.nativeplantcenter.org).

Although the cooperation of various stakeholders has resulted in legislation banning some exotic ornamental species within the next decade, the cultivation of existing plants and introduction of new species suggests that exotic ornamentals will continue to become naturalized and increasingly impact the Hudson River Valley landscape. There are very few national regulations controlling exotic ornamental species; the importation, purchase, and introduction of exotic species into the United States is legal so long as insects, pathogens, noxious weeds, or endangered species are not among them (Reichard and White 2001). What constitutes a "noxious weed," however, is not standardized and it is left to states to identify which species are harmful.

Much of the information presented here is available to the public through the U.S. Department of Agriculture Plants Database (http://plants.usda.gov), *Invasive Plants Field and Reference Guide* (Huebner et al. 2007), National Park Service and U.S. Fish and Wildlife Service's *Plant Invaders of Mid-Atlantic Natural Areas* (Swearingen et al. 2002), the New York Flora Atlas (Weldy and Werier 2009), *New York State Museum Circular* Number 57 (Mills et al. 1997), the New York State Invasive Species Clearinghouse (Cornell University 2008), and the Invasive Plant Council of New York State (2005). Although this last organization has disbanded, data remain available from its constituent Partnerships for Regional Invasive Species Management (PRISMs), which work with state agencies, the public, conservation organizations, and trade stakeholders to address invasive species and their impacts. The Hudson River Valley region is covered by three PRISMs: Capital-Mohawk, Catskill Regional Invasive Species Partnership, and Lower Hudson.

CONCLUSION

Some scientists believe opinion and ideology should be excluded from studies of exotic species, but introduction, regulation, and management are largely informed by social values. Ecological studies on invasive species are ever-increasing, yet few studies assess public attitudes toward eradication or other control measures for invasive species. After centuries of introductions some exotic ornamental plants remain established as a result of positive public perception and are so common that they are not widely known to be nonnative. In many cases, these species become accepted as part of the local flora and the cycle of introduction and naturalization continues.

Research on regional histories of exotic species is also necessary because the historic value of some species might be place-based. For example, period-sensitive conservation of historic properties and landscapes often calls for maintenance of gardens as they appeared during the target era. The potential harm caused by exotic ornamental plants should be factored into such plans because naturalization is related to length of establishment and frequency of plantings. It is also difficult to coordinate management efforts in a mosaic of private and public lands, but the growing combination of legislative control and education on alternatives suggests that landscape gardening may be generally moving away from exotic ornamentals and toward native species.

REFERENCES CITED

Adams, D. W. 2004. *Restoring American gardens: An encyclopedia of heirloom ornamental plants 1640–1940.* Portland: Timber Press.

Belote, R. T., and J. F. Weltzin. 2006: Interactions between two co-dominant, invasive plants in the understory of a temperate deciduous forest. *Biological Invasions* 8: 1629–41.

Bergman, D. L., M. D. Chandler, and A. Locklear. 2000: The economic impact of invasive species to wildlife services' cooperators. *Proceedings of the Third National Wildlife Research Center Special Symposium: Human Conflicts with Wildlife: Economic Considerations.* United States Department of Agriculture, 169–78.

Bunce, M. 1994. *The countryside ideal: Anglo-American images of landscape.* London: Routledge.

Chambers, D. 1991. Painting with living pencils: Lord Petre. *Garden History* 19(1): 60–76.

Clavero, M., and E. Garcia-Berthou. 2005. Invasive species are a leading cause of animal extinctions. *Trends in Ecology and Evolution* 20(3): 110.

Cornell University. 2008. New York State Invasive Species Clearinghouse. www.nyis.info/Resources/IS_Risk_Assessment.aspx.

Cullina, W. 2002. *Native trees, shrubs, and vines: A guide to using, growing, and propagating North American woody plants.* Boston: Houghton Mifflin.

Currie, C. K. 1989. Fishponds as garden features, c. 1550–1750. *Garden History* 18(1): 22–46.

Dehnen-Schmutz, K., J. Touza, C. Perrings, and M. Williamson. 2007. A century of ornamental plant trade and its impact on invasion success. *Diversity and Distributions* 13: 527–34.

DiTommaso, A., F. M. Lawlor, and S. J. Darbyshire. 2005. The biology of invasive alien plants in Canada. 2. *Cynanchum rossicum* (Kleopow) Borhidi [=*Vincetoxicum rossicum* (Kleopow) Barbar.] and *Cynanchum louisea* (L.) Kartesz & Gandhi [=*Vincetoxicum nigrum* (L.) Moench]. *Canadian Journal of Plant Sciences* 85: 243–63.

Doell, M. C. K. 1986. *Gardens of the Gilded Age: Nineteenth-century gardens and homegrounds of New York State.* Syracuse: Syracuse University Press.

Downing, A. J. 1991. *Landscape gardening and rural architecture.* Toronto: General Publishing Company. Reprint of Downing, A. J. 1865. *A treatise on the theory and practice of landscape gardening, adapted to North America; With a view to the improvement of country residences.* New York: Orange Judd Agricultural Book Publisher.

Duncan, J. S. 1973. Landscape taste as a symbol of group identity: A Westchester County village. *Geographical Review* 63(3): 334–55.

Ehrenfeld, J. G. 1997. Invasion of deciduous forest preserves in the New York Metropolitan Region by Japanese barberry (*Berberis thunbergii* DC.). *Journal of the Torrey Botanical Society* 124(2): 210–15.

———. 2003. Effects of exotic plant invasions on soil nutrient cycling processes. *Ecosystems* 6: 503–23.

Heisey, R. M. 1990. Allelopathic and herbicidal effects of extracts from tree of heaven (*Ailanthus altissima*). *American Journal of Botany* 77(5): 662–70.

Hu, S. Y. 1979. Ailanthus. *Arnoldia* 39(2): 29–50.

Huebner, C. D., C. Olson, and H. C. Smith. 2007. *Invasive plants field and reference guide: An ecological perspective of plant invaders of forests and woodlands.* United States Department of Agriculture.

Invasive Plant Council of New York State. 2005. New York State Early Detection Invasive Plants by Region, Assessment of Naturalized Invasive Plants. www.ipcnys.org.

Jarvis, P. J. 1973. North American plants and horticultural innovation in England, 1550–1700. *The Geographical Review* 63(4): 477–99.

Laba, M., R. Downs, S. Smith, S. Welsh, C. Neider, S. White, M. Richmond, W. Philpot, and P. Baveye. 2008. Mapping invasive wetland plants in the Hudson River National Estuarine Research Reserve using quickbird satellite imagery. *Remote Sensing of the Environment* 112: 286–300.

Leighton, A. 1987. *American gardens of the nineteenth century: "For comfort and affluence."* Amherst: University of Massachusetts Press.

Lonsdale, W. M. 1999. Global patterns of plant invasions and the concept of invasibility. *Ecology* 80(5): 1522–36.

Mack, R. 2003. Plant naturalizations and invasions in the Eastern United States: 1634–1860. *Annals of the Missouri Botanical Garden* 90(1): 77–90.

———, and M. Erneberg. 2002. The U.S. naturalized flora: Largely the product of deliberate introductions. *Annals of the Missouri Botanical Garden* 89: 176–89.

Manchester, S. J., and J. M. Bullock. 2000. The impacts of non-native species on UK biodiversity and the effectiveness of control. *Journal of Applied Ecology* 37: 845–64.

Martin, P. H. 1999. Norway maple (*Acer platanoides*) invasion of a natural forest stand: understory consequence and regeneration pattern. *Biological Invasions* 1: 215–22.

Mattila, H., and G. W. Otis. 2003. A comparison of the host preference of monarch butterlies (*Danaus plexippus*) for milkweed (*Asclepias syriaca*) over dog-strangler vine (*Vincetoxicum rossicum*). *Entomologia Experimentalis et Applicata* 107: 193–99.

McKinney, M. L. 2004. Citizens as propagules for exotic plants: measurement and management implications. *Weed Technology* 18: 1480–83.

Mickalus, P. 2002. Cultivating the Big Apple: The New York Horticultural Society, nineteenth-century New York botany, and the New York Botanical Garden. *New York History* 83(1): 34–54.

Mills, E. L., M. D. Scheurell, J. T. Carlton, and D. L. Strayer. 1997. Biological invasions in the Hudson River Basin: An inventory and historical analysis. *New York State Museum Circular* No. 57. The New York State Education Department.

Mitchell, C. E., and A. G. Power. 2003. Release of invasive plants from fungal and viral pathogens. *Nature* 421: 625–27.

Mooney, H. A., and E. E. Cleland. 2001. The evolutionary impact of invasive species. *Proceedings of the National Academy of Science* 98(10): 5446–51.

Niemeira, A. X.. and B. von Holle. 2009. Invasive plants and the ornamental plant industry. In *Management of Invasive Weeds*, ed. Inderjit. The Netherlands: Springer Netherlands.

Parks, D. 1983. The cultivation of flower city. *Rochester History* XLV(3 and 4): 25–47.

Pauchard, A., and K. Shea. 2006. Integrating the study of non-native plant invasions across spatial scales. *Biological Invasions* 8: 399–413.

Pimentel, D., R. Zuniga, and D. Morrison. 2005. Update on the environmental and economic costs associated with alien-invasive species in the United States. *Ecological Economics* 52: 273–88.

Plumptre, G. 1993. *The water garden: Styles, designs, and visions.* New York: Thames and Hudson.

Reichard, S. H., and P. White. 2001. Horticulture as a pathway of invasive plant introductions in the United States. *BioScience* 51(2): 103–13.

Sawyer, R. V., and E. H. Perkins 1928: *Water gardens and goldfish.* New York: A. T. De La Mare.

Schierenbeck, K. A. 2004. Japanese honeysuckle (*Lonicera japonica*) as an invasive species; history, ecology, and context. *Critical Reviews in Plant Sciences* 23(5): 391–400.

Shadel, W. P., and J. Molofsky. 2002. Habitat and population effects on the germination and early survival of the invasive weed, *Lythrum salicaria* L. (purple loosestrife). *Biological Invasions* 4: 413–23.

Sheeley, S. E., and Raynal, D. J. 1996. The distribution and status of species of *Vincetoxicum* in eastern North America. *Bulletin of the Torrey Botanical Club* 123: 148–56.

Silander, J. A., Jr., and D. M. Klepeis. 1999. The invasion of Japanese barberry (*Berberis thunbergii*) in the New England landscape. *Biological Invasions* 1: 189–201.

Silveri, A., P. W. Dunwiddie, and H. J. Michaels. 2001. Logging and edaphic factors in the invasion of an Asian woody vine in a mesic North American forest. *Biological Invasions* 3: 379–89.

Starfinger, U., I. Kowarik, M. Rode, and H. Schepker. 2003. From desirable ornamental plant to pest to accepted addition to the flora?—The perception of alien tree species through the centuries. *Biological Invasions* 5: 323–35.

Stohlgren, T. J., D. Binkley, G. W. Chong, M. A. Kalkhan, L. D. Schell, K. A. Bull, Y. Otsuki, G. Newman, M. Bashkin, and Y. Son. 1999. Exotic plant species invade hot spots of native plant diversity. *Ecological Monographs* 69(1): 25–46.

Stohlgren, T. J., D. Barnett, C. Flather, J. Kartesz, and B. Peterjohn. 2005. Plant species invasions along the latitudinal gradient in the United Sates. *Ecology* 86(9): 2298–2309.

Stohlgren, T. J., D. Barnett, C. Flather, P. Fuller, B. Peterjohn, J. Kartesz, and L. L. Master 2006: Species richness and patterns of invasion in plants, birds, and fishes in the United States. *Biological Invasions* 8: 427–47.

Stone, K. R. 2009. *Iris pseudacorus*. In Fire Effects Information System, Online. U.S. Department of Agriculture, Forest Service, Rocky Mountain Research Station, Fire Sciences Laboratory (Producer). www.fs.fed.us/database/feis/.

Swearingen, J., K. Reshetiloff, B. Slattery, and S. Zwicker. 2002. *Plant invaders of mid-Atlantic natural areas*. National Park Service and U.S. Fish and Wildlife Service.

Taylor, B. W., and R. E. Irwin. 2004. Linking economic activities to the distribution of exotic plants. *Proceedings of the National Academy of Sciences* 101(51): 17725–30.

Taylor, N. 1965: *The guide to garden shrubs and trees (including woody vines): Their identity and culture*. Boston: Houghton Mifflin.

Thompson, D. Q., R. L. Stuckey, and E. B. Thompson. 1987. Spread, impact, and control of purple loosestrife (*Lythrum salicaria*) in North American wetlands. U.S. Fish and Wildlife Service. Jamestown, ND: Northern Prairie Wildlife Research Center Online. www.npwrc.usgs.gov/resource/plants/loosstrf/index.htm (Version 04JUN99).

Thuiller, W., D. M. Richardson, M. Rouget, S. Proches, and J. R. U. Wilson. 2006. Interactions between environment, species traits, and human uses describe patterns of plant invasions. *Ecology* 87(7): 1755–69.

Todd, K. 2001. *Tinkering with Eden: A natural history of exotic species in America*. New York: W. W. Norton.

Ward, J. S., T. E. Worthley, P. J. Smallidge, and K. P. Bennett. 2006. *Northeastern forest regeneration handbook: A guide for forest owners, harvesting practitioners, and public officials*. Newtown Square, PA: USDA Forest Service.

Weldy, T., and D. Werier. 2009. New York flora atlas. Albany: New York Flora Association. http://newyork.plantatlas.usf.edu.

Whitney, G. G. 1994. *From coastal wilderness to fruited plain: A history of environmental change in temperate North America 1500 to the present*. Cambridge: Cambridge University Press.

Wyckoff, P. H., and S. L. Webb. 1996. Understory influence of the invasive Norway maple (*Acer platanoides*). *Bulletin of the Torrey Botanical Club* 123: 197–205.

PART III

RIVER OF COMMERCE

Robert E. Henshaw

> On this river there is great traffick in the skins of beavers . . . and the like.
> The land is excellent and agreeable, full of nobel forest trees and grape vines,
> and nothing is wanting but the labor and industry of man . . .
> —de Laet, 1625

> The New York of today is the product of big business. It is a state of mind.
> Nothing done there is motivated by historic traditions.
> —Gauss, 1938

ENTREPRENEURIAL ZEAL

Commercial exploitation is the driving force of the Hudson River today, and has been from the first arrival of European explorers. Fifteenth and sixteenth-century Dutch businessmen sponsored voyages of exploration around the world seeking products and materials from faraway places to sell in European markets. According to European custom then, all resources "found" in new corners of the world were available for the discoverer's exploitation (Lewis 2005). Henry Hudson's 1609 voyage, sponsored by the Dutch East India Company, sought a route to the Orient for commercial spice trade. Dutch merchants, upon receiving the 1611 reports of Dutch explorers Adriaen Block and Hendrick Christiaensen of a land rich in lush forests and furbearing animals, promptly created the New Netherlands Company to exploit these new resources that were apparently available for the taking. Poorly managed, the company lasted only a few years. It was followed by the Dutch West India Company, which replaced the Dutch East India Company, and was armed with a monopoly from the Crown for all fur trade from the Hudson River.

From the beginning the Dutch businessmen sought simply to exploit the natural riches, not to establish a new political entity. Their later sponsorship of colonization was simply an expedient to merchandise the products of the Hudson River. The settlements were tolerated by England because it was clear that the Dutch were there for commercial purposes only and not to create a new Netherlands which might compete politically or religiously with the Virginia and New England colonies. Dutch merchants sought settlers for their new holdings based simply on their willingness to enter into a contract to create and return wealth to the sponsors. They required none of the religious or political qualifications imposed in the English colonies.

The village of New Amsterdam, founded by Adriaen Block on the tip of Mannahatta Island, was cosmopolitan from the first because it was established and governed largely as a strategic trading post (Wikipedia 2010). Within a few years it was reported to have eighteen nationalities among its residents, and to be inhabited by "every sort of immigrant denizen or foreigner" (de Vries 1643). Petrus Stuyvesant, the newly appointed Director General, was disappointed to find the people "slovenly, drunken, dishonest, and with poor business management" (Wikipedia 2009). By 1661 a visiting Englishman could still characterize the small city as "Ye Towne of Mannadens . . . seated between England and Virginia, commodiously for trade . . . inhabited with several sorts of tradesmen and

195

marchants [*sic*] and mariners" (de Vries 1643). Later writers such as Washington Irving poked fun at the boisterous citizenry of the Hudson Valley, and the nickname "Knickerbockers" stuck. The new Dutch colony was, in short, a polyglot, and it has remained so since.

Earliest harvesting of forests and furs gave way to production of live stock and crops, particularly wheat, and later to extractive industries that exploited forests for charcoal and tanning bark, and mining of the abundant mineral resources. With the coming of the industrial age, manufacturing and industry settled along the shores. Some industries relied on the river as a source for raw materials, others for transport of supplies or products. Increasingly, industries with no reliance on, and no emotional attachment to the Hudson were established. For many, their only reason for their location on the river was to facilitate disposal of wastes. Often they were tolerated simply as providers of jobs. Albany and especially Troy, blessed with abundant water power and access to Adirondack forests for charcoal and iron deposits, became prosperous smelting iron and producing iron stoves, horseshoes, and other products—the first "Pittsburgh" in America. The growing metropolis at the mouth of the river needed a ready supply of bricks, stone, and wood, and the Hudson region became that reliable source. By the early twentieth century there were more than two dozen brickworks lining the river from Ossining to just south of Troy (Hutton 2003; Brickcollecting 2010). The last yard closed in the mid-1970s. Limestone was, and still is, quarried in the mid-Hudson region where the Allegheny Mountains cross the Hudson River rift (see "An Abbreviated Geography" in this volume), and brown stone was blasted from the Palisades cliffs opposite Yonkers and the city of New York. The people in the city required ice to preserve food and for cooling. In chapter 14 Harris and Pickman describe the robust ice harvest industry that grew along the Hudson. Clothing manufacture attracted large numbers of European immigrants. Troy is still known as the "Collar City" because of the invention and marketing there of a collar that could be removed for cleaning and starching.

Because of its ideal location and ice-free harbor, New York City—New Amsterdam renamed during English possession—became a key center for trade and transshipment of goods. Initially, sugar and slaves from the Caribbean, grain and forest products from the north and west, and later cotton from the southern colonies, all moved through New York City en route to Europe. Eventually, because of its deep water port, available slave labor, and especially because it had unique financial resources, the city transshipped virtually all of the cotton shipped from the eastern Confederate states (Barrows and Wallace 1999, 335–36). Cotton became the city's number one commodity, and the most lucrative. New York City received an estimated 40 percent of all cotton revenues because the city supplied insurance, shipping, and financing services and New York merchants sold goods to Southern planters (Mississippi History Now 2010). Thus, by 1860 New York City had become the de facto capital of the American South (Mississippi History Now 2010). In fact, New York City rose to its preeminent position as the commercial and financial center of America in large part because of its dominant role in the transshipment of cotton bound for Europe (Barrows and Wallace 1999, 335–36). During the twentieth century large heavy industries, such as automobile assembly and wire manufacturing became common. One modern industry completely dependent on a riverbank location was power generation because of the enormous amount of water needed to cool the huge steam generators. By the mid-nineteenth century, two-thirds of all American imports and one-third of all exports moved through New York City (Becker 1993, 17). Driven by its key location and ruled virtually only by business motivations, the importance of New York City for business has carried through to its modern designation as the "World Trade Center."

TRANSPORTATION ENABLED GROWTH

Much of the Hudson's history of the last four hundred years is tied to transportation. Early colonists adapted boat designs familiar to them in the Netherlands, and in so doing created a unique boat design, the Hudson River Sloop, with massive sail to capture the limited winds. Hudson River sloops were the principal freight haulers for more than a century. Because they were subject to the vagaries of winds, tides, and river flow, their journey from one end of the estuary to the other could take up to two weeks. As commerce grew in the Hudson River Valley (see Litten, ch. 11; and Panetta, ch. 17), pressure built to find a more reliable and faster way to get products to market in New York City.

Experimentation on steam navigation had begun on the River Seine in Paris at the end of the eighteenth century. Predicting its success here before even the creation of a steam-driven boat, Hudson resident and Patroon Robert R. Livingston secured from the New York Legislature a monopoly for all steam-propelled shipping on the Hudson River for himself and business partner, inventor Robert Fulton—a true Hudson River entrepreneurial coup! Fulton began building steamboats as rapidly as possible attempting to take advantage of their monopoly, but he was unable to meet demand. The life of the Hudson River freight-hauling sloop thus was extended another forty years. Travel between Albany and New York was reduced to overnight and became reliable, ushering in a glorious romantic era of Hudson River steamboat travel.

The first steam railroad in this country was constructed between Albany and Schenectady in 1831, just six years after the opening of the Erie Canal. With that success Erastus Corning, an Albany entrepreneur, taking advantage of the sea-level estuary, laid railroad tracks southward along first the west shore and then the east shore of the river in the early 1850s with a crossing just north of Albany at Troy (Wikipedia 2010). History tends to ignore the *original* "robber baron" and awards the dubious title to Cornelius Vanderbilt, who did not gain control of the rail lines until 1867 after service between Albany and New York City had begun. Vanderbilt added another railroad river crossing at Poughkeepsie in 1887. From the overnight journey of steamboats, rail reduced the trip between Albany and New York to a few hours and freight from the interior did not need to be off-loaded to ferries for the last leg into the city. Freight hauling, as Panetta shows in chapter 17, shifted to rail, hastening an end to the age of steamboats on the Hudson. Even today the original railroad tracks along both shorelines carry much of the freight bound for New York City from all parts of the United States. But the days of rail supremacy, like the steamboats before them, were numbered. During the twentieth century modern highways enabled trucks to carry an increasing percentage of the city's freight. Panetta provides in chapter 17 a full elaboration of the history of transportation in the Hudson River Valley. He demonstrates that in many ways it mirrors virtually all of the other human uses impacting the ecology and how the changing technologies of transportation modes forced huge changes in the subsequent ways that humans could use the Hudson River Valley.

Early siting of industries directly on the shoreline permitted ready access to the river for transportation. Later, shoreline siting increasingly included non-river-dependent industries, particularly for ready access to railroad transportation.

EFFECTS OF COMMERCE

As extractive, manufacturing, and construction industries in the Hudson region grew and prospered, the effects of those uses began to accumulate. Some earlier industries, for example, harvesting of river ice, leave virtually no vestiges today. The effects of other industries remain pervasive. Lumbering, tanning bark harvest, and charcoal production virtually eliminated original forests in the Hudson Region. Mining of iron and limestone leaves gaping quarries.

Railroad tracks, at first welcomed even in the center of cities as a sign of progress, caused important unintended consequences along the river shorelines. They linearized parts of the river's shoreline with rip rapped gravel rail beds, removing natural riparian habitat for aquatic species. Where the rail lines had to cross small embayments where the river once meandered, they were placed either on trestles or on elevated gravel beds with culverts beneath. The former embayments suffered reduced flushing effect of the river, causing many to become shallow marshes. Interestingly, modern regulatory oversight would never countenance partial blockage of embayments if the rails were sited today, but the marshes created by nineteenth-century construction are now valued and protected as tidally flushed fresh water wetlands for their uncommon character. These marshes provide refuge for early life stages of many riverine species. Their high productivity when flushed into the open river contributes to the food webs there.

As industries prospered, many owners became extremely wealthy. They created vast estates along the shores of the Hudson and, wishing to make them distinctive to impress their neighbors, they imported many species of exotic plants. Those plants subsequently escaped, with varying effects on native vegetation as discussed by Teale in chapter 13.

Not only did the river provide direct routes to markets, it also provided a ready means of disposal of the wastes and byproducts of commerce. Tonjes et al., in chapter 15, discuss struggles of New York City with sanitary wastes as the city grew. In chapter 16 Levinton details the disposal and aftermath of cadmium wastes from a factory near Cold Spring. In the Upper Hudson, waste polychlorinated biphenols ("PCBs") were discharged from an electrical capacitor factory; they contaminated both the Upper Hudson and the Lower Hudson. The assimilative capacity was presumed until, by the early part of the twentieth century, the waters, the shores, and the air above were filled with waste materials; the landscape cluttered. A cynical "joke" among the citizens of Tarrytown said they knew what color the automobile assembly plant was using that day by the color of the river. Although many of the former industries were disbanded and the factories razed, chemical residuals in the river bottom sediments remain, most (in)famously the PCBs.

Beginning in the 1950s, at a time when public interest in the Hudson was minimal due to its deteriorated state, large power plants were placed in the southern portion of the Lower Hudson for easy supply to New York City—a problematic location in the "biological center" center of the river (see "An Abbreviated Geography" in this volume). Generation of electricity produces vast quantities of heat, which the utilities relied on the river to receive. As the plants drew in enormous quantities of cooling water they unavoidably captured and killed massive numbers of fish. To avoid such "fishkills" less cooling water would have to be withdrawn and any corrective technologies would have to be commensurately large and/or expensive. At the time, this meant reduced generation during fish spawning season or use of gigantic natural draft cooling towers to dissipate the heat. Interestingly, this created a conflict between environmental qualities—protect fish and deteriorate the striking scenic quality, or permit killing of fish to preserve the scenic quality. Seeking solutions, in the 1960s utilities began a research program on whether the losses of fish in power plants endanger the fish populations. These studies, which continue today, represent the largest aquatic research effort on any river in the world. In the current decade, dry cooling in place of evaporative cooling has been proposed by regulators, but this is extremely expensive and reduces the net efficiency of the power plant. For these reasons utilities resist dry cooling. Despite the amount of research, answers are elusive, and results are highly contentious. In chapter 18 Young and Dey, who conducted some of those studies, attempt the difficult task of review and explanation of why interpretation is so contentious.

By the first half of the twentieth century New York had created a perfect combination of conditions for environmental decline: tacit public support of self-justifying commercialism; communities isolated from the river by railroads having lost their emotional attachment and managerial interest in the river; no legal framework to regulate despoliation of the environment; and perhaps most pervasive, an overwhelming sense of powerlessness as commerce swept through the Hudson River Valley. Decades of region-wide citizen apathy and grudging tolerance of the river-in-decline ensued.

But this was not the traditional way for the Hudson River Valley's entrepreneurial residents. New York had also created a mix countervailing forces. In the 1800s many of the wealthy industrialists had settled (or at least built summer places) in the Hudson River Valley. They chose the location because of the scenic and recreational qualities it offered; they sought to be (or to own) a part of the Hudson's valued "grandeur"; and they were accustomed to having their way. Their zeal surely rubbed off on the working-class residents who, themselves, had such a long history of entrepreneurism. Combining forces in the last quarter of the twentieth century, they found their legs to resist continued environmental despoliation. The unique commercial history of the Hudson River region all but dictated that it would be the birthplace of the environmental movement.

REFERENCES CITED

Bard College Field Station. 2009. The Hudson River National Estuarine Research Reserve. Annandale-on-Hudson, NY: Bard College. http://inside.bard.edu/archaeology/tivolibays/hrnerr.html; accessed Dec. 16, 2009.

Barrows, E. G., and M. Wallace. 1999. *Gotham: A history of New York City to 1898.* New York: Oxford University Press, 335–36.

Beckert, S. 1993. *The monied metropolis: New York City and the consolidation of the American bourgeoisie, 1856–1896.* Cambridge: Cambridge University Press, 17.

de Laet, J. 1625. New world. In *Narratives of New Netherland, 1609–1664,* ed. J. F. Jameson, 29–58. New York: Barnes and Noble, 1909.

de Vries, D. P. 1643. In *Narratives of New Netherland, 1609–1664,* ed. J. F. Jameson, 183–243. New York: Barnes and Noble, 1909.

Gauss, C. 1938. In *The historical atlas of New York: A visual atlas of New York City's history.* New York: Henry Holt, 113.

Hudson River Brickmaking. 2010. http://brickcollecting.com/hudson.htm.

Hutton, G. V. 2003. *The great Hudson river brick industry: Commemorating three and a half centuries of brickmaking.* Fleischmanns, NY: Purple Mountain Press, 240.

Mechaelius, Reverend D. J. 1628. Letter. In *Narratives of New Netherland, 1609–1664,* ed. J. F. Jameson, 117–31. New York: Barnes and Noble, 1909.

Mississippi History Now. 2010. http://mshistory.k12.ms.us/; accessed in 2010.

Wikipedia. 2010. History of New York City (1665–1783). http://en.wikipedia.org/wiki/History_of_New_York_City_(1665–1783); accessed in 2010.

———. 2010. History of railroads to New York City. http://en.wikipedia.org/wiki/New_York_Central_Railroad; accessed in 2010.

CHAPTER 14

THE RISE AND DEMISE OF THE HUDSON RIVER ICE HARVESTING INDUSTRY

Urban Needs and Rural Responses

Wendy E. Harris and Arnold Pickman

ABSTRACT

The origins of Hudson River ice harvesting can be traced to late-eighteenth-century New York City, where population growth, pollution, and technological advance resulted in the destruction of nearby water sources. By the second half of the nineteenth century, an enormous ice industry had developed along the rural shorelines of the Hudson River, north of Poughkeepsie. This unique industry involved the transformation of an important environmental resource—the waters of the Hudson River itself—into a commodity, represented by standard-sized blocks of ice that were stored in enormous ice houses until the summer months and then shipped by barge down the river to be consumed by urban dwellers. The interplay between the changing demands of distant urban markets and the technological capabilities of the ice industry had dramatic consequences—not only for the cultures and economies of local Hudson River communities but for shoreline appearance and morphology as well. Ironically, the demise of the Hudson River ice industry can be attributed to the same forces that drove its creation.

INTRODUCTION

The human desire for ice and the ability to harvest and store it date back at least four thousand years, to ancient Babylon and China (David 1995, xi). In the United States, the earliest documented ice harvesting occurred at Jamestown in the seventeenth century (Cotter and Hudson 1957, 10). By the eighteenth century it was widely practiced throughout the colonies (Belden 1983, 145–68; Funderberg 1995, 3–32). In New York City, at that time, the demand for ice was satisfied by the exploitation of water sources located on the city's immediate outskirts. In time, as a result of the city's expansion, these sources were destroyed and New Yorkers were forced to look elsewhere for their ice. By the third quarter of the nineteenth century, the city would come to rely upon the frozen waters of the mid- and upper Hudson River. During this period, ice was transformed into the greatest of the country's natural resources, and the Hudson River became America's greatest ice producer.

The story of the Hudson River ice industry is a story about the river's frozen waters, its shorelines, landscapes, and communities, but it is also the story about the enormous market for ice located at the river's mouth—New York City. "Ice," said Henry David Thoreau (1904, 231), "is an interesting subject for contemplation." To understand the significance of the Hudson River ice harvesting industry's history, we suggest that the reader follow Thoreau's suggestion and contemplate what ice must have meant to the generations of New Yorkers who lived their lives before fans, air conditioning, and electric-powered refrigerators. In the middle of the nineteenth century, a reporter—most likely an urban dweller—writing about what he termed the "consequences of ice," described:

> cold water on the table, cooling pillows for the sick, antiseptic layers for meats and fruits, cobblers, juleps, and smashes, neat plates of well-moulded butter, wines not flat but bright and sparkling. (*New York Times* 1865)

Ice helped New Yorkers endure the summer heat, allowed them to create such hot weather treats as water ices and ice cream, cooled their food and beverages, and brought them some relief during the periodic epidemics that swept through crowded neighborhoods killing thousands.

In the following discussion, we examine the relationships between changing patterns of consumption among the urban consumers who constituted the New York City market for ice and the Hudson River communities and landscapes that were the settings for ice production. We trace these relationships within the framework of a feedback model that considers how human uses and shifts in human uses affect the character of the environment, and how, in turn, an altered environment may drive subsequent human uses. Specifically, in the case of the ice industry, we focus upon the processes of population growth, industrial expansion, and technological advance. We will look at the roles these processes played in two distinct environmental, social and temporal settings—late-eighteenth/early-nineteenth-century Manhattan Island and the Hudson River in the late nineteenth/early twentieth century—and also how these processes shaped the rise as well as the demise of the once thriving ice industry.

EIGHTEENTH-CENTURY USE OF ICE

In the eighteenth century, when first encountered in New York City's historical record, commercially distributed ice and foods requiring ice for their preparation were luxuries enjoyed by the wealthy. The earliest documentary evidence suggesting that ice was being harvested, sold, and used commercially in New York City comes in the form of advertisements placed in newspapers by confectioners, many of French and Italian descent, who had emigrated to North America in order to ply their trade. Arriving here, they brought with them a great culinary novelty—ice cream—the consumption of which had lately become a craze among fashionable circles in Britain and on the Continent (David 1995, 316). Anglo-American elites soon developed a taste for ice cream and fruit-infused water ices, thus creating—for the first time in the city's history—a market for ice.

Among the earliest New York confectioners to sell ice cream was Philip Lenzi, who came to the city by way of London. On November 25, 1773, an advertisement in *Rivington's New York Gazetteer*, announced "Monsieur Lenzi's" arrival with an explanation of his many offerings and services. Based temporarily in a private home "near the Exchange," he planned to sell a wide range of European-style "sweetmeats." Ice cream appears in this announcement, buried in a long list between "brown sugar candy" and "sugar ornaments." By 1777, Lenzi had his own confectioner's shop at 517 Hanover Square and had placed ice cream in a more noticeable spot in his advertisement. "May be had everyday," his notice reads, "ice cream" (*New-York Gazette and the Weekly Mercury* 1777).

In the early 1780s, another French confectioner began advertising in the city's newspapers. Joseph Corre arrived in America during the Revolutionary War as the personal cook to a British officer (Garrett 1978, 201). Before the war's end, he had struck out on his own and opened a confectionary store, near Lenzi's shop, on Hanover Square. In a 1781 advertisement, Corre announces that he "continues to serve the Ladies and Gentleman of this garrison, upon the most reasonable terms, with ICE CREAM" (*New-York Gazette and the Weekly Mercury* 1781).

As seen in other late-eighteenth-century advertisements, a number of the city's confectioners began to focus exclusively on the sale of ice cream, suggesting the rise of what we would now recognize as ice cream shops. Of equal interest are the efforts made by at least one confectioner, Isherwood and Grieg, to ensure the patronage of the city's more genteel citizens. Their ad announced that a special room, next to their store, had been "fitted up" for ladies (*Commercial Advertiser* 1798).

Along with ice cream, the city's gentry adopted other aspects of English cuisine. Among these were elaborate desserts requiring ice for their creation and preservation. Such items included syllabubs, molded jellies, bombes, blancmanges, and flummeries. Serving and consuming these concoctions was to become a form of entertainment, as well as symbolic of one's membership among the elite (Belden 1983, 157, 165–68). From their earliest arrival on Manhattan's shores, the confectioners—sensing an opportunity—also offered their services as caterers. In his 1773 advertisement, Lenzi indicated his availability "to furnish any public entertainment, as he has had the management of several given at Balls, Masquerades &c. in most of the principal cities of Europe" (*Rivington's New York Gazetteer* 1773). Corre also advertised that he was able to "provide dinners or suppers at any private house in town for the convenience of ladies and gentlemen" (*New-York Morning Post and Daily Advertiser* 1785). Throughout this period, wealthy households relied upon confectioners to provision the receptions, assemblies, routs, dinner parties, and banquets that constituted society's seasonal social round.

At least four New York City confectioners—Joseph Corre, Jacques Madeline Delacroix, Peter Thorin, and John H. Contoit—are known to have turned their entrepreneurial talents to the creation and management of pleasure gardens—a bigger and grander venue for the sale of ice cream. As one European visitor remarked after visiting two such establishments, "They are both kept by French people who through the sale of ice cream alone have gained a large fortune" (Garrett 1978, 210). Much like the ice trade itself, pleasure gardens were a ubiquitous but now largely forgotten part of urban life. Predating public parks by more then a century, pleasure gardens were in fact gardens with manicured lawns, groves of shade trees, fountains, statues, gravel walkways, gazebos, pavilions, and muslin-draped outdoor "supper boxes." At the same time, however, they were also social centers where city dwellers could gather to see one another and be seen, and be treated to theatrical performances, concerts, fireworks, and other forms of entertainment. In 1800, the city's eight pleasure gardens represented its "most popular diversion" (Garrett 1978, 621).

One of the city's best-known pleasure gardens was Brannon's, located on the road to Greenwich Village. Recalling it in 1794, an English visitor remarked that "iced creams and iced liquors are much drank [*sic*] here during the hot weather" (Garrett 1978, 147). In fact, due to the involvement of so many confectioners, the consumption of ice cream and iced beverages became one of the more notable activities at the city's pleasure gardens—playing a major part in the promotion of these establishments and in the memories of those who frequented them. Delacroix's Columbia Garden was located adjacent to the Battery where it received pleasant waterfront breezes during the heat of summer. Elizabeth Bleeker remembered it as "a most romantic place," and wrote in her diary that she "had a charming glass of ice cream, which has chilled me ever since" (Garrett 1978, 204–205). As expressed by a young woman in 1803, in a letter to a friend:

> In the cool of the evening we walk down to the Battery and go into the garden. Sit half an hour, eat ice cream, drink lemonade, hear fine music, see a variety of people, and return home happy and refreshed. (Belden 1983, 168)

SALE AND STORAGE OF ICE

As the literature suggests, confectioners and pleasure gardens had to have an abundance of ice on hand in order to provide products to their customers. Writing of a famous early-nineteenth-century British confectioner/caterer, food historian Elizabeth David (1995, 325) observes that "the consumption of ice for cooling and freezing in [this establishment] was clearly quite considerable." In addition to using ice to prepare ice cream and desserts, confectioners also treated ice as a commodity to be bought and sold. Fairly early on in their New York careers, both Lenzi and Corre were selling ice from their shops. In June 1777, Lenzi ran an advertisement: "May be had everyday, ice cream; likewise ice for refreshing wine, &c." By 1788, Corre and an unnamed confectioner were also selling ice on a daily basis out of their Wall Street shops (*The New-York Morning Post and Daily Advertiser* 1788a; 1788b). Since the peak period for ice consumption was during the warmer months, the confectioners would have required spaces that were specially constructed to slow down the inevitable melting of ice. Their options would have included interior "cellars" built within the basements of their shops, exterior ice houses in their backyards that had storage space below ground accessible through some sort of superstructure, or else to purchase ice everyday from some nearby source.

At New York City's pleasure gardens, ice was apparently stored on the premises. Columbia Gardens, for example, most likely had an ice house somewhere on its grounds. In 1806, we find Corre, its owner, offering ice for sale here by subscription, from May through September (Stokes 1929 Vol. 5, 144). Additionally, on a number of occasions in which the pleasure gardens themselves came up for sale, ice houses are mentioned in the advertisements. For example, Peter Thorin, announcing the sale of New Vauxhall Garden, stated that the property contained "a large icehouse almost full of excellent ice" (Garrett 1978, 193). In 1804, when Mount Vernon Garden was offered for sale, the advertisement stated that the property contained an ice house along with several other buildings (*American Citizen* 1804). It should also be noted that at least two different late-eighteenth-century pleasure gardens took the name "Ice House Garden" (*Daily Advertiser* 1798; Garrett 1978, 170–91).

During this period, ice could also be purchased from commercial ice houses not associated with confectioners. According to a newspaper notice appearing in 1784, ice was available for delivery every morning and evening from an ice house that stood at the end of Wall Street near the Hudson River (*The New York Packet and the American Advertiser* 1784). Another outlet was a "cellar" located beneath the Government House (formerly standing near the tip of Manhattan, overlooking the harbor) where ice was sold every morning (Stokes 1929 Vol. 5, 1369). We know as well that wealthy families maintained private ice houses at their city townhouses and at their estates on the city's outskirts. For example, a 1795 advertisement for a thirteen-room house on Division Street, a "Spacious Ice House" is described as "under the Cellar" (*The American Minerva and the New-York Advertiser* 1795). In 1804, a "large elegant new HOUSE" being sold on Water Street had "a liquor vault and ice House" in its yard (*The Daily Advertiser* 1804).

Unfortunately there are no surviving images of New York City's eighteenth-century commercial or domestic ice houses, nor have any remains of these been identified by archaeologists. We can reconstruct their probable appearance and size, however, from documentary sources, from archaeological excavations that have been conducted elsewhere, and from ice houses preserved at historic properties such as Mount Vernon, Monticello, and the Van Cortlandt Manor (Dillon 1975, 5; Sack Heritage Group; Thomas Jefferson Foundation).

For a sense of how an interior ice "cellar" might be constructed, we can extrapolate from an early-nineteenth-century English description (McIntosh 1828, 262):

> The London confectioners, as well as most people on the continent, content themselves with keeping [ice] in cellars, surrounded with very thick walls, and without windows, being entered sometimes by straight and sometimes by crooked passages, secured by double and often treble doors, and the ice thickly covered by straw or mats.

Freestanding backyard ice houses, other sources tell us, included both aboveground and subterranean masonry components. The buried or shaft portion where the ice was stored was shaped like an "inverted cone" with a grate or gravel at the bottom for drainage. Access to the building was through the upper part of the structure, which was located at ground level. The latter was often round or octagon-shaped with a shingled, thatched, or bricked domed roof (David 1995, xiv–xvi; Thomas Jefferson Foundation; Webster 1845 (VIII, 2). In Philadelphia, the archaeological excavation of portions of a circa 1780s ice house indicated that its shaft portion was originally eighteen feet (5.49 meters) deep with a diameter of approximately thirteen feet (3.96 meters) (Yamin 2008, 40–43) (Fig. 14.1). An eighteenth-century exterior ice "well" excavated in Alexandria, Virginia, was estimated to have a capacity of sixty-eight tons (City of Alexandria).

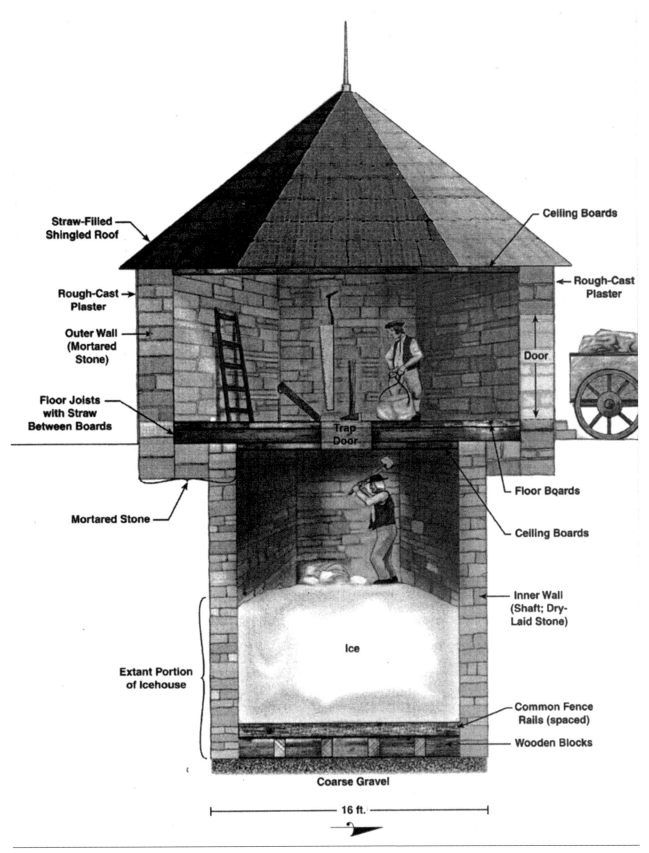

FIG. 14.1. Drawing of a ca. 1780s Ice House, probably belonging to Robert Morris—based on documentary sources and results of archaeological excavations conducted by John Milner Associates, Inc., at the site of the Liberty Bell Center in Philadelphia. Rendering by Todd Benedict and Rob Schultz, John Milner Associates, Inc. Source: Yamin (2008, 45, Fig. 3.6) (courtesy of Rebecca Yamin and John Milner Associates).

EARLY NYC ICE SOURCES

By 1800, New York City had a rapidly growing population of 60,500. Ten years earlier, it had surpassed Philadelphia as the largest city in the United States. Also, as stated above, during this period certain sectors of its citizenry apparently enjoyed a dependable and sizeable supply of ice. Where did it come from? At that time, the most readily available ice source would have been the Collect (from the Dutch "kolck" or "small body of water") also known as the Fresh Water Pond. Located on what was then the city's northern outskirts, near present-day Foley Square, it was most likely a glacial kettle pond, fed by springs and local streams (Fig. 14.2). The Collect was praised by Early Euro-American residents of New York as an excellent spot for fishing and snipe hunting. In the winter the Collect was a popular spot for ice skating. Gallows were erected along its shorelines during the summer, and crowds gathered here to watch public hangings. By the time ice suppliers and confectioners began harvesting ice commercially from the Collect, however, a group of so called "noxious" industries were located on its southern shore. At the close of the American Revolution, the pond, now considered a prime industrial water source, was surrounded by tanneries, breweries, distilleries, rope-walks, potteries, and furnaces. Contemporary accounts described it as a dumping ground for animal carcasses and human corpses. Ironically, the first documentary evidence we have that the Collect served as an ice source appears in 1806 on the eve of its abandonment for that purpose. A newspaper advertisement placed by Joseph Corre, the confectioner, ice supplier, and owner of a pleasure garden, informed the public that because the Collect had become so "putrid" he had (at great cost) "procured ice from a fresh spring about three miles from the city." By then, the city had begun the process of filling in portions of the Collect and leveling the hills that surrounded it (Hill and Waring 1897, 207–208; Koeppel 2000, 42, 52, 116; Stokes 1929, Vol. 5, 144).

Corre's new ice source may well have been Sunfish Pond, then located at a greater distance from the city near present-day Park Avenue South and

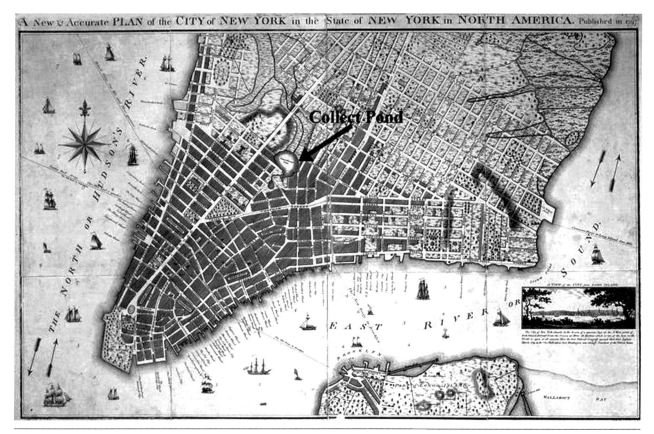

FIG. 14.2. 1797 map (Taylor 1797) showing location of early ice source, Collect Pond, at what was then the northern edge of New York City (courtesy of the Map Division, New York Public Library).

32nd Street. Streams originating near present-day Times Square fed the one-acre pond, which then drained into the East River's Kip Bay. Described as a favorite spot for fishing and muskratting, by the 1830s Sunfish Pond's purity too had been compromised as a result of a glue factory that had located along its northern shore. It was filled in by the 1840s (Depew 1895, 467; Koeppel 2000, 240).

CAUSES UNDERLYING THE RISE OF THE HUDSON RIVER ICE INDUSTRY

The discussion now moves nearly a century ahead and more than a hundred miles (over 160 kilometers) north of the island of Manhattan, to the reaches of the Hudson located between Poughkeepsie and Albany, where, by the 1880s, the shorelines of Ulster, Greene, Rensselaer, and Albany Counties had become the center of New York City's ice production. Here, far from the city's contaminated waters and safely above the estuary's saltwater front, 135 ice houses had been constructed. The Hudson River was now producing between 2,000,000 and 2,750,000 tons of ice during a good winter season, making it the largest producer in the United States (Hall 1884, 24–26).

The Hudson River ice industry's true beginnings, however, date to the 1830s and to only thirty miles (some 48 kilometers) upstream from New York City at Rockland Lake, where ice was cut and hauled down the side of Hook Mountain to the river. Here it was stored in a 1,500-ton ice house until it could be shipped to the city by steamboat (Stott 1979, 7–8). Demand for ice kept growing, and by the 1860s, the company that controlled production at Rockland Lake, the Knickerbocker Ice Company, had developed ice fields on the river almost as far north as Albany (Stott 1979, 9). A number of causes underlay this expansion, including the tremendous improvement in ice harvesting and storage technology that had occurred since ice was first taken from the Collect, New York City's unprecedented growth, and changes in the food distribution system.

Many of the innovations that would transform the ice trade were introduced by Nathaniel Jarvis Wyeth, a young associate of Frederick Tudor, the Boston man generally credited with the creation of North America's international ice trade. Tudor's business had begun in 1806, centering upon the shipment of Massachusetts ice to warm weather ports, including Charleston, Martinique, Cuba, and later to India. Wyeth, as an ice supplier and later Tudor's foreman, had grown frustrated with the slow and haphazard methods used in the hand-harvesting of ice from New England's frozen ponds. Among other innovations, in the 1820s he devised a horse-drawn ice plow with parallel blades that allowed workers to grid out and cut an ice field into easily removable standard-sized blocks of ice cut at right angles (Hall 1884, 2–4; Maclay 1895; Smith 1961; Stott 1979, 7).

Not only did Wyeth's invention allow greater quantities of ice to be cut at a faster rate, but the ice that was produced could be efficiently stacked within Wyeth's newly configured ice houses. Unlike their eighteenth-century predecessors, these were above ground, multiroomed, multistoried, and constructed of wood (Fig. 14.3). Construction of these ice houses was made feasible by other early-nineteenth-century technological innovations. One of these was the discovery of the insulating property of various materials (Weightman 2003). Most ice houses had double walls packed with such materials, including wood shavings, sawdust, and hay, which would also be packed around the ice blocks. Additionally, the development of steam-powered ice elevators enabled the ice houses to be easily filled. A large ice house could hold sixty thousand tons of ice.

In the spring, the ice was removed and shipped to market in specially constructed ice barges (Fig.

FIG. 14.3. Undated photograph of the Empire Number 2 Ice House at Catskill, N.Y. Note the powerhouse in the center of the photograph and the ice elevators in front each of the large vertical loading doors (courtesy of the Vedder Memorial Library, Greene County Historical Society).

FIG. 14.4. 1884 Illustration showing Hudson River ice barges (cover illustration, *New York Chronicle* 1993).

14.4). These barges are identifiable by the large windmills that extended above the deck that were used to pump out any meltwater that accumulated during the trip downriver, an invention that has been credited to Thomas Edison (de Noyelles 1982, 138; Dibner n.d., 17–20; Hall 1884, 17; Walsh 1983).

Because more ice could be harvested and stored, and less lost to meltage due to the advances in technology, greater amounts of ice were reaching New York City (Hall 1884, 9–10; Jones 1984, 800; Weightman 2003, 105–15). The city was now the largest ice market in the country, requiring nearly 1.5 million tons of ice yearly by the 1880s (Hall 1884, 24). It had seen tremendous growth in just a short period of time. From 1830 to 1880 more than 960,000 people were added to its population (New York State 2000, 2). Not only did the city require ever-increasing amounts of food to feed its citizens, but more ice was needed to keep food from spoiling as it moved through an expanding and increasingly complex food distribution system.

For more than two hundred years, beginning with the Dutch, New Yorkers had obtained fresh locally produced food at centrally located municipally regulated markets. Provisions arrived from throughout the neighboring region by various means: from the gardens of New Jersey and Long Island on market boats and farm wagons; by sloops from Hudson River Valley farmsteads; in fishing and oyster boats from Long Island Sound and Raritan Bay; and on the hoof as droves of cattle were herded overland. Some of the produce originated from gardens and farms on the portion of Manhattan Island immediately north of the city. As late as 1837, only one-sixth of the island was covered with buildings and streets. Although much of the remaining land was either wooded or enclosed by private estates, an extensive portion was still under cultivation (Lindner and Zacharias 1999, 340; Spann 1981: 124–28).

Thomas F. Devoe, butcher and market historian, chronicled practices at the city's twelve municipal markets during the 1820s. Because perishable items, such as fish, meat, milk, and butter, were usually purchased on a daily basis, marketmen used very limited amounts of ice. Devoe records only one case during this period in which any significant quantity of ice was stored at a city market. This was the Grand Street Market, where a customer, James Allaire, paid the costs for constructing and filling a subterranean ice house. His reason for doing this, according to Allaire, was so he might "now and then, have a good piece of corned beef through the warm weather" (Devoe 1865, 456). Beginning in 1839, with the introduction of iceboxes into the public markets, butchers and fishmongers began to keep ice on hand so that they might "keep pieces over." Some consumers had also acquired iceboxes, and they too needed a constant supply of ice to keep food cool (Burrows and Wallace 1999, 46, 451; Devoe 1865, 346–47, 456, 485).

Gradually, the entire system underwent a major transformation due to the loss of local food producers whose farmland had been swallowed by suburbanization, the proliferation of middlemen grocers, the rise of privately owned neighborhood grocery stores, and the development of refrigerated railroad cars that carried perishable food into the city from distant sources. As the Federal Trade Commission was to observe early in the twentieth century, "supplying the needs of a great city is no longer the casual affair of farmers with their farmers' wagons" (Lindner and Zacharias 1999, 281–82; Spann 1981, 124–28; Tangires 2003).

In general, underlying the city's growing market for ice was a change in consumption patterns favoring large industry. In the mid-nineteenth century the major consumers of ice had been families, hotels, saloons, and ice cream stores. As the century drew to a close, the major consumers became breweries, meat packers, and the railroads (Hall 1884, 5). Ice, once considered a luxury, was now a necessity.

Having discussed the factors driving the rise of the Hudson River ice industry, we will now examine the consequences—for the river's landscapes and its communities.

EFFECTS UPON THE LANDSCAPE

During the latter third of the nineteenth century and the opening decades of the twentieth, ice fields and ice houses became the dominant elements along portions of the river's shoreline north of Poughkeepsie. During the 1990s, as archaeologists working for the New York District Corps of Engineers, we undertook a series of investigations here as part of a habitat restoration study. We focused especially upon three ice house sites located on lower Schodack-Houghtaling Island, a seven-mile-long island just south of Albany. Our research suggests that not only did the ice industry greatly affect the economic and social history of the Hudson River but it also transformed the physical landscape (Harris and Pickman 1999; 2000; Huey 1998). Elsewhere we have examined how throughout the nineteenth century, the river's shorelines were greatly modified as a result of a series of engineering efforts that included navigation improvement projects, railroad construction, and the making of land to support industrial production (Harris et al. 1996). Landscape changes associated with ice harvesting fall within the latter category.

Modifications to the shoreline morphology began during the opening phases of an ice house's construction. Sediment was removed from the river bottom by steam dredges and then redeposited behind bulkheads in shallow water near the shore to form a projecting wharf directly in front of the ice house, providing a loading area for ice barges. This process would be followed at almost every ice house built along the river. In a contemporary aerial photograph (Fig. 14.5), the areas of shoreline terrain seen protruding into the river from its western bank are actually former ice house sites. This altered shoreline survives today as a highly visible feature of the riverine landscape.

Historic maps and newspapers indicate that an ice house building boom occurred on Schodack-Houghtaling Island and adjoining areas of the river in the 1880s. The construction of so many of these huge buildings drastically altered the appearance of less-populated reaches of the river as well as the waterfronts of small villages such as Catskill, Athens, New Baltimore, Coeymans, and Coxsackie. The map shown in Figure 14.6, indicating the locations of seventy-one ice houses that once stood on the

FIG. 14.5. A 1989 aerial photograph of the Hudson River's western shoreline clearly indicates the locations of former ice house wharves. Scale of original: 1 in = 480 ft [1 cm = 57.6 m] (courtesy of Col-East, Inc.).

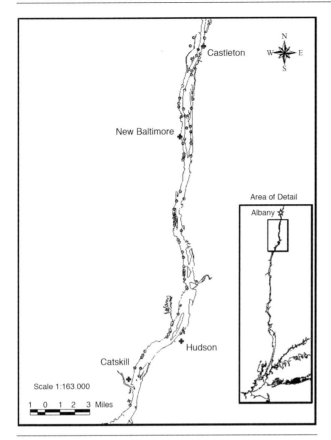

FIG. 14.6. Locations of Hudson River ice houses between Catskill and Castleton, N.Y., ca. 1890s. Circles indicate ice house locations, plotted using maps of that period (Beers 1891; USACE 1897), aerial photographs (Col-East Inc. 1989), and twentieth-century topographic maps (USGS 1953a, 1953b, 1953c, 1953d) (drawn by Dag Madara).

riverbanks between Catskill and Castleton, should provide readers with a sense of the density of these industrial structures. What the map cannot convey is the visual impact of individual ice houses. Painted a brilliant white in order to repel the sun's rays and forestall the melting of their contents, most icehouses were more than three stories high, thus dwarfing all other aspects of the river's built environment. The remains of one of the ice houses we investigated extended some four hundred feet (approximately 122 meters) along the shoreline.

For those viewing the shorelines from a distance, such as passengers on the railroad or on steamboats traveling between New York City and Albany, the newly constructed clusters of ice house complexes may have been experienced as glaring intrusions upon the landscape's scenic beauty. Indeed, one Hudson River guidebook referred to the ice houses as "immense storehouses that line the banks of the river . . . all the way to the head of navigation, and which form a feature of the scenery more conspicuous than ornamental" (Ingersoll 1893, 129). Another guidebook praised the river's pastoral views but made an exception of what the writer termed "the great unattractive whitewashed ice houses perched on the river banks, suggesting the outreaching grasps of the monopolistic ice barons" (Buckman 1909, 122) (Fig. 14.7).

EFFECTS ON HUDSON RIVER COMMUNITIES AND WORKERS

Although the available evidence indicates that outsiders responded to the transformed landscape with ambivalence, for those living and working in Hudson River communities the economic benefits of the developing ice harvesting industry far outweighed changes to the familiar landscape, and they eagerly embraced it. Local newspapers reported enthusiastically upon the construction and expansion of individual ice houses, providing readers with progress reports on dredging and filling of mudflats, pile driving, bricklaying, framing, roofing, and the in-

FIG. 14.7. View of ice houses from the river illustrates impact of these large structures on the landscape (Source: Bruce 1903).

stallation of engine rooms, smoke stacks, and elevators. In a village such as Coeymans in 1892, the construction of a huge new ice house was "astonishing" and a source of wonder rather than dismay. Local residents were encouraged to visit the construction site and reassured that "if a favorable winter ensues, an increased force of help will find employment in the ice harvest at this point" (*Coeymans Herald* 1892).

Watching the ice workers from his home on the banks of the Hudson, the naturalist John Burroughs (1886, 202), wrote:

> On a stern winter night, it is a pleasant thought that a harvest is growing down there on those desolate plains which will bring work to many needy hands by and by, and health and comfort to the great cities some months later.

As Burroughs's observations suggest, the ice harvesting industry involved the transformation of the river's frozen surface into a site of human labor. Despite the use of horse power to mark out and cut ice blocks, and steam power to load the block into ice houses, Hudson River ice harvesting was essentially a labor intensive industry (Fig. 14.8).

Estimates place the size of the Hudson River ice industry seasonal work force at up to twenty thousand workers (Hall 1884, 26). Like the industry they labored in, the history and culture of Hudson River ice house workers have been largely forgotten. As part of our research on the Schodack-Houghtaling Island ice houses, we tried to learn more about these people, their origins, and their experiences.

During the third and fourth quarters of the nineteenth century, the Hudson River ice houses became a major factor in local economies, and for workers in river communities their construction was a welcome development. Most prospective ice workers engaged in seasonal pursuits such as agriculture, logging, fishing, ship building, brickyard work, and river transportation—all of which ceased during the winter months (Hall 1884, 26; Post n.d.). A local newspaper column noted that "several hundred men hereabouts annually make calculation on the income derived from two or three weeks' employment in the ice harvest to tide them over to the settled Spring work" (Anonymous n.d.).

FIG. 14.8. 1883 photograph of Hudson River ice crew scraping the ice prior to cutting, near Athens, N.Y. (courtesy of the Vedder Memorial Library, Greene County Historical Society).

Work in the ice industry could also extend beyond winter. Commenting on Greene County's forty ice houses, Beers (1884, 58) noted: "The business gives employment to a large number of men, both in harvesting the ice in the winter and breaking it out and loading barges in the summer." Thus, not only did the ice houses provide a new source of income to the region's farmers, artisans, tradesmen, and laborers, but one that could potentially provide a year-round basis of support for themselves and their families.

To further explore the composition of the ice house workforce, and to confirm the data contained in secondary sources and in newspapers of the period, we examined the records of the Van Orden, Vanderpool, and Sherman Ice House, constructed in 1881 on the western shoreline of Houghtaling Island opposite the village of New Baltimore. Weekly payrolls from the 1889 harvesting season list sixty-six persons, a figure that apparently included both year-round and seasonal workers (Sherman 1889b; 1889c). The names of twenty of these workers also appear in the 1892 New York State census records and directories for two adjacent villages on the west bank of the Hudson (Lant 1892; New York State 1892). Six of these workers were described as farmers, while three others were river pilots or boatmen. The other workers included two carpenters (one a ship carpenter), a painter, a stonemason, and a butcher. Another, described as an engineer, apparently operated the ice house steam engine, and may have been a year-round employee. Only five of these ice house workers were described as laborers. Two of the ice house employees were substantial landowners—one of the farmers having 139 acres, and one of the carpenters, 92 acres. The butcher also owned an acre of land. This small sample supports the inference that many ice house employees were unaccustomed to working as industrial laborers.

The development of the ice industry brought communities into a new relationship with the frozen waters of the Hudson. In addition to the funds that flowed directly to the workers, the economic power of the ice industry derived from its indirect impact upon local villages as ice house employees spent their money in hotels, boardinghouses, restaurants, saloons, clothing stores, and other retail establishments (Beecher 1991,79). Between wages spent and wages pocketed, many river communities became economically dependent upon the ice industry. Newspapers such as the *Catskill Examiner* ran special columns during the winter months devoted wholly to the progress of the harvest. One column proclaimed that "ice is the only the only thing talked about in New Baltimore now" (*Catskill Examiner* 1883a). During warm winters, when the ice harvest was poor, the columns chronicled the dismal mood of the villages:

> The ice grows less and less encouraging. We have had and are having uniform spring weather . . . up the river the ice men have done nothing and below us it is of course the same . . . the laboring class feel the loss of their work on the ice very severely and when they suffer, the interests of the business community are seriously affected. (*Catskill Examiner* 1880)

During good harvest seasons, when the winters were cold, the workers had employment, but this involved exposure to the harsh, and often dangerous, working conditions on the ice fields. The diary of ice house owner Augustus Sherman attests to numerous days of sub-zero temperatures, and days when the wind and the temperature created conditions so severe that work became impossible (see, e.g., Sherman 1882; 1889a; 1895). Both secondary accounts and journals of the ice men indicate that working on the ice had other hazards that sometimes led to severe injury or even death. Accidents recounted in the local newspapers and elsewhere include falling through the ice or open channels into the freezing water, being struck by falling ice cakes which weighed up to several hundred pounds, and being ensnared in the ice house elevating machinery (Beecher 1979, 3; Rothra 1988, 18).

While the wages earned in the ice fields provided a needed supplement to local incomes, ice workers also encountered, possibly for the first time, relations of production typical of industrial capitalism. The process of being incorporated into the wage labor system was not always a smooth one, as suggested by the many accounts of strikes on the Hudson River ice fields. Some affected single ice houses and were quickly resolved. Others were more widespread and involved violence and threats of violence (*Catskill Examiner* 1875; 1876; 1879; 1883a; 1883b; *Coeymans Herald* 1879; 1881; 1882). Thus, for the farmers, artisans, and tradesmen listed in the Vanderpool, Van Orden, and Sherman payroll, life in the ice fields may have provided an initial personal encounter with labor strife. Participation in the ice industry workforce also brought many workers into contact with men and women of other ethnic and cultural backgrounds. Contemporary newspaper accounts note that the ice industry's labor force included African Americans and women, as well as Irish and Italian immigrants (*Catskill Examiner* 1875; 1878).

The following quote is from an atypically pro-labor local newspaper account of an 1875 strike on the ice fields:

> By 10 o'clock the crowd numbered about 500 tough and determined men, many of whom had come from points 8 to 10 miles [13 to 16 km] distant to get work, and they formed a line and marched up and down Main Street. . . . The procession comprised all nationalities, including a liberal infusion of the Hibernian element—fairly spoiling for a fight—and was peppered with Anglo-Africans. . . . Pale faces and darkies met in peace on the platform of "fourteen shillings a day." (*Catskill Examiner* 1875)

While this account reflects racial and ethnic attitudes typical of the period, it also indicates the workers' solidarity in the face of what they perceived as economic exploitation by the ice house owners.

Thus, within the larger Hudson River landscape, ice fields and ice houses became sites of both human conflict and accommodation as a generation of workers were absorbed into the culture of the new industrial society.

THE DEMISE OF THE HUDSON RIVER ICE INDUSTRY

Beginning in the first decade of the twentieth century, New York consumers began to question the purity of Hudson River ice. It was alleged that raw sewage was pouring into the river from towns directly adjacent to the ice fields (*The New York Times* 1903a; 1903b). After testing the water, the city issued a statement saying that typhoid fever may have been caused by "the present condition" of Hudson River ice (*The New York Times* 1907). The ice industry fought back, claiming convincingly that typhoid bacteria were killed by long-term freezing and citing as proof the city's low death rate from intestinal diseases (*The New York Times* 1903c; 1913; *Albany Telegram* 1910). By then, though, other factors were beginning to take a toll on New Yorkers' desire for natural ice. Chief among these was the high cost of ice, caused primarily by the development of ice monopolies and the inability of the Hudson River ice industry to meet urban demand (*The Troy Weekly Times* 1889; *The New York Tribune* 1900). In 1920, the *New York Times* carried a story noting that decreasing amounts of Hudson River ice were being harvested and that as a consequence ice houses were being abandoned.

FIG. 14.9. Visible Remains of Hudson River Ice Houses. Clockwise from upper left: Brick Structure that probably housed the steam engine and firebox—P. McCabe & Co. Ice House, Schodack-Houghtaling Island; Powerhouse—Scott Brothers Ice House, Nutten Hook; Eroding Foundation Wall—Van Orden, Vanderpoel & Sherman Ice House, Schodack-Houghtaling Island; Barge Remains—P. McCabe & Co. Ice House, Schodack-Houghtaling Island; Brick Chimney—Miller & Whitbeck Ice House, Schodack-Houghtaling Island (photographs by Wendy Elizabeth Harris, 1998).

Coinciding with these trends and contributing to the industry's decline was the increasing availability of technology for cooling air and manufacturing ice (Gosnell 2005, 378). Unfortunately, this involved the use of large machinery and dangerous gases such as ammonia. By the 1930s, however, these obstacles had been overcome and millions of electric refrigerators were in use in American homes (Rogers n.d; Weightman 2003). The era of the natural ice industry had come to an end. As one Catskill resident observed at the time: "The ice house has followed the livery stable to oblivion" (Anonymous 1931).

The falloff and ultimate disappearance of the ice trade severely disrupted the economies of such industry centers as New Baltimore, Coeymans, and Catskill. Some of these communities would never recover the vitality they had known during the ice industry's brief existence (Bush 2009, pers. comm.).

Within the reaches of the river that we examined as archaeologists, the landscape and economy that was created by the ice industry lasted no more than fifty years. By the time we encountered the industry's infrastructure—at the end of the twentieth century—it existed as ruins (Fig. 14.9). The frozen waters of the Hudson had long ago ceased to have any value for urban consumers.

CONCLUSIONS

What observations can we make when confronted with the history of this industry? Looking back from the vantage point of the twenty-first century, we see a regionally focused and sustainable extractive industry that both flourished and collapsed as a result of the same forces. Population growth, pollution, economic change, and technological advance pushed the ice trade off the island of Manhattan, up to the reaches of the mid-and upper Hudson, where it became an enormous industry. In the end, these forces followed the industry, and became its undoing.

REFERENCES CITED

Albany Telegram. 1910. Natural ice men prepared to fight manufacturers of artificial ice, July 31. Ice Industry Folder, Vertical Files, Greene County Historical Society, Vedder Memorial Library, Coxsackie, NY.

American Citizen. 1804. Advertisement, Vol. 5, Issue 1352 (Aug. 4), 3. New York: D. Denniston and J. Cheetham. Collection of the New York Public Library.

American Minerva and the New-York Advertiser. 1795. Advertisement, Vol. II, Issue 587 (Nov. 23), 1. New York: G. Bunce. Collection of the New York Public Library.

Anonymous. n.d. The ice prospects. Unidentified newspaper article in Ice Industry Folder, Vertical Files, Greene County Historical Society, Vedder Memorial Library, Coxsackie, NY.

———. 1931. Deplores passing of a once great industry. Unattributed newspaper (Feb. 5). Ice Industry Folder, Vertical Files, Greene County Historical Society, Vedder Memorial Library, Coxsackie, N.Y.

Beecher, R. 1979. A winter's ice harvest, 1900–1901. *Greene County Historical Society Journal* 3(4): 1–3.

———. 1988. The Shermans of New Baltimore: The fourth generation's Augustus Sherman. *Greene County Historical Society Journal* 12(3): 21–28.

———. 1991. *Under three flags: A Hudson River history, Coxsackie and Clinton and settlements.* Hendersonville, NY: Black Dome Press.

Beers, F. W. 1891. *Atlas of the Hudson River Valley from New York City to Troy.* New York: Watson and Co.

Beers, J. B. and Co. 1884. *History of Greene County, New York.* 1969 reprint ed. Cornwallville, NY: Hope Farm Press.

Belden, L. C. 1983. *The festive tradition, table decoration, and desserts In America, 1650–1900.* New York: W. W. Norton.

Bruce, W. 1903. *Panorama of the Hudson showing both sides of the river from New York to Albany.* Photographs by G. Willard Shear. New York: Bryant Literary Union.

Buckman, D. L. 1909. *Old steamboat days on the Hudson River.* New York: The Grafton Press.

Burroughs, J. 1886. A river view. In *The writings of John Burroughs, Volume VII. Signs and seasons,* 195–213. Cambridge: The Riverside Press.

Burrows, E. G., and M. Wallace. 1999. *Gotham: A*

history of New York City to 1898. New York: Oxford University Press.

Bush, C. S. 2009. Personal Communication. Town Historian, Town of New Baltimore, Greene County, N.Y.; July.

Catskill Examiner. 1875. The strike, January. Ice Industry Folder, Vertical Files, Greene County Historical Society, Vedder Memorial Library, Coxsackie, NY.

———. 1876. The ice business, n.d. Ice Industry Folder, Vertical Files, Greene County Historical Society, Vedder Memorial Library, Coxsackie, NY.

———. 1878. The ice harvest, Feb. 23. Ice Industry Folder, Vertical Files, Greene County Historical Society, Vedder Memorial Library, Coxsackie, NY.

———. 1879. The ice harvest, Jan. 18. Ice Industry Folder, Vertical Files, Greene County Historical Society, Vedder Memorial Library, Coxsackie, NY.

———. 1880. The ice harvest. Jan. 24. Ice Industry Folder, Vertical Files, Greene County Historical Society, Vedder Memorial Library, Coxsackie, NY.

———. 1883a. The ice harvest, Jan. 20. Ice Industry Folder, Vertical Files, Greene County Historical Society, Vedder Memorial Library, Coxsackie, NY.

———. 1883b. The ice harvest, Jan. 27. Ice Industry Folder, Vertical Files, Greene County Historical Society, Vedder Memorial Library, Coxsackie, NY.

City of Alexandria. Gadsby's Tavern Museum. The ice well. http://gadsbys.home.att.net/icewellscr.pdf; accessed July 2009.

Coeymans Herald. 1879. The meeting of the workingmen, June 18. New York State Library, Albany, NY.

———. 1881. Local brevities, July 20. New York State Library, Albany, NY.

———. 1882. Coeymans, June 15. New York State Library, Albany, NY.

———. 1892. Coeymans, Oct. 5. New York State Library. Albany. NY.

Col-East Inc., North Adams, Massachusetts. 1989. Aerial Photograph Series of the Hudson River. April 20.

Commercial Advertiser. 1798. Advertisement for Isherwood and Greig, Confectioners. Vol. I(186) (May 5): 2.

Cotter, J. L., and J. P. Hudson. 1957. *New discoveries at Jamestown: Site of the first successful English settlement in America.* Washington, DC: National Park Service, United States Department of the Interior. http://www.nps.gov/history/history/online_books/jame/discovery.pdf; accessed July 2009.

Daily Advertiser. 1798. Advertisement for "Ice House Garden" Vol. 4, Issue 4263 (Oct. 6), 4. New York: Collection of the New York Public Library.

———. 1804. Advertisement. Vol. 20, Issue 5863 (Jan. 13), 3. New York: Francis Childs. Collection of the New York Public Library.

David, E. 1995. *Harvest of the cold months: The social history of ice.* New York: Viking Penguin.

deNoyelles, D. 1982. *Within these gates.* Privately published.

Depew, C. M. 1895. *1795–1895. One hundred years of American commerce.* New York: D.O. Haynes.

Devoe, T. F. 1862. *The market book: A history of the public markets in the City of New York.* New York: The Author.

Dibner, B. n.d. The other end of the hawser: The history and technology of the barges that tugs towed . . . cold cargoes . . . *Tugbitts: The Quarterly Journal of the Tugboat Enthusiasts Society of the Americas.* South Street Seaport Library Vertical Files.

Dillon, J. 1975. National Register of Historic Places Nomination Form, Van Cortlandt Manor. Washington DC: United States Department of Interior, National Park Service.

Funderburg, A. C. 1995. *Chocolate, strawberry, and vanilla: A history of American ice cream.* Bowling Green: Bowling Green State University Press.

Garrett, T. M. 1978. A history of pleasure gardens in New York City, 1700–1865. Dissertation, New York University. Ann Arbor: University Microfilms International.

Gosnell, M. 2005. *Ice: The nature, the history, and the uses of an astonishing substance.* Chicago: The University of Chicago Press.

Hall, H. 1884. The ice industry of the United States. *Tenth U.S. Census,* 1880, Volume 22. Washington, DC: U.S. Census Office.

Harris, W. E., and A. Pickman. 1999. Cultural resources investigation, ice harvesting industry remains located on property belonging to the U.S. Army Corps of Engineers, New York District, Schodack-Houghtaling Island, Town of New Baltimore, Greene County, New York. Prepared for New York District Army Corps of Engineers. On file New York District Army Corps of Engineers and New York Office of Parks, Recreation, and Historic Preservation, Waterford.

———. 2000. Towards an archaeology of the Hudson River ice harvesting industry. *Northeast Historical Archaeology* 29: 49–82.

Harris, W. E., A. Pickman, and N. Rothschild. 1996. *Landscape, land use, and the iconography of space in the Hudson River Valley: From prehistory to the present.* Paper presented at the New York Academy of Sciences, Dec. 9, New York City.

Hill, G.E. and G. E. Waring. 1897. Old wells and water-courses on the isle of Manhattan, Part II, In *Historic New York: Being the first series of the Half Moon papers*, ed. M. W. Goodwin, A. C. Royce, and R. Putnam, New York: G. P. Putnam's Sons.

Huey, P. R. 1998. Historical and archaeological resources of Castleton Island State Park, towns of Stuyvesant, Columbia County, New Baltimore, Greene County, and Schodack, Rensselaer County, New York: A preliminary phase I cultural resources assessment. New York State Office of Parks, Recreation and Historic Preservation.

Ingersoll, E. 1893. *Illustrated guide to the Hudson River and Catskill Mountains.* Chicago and New York: Rand, McNally.

Jones, J. C. 1984. *America's icemen: An illustrated history of the United States ice industry trade, 1665–1925.* Humble, TX: Jobeco Books.

Koeppel, G. T. 2000. *Water for Gotham: A history.* Princeton: Princeton University Press.

Lant, J. H. 1892. *Greene County directory for 1892: Containing the names of the inhabitants of Catskill and Coxsackie.* Hudson, NY: Bryan and Wren.

Linder, M., and L. S. Zacharias. 1999. *Of cabbages and Kings County.* Iowa City: University of Iowa Press.

Maclay, R. 1895. *The ice industry.* In *One hundred years of American commerce*, Vol. II, ed. Chauncey Depew, 466–69. New York: D. O. Haynes.

McIntosh, C. 1828. *The practical gardener and modern horticulturist.* London: Thomas Kelly. http://books.google.com/books.

New York Chronicle. 1993. Bringing ice to New York City in 1884. Reprint from 1884 *Harpers Weekly*. Vol. 6, No 2, Cover Illustration. Copy in South Street Seaport Museum Library Vertical Files.

New York Gazette and Weekly Mercury. 1777. Advertisement for Philip Lenzi. Issue 1336 (June 2): 4. Collection of the New York Public Library.

———. 1781. Advertisement for Joseph Corre, Confectioner. Issue 1553 (July 23): 3.

New-York Morning Post and Daily Advertiser. 1785. Advertisement for Joseph Corre, Confectioner. Issue 316 (March 24): 1.

———. 1788a. Advertisement for Joseph Corre, Confectioner. Issue 1230 (June 19): 4.

———. 1788b. Advertisement for "Ice & Ice Cream, Mead & Cakes." Issue 1262 (July 26): 3.

The New York Packet, and the American Advertiser. 1784. Advertisement. Issue 392 (June 10): 2. New York: Samuel Loudon. Collection of the New York Public Library.

New York State. 1892. State census records for Greene County, New York. Microfilm copy in collection of Greene County Historical Society, Vedder Memorial Library, Coxsackie, NY.

———. 2000. Population of New York State by county: 1790 to 1990. Prepared by the Department of Economic Development, State Data Center. http://www.empire.state.ny.us/nysdc/StateCountyPopests/CountyPopHistory.PDF; accessed July 2009.

The New York Times. 1865. The ice trade. May 3, 8.

———. 1903a. Ice water. April 28, 8.

———. 1903b. Ice. May 31, 6.

———. 1903c. Concerning ice. July 20, 6.

———. 1907. Sees danger in Hudson ice. March 6, 6.

———. 1913. Natural ice superior. Jan. 24. Ice Industry Folder, Vertical Files, Greene County

Historical Society, Vedder Memorial Library, Coxsackie, NY.

———. 1920. Hudson River ice on decline." Ice Industry Folder, Vertical Files, Greene County Historical Society, Vedder Memorial Library, Coxsackie, NY.

New York Tribune. 1900. An ice famine imminent. March 14, 6.

Post, P. n.d. Ice harvesting, A once important area industry. Clipping in Ice Industry Folder, Vertical Files. Greene County Historical Society, Vedder Memorial Library, Coxsackie, NY.

Rivington's New-York Gazeteer. 1773. Advertisement for Philip Lenzi. Issue 32 (Nov. 25): 4. Collection of the New York Public Library.

Rogers Refrigeration. n.d. Refrigeration History. http://www.rogersrefrig.com/ history.html; accessed July 2009.

Rothra, W. H. 1988. *Two in a bucket: A personal account of ice harvesting.* Rev. ed. Published by the author. Printed by Overlake Enterprises, Stowe, NY.

Sack Heritage Group. George Washington's ice house. http://www.sackheritagegroup.com/articles/articles.php?articleID=142; accessed July 2009.

Sherman, A. 1882. Notebooks. Entry for Jan. 27, 1882. Edward Ely Sherman Memorial Collection, Box 147. Greene County Historical Society, Vedder Memorial Library, Coxsackie, NY.

———. 1889a. Notebooks. Entry for Feb. 4, 1889. Edward Ely Sherman Memorial Collection, Box 147. Greene County Historical Society, Vedder Memorial Library, Coxsackie, NY.

———. 1889b. Ice house account book. Amounts paid Feb. 22, 1889. Edward Ely Sherman Memorial Collection, Box 138. Greene County Historical Society, Vedder Memorial Library, Coxsackie, NY.

———. 1889c. Ice house account book. Amounts paid March 2, 1889. Edward Ely Sherman Memorial Collection, Box 138. Greene County Historical Society, Vedder Memorial Library, Coxsackie, NY.

———. 1895. Notebooks. Entry for Jan. 5, 1895. Edward Ely Sherman Memorial Collection, Box 147. Greene County Historical Society, Vedder Memorial Library, Coxsackie, NY.

Smith, P. C. F. 1961. Crystal blocks of Yankee coldness. *Essex Institute Historical Collections* 97(3): 197–232.

Spann, E. K. 1981. *The new metropolis, New York City, 1840–1857.* New York: Columbia University Press.

Stokes, I. N. P. 1929. *The iconography of Manhattan Island, 1498–1909.* New York: Robert H. Dodd.

Stott, P. 1979. The Knickerbocker Ice Company and Inclined Railway. *The Journal of the Society for Industrial Archaeology* 5(1): 7–18.

Tangires, H. 2002. Public markets and civic culture in nineteenth century America. Baltimore: Johns Hopkins University Press.

Taylor, B. A. 1797. *A new and accurate plan of the city of New York, in the State of New York in North America.* Collection of the New York Public Library, Map Division.

The Thomas Jefferson Foundation. n.d. The ice house. *Jefferson Encyclopedia.* http://wiki.monticello.org/ mediawiki/index.php/Ice House.

Thoreau, H. D. 1904. *Walden.* New York: E. P. Dutton.

Troy Weekly Times. 1889. The hermit's letter, ice items. Vol. XXXIV(2)(Aug. 8): 3.

United States Army Corps of Engineers (USACE), New York District. 1897. Improvement of the Hudson River, New York, in charge of Lieut. Col. William Ludlow and Major A. N. Miller, Corps of Engineers, U.S.A. Map on file U.S. Army Corps of Engineers, New York District, Albany Field Office, Troy, NY.

United States Geological Survey, Department of Interior. 1953a. Delmar Quadrangle, New York. 7.5 Minute Series (Topographic), Photorevised 1980. Digital Raster Graphic (DRG) File 1998.

———. 1953b. Hudson North Quadrangle. 7.5 Minute Series (Topographic), Photorevised 1980. Digital Raster Graphic (DRG) File 1998.

———. 1953c. Hudson South Quadrangle. 7.5 Minute Series (Topographic), Photorevised 1980. Digital Raster Graphic (DRG) File 1998.

———. 1953d. Ravena Quadrangle, New York, 7.5 Minute (Topographic), Scale 1: 24,000. Digital Raster Graphic (DRG) File 1998.

Vedder Memorial Library. n.d. One of the two large

ice houses north of the Rip Van Winkle Bridge, Catskill, New York, Collection of Greene County Historical Society, Vedder Memorial Library, Ice Industry Photographic Files.

———. 1883. Men and ice fields, Lower end of Athens, January 27, 1883. Collection of Greene County Historical Society, Vedder Memorial Library, Ice Industry Photographic Files (Oversize Box 10).

Walsh, A. 1983. Ice harvesting on the Hudson River. Articles appearing in the *Chatham Courier*, Chatham, New York, March 3 and March 10.

Webster, T. (assisted by the late Mrs. Parks). 1852. *An encyclopedia of domestic economy*. Boston: Little, Brown. http://books.google.com/books.

Weightman, G. 2003. *The frozen-water trade*. New York: Hyperion.

Yamin, R. 2008. *Digging in the city of brotherly love*. New Haven: Yale University Press.

CHAPTER 15

HUMAN SANITARY WASTES AND WASTE TREATMENT IN NEW YORK CITY

David J. Tonjes, Christine A. O'Connell, Omkar Aphale, and R. L. Swanson

ABSTRACT

Henry Hudson first sailed to New York harbor four hundred years ago. Since then, New York City has both affected and been affected by water quality in greater New York Harbor. In this chapter, we focus on sewers, sewerage, and sewage treatment in Manhattan and their effects on the Hudson River. It is clear that feedbacks among drinking water quality and quantity, population, public perceptions, regulations, and estuarine water quality exist, although their strength and character have varied over time.

Early land uses damaged local water supplies found on Manhattan Island. New York then began to exploit the large fresh water resources available to its north, which helped the city to expand more rapidly. Water availability also allowed for water carriage sanitary practices, increasing discharges of wastes through a growing sewer network into local waters. The discharge of wastes degraded water quality, affecting natural resources in the harbor. Untreated wastes led to disease from contaminated seafood, and also more generalized effects on public health. Overall, New York lifestyles became largely detached from its shoreline, partly due to the industrial character of the waterfront, and partly because of odors and visual blight from pollution. Growing public distaste over poor harbor water quality, especially in the early twentieth century, led to some sewage treatment. More and more comprehensive treatment followed regulatory and legal actions, beginning in mid-twentieth century. Concurrently, maritime commerce declined, and the waterfront became underutilized. However, in the twenty-first century, natural resources are recovering, and New York City citizens once again flock to the shores of the Hudson River, to new and revitalized parks, new areas of development and older areas undergoing transformation, and into the harbor, now largely cleaned of its fouling from sanitary waste disposal. Today New York City public life has a much greater orientation toward the waterfront, which certainly was fostered by improved harbor water quality and the opportunities for growth that were available with the disappearance of the City's maritime industries.

Thus, there has been a complicated relationship between the city and its rivers and harbor. One aspect has been continuing use of local water bodies as receptacles for wastes, which has benefited those living in the city. Gaining these benefits has had continuing costs, however. Marine resources were damaged and some were lost, and quality of life on land was affected. Trying to undo the impacts, which has required great effort and much capital, has been hampered by technology decisions that appear suboptimal with the advantage of more than one hundred years of hindsight. Still, modern sewage treatment, initiated by local efforts and concerns, but spurred on to completion by the forces unleashed by the great environmental awakening of the 1960s and 1970s, has made it possible for the citizens of New York to again fish, boat, and even swim in city waters.

NEW YORK, DRINKING WATER, AND POPULATION GROWTH

In 1609, Manhattan had several large ponds and some streams, predominantly fed by groundwater. Drinking water of good quality was thus available (Sanderson 2009). As the population of the city grew, the common practices of the day led to impacts to the shallow groundwater system and associated surface water bodies. Human sanitary wastes, for instance, were managed through permeable and solid wall privies. The first regulations regarding acceptable design for these devices were promulgated by the Dutch in 1657 (Loop 1964). The preferred and approved design was for impermeable pits, although it is understandable that not all privy pits proved to be watertight. In addition, there was no organized management of solid wastes. These were disposed onto streets, or into marshes, ponds, and other low-lying areas where fill might create more usable land, and sometimes directly into the surrounding waters of the Hudson and East rivers (Melosi 1981). These practices affected local drinking water quality (Koeppel 2000).

Continuing increases in population (and population density) created greater impacts on local water supplies. Human wastes from privies, solid wastes, and wastes from various businesses and industries were directly disposed in bodies of water that also supplied drinking water. Wastes released into the subsurface directly contaminated groundwater-fed wells, and indirectly affected surface water bodies through groundwater discharges. The effects became greater in degree and geographical scope as the population growth rate increased in the mid-1700s (by 1800 the population of Manhattan exceeded fifty thousand) (Fig. 15.1) (data from Goldman 1997; Loop 1964; Burrows undated; New York City Department of Planning undated). Continued growth of the city was thought to be threatened by potable water shortages and the absence of a water system capable of supporting fire suppression. Entrepreneurial efforts to provide water from outside of the developed area of the city, and to construct distribution networks, were therefore encouraged by the city, beginning about the time of the Revolutionary War. The Manhattan Company, for instance, built a small reservoir around 1800 (this company is better known for

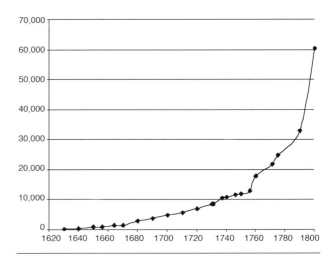

FIG. 15.1. Population of Manhattan, 1600–1800 (not including aboriginal peoples).

being used by Aaron Burr for political purposes). Other small systems were also constructed, but the need for a city-wide system was recognized by both city and state governments. The state legislature chartered the Croton Aqueduct Board in 1833, and the Croton Aqueduct was completed in 1842, along with a nascent distribution network that radiated from the central reservoir at 42nd St. and 5th Ave. The Croton system could deliver up to seventy-five million gallons of water each day to the city, although subscribers for home delivery were slow to be added at first (most still took water from central distribution points such as fountains and hydrants) (Koeppel 2000).

This development unleashed an unforeseen effect, an outgrowth of the perception of an unlimited water supply, and the development of new technologies to manage human septic wastes. These changes meant population growth no longer impacted city drinking water quality, but, rather, increasingly affected water quality in its surrounding water bodies. This was the result of the installation of sewers, and changes in their use.

SANITARY WASTE MANAGEMENT IN THE 1600s, 1700s, AND EARLY 1800s

Dutch colonists began to recreate familiar urban infrastructure from Holland within twenty-five years of settling New Amsterdam, including street gutters to convey stormwater to nearby rivers, and canals. The first true underground sewer was created under the English in the 1680s when the Broad Street Canal was covered over. In 1703, it was classified as a "common sewer"—a portal for many sources of wastewater, which was to be managed by local government. Common sewers were intended to be used only for stormwater, not human wastes (Loop 1964).

Sewer construction continued through the 1700s as impervious surfaces increased (Loop 1964). At first, sewering was an entirely private enterprise, where open trenches were dug to the closest shoreline, but by the middle of the century the trenches were replaced by underground pipes. These pipes were made both of wood and fired clay. Clay pipes needed to be preformed, and were less expensive when many were fabricated at one time. Wooden pipes were easier to create for custom jobs, but required greater skill to fit lengths together (Goldman 1997).

By the 1800s formal procedures were established for new sewer construction. The applicant, generally a group of landowners in a particular area, would petition for a project to the City Common Council. The Common Council would hold a hearing to determine if there were objections from other residents to the proposed sewer. If objections were limited, the council would approve the project. The construction process included the city soliciting bids and then contracting for approved materials and labor from private sector sources. The city used municipal staff to oversee construction. The participating property owners were billed following project completion for all contract costs (Goldman 1997).

Thus, underground piping was installed unsystematically, usually only in wealthier neighborhoods. Sometimes, multiple pipelines were set down in the same street. Until the 1840s, these sewers were intended to drain stormwater from property, or sometimes to dewater groundwater; human sanitary wastes were explicitly banned (Goldman 1997).

Impermeable cesspits continued to be the preferred means of managing human wastes. Household wastes were collected in chamber pots, or from enclosed water closets, and brought to these cellar or backyard structures (Loop 1964). Wastes were cleaned from the cesspits as needed, although these intervals were widely spaced, because water-carrier technologies for wastes were not used. The contents of the pits were only human wastes; other organic materials, such as kitchen wastes and household slops, were managed separately. Thus, the cesspits were much slower to fill than would be the case today. Wastes collected from cesspits were sometimes dumped into the closest river. At times, the city contracted with collection companies so that cesspit wastes could be sold as fertilizer. These contracts specified that the city would make sure wastes were set out curbside for collection (Goldman 1997).

By the late 1700s, it was fairly clear that the privy system had not protected drinking water supplies from contamination (Koeppel 2000). However, it was not until 1820 that city government formally took notice of soil pollution from privies, and the associated pollution of groundwater drinking supplies (Loop 1964).

WASTE CRISIS CAUSED BY ABUNDANT WATER

The Croton Aqueduct began delivering water to Manhattan in 1842 (Koeppel 2000). Unlimited, widely distributed water radically changed sanitary practices and led to a waste management crisis. Water closets and sinks had been rare because city regulations forbade the use of sewers for sanitary wastes. A mechanism to enforce this ban was that all household lines were required to have screens where they connected to street sewers, thus creating barriers to the transport of solid materials. Also, it had been difficult for most households to provide enough water to make "water carriage" systems practicable. However, with seemingly unlimited water supplies, installation of these household technologies was rapid. Although flush toilets were not invented for another twenty-five years, large quantities of household water now made it possible to carry human wastes away from living quarters rapidly and efficiently. This, in turn, quickly overwhelmed the holding capacity of privy systems (Melosi 2000). Therefore, in defiance of city regulations, many homeowners connected their new waste lines to household stormwater sewers. Only three years after the opening of the Croton water system (1845), new regulations allowed sanitary wastes to be sent through sewer systems. This led to a 50 percent expansion of sewers over the next decade, installed through the permit process discussed above. Thus, some streets had multiple lines, and others had no service (Goldman 1997).

Relatively few of the effluent pipes were extended as far as the end of the piers that surrounded lower Manhattan, and so wastes were discharged close to shore. This enhanced sedimentation in the berthing areas, and led to accumulations of wastes along the shoreline and in and around ships. Regulations in 1849 ordered outlet extensions to open waters to try to minimize these shoreline impacts. The regulatory revisions, however, did not address other technological issues such as the grates on household lines, right-angled turns that clogged with solids, and over-capacity pipes lacking sufficient gradients to flush, especially when battling tidal ebb and flow. These unaddressed problems led to many odor and overflow problems in the early sanitary sewers (Goldman 1997).

Although the city had regulated sewers and managed their use since the late 1600s in various ways, it was not until 1870 that the city assumed ownership and complete responsibility. This change was part of a general reform of city institutions, but also it was in response to public health concerns. The perceived need to convey wastes away from people grew as the miasma theory of disease gained wider acceptance ("miasmas," or vapors and gases, were the cause of illness, and septic wastes clearly emanated vapors). In addition, in poorer areas of town, tenements still dumped sanitary wastewater directly onto streets, because there had been no private enterprise to install sewers (Goldman 1997).

As a result, sewers were extended into many parts of Manhattan, and the existing pipe jumble was simplified. Many outfalls were extended and otherwise modified to try to address shoreline issues (Goldman 1997). Manhattan privy counts fell from fifteen thousand in 1875 to less than one thousand in 1891. Still, much work remained in older sections of the city, so that when the subway building boom in the early twentieth century occurred, another priority was to rebuild the sewers downtown (Loop 1964).

Few complaints regarding harbor water quality near Manhattan were recorded until well after freshwater supplies began to become unpotable from pollution in the 1700s (Koeppel 2000). However, there is indirect evidence that sewers, and the septage emanating from them, caused disagreeable water quality, even before human wastes were allowed to be disposed through them. Throughout the early 1800s, some new sewers were opposed at public hearings; testimony was presented on odors and explosions. In the 1830s, the design of outfalls was codified to encourage flushing by tides. In 1841, certain industries were forced to disconnect from the sewers, because the sewers they used impacted local air quality and discharged especially objectionable wastes.

The use of sewers for human wastes increased shoreline impacts. When the Common Council considered this change, among the comments was a concern for impacts to shoreline fish populations if there was insufficient tidal flushing. By 1849, complaints from residents led to new rules requiring outfalls to be extended past the pier line. In 1864, "pools of decomposing animal and vegetable offal" were described at the shoreline and by 1870 sewage created "white stringy slimes" and gray films near the shore (Goldman 1997). In 1875, the *New York Herald* opined it was "fallacious to assume that the discharge of sewage to rivers was borne away to the ocean" (Loop 1964).

Another source of contamination to the city's surface waters was garbage and other solid waste. Before the 1860s there was no organized, municipal solid waste collection system. Then, the city began to experiment with various schemes to harness entrepreneurial skills. These early efforts never entirely succeeded, mostly because they depended on implementing grand, complicated technologies. Although they all eventually failed, most of the companies removed some of the solid wastes accumulating at residences, businesses, and on streets. It was not until late in the century that city management of solid waste removed this loading from the stormwaters that ran into the sewers, and then into the harbor (Miller 2000).

The city's animal populations, including pigs which ran free eating trash, cattle, oxen, and especially horses, also contributed to the pollutant loading on city streets. Some of these wastes were collected from streets for household gardens, but the remainder washed into the storm sewers. The number of horses per person increased with the introduction of streetcars in the early 1800s, and again with expansions of freight transport in the middle portion of the century, due to growth of railroads (McShane and Tarr 1997). This led to there being more than 125,000 horses in the city around 1900—one for every twenty-five people (Tarr and McShane 2005). Each horse produced fifteen to thirty pounds of manure and a quart of urine each day, and only lived two to three years. It was easier to remove a horse when it could be disarticulated after rotting a little (Morris 2002); thus, even when dead, horses continued to affect the quality of urban runoff.

THE IMPACTS INCREASE AND REACH A NADIR

In the late 1800s, the population growth rate (Fig. 15.2), housing densities, and industrialization of Manhattan increased, causing growing effects on harbor water quality from additional wastewaters. In 1891 beaches and open waters were called "unsightly" and the "stench was unbearable." Proposed re-routing of sewage outfalls from the Passaic River in New Jersey to New York Harbor caused the New York Legislature to create the New York Pollution Commission. In 1906, it found that the harbor was "heavily polluted," with navigation obstacles and "local nuisance conditions," because dispersion and diffusion of sewage was incomplete. There were only three small chemical precipitation plants for the wastes of ten million people in the harbor basin.

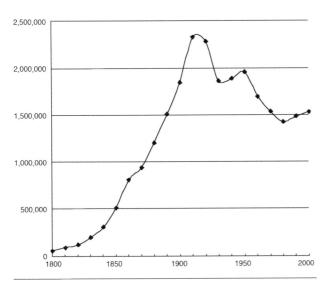

FIG. 15.2. Population of Manhattan, 1800–2000.

The commission found that dissolved oxygen levels throughout the harbor were insufficient to support oxidative degradation of wastes (Loop 1964).

The New York State Legislature created the Metropolitan Sewerage Commission in 1906 for follow-up work, replacing the Pollution Commission. The Sewerage Commission thoroughly described water conditions, qualitatively and quantitatively. The harbor above the Narrows was called "dangerous to health," local nuisance conditions were "innumerable," and several waterways were "open sewers." Impacts included declines in fisheries and shellfisheries so that the commission advocated abandonment of the local oyster industry. There was also contamination of municipal baths (which were located along the shore and which used the ambient river and harbor waters). Visible garbage, offal, and solid matter were present throughout the harbor. The waters of the harbor generally were found to be discolored, turbid, effervescent, oily, and odorous (Loop 1964).

The commission found that a dissolved oxygen concentration of three milligrams per liter (mg/l) was a critical concern (Loop 1964). There is some evidence that low dissolved oxygen concentrations were affecting fish populations (Limburg et al. 2006), because low dissolved oxygen levels make it impossible for fishes and shellfish to respire (U.S. Environmental Protection Agency 2000). In addition, it was understood that if the harbor was to provide waste treatment, higher oxygen levels were important because aerobic decomposition of matter is much more efficient than anaerobic decay processes, and so as oxygen levels decreased there was less biological decomposition of wastes (Loop 1964). There was such robust debate, however, on whether to enshrine the 3 mg/l level as a standard that no official action was taken.

Entangled in the debate over the point to establish a standard, and associated with the survival of fishes and other macro-organisms, was the issue of defining water quality impacts based on saturation concentrations or absolute concentrations of dissolved oxygen. Some declines in absolute oxygen levels result only from seasonal or tidal changes in temperature and salinity that limit the amount of oxygen the water can contain. In the early days of the twentieth century, one proposed indicator of major water quality impacts was depletion of dissolved oxygen to 50 percent of saturation. Impacts to this level were measured in some parts of the harbor (the Harlem River, parts of the East River, and certain embayments) in summer sampling conducted at and around 1910, but generally most areas were not impacted to that degree (Loop 1964). Conditions worsened, however, and the lowest dissolved oxygen levels were measured in the late 1920s through the mid-1930s (Suszkowski 1973).

The Metropolitan Sewerage Commission advocated for waste treatment, because it was clear that sections of the harbor, such as the lower East River, could not reach adequate waste treatment in situ. The level of treatment would have to be sufficient to support desired end uses of the water bodies. An interstate commission to administer the plans was recommended, as problems crossed state lines, and activities in New York affected New Jersey, and vice versa. No comprehensive action followed, so that although mitigations were prescribed before World War I, conditions continued to deteriorate into the 1920s as the city grew and discharges increased. In 1925, co-incident with the imposition of national standards, New York State closed all shellfish beds (Loop 1964) (New Jersey kept some areas open but under strict supervision). By 1928, the five-year, running average of summer, bottom-water dissolved oxygen was only 35 percent of saturation for Hudson River monitoring stations. Hudson River and inner harbor areas reached the lowest dissolved oxygen values then, although water quality in the East and Harlem rivers continued to decline into the mid-1930s (Suszkowski 1973).

Some actions to provide treatment were made in the 1920s. Screening plants in Manhattan removed gross contamination from nearly 20 percent of 150 million gallons of Manhattan sewage each day, and nine other screening plants operated elsewhere in the harbor. These screening plants removed the more visible indicators of waste discharges. In doing so, they improved aesthetics slightly and also removed some of the organic matter formerly loaded into the waters (Loop 1964).

As of 1930, nearly 1.5 billion gallons of sewage were discharged from the city and other areas fronting on New York Harbor, receiving no treatment, except for the fraction treated by screening plants. Thus, there were solids and visible turbidity traceable to sewage throughout the harbor, and slack waters were gassy and black. The City Department of Health banned swimming from the mouth of the harbor northward throughout its jurisdiction (Loop 1964). The population of Manhattan leveled off and began to decrease about this time (Fig. 15.2), because of changes in immigration law, the early stirrings of suburbanization, and the Great Depression. Smaller numbers of people in Manhattan, coupled with greater waste treatment levels, signaled the end of the long declining trend in water quality, because septic waste generation is generally proportional to population.

MODERN SEWAGE TREATMENT

Modern treatment methods for sanitary wastes brought about the recovery of water quality measured in the latter part of the twentieth century, although it took more than fifty years to build enough facilities to cover all of New York City. Ward's Island was the first. Its construction begun in 1931, but it did not achieve full operational status until 1937, because of city financial difficulties caused by the Depression (Gould 1951). Ward's Island was designed to use "activated sludge" technology (Loop 1964), which is the predominant process in use at large sewage treatment plants in the twenty-first century. Activated sludge treatment generally results in 80 percent less consumption of oxygen in receiving waters affected by effluent (biological oxygen demand, BOD), and approximately 80 percent of dissolved and settlable solids are also removed (total suspended solids, TSS). This level of treatment level is known as "secondary treatment," because it employs a biological process as well as the physical process screening and settling solids from sewage (Nathanson 2007).

In 1936, the Interstate Sanitation Commission (ISC) began to regulate sewage impacts on the harbor. With Ward's Island beginning to operate then, the city was treating 13 percent of its sewage flow, removing about 1 percent of the total amount of dissolved solids in influents citywide (treatment levels were so low because most wastewater being treated was only being screened). By the beginning of World War II, with a total of three plants online, treatment resulted in 32 percent of dissolved solids being removed (Loop 1964).

In 1948, Congress updated the 1899 Federal Rivers and Harbors Act (which prohibited the dumping of garbage into navigable waterways). This allowed the U.S. Public Health Service to monitor and assist in situations where there was interstate pollution, and authorized financial assistance to municipalities that voluntarily participated in such programs (Melosi 2000). It gave an impetus for the city to sign an "order on consent" with the ISC to "virtual[ly] eliminate pollution" in Class A recreational waters by 1953 (Loop 1964). New York State had codified uses of waterways, and created differential water quality standards to allow those uses, meeting another of the goals of the Metropolitan Sewerage Commission. In New York State, "Class A" waters were the "highest and best" use waterways, suitable for fishing, swimming, and shellfishing.

The reform of sewage treatment financing accomplished in 1950 was a breakthrough, essential as a means for the city to fund its plans. A dedicated funding source was created by explicitly linking sewage fees for system users to water usage. Although the measurements of water use were only approximate, based on building size and tenancy (until water meters were required in the 1990s), sewage plant operational monies, and, more importantly, capital expenses for plant construction, had been made independent of other city taxes and fees. Five new major projects, expansions at other plants, and upgraded sewer infrastructure were quickly accomplished, because construction bonds were no longer limited by city debt limits (Gould 1951; Loop 1964).

Later in the 1950s, the city had difficulty meeting all requirements set by the ISC. Industrial wastewater inputs resulted in plant process failures, because secondary treatment requires healthy microorganisms, and many of the chemicals dumped into the municipal sewer system by factories were toxic. Newtown Creek, the largest plant constructed in New York, used "modified aeration treatment," a less effective process than full activated sludge treatment used in an attempt to reduce plant size and overall construction costs. Newtown Creek, as a result, does not achieve the standard 80 percent reductions in BOD and TSS (Loop 1964) and has been targeted for an upgrade ever since it began operations in 1967 (construction began in the middle 2000s) (ISC, 2009).

In 1972, Congress passed the landmark Clean Water Act amendments. An important element of the act was a requirement that all discharged sewage needed to meet secondary treatment levels for BOD and TSS (with very few exceptions). Over 50 percent of wastewaters discharged to open waters in New York City already met the standard, and 75 percent of city wastewaters regularly treated because of the previous forty years of effort (Gross 1974).

New York City's fiscal crisis of the 1970s prevented completion of its sewage treatment system immediately following the 1972 legislation. This meant that for many years a large proportion of discharged city wastewaters did not meet standards. In the 1980s, the last two large city treatment plants, North River and Red Hook, were built. General upgrades and expansions of the systems meant all of the city (except for parts of Staten Island) have sanitary sewers. The wastes from sewered areas, with the exception of those treated by the Newtown Creek plant, all receive secondary treatment (Brosnan and O'Shea 1996).

In 2009, there were a total of fourteen sewage treatment plants in New York City (Fig. 15.3, adapted from Swanson et al. 2000), nine of which discharge to the inner harbor and the Hudson River, or tributaries to the inner harbor such as the East River and Kill van Kull (Adamski and Deur 1996) (Table 15.1) (Tonjes 2005).

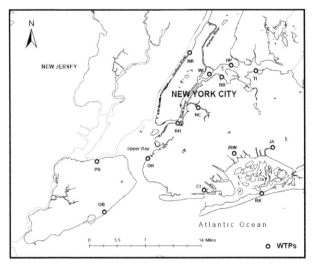

FIG. 15.3. Wastewater Treatment Plants (WTPs) in New York City: BB = Bowery Bay, CI = Coney Island, HP = Hunts Point, JA = Jamaica, NC = Newtown Creek, NR = North River, OB = Oakwood Beach, OH = Owls Head, PR = Port Richmond, RH = Red Hook, RK = Rockaway, TI = Tallman Island, WI = Wards Island, 26W = 26th Ward.

Table 15.1. New York City Wastewater Treatment Plants (directly or indirectly discharging to the Hudson River)

WPCP	Primary Treatment	Secondary Treatment	Last Upgrade	Current Capacity (MGD)
Ward's Island	1937	1937	1998	275
Bowery Bay	1939	1942	1973	150
Tallman Island	1939	1939	1976	80
Hunts Point	1952	1952	1979	200
Owls Head	1952	1952	1995	120
Port Richmond	1953	1978	1979	60
Newtown Creek	1967	2009*	Ongoing	310
North River	1986	1991	1991	170
Red Hook	1987	1989	1990	60

*Upgrades to full secondary treatment were to be completed by February 2009 (ISC, 2009) but no official notice of the project completion could be found

Water quality was still not good in 1980s in the Hudson, as illustrated by average summer dissolved oxygen concentrations less than 4 mg/l some years, and generally high fecal coliform bacteria counts. Fecal coliform, used an indicator of human pathogen contamination, decreased geometrically at the 42nd St. monitoring point in the river as the North River sewage treatment plant became operational in 1985–86, for instance (Fig. 15.4). (Swanson et al. 2000). Fecal coliform concentrations are reduced partly due to the biological activity of a sewage treatment plant, but primarily because of disinfection practices at the plant outfall (Nathanson 2007). Dissolved oxygen concentrations for bottom waters just south of the North

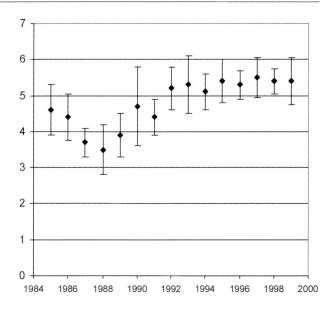

FIG. 15.5. Dissolved oxygen concentrations (summer bottom water means), 1985–1999, Station N3 (W. 155th St., Hudson River).

River treatment plant improved as the plant increased its treatment level through the late 1980s (Fig. 15.5) (Swanson et al. 2000). Generally, similar trends are found across almost all New York City waters, because all dry weather flows go through sewage treatment plants under normal operating conditions, engineering improvements and regulatory reforms on discharges are in place, and the city has made a clear commitment to other practices that result in improved harbor water conditions (Brosnan and O'Shea 1996; NYCDEP 2009).

CHRONIC PROBLEMS

Much municipal infrastructure is maintained and kept in good operating order. However, this ensures it never requires replacement or is supplanted by newer models, even though it was built to outmoded designs. Much important infrastructure is never determined to be "obsolete," and required to be replaced. Therefore, early, long-lived decisions result in technology lock-ins where changes to meet new conditions or address uncovered problems are difficult to implement.

For instance, there was debate in the nineteenth century over whether to install separate stormwater and wastewater sewer systems, or keep the combined approach. Combined sewers had a perceived economic benefit, because separate systems required

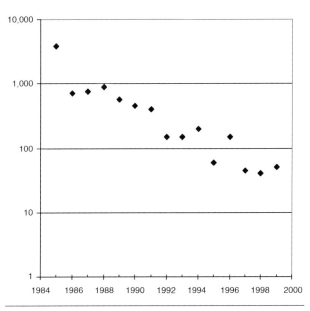

FIG. 15.4. Fecal coliform counts (summer geometric means), 1985–1999, Station N4 (W. 42nd St., Hudson River).

installing two sets of pipes. A specific analysis for Memphis, Tennessee, after the Civil War actually forecast slightly lower costs for the separate sewer systems. Most public health advocates also favored separate systems. Separate sewers resulted in smaller pipes, and especially for sanitary systems, less airspace in the pipes (large pipes were needed for combined systems to manage unusual storm events on top of the daily production of septic wastes). The miasma theory of disease, which had more adherents than the competing germ theory, was based on vapors and gases transmitting illness. Vapors and gases could be minimized in the septic sewers if the pipes were smaller and better fitted to typical volumes, and it was anticipated storm sewers would mostly be empty (except when it rained). Despite these analyses that seemingly favored separate systems, New York, like all other large eastern North American cities (including Memphis), chose to stay with combined sewers, even as it greatly expanded its system (Melosi 2000).

Combined sewers presented technical and economic problems as treatment plants came on line, as it was impractical to size plants to meet maximal flows. Generally, plant designs called for a capacity of two times dry weather flow (Loop 1964). Thus, small amounts of rainfall lead to diversions of flow. As little as 0.1 inch of heavy rain can cause diversions of sewage in modern-day New York City (New York City 2007). One report prepared for the New York City in the 1960s estimated that as much as 30 percent of annual loadings associated with sewage may bypass treatment because of wet weather overflows; although the report claimed it only rains 3 percent of the time in the northeast United States, wet weather causes outsized effects because stormwater carries its own pollutants and washes out accumulated sedimentation (Loop 1964). Thus, water quality tends to be better in many areas of the harbor in drier summers (Swanson and Tonjes 2001a), which can be a factor especially in interpreting long-term trends for fecal coliform (such as Fig. 15.4). For the past seventy-five years, the city has faced the challenge, as combined sewers were connected to treatment plants, to create enough capacity during wet weather to store the combination of human wastewater and stormwater to allow treatment after rains end (IEC 2009). One incremental step was to adopt water conservation programs in the 1990s, which reduced influents by nearly 20 percent and allowed for stormwater treatment rates citywide to exceed 50 percent (Swanson and Tonjes 2001b). By 2000, the city estimated that 60 to 80 percent of all wet weather flows in Manhattan were treated (Swanson and Tonjes 2001a).

Eutrophication of coastal waters is most often associated with effluents discharge (Cleorn 2001). Eutrophication problems in Long Island Sound have been closely tied to New York City wastewater nutrient releases (Long Island Sound Study 2008). Nutrient removal from wastewater is well understood, and the technology is well tested (Nathanson 2007). However, space limitations at existing city sewage treatment plants, and cost projections that range from $500 million to several billion dollars, have made the city slow to implement these additional treatment steps (Andersen 2002).

Solids in sewage are removed from the influent, but this creates a solid waste (sewage sludge), which then requires disposal. The city relied on ocean dumping to mitigate sewage sludge disposal effects on local waters. The wastes were first dumped twelve miles southeast of the harbor entrance until the 1980s. In 1986, a site 106 miles southeast of the harbor entrance was adopted, and it was used until 1992, following the passage of the Ocean Dumping Ban Act. The ban on ocean dumping of sewage sludge has resulted in additional treatment of sludge in the treatment plants to make the sludge more amenable to transport by truck. This sludge dewatering resulted in approximately 30 percent more nitrogen loadings in plant discharges, which makes the task of overall nitrogen removal that much more difficult (Swanson et al. 2004). This increase in nitrogen releases in the 1990s probably contributed to some areas continuing to experience low dissolved oxygen concentrations, particularly in western Long Island Sound (Wilson et al. 2008). There is no mention of nitrogen removal in the environmentally oriented, twenty-year "PlaNYC" (New York City 2007), although in 2010 the city reached an agreement with the Natural Resources Defense Council to implement nutrient removal at its four Jamaica Bay treatment plants (NRDC 2010).

Sewage also has contributed to poor harbor sediment quality and toxicity. This affects the literal base of the marine food chain (Long et al. 1995).

Concentrations of contaminants in the water column, especially metals, have declined (Sanudo-Wilhemy and Gill 1999). The Clean Water Act specified that generators of industrial effluents needed to pre-treat wastes prior to release into sewer systems. Pre-treatment programs have been very effective. In addition, New York City has a "trackdown" program—painstaking efforts to find the source of metals in plant influents by testing in the sewer lines. When sources are determined, modifications of practices follow to eliminate the inputs (Swanson et al. 2000). Removing contamination inputs is one element in the overall program to remediate harbor sediments.

Recently, advances in analytical chemistry have allowed the detection of "organic wastewater contaminants" (OWCs) in aqueous samples, including pharmaceuticals and personal care and other household products that are not entirely degraded in wastewater systems. The concentrations of most OWCs are well below therapeutic levels, so that direct human health concerns appear to be unlikely (Benotti et al. 2009). However, OWCs have caused endocrine effects in marine organisms, because many are hormonally active substances, or are functionally similar to them. Measured impacts include gravely skewed sex ratios or developmental problems in fishes where concentrations of these compounds are highest (Sumpter 1995). Treatment does not affect certain OWCs, and may in fact make some compounds more potent endocrine disruptors (Auriol et al. 2005). Endocrine system responses and changes in those responses have been measured in fishes exposed to New York City effluents (Todorov et al. 2002). In areas like the harbor that receive so much sewage effluent, it is probable more effects will be detected as analyses continue. Thus, it is likely that OWCs will come under more regulation, leading either to societal changes to reduce influent quantities or major treatment process modifications.

One intent of the Clean Water Act is to restore "biologic integrity" to impacted waterways (Karr 1991), and it seems unlikely that harbor ecosystems will ever be restored to that degree. In practice, compliance with regulations is determined by measuring water quality indicators, a process assumed to ensure ecological quality. Most water bodies that are not used as drinking water supplies have routine indicator testing for the two obvious measures of water quality: dissolved oxygen and fecal coliform. Most of New York Harbor, under average conditions, now meets standards. However, complete compliance with water quality regulations requires that all regulated contaminants conform to regulated levels. Routine comprehensive testing is rarely required due to the costs; if such testing were made regularly, it is unlikely that the harbor would ever achieve full compliance, because of contaminated sediments, continuing combined sewer overflows under wet weather conditions, and the inability of standard sewage treatment to remove nutrients and many OWCs.

POLLUTED RIVER IMPACTS ON NEW YORK CITY

The focus thus far has been on the impact of people on the river, and not the effect of the impacted river on city residents. The colonial-era settlers of New York ate local shellfish and fishes. Diamond Jim Brady was famous for his oversized oyster feasts and the city was also well known for other seafood (Boyle 1969). But by the mid-1800s, pollution of the harbor had brought disease, typhus and cholera. As early as the 1860s New York death rates were documented to be higher than other cities with better sewers, an indirect indictment of the effect of rising pollution of an important food source (Melosi 2000). By 1900, people were able to make connections among water pollution, shellfish, and intestinal illnesses, and so harbor fisheries declined (Andersen 2002). Still, at least some city residents continued to exploit available resources, and Robert Boyle made a vivid, unsettling description of striped bass fishing at the 42nd Street untreated sewage outfall in the 1960s, with fish being caught and eaten despite "an oily flavor" (Boyle 1969).

Maritime businesses were certainly aware that sewer effluents caused shoaling, and nuisance odors and other aesthetic concerns. Wood waste gathered from the harbor around 1900 often had an inch or more of accreted sewage on it, an illustration of potential effects (Loop 1964). Certainly the luxury passenger lines with dockage at West Side piers in the mid-twentieth century must not have relished collecting upscale passengers while nestled in among the raw sewage outfalls there (these outfalls were used into the 1980s, and still serve as outlets for untreated wastes when the system receives too much rain today).

Sewers affected everyday life in the city, from earliest times. Odor complaints were raised in the 1830s, and shoreline odors disturbed residents enough in the 1840s that the matter was brought to the Common Council (Goldman 1997). In the early 1900s, the Metropolitan Sewerage Commission decried "objectionable conditions" in the harbor, making it clear that citizens were disturbed by how bad water quality was. Activities such as sanitary bathing and recreational swimming in city rivers were banned (Loop 1964).

Although the overall lack of residential or recreational use of the waterfront was mostly due to commercial appropriations, there was, for example, growing residential development along the East River at Tudor City and north in the 1930s (Loop 1964), including luxury apartments such as the River House at East 52nd St., which was said to be the "best apartment house" in the city when it was built (Bower 2009). This kind of development was said to be an added impetus for sewage treatment (Loop 1964).

By 1936, as reported by Loop (1964), the need for treatment was justified because of "common standards of decency," such that citizens along the waterfront should not be exposed to recognizable human wastes. In the 1940s, the ISC had a clear goal to recover city waters for boating, fishing, and even shellfishing and swimming (Loop 1964). With the development of a comprehensive sewage treatment system, most waters in the city met dissolved oxygen and pathogen standards by the 1990s (Swanson et al. 2000). Fish populations and other marine life had made notable recoveries (Waldman 1999), and there have even been pilot programs to try to restore oysters in some areas of the harbor (Swanson and Tonjes 2001a).

Advancing levels of sewage treatment have clearly coincided with decisions to expand waterfront uses and access for both city residents and visitors. Growing citizen distaste with the condition of New York City water led to municipal action; the potential for further improvement has continued to change perceptions. Urban waterways are no longer considered suitable for untreated waste or as treatment facilities. As water quality improves, potential use of the shoreline and waters increase. And, as more people use and appreciate the shoreline, there is more support for programs that increase the scope of use of the harbor. In the 1960s and 1970s, newspaper stories about swimmers or kayakers or fishers in New York City waters were novelty feature items. Now these activities, while not commonplace for most New Yorkers, are widely advertised, and no longer qualify as news.

As commercial use of waterfront areas has declined, projects such as Battery Park City and the South Street Seaport have received much attention and achieved commercial success, and areas such as Battery Park, Riverside Park, and the Brooklyn Promenade have regained lost luster (Freudenberg and Pirani 2007). Although New York still is not a city like Paris or London where its riverside is an essential part of its image and appeal, there is growing awareness of the magnificent New York shoreline. This appreciation will grow as the Hudson River Park develops, and water-oriented construction, especially along the West Side, continues. New York City has already rezoned (or is planning to rezone) most of the west side waterfront and the shoreline along the Harlem River to increase densities and mixed residential-commercial uses, for instance (New York City 2007). It is difficult to imagine these projects being pursued if there were not effective treatment of the city's wastes. Nonetheless, although PlaNYC discusses the need (and State requirements) to upgrade capture rates and decrease the generation of stormwater to reduce the amount of untreated sewage released under wet weather conditions, it does not address the potential for additional treatment of nutrients or OWCs (New York City 2007).

CONCLUSIONS

In its earliest days, the relatively small population of New York City could manage wastes and minimize impacts to the environment, especially key natural resources that people depended on for daily life. As the population grew, however, waste management practices especially impacted important water resources. Declining drinking water quality and increasing demand for water supplies forced the city to create a distribution system based on supplies from outside its own borders. The availability of the harbor as receiving waters for wastes, with tides and currents that made many of the wastes "disappear" as they were discharged, made it a natural catchment for the great increase in wastewater disposal needs that occurred with the public water distribution system in the 1840s. The use of stormwater sewers for sanitary wastes alleviated impacts on people from growing pollution levels around their dwellings. However, the increasing city population increased its use of sanitary sewers, and the associated waste burden on harbor waters. Although it was slow to be recognized, eventually this waste loading also affected human health, albeit not enough to slow city growth rates.

The long rehabilitation of the harbor only began when public distaste for its degradation forced the initiation of waste treatment. Although New York began work on its sewage treatment system before many other American cities, its slow progress (fifty years of construction following thirty years of planning) appears to be an indictment of the degree its citizens were disengaged from its shoreline. City engineers deserve credit, however, for once the plants began to be built, the selected technologies reached treatment levels that were not mandated until the Clean Water Act was passed in 1972.

Although Manhattan is an island, for much of the twentieth century it had few public spaces and little public activity along the banks of its rivers; waterfront uses were largely restricted to shipping and related commerce. But these vast industries declined, and by 2000 they had essentially vanished. As water quality has improved, perceptions of the harbor have changed from a waste receptacle to a natural resource (once again). Certainly, many more elements of New York life, such as parks and recreation, include the harbor and there is much recent commercial and housing redevelopment, replacing empty and underused spaces that appeared as the maritime industries withered away. The malodorous, fouled waters of the harbor circa 1920 would not support the civic life that now is found using the recovering shoreline.

ACKNOWLEDGMENTS

We received many helpful comments on a draft of this manuscript from editor Robert Henshaw, the conference committee, and anonymous peer reviewers, which were used to improve the paper immensely.

REFERENCES CITED

Adamski, R. E., and A. A. Deur. 1996. History of New York City's wastewater treatment. In *The Proceedings of the International Symposium on Sewage Works*, 59–74. Tokyo: Japan Sewage Works Association.

Andersen, T. 2002. *This fine piece of water*. New Haven: Yale University Press.

Auriol, M., Y. Filali-Meknassi, R. D. Tyagi, C. D. Adams, and R. Y. Surampelli. 2005. Endocrine disrupting compounds removal from wastewater, a new challenge. *Process Biochemistry* 41(3): 525–39.

Benotti, M. J., R. A. Trenholm, B. J. Vanderford, J. C. Holady, B. D. Stanford, and S. A. Snyder. 2009. Pharmaceuticals and endocrine disrupting compounds in US drinking water. *Environmental Science and Technology* 43: 597–603.

Bower, M. (picture editor). 2009. *New York 400*. Ed. J. Thorn. Philadelphia: Running Press.

Boyle, R. H. 1969. *The Hudson River*. New York: W. W. Norton.

Brosnan, T. M., and M. L. O'Shea. 1996. Sewage abatement and coliform bacteria trends in the lower Hudson-Raritan Estuary since passage of the Clean Water Act. *Water Environment Research* 68(1): 25–35.

Burrows, E. G. Undated. Table I: Population of North American ports to 1775. Table II: The black population of New Netherland and New York colony. Table III: White and black population of colonial New York City. http://www.acadmic.brooklyn.cuny.edu/history/burrows/demog.htm; accessed July 20, 2009.

Cloern, J. E. 2001. Our evolving conceptual model of the coastal eutrophication problem. *Marine Ecology Progress Series* 210: 223–53.

Freudenberg, R., and R. Pirani. 2007. *On the verge: Caring for New York City's emerging waterfront parks and public spaces*. New York: Regional Plan Association.

Goldman, J. A. 1997. *Building New York's sewers: Developing mechanisms of urban management.*

West Lafayette, IN: Purdue University Press.
Gould, R. H. 1951. Progress on sewer rentals for New York City. *Sewage and Industrial Wastes* 23(7): 849–52.
Gross, G. 1974. Sediment and waste deposition in New York Harbor. In *Hudson River Colloquium*, ed. O. A. Roels, 112–28. New York: Annals of the New York Academy of Science (V. 250).
IEC. 2009. *2008 Annual Report.* New York: Interstate Environmental Commission.
Karr, J. R. 1991. Biological integrity: A long-neglected aspect of water resource management. *Ecological Applications* 1(1): 66–84.
Koeppel, G, T. 2000. *Water for Gotham: A history.* Princeton: Princeton University Press.
Limburg, K. E., K. Ahattala, A. W. Kahnle, and J. R. Waldman. 2006. Fisheries of the Hudson River Estuary. In: *The Hudson River Estuary*, ed. J. S. Levinton and J. R. Waldman, 189–204. New York: Cambridge University Press.
Long, E. R., D. A. Wolfe, K. J. Scott, G. B. Thursby, E. A. Stern, C. Peven, and T. Schwartz. 1995. *Magnitude and extent of sediment toxicity in the Hudson River Estuary.* NOAA Technical Memorandum NOS ORCA 88. Rockville, MD: National Oceanic and Atmospheric Administration.
Long Island Sound Study. 2008. *Sound health 2008.* Long Island Sound Study, Stony Brook, NY.
Loop, A. S. 1964. *History and development of sewage treatment in New York City.* New York: New York City Department of Health.
McShane, C., and J. A. Tarr. 1997. The centrality of the horse in the 19th century American city. In *The making of urban America*, ed. R. A. Mohl, 105–30. Lanham, MD: Rowan and Littlefield.
Melosi, M. V. 1981. *Garbage in the cities: Refuse, reform, and the environment, 1880–1980.* College Station, TX: Texas A&M University Press.
———. 2000. *The sanitary city: Urban infrastructure in America from colonial times to the present.* Baltimore: The Johns Hopkins University Press.
Miller, B. 2000. *Fat of the land: Garbage in New York the last two hundred years.* New York: Four Walls Eight Windows.
Morris, E. 2002. *From horse power to horsepower.* http://www.uctc.net/access/30/Access%2030%20-%2002%20-%20Horse%20Power.pdf.; accessed July 15, 2009.
Nathanson, J. A. 2007. *Basic environmental technologies.* 5th Ed. Upper Saddle River, NJ: Pearson Prentice-Hall.
NRDC. 2010. City commits to major water quality improvements in Jamaica Bay. Press release, http://www.nrdc.org/media/2010/100225.asp., dated February 25, 2010; accessed March 16, 2010.
NYCDEP. 2009. *2008 New York Harbor water quality report.* New York City Department of Environmental Protection. http://www.nyc.gov/html/dep/pdf/hwqs2008.pdf; accessed July 23, 2009.
New York City. 2007. *PlaNYC.* http://www.nyc.gov/html/planyc2030/html/downloads/the-plan.shtml; accessed December 9, 2009.
New York City Department of Planning. Undated. Total and foreign-born population, New York City, 1790–2000. http://www.nyc.gov/html/dcp/pdf/census/1790-2000_nyc_total_foreign_birth.pdf; accessed July 20, 2009.
Sanderson, E. W. 2009. *Manhatta: A natural history of New York City.* New York: Harry N. Abrams.
Sañudo-Wilhelmy, S. A., and G. A. Gill. 1999. Impact of the Clean Water Act on the levels of toxic metals in urban estuaries: The Hudson River Estuary revisited. *Environmental Science and Technology* 33: 3477–81.
Sumpter, J. P. 1995. Feminized responses in fish to environmental estrogens. *Toxicology Letters* 82–83: 737–42.
Suszkowski, D. J. 1973. Sewage pollution in New York Harbor: A historical perspective. MS Thesis, State University of New York at Stony Brook.
Swanson, R. L., M. L. Bortman, T. P. O'Connor, and H. M. Stanford. 2004. Science, policy, and the management of sewage materials: The New York City experience. *Marine Pollution Bulletin* 49: 679–87.
Swanson, R. L., and D. J. Tonjes. 2001a. *New York City 2000 regional harbor survey.* Ed. P. Heckler, B. Ranheim, N.-J. Yao, and B. W. Stephens. New York: New York City Department of Environmental Protection.
———. 2001b. Water conservation cleans Long Island Sound. *Clearwaters* 31(2): 8.

Swanson, R. L., D. J. Tonjes, N. Georgas, and B. W. Stephens. 2000. *New York City 1999 regional harbor survey*. Ed. P. Heckler, A. I. Stubin, and N.-J. Yao. New York: New York City Department of Environmental Protection.

Tarr, J. A., and C. McShane. 2005. Urban horses and the changing city-hinterland relationships in the United States. In *Resources of the city: Contributions to an environmental history of Europe*, ed. D. Shott, B. Luckin, and G. Massard-Guilband, 48–62. Aldershot, UK: Ashgate.

Todorov, J. R., A. A. Elskus, D. Schlenk, P. L. Ferguson, B. J. Brownawell, and A. E. McElroy. 2002. Estrogenic responses of larval sunshine bass (*Morone saxatilis* x *M. chrysops*) exposed to New York City sewage effluent. *Marine Environmental Research* 54: 691–95.

Tonjes, D. J. 2005. The New York City harbor survey. In *Water encyclopedia*, ed. J. Lehr and J. Kelley, DOI: 10.1002/047147844X.ww144. Online product: John Wiley and Sons (www3.interscience.wiley.com/cgi-bin/mrwhome/110431379/HOME).

U.S. Census Bureau. Undated. Population of New York by Counties 1900–1990. http://www.census.gov/population/cencounts/ny190090.txt; accessed July 20, 2009; and, 2000 Census, New York County, New York. http://factfinder.census.gov/servlet/GCTTable?_bm=y&-context=gct&ds_name=DEC_2000_PL_U&-CONTEXT=gct&mt_name=DEC_2000_PL_U_GCTPL_ST2&-tree_id=400&-redoLog=true&-caller=geoselect&-geo_id=05000US36061&-format=CO-1&-_lang=en; accessed July 20, 2009.

U.S. Environmental Protection Agency. 2000. *Ambient aquatic life water quality criteria for dissolved oxygen (saltwater)*. EPA-822-R-00-012. Washington, DC: U.S. Environmental Protection Agency.

Waldman, J. 1999. *Heartbeats in the muck*. New York: Lyons Press.

Wilson, R. E., R. L. Swanson, and H. A. Crowley. 2008. Perspectives on long-term variation in hypoxic conditions in western Long Island Sound. *Journal of Geophysical Research* 113:C12011, doi:10.1029/2007JC004693.

CHAPTER 16

FOUNDRY COVE

Icon of the Interaction of Industry with Aquatic Life

Jeffrey S. Levinton

ABSTRACT

The history of the Hudson River has paradoxically combined its natural beauty with its industrial and commercial utility. Foundry Cove is an iconic example of this intersection, being located in the exquisite Hudson Highlands but also being a locus of industrial activity since a large forge was located there in 1817. A major nickel-cadmium battery factory operated from the early 1950s to the 1970s, and releases of nickel-cadmium wastes filled the cove and were released into the open Lower Hudson River estuary and taken up by the blue crab, which created a health hazard for fishers. Within the cove, the cadmium-laden deposits caused a dominant benthic invertebrate species to evolve resistance to cadmium through Darwinian natural selection, which allowed trophic transfer to higher levels of the food web. Foundry Cove was declared a Superfund site in 1980 and its cleanup was largely successful, which is unusual for major dredging projects. Following the cleanup, the cadmium-tolerant species lost its resistance in only about nine generations. The cleanup also resulted in a greatly reduced release of cadmium into the Hudson, and a strong reduction in cadmium in the blue crab. Restoration of Foundry Cove itself was variably successful with an only partially restored marsh and strong effects on the soft-bottom benthic community. The success of the cleanup and restoration owes a great deal to a wide range of participation by government agencies, conservation groups, and research scientists.

INTRODUCTION

The Lower Hudson River estuary is an estuary with a long history of industrial impact. Ironically, the birth of the Industrial Revolution was the seed of the estuary's development into a national icon of natural beauty and also its degradation into one of the most impacted large rivers in America. The steamboat allowed many thousands of tourists to see the astounding beauty of the waters and vistas of the Hudson, but it also turned the Hudson into part of a major conduit for trade to the western interior of eastern North America. This conduit presented many opportunities to those entrepreneurs who developed manufacturing and processing operations along the lower Hudson shores.

The intersection of the Hudson River with its highlands, near West Point, is a location where natural beauty, ecological function, and human impact have intersected many times. In the Revolutionary War, the area was strategically crucial, and a small forge built at West Point helped the completion of a large chain, which was stretched across the Hudson to impede the northerly advancement of British forces (Diamant 1989). Between 1817 and 1911, a major foundry was established and operated in East Foundry Cove, where the first railroad steam engine in New York was constructed. Later this foundry became the crucial site of construction of guns and ammunition for the Union in the Civil War. This activity must have released a variety of wastes, but the impact on aquatic life is unknown. In 1837, an ill-fated attempt to produce wild rice resulted in the impoundment of a major area of marsh, now known as Constitution Marsh. After a relatively long quiet period, this area was challenged yet again. A major factory for the production of batteries was installed in the early 1950s in the village of Cold Spring, New York, which will be the major focus of this chapter. Finally, a major pump-storage plant was suggested and nearly built in the walls and upper parts of Storm King Mountain on the west side of the Hudson. A major protest regarding aesthetics and the impact on Hudson River fisheries stalled and finally killed this proposal. This episode resulted in the first accorded right of environmental groups to sue against proposed projects that might harm the ecological function or aesthetic features of the environment (Boyle 1979; Suszkowski and D'Elia 2006).

We will focus on the last ignominious phase of the Hudson Highland's industrial pollution history—one of the greatest sources of metal pollution

FIG. 16.1. Air photo taken in 1997 showing field sites at Foundry and South Cove, near Cold Spring, New York.

in world history—an NiCd battery factory located at Cold Spring, New York, adjacent to Foundry Cove. Foundry Cove (Fig. 16.1) is a tidal bay that consists of fresh water for nearly the whole year, except in summer when Hudson River discharge is very low and local salinities reach 3–6 psu. It is on the east side of the Hudson River about ninety kilometers north of the Battery, which is km 0 at the southern tip of Manhattan Island. A railroad causeway divides the cove into western and eastern sections connected by a channel about twenty-five meters wide. The railroad tracks move north-south through a rocky Constitution Island. Constitution Marsh, a cattail marsh rich with waterbird species, borders East Foundry Cove on the south.

THE BATTERY FACTORY AND METAL RELEASE

The establishment of the battery plant in Cold Spring emerged as a response to a need generated by the cold war. The Nike missile system was designed and deployed to protect the United States from attack by airplanes possibly carrying nuclear weapons. These line-of-sight antiaircraft missiles, aided by radar detection, were installed throughout the United States in the period 1953–1962, and the missiles had battery-powered guidance systems, whose nickel-cadmium batteries were built in the factory at Cold Spring. The factory was constructed by the United States Army in 1952, and used for the manufacture of batteries through 1979 (Knutson et al. 1987).

The factory wastewater contained cadmium and nickel in the form of metallic hydroxides and was released mainly as suspended solids (cobalt was briefly used as a stabilizer and also was released). Initially, waste materials were released through the Cold Spring village sewer system directly into the main stream of the Hudson, but a small amount was diverted through a pipe that entered East Foundry Cove (Fig. 16.1). The manufacturer was ordered to disconnect from the sewer system leading directly to the Hudson in 1965, and all wastes were subsequently discharged into East Foundry Cove. After a treatment system was installed in 1971, waste was again discharged via the Cold Spring sewer system into the Hudson River, at the Cold Spring town pier. Battery manufacture ceased in 1979.

METAL LOADS IN THE SEDIMENT AND METAL EXPORT TO THE HUDSON RIVER

What was the environmental chemical consequence of this long-term industrial activity? About 179 MT of total waste was discharged; 51 MT of solids and 1.6 MT of soluble cadmium had entered East Foundry Cove. Cadmium and nickel were concentrated in surface sediments, but at a depth of 20 cm, these metals occurred at preindustrial levels. Outside East Foundry Cove, Cd and Ni dropped by three orders of magnitude in West Foundry Cove sediments (Bower et al. 1978). East Foundry Cove became the most metal-polluted estuarine site in the world, and hot spots of cadmium and nickel also occurred in open Hudson River sediments (Bower et al. 1978).

By the early 1970s, the United States Clean Water Act had been enacted and New York State pollution standards were in force. A New York State court order in 1972 dealt with the failure of the Marathon Battery Company to meet state discharge standards and stipulated that the most heavily contaminated area (> 900 ppm Cd/wet weight of sediment) be dredged to a depth of 30 cm. The eastern end of East Foundry Cove was dredged in 1972 and 1973, which only removed about 10 percent of the cadmium (Engineering 1983). A sediment survey in 1974 and 1975 showed that concentrations were still very high and no appreciable improvement was made in the areas with highest sediment Cd concentrations (Kneip and Hazen 1979). In 1983 another survey still showed very high concentrations in East Foundry Cove but demonstrated a reduction in surface sediment concentrations, relative to previous surveys. Sediment at the surface averaged ca. 4,000 ppm Cd with a range of 12–39,500. But subsurface peaks in cadmium and nickel were also found, suggesting that some of the waste was now being buried by sediments of lower concentration (Knutson et al. 1987). Nevertheless, hotspots of 10,000 ppm Cd still persisted and a small area near the outfall pipe had sediments of about 25 percent Cd. Overall, there was no evidence that total cadmium load had declined from previous surveys.

Cadmium concentration within East Foundry cove declined with distance from the outfall and was positively correlated with the silt-clay fraction of the sediment (Knutson et al. 1987).

Aside from the extreme pollution of sediments within East Foundry Cove, there was strong tidal exchange between the cove and the open Hudson River through the railroad trestle opening at the western end, with currents approaching 2 m s^{-1} during full tidal strength (J. Levinton, unpublished data). Metal-rich particulates were leaving East Foundry Cove at tidal ebb. A simple means of estimating export to the Hudson involves regular collection of water samples over a tidal cycle, filtering particulates, measuring cadmium concentrations of both filtered water and particulates using atomic absorption spectroscopy, and comparing cadmium entering and leaving Foundry Cove through the railroad trestle pass. We made such measurements in 1994 and found that nearly all of the cadmium was in particulate form. Extrapolating the exchange of these two cycles suggested that East Foundry Cove was exporting ca. 0.5–1.0 MT y^{-1} to the Hudson River (R. Young, unpublished). This estimate is consistent with two other estimates of 0.7 and 0.3 MT y^{-1} in 1974 and 1976, respectively (Hazen and Kneip 1980). While this export was not materially reducing the overall load within the system, it was affecting the region, as we discuss below.

BIOLOGICAL EFFECTS OF CADMIUM WITHIN FOUNDRY COVE

Cadmium Toxicity

Cadmium is a common aquatic pollutant and is taken up by a large number of marine plant, algal, and animal species (Klerks 1987; Rainbow 2002). It has a wide range of effects, resulting from industrial exposure, on humans, including impairment of renal function, cancer, and arteriosclerosis, among others (Flick et al. 1971). Cadmium is found commonly in estuaries, but is often complexed with chloride as salinity increases down-estuary (Windom 1989). Cadmium is usually more toxic in fresh water as it is available in dissolved form and is taken up readily by a variety of aquatic invertebrates. The complexing with chloride makes it less toxic as salinity increases (Sunda et al. 1978). Cadmium has a wide variety of toxic effects on aquatic invertebrates, such as oligochaetes, which may inhibit growth and reproduction, and increase mortality (Bouchet et al. 2000; Klerks and Levinton 1989a). Cadmium binds to metal-binding proteins and may be readily taken up by aquatic invertebrates and passed up the food web (Wallace and Lopez 1997).

Effects on the Benthos

The pollution at Foundry Cove stimulated a great deal of interest in possible biological effects. While there was evidence of export of cadmium mainly in the particulate phase into the Hudson River, and a possible mechanism of de-adsorption from particles and complexing with chloride ions, the evidence in Foundry Cove suggested a very low rate of release from particulates to the water (Hazen and Kneip 1980). Still, elevated cadmium concentrations were found in a wide variety of organisms living in Foundry Cove sediments and in the adjacent marshes (Klerks and Levinton 1989a; Kneip and Hazen 1979). Unfortunately, no extensive experimentation was done on the components of the community, but reduced abundance was found in the 1970s, in association with high cadmium concentrations in the sediment.

The benthos of East Foundry Cove was sampled before the cleanup and was dominated by oligochaetes and a number of species of chironomid fly larvae. While Kneip and Hazen (1979) found some reduced abundance at high cadmium levels, in 1983 no differences in overall population density of dominant groups could be found over a wide range of cadmium concentrations (Klerks and Levinton 1989a). Differences in diversity were also not found over this same range, although a later study found a drop in species richness very close to the outfall where concentrations are very high, reaching 25 percent (Levinton et al. 1999). The broad overlap, in community composition (Fig. 16.2) over a wide range of cadmium concentration within East Foundry Cove and with benthos from a nearby much cleaner South Cove was striking (Kelaher et al. 2003a).

FIG. 16.2. Two-dimensional multidimensional-scaling (MDS) ordination plots comparing faunal assemblages using the Bray-Curtis dissimilarity coefficient in locations in Foundry Cove with low (unfilled symbols) and high (unfilled dashed symbols) levels of cadmium and a reference location in South Cove with low cadmium (grey-filled symbols). MDS plots take the multidimensional data for many species from different sites and render it into a two-dimensional plot. Similarity of location denotes similarity among faunal assemblages. The overlap of Foundry Cove and South Cove samples before the cleanup (left diagram) shows that the two sites cannot be distinguished, in contrast to the complete separation of samples from the two sites after the cleanup (right diagram). Differently shaped symbols indicate different times of sampling: Squares = May/June; inverted triangle = August; triangle = October; circle = December (from Kelaher et al. 2003).

Cadmium Pollution and Natural Selection

The apparent lack of effect on soft-sediment community composition of such a high concentration of cadmium in the sediment was unexpected, given the expectation that cadmium would be highly toxic. Perhaps despite the high cadmium sediment surface concentrations, the exposure to benthos was minimal. Unfortunately, no measurements of dissolved cadmium in pore waters were made. We chose the most common species in soft sediments, the oligochaete *Limnodrilus hoffmeisteri* for further study. *L. hoffmeisteri* is a deposit-feeding infaunal worm, a simultaneous hermaphrodite, and reproduces without selfing. It is small, rarely exceeding 1–2 cm in length. Young are deposited into a cocoon, which is supplied with nutrients for the embryo, and juveniles crawl away from the cocoon into the sediment.

We found that Foundry Cove worms survived well when placed in Foundry Cove high-cadmium sediments as expected, but worms from the nearby clean South Cove died immediately when exposed similarly (Klerks and Levinton 1989b). Offspring of Foundry Cove worms raised for two generations in clean sediment were still far more resistant to cadmium than worms collected from South Cove (Fig. 16.3). Estimates of family resemblance demonstrated that nearly all of the variation in resistance among individuals was genetically based, and it was likely that resistance was explained by a single segregating genetic element (Martinez and Levinton 1996). The resistance was apparently facilitated by the ability of Foundry Cove worms to synthesize a metal-binding protein of ca. 16,000 daltons molecular mass, which could not be synthesized to nearly the same degree by South Cove worms when exposed to the same level of cadmium (Klerks and Bartholomew 1991). The abundance of *L. hoffmeisteri* despite high cadmium levels could therefore be explained by intense Darwinian natural selection for resistance. Cadmium was therefore actively toxic in Foundry Cove and having interesting biological effects.

The evolution of resistance may appear to be a somewhat abstruse result of little interest to environmental considerations, but that is far from the case. The mechanism of resistance involves binding and detoxification of cadmium by binding with a metal-binding protein or immobilization in cadmium-rich granules, concentrated in the chloragog tissue surrounding the gut (Klerks and Bartholomew 1991). These mechanisms increase the body burden of cadmium in worm tissues, which makes abundant benthic prey species available for trophic

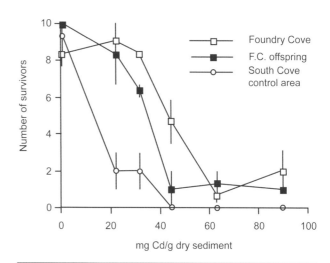

FIG. 16.3. Numbers of *L. hoffmeisteri* surviving a 28-day exposure to sediments with different cadmium concentrations including field collected worms from Foundry Cove, South Cove, and second-generation offspring of Foundry Cove worms reared in the laboratory in clean sediment (From Klerks and Levinton 1989b).

transfer of cadmium to higher trophic levels. Transfer to a predatory shrimp was especially efficient when the cadmium was bound to the metal-binding protein in the cytosol, but was inefficient when found in the metal-rich granules, which are relatively insoluble (Wallace et al. 1998). Thus, rapid evolution of resistance reorganized the community into a source of cadmium for local and mobile predators that might enter and leave the cove, such as fish and crabs.

The regional impact of cadmium release at Foundry Cove was limited to the finding of high cadmium concentrations in the blue crab, *Callinectes sapidus*, which migrates up-estuary into the lower Hudson, sometimes nearly to Albany. A longitudinal survey of blue crabs in the Lower Hudson demonstrated high cadmium concentrations, especially in the hepatopancreas. Though the distribution was broad, a regional high occurred in the vicinity of Foundry Cove, with declines of body burden both up and downriver (Sloan and Karcher 1984). A health advisory was issued by the New York State Department of Health suggesting a restriction of dietary intake of blue crabs.

THE CLEANUP

Citizen groups, state agencies, and federal agencies recognized the pollution at Foundry Cove. The site was placed on the National Priorities List of the United States Environmental Protection Agency (USEPA). A series of records of decision was published in the late 1980s, consistent with the 1980 U.S. federal "Superfund" Act, to clean up the factory site and to remove most of the cadmium-laden sediments from East Foundry Cove. Six areas, including three within East Foundry Cove, were included in the total plan (USEPA 1989). Preparation commenced in 1993, including construction of a haul road, dike, and treatment facilities(USEPA 2003). The cleanup was performed in 1994–95 at an estimated cost of 91 million dollars (USEPA 1994). Dredging of the marsh and open bottom sediments of East Foundry Cove was completed by February 1994. An objective of reduction to 10 mg kg^{-1} was set, which involved dredging the sediments of East Foundry Cove to a depth of 30 cm. East Foundry Cove marsh was dug out with the objective of achieving concentrations of 100 mg kg^{-1}. A bentonite and geotextile cap was installed, which turn was covered by sedimentary material that was planted with native plants that had been started in a greenhouse. During the cleanup, a wall prevented tidal exchange with West Foundry Cove and the Hudson. A large tubular bladder was installed around the site of the marsh to prevent cadmium release from the marsh area to the open area of East Foundry Cove. Dredged sediments were treated on site and transported away on railroad cars, which traveled on a railroad spur, rebuilt on a former railroad track that connected with the Metro-North Railroad tracks at Cold Spring, New York. Material was transported off-site. Overall, a total of 189,000 MT of contaminated sediments and soils were removed from the Foundry Cove site, stockpiled, cured, and tested for leaching before transport to City Management Landfill in Michigan (USEPA 2003).

THE EFFECT OF THE CLEANUP ON SEDIMENT CADMIUM CONCENTRATIONS AND EXPORT TO THE HUDSON

The aftermath of the cleanup was studied by Advanced Geoservices Corporation (AGC) and by other subcontracting and independent groups. An investigation following the cleanup demonstrated general compliance with the 10 mg kg^{-1} limit in open East Foundry Cove sediments (AGC 2001). However, hot spots higher than this concentration were found in several sites most proximate to the former outfall within East Foundry Cove (Mackie et al. 2007).

Overall, relative to the very high levels of cadmium before the cleanup, we can consider the dredging project a very salutary event. Indeed, a major National Research Council summary recognized the Foundry Cove cleanup as a major success relative to the majority of other Superfund dredging cleanups (Council 2007). This success however, must be regarded in context. After the dredging, there still was a significant Cd sediment concentration gradient from the outfall with concentrations declining with increasing distance to outside of East Foundry Cove. Cd contamination remaining near the former outfall within East Foundry Cove (60 mg kg^{-1} Cd) and proximal marsh (102 mg kg^{-1}) is still high in a wider geographic context. Concentrations in 2002 (Mackie et al. 2007) still exceeded the consensus-based threshold, at which negative effects on standard test organisms are expected when exposed to freshwater sediments, namely, 5.0 mg kg^{-1} Cd (Crane et al. 2002).

As mentioned above, before the cleanup Foundry Cove exported a significant amount of cadmium by means of tidal flow to the open Hudson. A Stony Brook University class project was done in September 1995 to assess the effectiveness of the cleanup on this export (Sokol et al. 1996). Previous estimates demonstrated export of 350–1,600 g in an ebbing tidal cycle (Mackie et al. 2007), whereas the 1995 estimate was only 22.9 g, a drop of ca. 98 percent (Fig. 16.4). A later study in 2002 demonstrated that in waters exiting on the outgoing tide near the railroad trestle, both dissolved and particulate metals (Cd, Co, Cu, Pd, Ni, Ag) were lower in concentration than in the waters of New York Harbor. Tidal export of Cd in one cycle was estimated to be only 1.6 g, or a decline of more than 300 times in concentration than before the cleanup, suggesting that the dredging has successfully reduced output of Cd to the larger Hudson River system (Mackie et al. 2007).

BIOLOGICAL IMPACT OF THE CLEANUP

Cadmium Body Burdens and Uptake

Within East Foundry Cove, the most notable outcome of the cleanup was a major drop in cadmium concentrations of benthic species. One study used a mix of small-bodied benthic species and demonstrated a substantial reduction in body burdens (AGC 2001). A study focused on the dominant species, the above-mentioned common oligochaete *L. hoffmeisteri*, demonstrated a drop of body concentration (Fig. 16.5) from an average of nearly 600 μg g^{-1} Cd in 1984 before the cleanup to < 10 μg g^{-1} Cd in 2002 after the cleanup (Levinton et al. 2003). Cadmium uptake by laboratory model

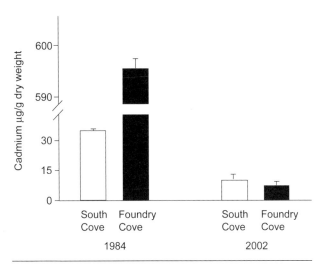

FIG. 16.5. Concentrations of cadmium in *L. hoffmeisteri* in Foundry and South Coves, before (1984) and after (2002) the Superfund cleanup (from Levinton et al. 2003).

organisms used in toxicology assessment (killifish, *Fundulus* sp. and crayfish, *Orconectes* sp.) was highly reduced when placed in waters of East Foundry Cove in the five years following the cleanup (AGC 2001). The reduction of sediment concentrations clearly resulted in a corresponding reduction of exposure and uptake of organisms living within East Foundry Cove sediments. This reduction extended to other organisms including plants and vertebrates collected in a five-year study following the cleanup (AGC 2001).

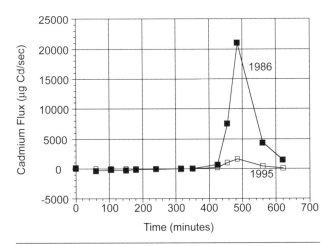

FIG. 16.4. Flux of cadmium through the railroad channel at Foundry Cove, comparing measurements made before (1986) and after (1995) the Superfund cleanup. Measurements in each case are based on a single tidal cycle (after Sokol et al. 1996).

Regional Impact on the Blue Crab

As mentioned above, cadmium concentrations of muscle and hepatopancreas were elevated in the blue crab throughout the Lower Hudson, but appeared to decline in crabs collected north of Foundry Cove (Sloan and Karcher 1984). This suggested that Foundry Cove was the source of the elevated concentrations. The gills of both field and laboratory-investigated blue crabs from Foundry Cove and Haverstraw Bay had metal-binding proteins that bound cadmium, when presented in dissolved form (Engel and Brouwer 1984).

Unfortunately, this species was not included in the approved USEPA five-year follow-up study. However, in 2004 Michael Kane of the New York State Department of Environmental Conservation (pers. comm.) conducted a collection survey and cadmium analysis of the blue crab throughout the Lower Hudson. While the localities were not exactly the same as the earlier survey (Sloan and Karcher 1984), their range allowed a comparative assessment of changes in cadmium concentration in blue crabs before and after the Foundry Cove Superfund cleanup (Levinton et al. 2006). Cadmium concentrations in crabs collected near Foundry Cove had declined substantially. Even more striking was a four to fivefold reduction of cadmium concentrations in both claw muscle and hepatopancreas, and also a strong reduction of variation in Cd concentrations in crabs collected throughout the estuary. The results before and after reveal a major problem: toxic substances can be exported from a point source and can be transported widely by tidal transport, river flow, and wind mixing. Blue crabs can move great distances merely in a tidal rise or ebb (Geyer and Chant 2006), and females are known to exploit localized tidal jets to enhance migration distances (Tankersley et al. 1998). Thus, the combination of organism locomotion and water motion can spread toxic substances great distances (Levinton and Pochron 2008).

Ecological Community Responses Following the Cleanup

The marsh dredging, restoration, and planting was studied mainly from a qualitative point of view. Survival of plantings of cattails, *Typha* spp., and other species was generally good at first, although some of the areas restored were filled with sediment to a relatively high elevation, resulting in the growth of trees and shrubs instead of plants typical of a cattail marsh. Invasive species such as purple loosestrife (*Lythrum salicaria*) were notably abundant, but it is not clear that the dredging and replanting was responsible for a higher vulnerability to invasiveness. After the initial planting, Canada geese, *Branta canadensis*, grazed many new plantings, and it has been difficult to prevent their continued damage to this day. A large proportion of the marsh area is bare, but substantial areas are completely grown in with cattails. Much of the artificial dredged channel has been colonized by water chestnut, *Trapa natans*, which was true of much of the open area of East Foundry Cove, before and after the dredging (E. Lind, unpublished reports).

The benthic invertebrate community of East Foundry Cove in 1984, before the cleanup, was overall surprisingly similar in population densities and biodiversity to South Cove, a comparable nearby tidal freshwater cove to the south (Klerks and Levinton 1989a). The cleanup involved, however, dredging to a depth of 30 cm, which must have severely disturbed the communities. A cursory examination of five grab samples showed at least that in East Foundry Cove, in 1995 just after the dredging, the major species were still present (Sokol et al. 1996). In 2001, we re-sampled East Foundry Cove in May, August, October, and December, along with South Cove and another cove, North Cove, to the north of Cold Spring, with additional sampling stations within each cove, relative to the previous 1984 sampling study (Kelaher et al. 2003b). While the same taxa were found in East Foundry Cove as had been observed in the earlier sampling, there was substantial change in the relative abundances (Fig. 16.2), either in comparison with 2001 reference sites or the 1984 sites. The formerly very abundant *L. hoffmeisteri* was much lower in abundance within East Foundry Cove and nematodes were greatly increased in abundance in South Cove. Chironomids and gastropods showed no consistent pattern of difference either spatially or temporally. The most notable changed parameter within Foundry Cove was greatly increased compaction of the sediment. The upper layers of sediment at East Foundry Cove were more compact than the soft mud at the surface in cores taken at the two reference coves (Kelaher et al. 2003b). This may have been caused by removal of relatively soft overlying sediment in the dredging, or the low abundance of *L. hoffmeisteri* might have resulted in reduced burrowing, which tends to increase water content in soft sediments (Levinton 1995).

Devolution of Resistance to Cadmium

Foundry Cove had extraordinary concentrations of cadmium, which were now reduced to those below which we would be able to detect resistance before the cleanup. So the obvious question arose: Would the population of *L. hoffmeisteri* lose resistance? Resistance might be lost for two distinct reasons. First, there might have been a cost to maintaining resistance. Genotypes with resistance might have lower fitness than those that did not have to maintain the ability to detoxify Cd and the former might therefore lose out in reproductive competition with the latter. It was also possible that after the barrier to the open Hudson was lifted, nonresistant genotypes might have entered East Foundry Cove and become the dominant worms, since nearly all the local adapted worms had presumably been dredged and removed.

We used a simple metric of resistance: survival, or the time for 50 percent of the population to succumb to exposure to high dissolved Cd. We followed survival after the cleanup and found that at first, Foundry Cove worms maintained their higher resistance to Cd, relative to South Cove worms. But a steady decline of resistance (Fig. 16.6) within East Foundry Cove proceeded to eventually cause the Foundry Cove worms to converge with South Cove worms by 2002, a period of nine years or nine generations (Levinton et al. 2003). The Foundry Cove worms had apparently lost their resistance. We have recently investigated the mechanism of loss of resistance and failed to find evidence for an evolutionary cost of resistance. Instead, the regional distribution of DNA haplotypes (based on 16S rDNA) suggest that the loss of resistance was due to a the dispersal of worms from the nearby Hudson River (Mackie et al. 2010). Still, an experiment selecting for resistance demonstrated that the Foundry Cove worm population evolved more resistance than a reference population from South Cove and evolved a greater degree of resistance when challenged with dissolved cadmium over several generations. Apparently, even though the whole East Foundry Cove population was now on average no different in resistance than the South Cove population, there still was a minority of resistant genotypes that had survived the cleanup and the recolonization.

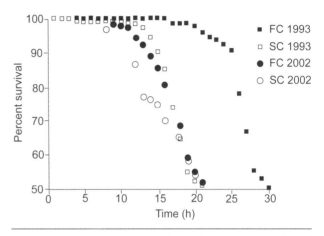

FIG. 16.6. Comparison of survival of *L. hoffmeisteri* from Foundry and South Cove before (1993) and after (2002) the cleanup. Note the similarity of worms collected in 2002 from Foundry Cove with survival of South Cove worms in both years (from Levinton et al. 2003).

LESSONS LEARNED

Effects on Several Scales

In one important sense, the cadmium pollution was biologically momentous. Never before had any bottom environment been so challenged with such high cadmium concentrations. Given the long time before remedial action was taken, it might have been best if the cadmium was very toxic and localized. It was originally thought that the rate of release of cadmium from the solid phase to possible biological uptake was very limited although there was certainly evidence for elevated concentrations of cadmium in many organisms (Kneip and Hazen 1979). But clearly cadmium was being released and taken up by plants and animals at relatively high rates, resulting in high body burdens of cadmium. This was likely occurring at small spatial scales, from cadmium hydroxides, to the sediment pore water, to the organisms exposed to pore water. On the other hand, large amounts of cadmium were being exported by tidal exchange to the open Hudson River. Given the effect on blue crabs, this export was highly detrimental to the Hudson River ecosystem.

Even more surprising was the apparent lack of effect of the high cadmium concentrations on the biota. Indeed, the relative abundances and diversity of soft bottom benthos in East Foundry Cove could not be distinguished statistically from a nearby clean reference site. In the most common species, the oligochaete *L. hoffmeisteri,* it was apparent that strong natural selection had caused the evolution of resistance to cadmium toxicity by up-regulation of the ability to produce large amounts of a metal-binding protein. As a result, the worms' mechanism of resistance resulted in binding of Cd to the cytosol and increased uptake of Cd, relative to nonevolved populations of worms. Thus, the opportunity existed for trophic transfer of Cd to higher levels of the food web.

Another equally important mechanism of Cd release occurred at a higher spatial level of organization. With every ebbing tide, East Foundry Cove exported substantial amounts of Cd, the majority in the solid phase. As the material moved downstream into saltier water some of the Cd was complexed with chloride, but some must have been released into solution. We do not know how much uptake can be explained by this or by direct feeding on cadmium-laden prey, but the blue crab *Callinectes sapidus* was clearly affected for tens of kilometers around Foundry Cove. Both water motion carrying Cd directly, and movement of crabs and perhaps other species spread Cd over a wide reach of the Lower Hudson.

It is therefore clear that the human use of Foundry Cove has had major impacts on different spatial scales. On the localized scale, strong natural selection took advantage of microscale release of cadmium into pore waters, and produced a population of at least one species with enhanced individual capacity to bind Cd and take it up, producing very high body burdens. On this scale, Cd could move through benthic food webs, eventually increasing exposure to human consumers. On the larger spatial scale, Cd was spread throughout the Lower Hudson, providing no major apparent toxic effect on the blue crab carriers, but presenting a health threat to human consumers.

Changing Ecology that Forced Changes in Human Uses

The installation of the battery factory in Cold Spring was an important component of a larger network devoted to our national defense. The release of a toxic metal used in an industrial process is a rather typical way that strong pollution enters an aquatic system. While mercury enters confined inland waters often from atmospheric deposition, a number of industrial inputs and burning of mercury-laden waste are major localized sources that have poisoned our waterways. In the case of Foundry Cove, a localized input resulted in high body burdens in a variety of organisms and additional risks from cadmium-laden materials that existed at the factory and entered into soils.

In effect, the installation of a battery factory and the release of waste was a human experiment in toxicity, uptake of toxic substances, transfer of metals through food webs, and transport through a larger system. Unfortunately, this experiment was not monitored very well, and it took a number of years before the dangers were appreciated. This is a model for many such inputs of toxic materials into water bodies, such as the release and spread of PCBs through the Hudson River. In this case, a factory released a class of deadly substances over a number of decades, and the removal of a dam allowed the spread of the material throughout the Lower Hudson. Phytoplankton, fish, and benthos all bioconcentrated PCBs (Baker et al. 2006; Limburg 1986), which led to closures of major fisheries. The pattern was similar to the cadmium example we discussed above but on a larger scale. Indeed, this input resulted in the largest Superfund site in the United States. The difficulty of dredging this site is much greater, and indeed many dredging operations in the United States have demonstrated strong difficulties (National Research Council 2007).

The Foundry Cove story had a relatively happy ending, because it was possible to isolate the system by walling it from the main course of the Lower Hudson. The cleanup plan, from the point of view of removal of toxics, was a success (AGC 2001;

Levinton et al. 2003). But even here, as discussed above, the necessary dredging led to a less than totally successful marsh restoration. The dredging of the open part of the cove must have severely disrupted the benthic communities, and we have evidence for this in the study of the aftermath (Kelaher et al. 2003). The deposition of cadmium and the widespread uptake of cadmium by many organisms testify to the pervasive effects of this point source of pollution. The prescribed remedy for such a pollution event, namely, massive dredging, can only have profound effects on marsh and cove systems. Indeed, hot spots within Constitution Marsh were left undredged because the impacts on water bird populations were deemed too costly relative to the benefits of toxic removal. Success, even in this instance must be muted with the enormous disruption that was imposed on this ecosystem.

Vigilance and Swift Response

The pollution caused by a battery factory at Foundry Cove, combined with release of toxic materials and wastewaters into the Hudson and East Foundry Cove demonstrates that we must maintain vigilance on the effects of releases into our coastal waters, combined with sound science on the toxic effects of substances, their biogeochemical properties, and their uptake. Along with this vigilance must be the appreciation that Darwinian natural selection is at work even if the challenges involve compounds not encountered before by aquatic species. A range of biological responses, including evolution, can drastically affect the movement of toxic substances through ecosystems. It is very likely that such adaptation, already found widely in the introduction of metals into the environment (Klerks and Weis 1987) can be found widely for other toxic substances, including PCBs, PAHs, and other organic toxins.

Swift and efficient cleanups are rarely the rule at industrial sites. Delays in cleanup greatly exacerbate the degree of uptake and the spread of toxic substances through ecosystems. At Foundry Cove, it took a number of years before an ineffectual dredging occurred, and even more years before a comprehensive plan was enacted to clean up and restore this region of the Hudson. Even though it was known early that large amounts of cadmium were being exported into the Hudson and that there was a cadmium-related health advisory for a fishery, no action was taken to reduce such export. Such delays are commonplace, and are elongated by lack of appreciation of the degree of toxicity, inexperience in the ecological and physical forces that spread toxic substances through aquatic ecosystems, and even legal battles by industrial corporations that ultimately would do damage by increasing the time from perception of a problem to cleanup. The Superfund law has not been a major accelerant of solutions, partially because the solutions are not always so effective by the time the various scientific, management, and legal hurdles are crossed. We will solve this general problem only when good monitoring, sound scientific research on effects and spread of toxicants, and swift cleanup when problems are perceived, are all achieved.

ACKNOWLEDGMENTS

This work could not have been accomplished without the work of many colleagues, graduate students and undergraduates who helped with field work, advice, and many types of information. I would like especially to cite the late James Rod for his unfailing support of our work and his many efforts at making our work easier and more successful. Ron Sloan of the New York State Department of Environmental Conservation was also of great help in my understanding of Foundry Cove. Eric Lind and colleagues of the Constitution Marsh Audubon sanctuary and field station have continued to facilitate our work and help in so many ways. I also thank Eric Lind for giving me access to unpublished reports recording studies of the aftermath of the cleanup. The Hudson River Foundation supported much of this work. Those interested in the story of Foundry Cove pollution will find useful information and videos at http://life.bio.sunysb.edu/marinebio/foundryframe.html.

REFERENCES CITED

AGC. 2001. *Five year review: Long term monitoring program, Marathon Remediation Site.* Ed. P. F.

Marano. Chadds Ford PA: Advanced Geoservices Corporation.

Baker, J. E., W. F. Bohlen, R. F. Bopp, B. Brownawell, T. K. Collier, K. J. Farley, W. R. Geyer, et al. 2006. PCBs in the Upper and tidal freshwater Hudson River estuary: The science behind the dredging. In *The Hudson River Estuary,* ed. J. S. Levinton, and J. R. Waldman, 349–67. New York: Cambridge University Press.

Bouchet, M.-L., S. Habits, S. Biagianti-Risbourg, and G. Vernet. 2000. Toxic effects and bioaccumulation of cadmium in the aquatic oligochaete *Tubifex tubifex. Ecotoxicology and Environmental Safety* 46: 246–51.

Bower, P. M., H. J. Simpson, S. C. Williams, and Y. H. Li. 1978. Heavy metals in the sediments of Foundry Cove, Cold Spring, New York. *Environmental Science and Technology* 12: 683–87.

Boyle, R. H. 1979, *The Hudson: A natural and unnatural history.* New York, W.W. Norton.

Crane, J. L., D. D. MacDonald, C. G. Ingersoll, D. E. Smorong, R. A. Lindskoog, C. G. Severn, T. A. Berger, et al. 2002. Evaluation of numerical sediment quality targets for the St. Louis River area of concern. *Archives of Environmental Contamination and Toxicology* 43: 1–10.

Diamant, L. 1989. *Chaining the Hudson: The fight for the river in the American Revolution.* New York, Lyle Stuart.

Engel, D. W., and M. Brouwer. 1984. Cadmium-binding proteins in the blue crab *Callinectes sapidus*: Laboratory-field comparison. *Marine Environmental Research* 14: 139–51.

Flick, D. F., H. F. Kraybill, and J. M. Dimitroff. 1971. Toxic effects of cadmium: A review. *Environmental Research* 4 :71–85.

Geyer, W. R., and R. Chant. 2006. The physical oceanography processes in the Hudson River estuary. In *The Hudson River Estuary*, ed. J. Levinton, and J. Waldman, 24–38. New York: Cambridge University Press.

Hazen, R. E., and T. J. Kneip. 1980. Biogeochemical cycling of cadmium in a marsh ecosystem. In *Cadmium in the Environment. Part I: Ecological Cycling,* ed. J. O. Nriagu, 399–424. New York: John Wiley and Sons.

Kelaher, B. H., J. S. Levinton, J. Oomen, B. J. Allen, and W. H. Wong. 2003. Changes in benthos following the clean-up of a severely metal-polluted cove in the Hudson River estuary: Environmental restoration or ecological disturbance? *Estuaries* 26:1505–16.

Klerks, P. L. 1987. *Adaptation to metals in benthic macrofauna.* Stony Brook NY: State University of New York at Stony Brook.

———, and P. R. Bartholomew. 1991. Cadmium accumulation and detoxification in a Cd-resistant population of the oligochaete *Limnodrilus hoffmeisteri. Aquatic Toxicology* 19: 97–112.

Klerks, P. L., and J. S. Levinton. 1989a. Effects of heavy metals in a polluted aquatic ecosystem. In *Ecotoxicology: Problems and approaches,* ed. S. A. Levin, J. R. Kelley, and M. A. Harvell, 41–67. Berlin: Springer-Verlag.

———. 1989b. Rapid evolution of resistance to extreme metal pollution in a benthic oligochaete. *Biological Bulletin* 176: 135–41.

Klerks, P. L., and J. S. Weis. 1987. Genetic adaptation to heavy metals in aquatic organisms: A review. *Environmental Pollution* 45 :173–205.

Kneip, T. J., and R. E. Hazen. 1979. Deposit and mobility of cadmium in a marsh-cove ecosystem and the relation to cadmium concentration in biota. *Environmental Health Perspectives* 28: 67–73.

Knutson, A. B., P. L. Klerks, and J. S. Levinton. 1987. The fate of metal contaminated sediments in Foundry Cove, New York. *Environmental Pollution* 45: 291–304.

Levinton, J. S. 1995. Bioturbators as ecosystem engineers: Control of the sediment fabric, interindividual interactions, and material fluxes. In *Linking species and ecosystems,* ed. C. G. Jones, and J. H. Lawton, 29–36. New York: Chapman and Hall.

Levinton, J. S., P. Klerks, D. E. Martinez, C. Montero, C. Sturmbauer, L. Suatoni, and W. Wallace. 1999. Running the gauntlet: Pollution, evolution, and reclamation of an estuarine bay and its significance in understanding the population biology of toxicology and food web transfer. In *Aquatic life cycle strategies,* ed. M. Whitfield. Plymouth, UK: The Marine Biological Association.

Levinton, J. S., and S. T. Pochron. 2008. Temporal and geographic trends in mercury concentrations in muscle tissue in five species of Hudson River fish. *Environmental Toxicology and Chemistry* 27: 1691–97.

Levinton, J. S., S. T. Pochron, and M. W. Kane. 2006. Superfund dredging restoration results in widespread regional reduction in cadmium in blue crabs. *Environmental Science and Technology* 40: 7597–7601.

Levinton, J. S., E. Suatoni, W. Wallace, R. Junkins, B. P. Kelaher, and B. J. Allen. 2003. Rapid loss of genetically based resistance to metals after the cleanup of a Superfund site. *Proceedings of the National Academy of Science* 100: 9889–91.

Limburg, K. E. 1986. PCBs in the Hudson. In *The Hudson River Ecosystem,* ed. K. E. Limburg, M. A. Moran, and W. H. McDowell, 83–130. New York: Springer-Verlag.

Mackie, J. A., J. E. Levinton, R. Przelawski, D. DeLambert, and W. Wallace. 2010. Loss of evolutionary resistance by the oligochaete *Limnodrilus hoffmeisteri* to a toxic substance—Cost or gene flow? *Evolution* 64 :152–65.

Mackie, J. A., S. M. Natali, J. S. Levinton, and S. Sanudo-Wilhelmy. 2007. Declining metal levels at Foundry Cove (Hudson River, New York): Response to localized dredging of contaminated sediments. *Environmental Pollution* 149: 141–48.

Martinez, D. E., and J. S. Levinton. 1996. Adaptation to Cadmium in the aquatic oligochaete *Limnodrilus hoffmeisteri*: Evidence for control by one gene. *Evolution* 50: 1339–43.

National Research Council, 2007. *Sediment dredging at Superfund megasites.* Washington, DC: National Research Council, Pages 1–294.

Rainbow, P. S. 2002. Trace metal concentrations in aquatic invertebrates: Why and so what? *Environmental Pollution* 120: 497–507.

Resource Engineering. 1983. Preliminary site background data analysis of Foundry Cove, prepared for Vinson and Elkins. Houston.

Sloan, R., and R. Karcher. 1984. On the origins of high cadmium concentrations in Hudson River Blue Crab (*Callinectes sapidus Rathbun*). *Northeast Environmental Science* 3: 222–32.

Sokol, R. A. Jr., R. C. Magiore, N. J. Alonsozana, J. L. Gulizea, J. G. Hopkins, M. T. Hyland, L. A. Ingoglia, et al. 1996. Restoration and recovery of an ecosystem polluted by cadmium. *Journal of Undergrad. Res.* 3 :115–27.

Sunda, W. G., D. W. Engel, and R. M. Thuotte. 1978. Effect of chemical speciation on toxicity of cadmium to grass shrimp *Palaemonetes pugio*: Importance of free cadmium ion. *Environmental Science and Technology* 12: 409–13.

Suszkowski, D. J., and C. F. D'Elia. 2006. The History and science of managing the Hudson River. In *The Hudson River Estuary,* ed. J. S. Levinton, and J. R. Waldman, 313–34. New York: Cambridge University Press.

Tankersley, R. A., M. G. Wieber, M. A. Sigala, and K. A. Kachurak. 1998. Migratory behavior of ovigerous blue crabs *Callinectes sapidus*: Evidence for selective tidal stream transport. *Biological Bulletin* 198: 168–73.

USEPA. 1989. Record of decision: Marathon Battery Corp. EPA ID: NYD010959757. OU 02, Cold Springs, NY. EPA/ROD/R02-89/097. Washington, DC: U.S. Environmental Protection Agency. Superfund Information Systems, U.S. Environmental Protection Agency [online]. Pages http://www.epa.gov/superfund/sites/rods/fulltext/r0289097.pdf.

———. 1994. Marathon Battery Co. site profile. Washington, DC, United States Environmental Protection Agency.

———. 2003. Five year review report: Marathon Battery Company Superfund site, Village of Cold Spring, Putnam County, New York. New York: United States Environmental Protection Agency, Region 2, 1–14, T11-T14.

Wallace, W. G., and G. R. Lopez. 1997. Bioavailability of biologically sequestered cadmium and the implications of metal detoxification. *Marine Ecology Progress Series* 147: 149–57.

Wallace, W. G., G. R. Lopez, and J. S. Levinton. 1998. Cadmium resistance in an oligochaete and its effect on cadmium trophic transfer to an omnivorous shrimp. *Marine Ecology Progress Series* 172: 225–37.

Windom, H. L. et al. 1989. Natural trace metal concentrations in estuarine and coastal marine sediments of the southeastern United States. *Environmental Science and Technology* 23: 314–20.

CHAPTER 17

RIVER CITY

Transporting Commerce and Culture

Roger Panetta

ABSTRACT

Transportation has been the key force the shaping the Hudson River region's settlement patterns and economic growth, historically linking city and country, nature and technology, and serving as a powerful shaper of the riverscape.

This chapter will provide a historical overview of river-based transportation including both primary and secondary connections to the Hudson. The evolution of the forms of transportation, their impact on the river and the valley, the response of the public and artists, and the shape of the metropolitan network they created are all part of this exploration.

THE NINETEENTH-CENTURY transportation revolution engineered a city-country nexus linking the river valley to New York City via the Hudson corridor with the support of a system of turnpikes, steamboats, railroads, tunnels, bridges and highways. An examination of the development of this vital connection implies a reevaluation of the standard geographic focus in the study of the Hudson River.

For the past quarter-century, the study of the Hudson River has concentrated on aesthetic and scientific dimensions, with a geographic concentration in the Mid-Hudson region. The history of transportation and commerce, the twin engines of economic development, and key forces in shaping the river and the valley, have been marginalized by scholars; neither has been the subject of a contemporary integrating narrative history. Indeed, the subjects of transportation and commerce lack the glitz of the Hudson River painters and the passion of the environmental battles of recent days. This intellectual oversight has not only diminished our appreciation of the critical role of transportation in the Hudson Valley but also skewed our view of the region in profound and fundamental ways. Transportation has been the key force shaping the region's settlement patterns and economic growth, historically linking city and country, nature and technology, and serving as a powerful shaper of the riverscape. Indeed, we have only just begun to document the impact of the successive waves of transportation on the ecology of the river and the valley.

Scholars and writers have tended to compartmentalize the Hudson's history into simple dichotomies—urban and rural, city and country, and culture and nature. This division has been a byproduct of the undervaluing of the role of the Hudson as a transportation corridor and its centrality to the creation of an integrated and interdependent metropolitan network. While most writers and historians acknowledge this role, it usually receives only cursory treatment as the preamble to the "true" history of the Hudson, which is to be found upriver in the aesthetic and political domains.

In this chapter I examine the ways humans connect to and manipulate nature with an emphasis on the historical dimension of this dynamic relationship. An examination of the history of river transportation will document the ways this ignored pathway connected us to the river, to nature, and to regional identity. While transportation is fundamentally shaped by technology, its interaction with nature is equally important. As machines of movement are successively introduced, each innovation, while at first celebrated, is quickly scrutinized and soon thereafter judged an ominous and powerful intruder in the landscape. How and why have these transportation innovations changed our perception of and relationship to nature?

When Europeans arrived in the Hudson Valley they detected the outlines of an inchoate highway system linked to the Hudson River corridor. For generations the Lenai Lenapi and the Mahicans had marked numerous trails, twelve to eighteen inches in width, which served as the cornerstone of land travel for the new European arrivals (Dunbar 1915, I: 19). Many of these local roads were linked to streams and tributaries, which, with the aid of dugout canoes, mapped a north-south Hudson axis, their main highway for trade and communication. European migration patterns followed these roads and natural waterways, which continued to be the distinguishing characteristic of regional travel. Many of these roads joined the Hudson River where travelers in turn could navigate south or north and cross the river at key points. Throngs passed through; many crossing the Hudson at Catskill and continuing westward (Dunbar 1915, I: 224).

EARLY ROADS

Europeans recognized early on that the native peoples were economic strategists who were living in localities selected because of their geographic advantages proximate to the Hudson and its tributaries. Thus, the Europeans followed the precedent of native peoples and settled on waterways, constructing bateaux and other small sailing vessels suitable for navigating the streams, bays, and rivers that both separated and connected them (Dunbar 1915, I: 29). Settlers who moved into the interior of New York from the lower part of the state and from New Jersey made their way up the Hudson in sailing boats; some pushed farther inland along the Mohawk River in bateaux, all in search of new possibilities (Dunbar 1915, I: 312).

Sailing vessels were the dominant mode of transportation on all these riverways in the seventeenth and eighteenth centuries.

SLOOPS

In 1639 Dutch colonists left New Amsterdam and occupied distant points along the Hudson River. They sailed the Dutch *sloep* with a fore-and-aft and single sail rig set quite forward. Its shallow draft accommodated many of the Hudson's tributaries while its ample beam offered considerable cargo space. For much of the eighteenth century it was the most reliable, comfortable, and efficient form of transportation (Fontenoy 1994, 27). Peter Kalm, a Swedish traveler, noted that the southbound sloops "bring Albany boards or plancks and also all sorts of timber, flour, peas, and furs" On their northern return, "they come almost empty, and bring a few kinds of mechandise the chief of which is rum" (quoted in Fontenoy 1994, 32). Even in these early accounts we can see the outlines of a reciprocal commercial exchange between Albany and New York: essential New World products to the city and specialty merchandise upriver.

For many farmers, sloop landings, usually at the conjunction of a tributary and the Hudson, served as embryonic market centers where the exchange of goods and news linked the rural with the urban. Englishman James Birket observed:

> This City [New York] also reaps great Advantages by the navigation of the Hudson or North River. . . . This city is well Scituated [*sic*] for business having the Advantage of all trade . . . (to)the Inland Country at least two hundred miles by the North river large Sloops go as far as Albany which is 166 miles. (quoted in Fontenoy 1994, 34)

Thomas Pownell, a British colonial official, also recognized the importance of the Hudson, which he described as "the most perfect navigation," and, "of great value in opening Communication with the inland parts of the Continent"(quoted in Fontenoy 1994, 34).

Paul Fontenoy correlated the economic prosperity of the eighteenth-century Hudson Valley merchants with the size of the sloop fleet. He noted the prerevolutionary increase in grain production was correlated with increases in the number of sloops—the two factors, he argued, that were the keys to New York State's prosperity (Fontenoy 1994, 37–38). But the cornerstone of that prosperity was at the intersection of the Hudson and the Atlantic, of the valley and Europe, where New York City emerged as a major commercial hub for the region.[1]

CITY AND COUNTRY

The connection between hinterland and city is the focus of William Cronon's *Nature's Metropolis: Chicago and The Great West*. In this 1991 work Cronon challenges the long-held American notion that city and country are separate places—partitioned entities. He believes they are in fact part of a unified narrative—one story—sharing a common history (Cronon 1991, xvi). Among the key links in this connection are the commodity flows that yoked marginal frontier zones with the metropolitan economy. Cronin recognized the problematic use of the term *frontier* to describe Chicago's hinterland and the interpretive morass surrounding the use of word *nature*. He intended to challenge the hard line drawn between human and nonhuman actions. Cronin believed this boundary was permeable and should be reimagined to include first nature—the prehuman nature—and second nature—the artificial world people erect atop the first. He argues that the nature we inhabit is an amalgam of the two (Cronon 1991, xvi–xix). One of the basic arguments in Cronon's work is that "environmental history of a single city made little sense if written in isolation from the countryside around it" (Cronon 1991, xix). He suggested that city and country depended on each other for meaning and shared a material bond. Cronon cautioned against locating the city outside of nature. He noted that the historical journeys between city and country through time and space will help us not only understand the city's place in nature and but appreciate that at journey's end one arrived at the historic possibilities of America's future (Cronon 1991, 9). He rejects the polarized concepts of man and nature, rural beauty and urban ugliness, pastoral simplicity and cosmopolitan sophistication. These dichotomous frames were the underpinning for the abstractions of city and country which Cronon directly challenged by connecting Chicago with the great west and arguing for a single region with interdependent relationships and a common past (Cronon 1991, 19).

Cronon's environmental history, like all good new interpretations, challenges scholars to rethink their own conceptual frames. Analogies between Cronon's Chicago and the Great West and New York and the Hudson Valley are not exact, yet a comparison is suggestive and offers transferable insights for the historical study of the Hudson's transportation network and the connection between city and country.

By the nineteenth century, Hudson River landings had become transportation nodes where farmers not only sold their produce, bartered with local merchants and discussed prices with sloop captains, but also gathered information from New York City— "The downstream metropolis had evolved into a pole that staked out the compass of, and helped structure daily activities in the countryside" (Bruegel 2002, 2-3). These river nodes connected the city with the hinterland (Bruegel 2002, 57). The city's growth fueled the growth of the upriver economy, which, in turn, had a reciprocal impact on the city.

TURNPIKES

The expansion of New York City's population in the first quarter of the nineteenth century compelled merchants to reach into the agricultural hinterland to satisfy the needs of the booming central city. Turnpike construction was one response to this new set of economic facts. The American Turnpike era, dating from 1785 to the 1830s, represented an effort to improve the older primitive natural paths with man-made artificial roads, open new links to New York State's frontier, and compete with Baltimore and Montreal. These new roads were often designed by local engineers to serve migrating populations and expanding agricultural markets. Privately financed turnpike companies sought straighter and shorter journeys to the inland trade.

While Albany benefited from this new construction, there was fierce competition from Hudson River port towns such as Hudson, Kingston, and Newburgh, all with connections to New England turnpikes (Thompson 1977, 156–57). Of course, other towns along the river, Athens, Catskill, and Poughkeepsie, became important transshipping points to New York City. Beginning with the Cherry Valley Turnpike in 1799 running west to Albany, the Ulster and Delaware Turnpike in Delaware County invited western migration from Connecticut while the Susquehanna Turnpike and the Newburg & Cochecton reached out to Pennsylvania (Adams 2006, 35).

These new turnpike roads extending out from river landings resembled, according to Benjamin DeWitt, the human body's circulation of the blood. In 1807 he wrote,

> the City of New York as the center of commerce, or the heart of the state, Hudson's river as the main artery, the turnpike roads leading from it as so many great branches extending the extremities, from which diverge innumerable small ramifications or common roads into the whole body and substance; these again send off capillary branches, or private roads, to all the individual farms, which may be considered as the secretary organs, generating the produce and wealth of the state. (quoted in Bruegel 2002, 72)

This biological metaphor suggests how vividly De Witt and others saw not only the deepening connection of the river valley and the city but the way they conceptualized this relationship using language that suggested a symbiosis. A system of economic circulation emanating from the city, conveyed by sloop to river landings and then by turnpikes to New York's western hinterland and completing the cycle returning back to the Hudson, the central connecting element.

STEAMBOATS

In navigating the Hudson and calculating river distances, the markers of nature played a key role. Sloop captains used river reaches, the stretch of water visible between bends in the Hudson, to determine distance and location. But a great deal of river travel remained imprecise and uncertain, dependant on winds and tides, reminding its users that the Hudson was still the creature of nature. Long before Fulton's *North River* steamed up the Hudson on August 17, 1807, there was a growing preoccupation with the possibility of steam navigation. The early work of John Fitch and John Stevens had laid the foundation and began to answer the eighteenth-century call emanating from the Atlantic world for "periodicity and regularity as the primary features of every sort of traffic that moves from one place to another throughout the continent." This attitude reflected a view deeply held by many of the Revolutionary generation, that transportation methods and their improvement and expansion, rather than politics or wars, were the prerequisite for a durable national unity (Dunbar 1915, 173–74). It reminds us early on of the founders' faith in the capacity of technology to advance the promise of republicanism (Kasson 1999).

Robert Fulton shared with George Washington a confidence in the potential of rivers as ligatures of national union and, with the aid of technology, paths to prosperity. They envisioned a republic yoked by a transportation network shaped by engineers. They imagined America as a product of commerce and art. It is not surprising that Fulton's innovative mind embraced both steamboats and canals, recognizing that the challenge was systemic and required new thinking that went beyond a single innovation to the larger national agenda of unifying and developing a new country through transportation. The American application of steam engine technology was rapid and widespread, and when it reached the rivers of the Great West worked to extend the reach of the Hudson and finances of New York City into new territories.

Fulton, whose entrepreneurial impulses were a perfect fit for the Hudson, quickly moved from the launching of the first steamboat, the *North River*, built in New York and fitted with British steam engines, on August 17, 1807, to scheduled service in early September. In partnership with Chancellor Robert R. Livingston and with the advantage of their New York steamboat monopoly, which secured an exclusive right to navigation on the Hudson

River, Fulton anticipated great profits. Other business interests quickly recognized new opportunities and challenged the Fulton-Livingston charter. The increase in river traffic, the anticipated opening of the Erie Canal in 1825, and the dual-state character of the Hudson between New York and New Jersey opened the river to navigation by all. The monopoly was legally dissolved in 1824 in *Gibbons v. Ogden* when the Supreme Court overturned the Livingston charter, which had effectively limited the number of steamboats on the Hudson to eleven. Within a decade of the Court's decision more than one hundred steamboats plied the river.

As competition intensified, steamboats increased in size, and improved speed and reduced fares enticed large numbers of the middle class to river excursions. The public was traveling on a far more regular basis not only for business but increasingly for recreation and leisure as well.

Steam ferryboats quickly replaced the old horse boat ferries and spurred the growth of river suburbs such as Brooklyn and Hoboken. These cross-river journeys, moving east to west, intersected the dominant north-south matrix. As the longitudinal connections matured they served as catalyst for the latitudinal bonds, which, by century's end, became part of a complex highly differentiated network of river crossings.

Nothing better documents the steamboat's conquest of time than the printed schedule, which measured a journey's distance not in miles but hours and minutes. The schedule challenged the uncertainties of the tides and the winds, instead offering predictability and speed, at least for eight to nine months of the year. The schedule also documented the relationship of the city and the river towns, and, with New York City serving as the point of origin, travel was measured in increments of time as one moved north from the city. The interdependence of city and valley was crystallized in these printed reminders, which provided travelers and traders with a new cartography.

Steamboat owners celebrated and flaunted the speed of their vessels. Impromptu river races endangered travelers trapped in the grasp of an irresponsible captain who responded to a challenge to race, often ending the lives of unsuspecting passengers . Racing served as a catalyst for real gains in speed: the run of the *North River* to Albany in 1807 took thirty-two hours; by 1852, the year legislation outlawed steamboat racing, the *Francis Skiddy* needed only seven and one-half hours to travel the same distance (Buckman 1907, 66). In this obsession, steamboat captains joined the country's moral "quest for speed," which promised to reduce the wasteful use of time and embraced the imperative of technological progress (Bennett 1998). But this notion also embodied the commercial and proto-capitalist demands for predictability and timeliness—the discipline of economic order now anchored in New York City. The clock had come to the valley and to its towns, villages, and farms.

But not all was commerce and trade with the steamboat. It offered the newly minted middle class of the city and the valley access to the river and its scenic places, historic monuments, healthful resorts, and summer vacations. Travel upriver became the antidote for the corruptions of downriver urban life. This bifurcation had the effect of seemingly creating two distinct worlds—one natural, one commercial—a division that, I argue, is polarizing when in fact they were both elements of one metropolitan universe.

The development of the Hudson's recreational landscape was aided by the publication of steamboat guidebooks, which initiated the novice traveler, codified the rules for appearance and behavior, and framed the river experience as a south-north journey from New York to Albany. Steamboats, often referred to as "beautiful machines," combined aesthetic and technological principles in their graceful architecture and the renowned speed (Seelye 1991, 361). These machines of transportation caught the eye of the artists and writers working to make sense out of the confrontation of the elemental forces of nature and technology. Were they oppositional or reconcilable, dissonant or harmonious forces? The river and the riverscape serve as the backdrop for this encounter and in the visual representations of steamboats we get a clue to the thinking of nineteenth-century American painters. Marine artists James and John Bard, twin brothers born in New York City in 1815, painted dozens of Hudson River sail and steamboats, in mostly upriver settings, as they speedily headed south toward New York City. The Bard brothers depicted efficient machines gliding over the Hudson with little resistance from a river yielding to Americans' "quest for speed." The river and the vessel are set in a linear frame emphasizing the longitudinal and celebrating technological efficiency. Steamboat and river are presented in harmony one with the other; nature and technology reconciled in a new kind of synergy. Even the clouds of engine smoke are quickly dissipated by the wind and are minimized by the taut flags and banners which one can almost hear snapping in the wind (Peluso 1991, 362).

The Bard paintings constitute a vivid example of the extension of the notion of the sublime beyond nature to the processes and engines of industrialization (Kasson 1999, 160). This variant of the term *sublime* or the technological sublime, as it came to be called, challenges the dichotomy between nature and culture, and by extension between upriver and downriver and country and city.

CANALS

In 1810 in one of the earliest recollection of a Hudson River steamboat journey Dewitt Clinton, then a member of the commission charged with the task of surveying a potential western canal, commented about the "warm weather" and the "crowded boat" (quoted in Seelye 1991, 223). While some may see this quote as an indication of the ready acceptance of steamboat travel we may miss the larger significance of his decision to begin the survey with a trip up the Hudson. Clinton's 1810 journey reminds us how essential the river was to any canal scheme. Indeed, the Hudson River with its Albany connection, contained a geographic imperative, providing the only place to the west where the Appalachians could be taken "in flank" (Albion 1939, 84). The river had predisposed New Yorkers to think westward and as businessmen and migrants sought more efficient travel the urgency for an overland waterway increased. There is no doubt that the ease of the southern leg on the Hudson River of this canal scheme is what made the Erie Canal proposal so attractive.

For many Americans rivers were seen as "national conduits of an advancing and cohering empire" which demanded the creation of man-made equivalents—roads and canals—that would " complete the great god-given diagram" (Seelye 1991, 382–83). As early as 1784 George Washington advised the new nation that it "must look to progressive improvement of interior communications and waterways." Linking the maritime strength of the Atlantic seaboard to the west was essential for national unity (Seelye 1991, 252). Washington, like others who pushed canal improvements, began with rivers and, not surprisingly, he envisioned the Potomac as the cornerstone of such a national project. He recognized early on that the Hudson was his principal competitor.

One of the chief catalysts for canal building was the turnpike movement. Turnpikes, roads financed by tolls, came late to New York State; the first one was only chartered in 1797. In New York they grew rapidly, more the three thousand miles in the next quarter-century, concentrated between the Hudson and the Genesee Rivers. Turnpike networks were "highly organized systems that sought to find the most efficient way of connecting eastern cities with western markets." The turnpike developers recognized the critical importance of the natural opening through the Appalachians and planned a system of turnpikes connecting Albany to Syracuse and beyond (Baer et al. 1993, 191–209). The New York and Albany Post Road was turnpiked and the towns on the east side of the river were now linked to New England by new roads running west to east. Fierce competition for the western trade pushed several Hudson River ports to press for new turnpike connections. The rapid growth of the Ohio Valley also caught the attention of New Yorkers. Turnpike development in the region increasingly focused on business connections in the west; yet many New Yorkers remained uncertain about how far to extend the reach of turnpikes and anticipated the Erie Canal with uncertainty, fearing its challenge to their livelihoods (Baer 2005, 1588–89).

By the beginning of the nineteenth century the drive to connect to the west was irresistible. Albert Gallatin. Jefferson's Secretary of the Treasury, in 1810 announced that good roads and canals were in the national interest (Seelye 1991, 254). John C. Calhoun, writing during his nationalist phase, called on the government to build a "connection between the Hudson and the Great Lakes" (Seelye 1991, 262). And build it they did, not with federal funds but with state financing.

With the end the War of 1812 and the return of peace, the talk of the canal was revived; public meetings in support of the enterprise were held in New York City and elsewhere. The legislature was petitioned, a board of commissioners appointed, and in 1817 legislation passed that provided for a system of canals and other internal improvements at the state's expense. Construction work was begun on the Fourth of July in 1817 and the Erie Canal was completed in 1825. In the "wedding of the waters" in New York harbor Dr. Samuel Mitchell arranged for a more compelling mixing in which he added contributions from the world's great rivers to Clinton's Erie donation and proclaimed the "circumfluent ocean republicanized." Colonel Stone then baptized the canal as "a new and additional river" (Seelye 1991, 347). The canal has often been described an artificial river embodying technology and nature—a remarkable republican achievement. At first glance we may see these as oppositional; nature subdued and sacrificed in the name of progress (Sheriff 1996, 59). But if we translate technology, the practical application of scientific knowledge, back to its nineteenth-century understanding we come to the word *arts,* which in its historical usage implied not tension but complementarity. Indeed the canal takes its inspiration from rivers, in this case the Hudson, and attempts to mimic nature and add to the gifts of God.

In *The Organic Machine,* a study of the Columbia River, Richard White argues that we should see the flow of rivers as a kind of energy. Like us, rivers work and rearrange the world. Indeed, the "river channels are historical products of their own past history" (White 1996, 12). Rivers represent the most efficient means of energy utilization working to eliminate obstructions to navigation.

For White, nature is made to serve and accept human manipulation for "the mechanical was not the antithesis of nature but its realization." Here he recognizes Emerson's notion that putting land or water to work was "opening a . . . new access to nature" (White 1996, 35). In the Emersonian vision of the machine as a force of nature, White finds a reiteration of the Western dream of liberation from labor (White 1996, 48). In applying these concepts to the history of the Columbia River, he notes that the dam builders argued that the project "depended on deciphering the 'but little understood schemes of nature' . . . by which nature made the great canal for the project." He believes that "what nature has so artfully arranged it would be criminal for humans to neglect to improve and finish. The dam was the final piece necessary to reveal nature's harmony." Thus, following White's lead, I suggested above that the idea of the Erie Canal was embedded in the Hudson River. The canal represented a deciphering of nature and a fulfillment of the river's historical function as a source of energy and movement, thus justifying the term *Hudson–Erie Canal System.* In the context of this argument, steamboats, railroads, and canals mediate between humans and the river with ties to both nature and human labor.

With the opening of the Champlain Canal in 1819 the Hudson collected the bulk of the traffic from the north and the west. By the 1830s the river was crowded with floating towns, long flotillas of canal barges lashed together four and five abreast and extending for almost a half-mile, making their way to the city (Buckman 1907, 84–85). This was a lucrative business with fierce competition among a few companies. Steamers that pulled these tows were recycled passenger boats capable of hauling sixty to eighty canal boats. River towns provided service to these floating villages and the locals helped maintain and repair these recycled steamers.

The canal also reshaped the agricultural and mercantile landscape of the river valley. Wheat had taken hold after the Revolution, and the area was known as the granary of the nation. Cattle, sheep, and poultry were added to butter and cheese and shipped south to feed the growing population of the city. The Erie Canal shifted wheat production westward, where better soils produced higher yields. Faced with this new reality valley farmers looked more closely at the needs of the New York market and began to specialize in dairying and orcharding. By the middle of the nineteenth century New York was the leading producer of oats, barley, and buckwheat, butter and cheese, and orchard and garden products (Thompson 1977, 161). A by-product of the relocation of wheat production westward was that New York City became the principal flour market of the country and began to monopolize most of that trade (Albion 1939, 92–93).

As farmers moved out of the eighteenth century world of self-sufficiency to the nineteenth-century world of commercial farming, in which more than half produced a surplus, their appetite for consumer goods, machines, and mobility increased (Parkerson 1995, 79–104). This change opened the door for New York City's merchants who, with the aid of the river and the canal, used the moment to extend their reach and accelerate exports to the valley and the hinterland (Albion 1939, 91).

The Erie spawned a "canal fever," which thickened the web of connections and increased the number of small feeder canals to the Hudson-Erie network. It also stimulated industrial production—large sawmills and flour mills began to appear in the valley and ironworks were established at Troy. An expanded and revamped lumbering industry emerged in the Upper Hudson in the Albany–Glens Falls region. Mills in New York City were able to draw on this supply via the Hudson (Thompson 1977, 161). The Erie Canal solidified the Mohawk-Hudson axis, suppressed competition from other canal systems, reduced the cost of transporting bulk cargo, and, with the aid of steam, turned the Hudson into a commercial superhighway.

In securing New York City's trading dominance over Baltimore, Boston, and Philadelphia the Erie Canal created a commercial vortex that sucked the economic world, if not the daily lives, of the Hudson River Valley farmers and the western hinterland into the Empire City. Albany, now secure as the gateway city, was, along with its downriver metropolis, key to making New York the Empire State. According the John Seeyle, "the Erie Canal was so latent with the promise of the future, that [it] gave the lordly river its American dimension, a linear diagonal pointing into the west" (1991, 315).

RAILROADS

The transportation revolution now began to accelerate, extending from the steamboat and the canal to the railroad. New York's first railroad, the Mohawk and Hudson, linking Albany and Schenectady, opened in 1831. Smaller trunk lines had been consolidated with larger carriers and railroads were thought of as adjuncts to the canals (Thompson 1977, 162). By mid-century that had changed when many feeder railroads to the west were linked to the Hudson River. However, no direct rail connection from Albany to New York existed. Competition from Boston for Albany's gateway traffic increased. At first there was confidence in the ability of steamboats to handle the increasing passenger and bulk cargo. But the new expectations for speed and regularity underscored the significance of the winter shutdown. Indeed, engineer R. P. Morgan in his report to the Hudson River Railroad Committee pointed out that the three-month obstruction of steamboats by ice would give the railroad a decided advantage. But he cautioned that railroads would not succeed only on the merits of year-round travel. His comparative study of speed and cost favored the railroad, which, he suggested, would attract those three to four thousand daily steamboat travelers "who value time, those who like to travel fast . . . those who prefer traveling cheaply and at least nine-tenths of the way-passengers on the east side the river" (Morgan 1842, 12). He also envisioned a large volume of traffic that would be of "immense importance to the city of New York and the community." Certain products, such as vegetables, meats, milk, butter, and eggs, would receive added value from speedy railroad transportation. And in turn exports from the city supplying 150,000 individuals with dry goods, groceries, cotton "may be safely described as the business of the railroad" (Morgan 1842, 14).

The Hudson River Railroad, completed 1851, extended 143 miles from Chamber Street in New York to Albany, and was constructed along the eastern bank of the river, five feet above high tide. The road maintained a "degree of directness" that was efficient and cost effective, and avoided "the sinuosities of the stream" by cutting through land and extending the line into shallow water (Anonymous 1851, 3). The selection of the river level grade provided a road that was astonishingly regular. One hundred and fourteen miles are dead level; the total rise and fall is only 213 feet. While some rock cutting and tunnel blasting was necessary this was a small trade-off for the ease of ride and the absence of hills, which were so taxing on the engines of the era. These and other technical challenges in building the Hudson River Railroad fell to civil engineer John B. Jervis, who was invited to survey the right of way following the rejection of the Morgan proposal by the steamboat interests in the legislature. He followed Morgan's example by conducting a cost analysis to demonstrate that the Hudson railroad line could be profitable. He wooed the legislature, won their approval, joined the board of directors and helped raise the necessary capital.

Jervis drew on his broad experience with the Erie Canal, the Delaware and Hudson Canal, the Mohawk and Hudson Railroad, and the Croton Aqueduct. Thus this self-taught engineer was responsible for a good deal of New York's early transportation infrastructure. Much of his work strengthened the north-south longitudinal axis linking valley and city. He almost singlehandedly reinforced the paradigm of travel established by sloops and retraced by steamboats. Jervis fully understood the importance of this north-south axis. However, Jervis's defense of the riverside location of the tracks went beyond efficiency and appealed to aesthetics. He agreed with Morgan's 1842 view that the "beauty of the scenery on the river is not diminished in any respect, while the country above the Highlands, has a more striking appearance from the railroad" (Morgan 1842, 13). Jervis had explored this idea earlier while working on the Croton Aqueduct, where his chief engineer Fayette B. Tower suggested that the aqueduct trail provided new viewing areas for all to enjoy the Hudson River. This is more that an engineer's footnote: it leads us to the central intellectual question of the "machine in the garden." Nothing more fully represented the nineteenth-century machine than the railroad and most especially the locomotive—the most sensuous of technological innovations: you could see it, feel it, hear it, and smell it. It fully marked the landscape it traversed. In his classic study *The Machine in the Garden,* Leo Marx recreated the literary discourse about the clash of these fundamental forces—nature and technology, the organic and the inorganic (Marx 1964). He outlined the apprehensions of the elite about the "intruder" while countering this negative view with the enthusiasm, and open excitement of the public. The danger of this debate is in its potential to bifurcate the nineteenth-century world and the history of the Hudson River, where many of these forces were in play, into a fundamental struggle. The byproduct of this intellectual divide is to see the Hudson being pulled in two opposite directions—the organic and inorganic, the natural and the artificial. This fundamental debate has served as a kind of boilerplate for the upriver-downriver, valley-city dichotomy that has framed if not skewed, much of our understanding of the Hudson.

The railroad, like other forms of transportation, dynamically linked two interdependent worlds. The speed and power of trains created a new order in which one easily moved from one domain to the other on a continuum of experiences. While it is true that the Hudson River Railroad hardened the river's edge, created an industrial zone, and blocked public access, it also published celebratory guidebooks, encouraged summer river travel, and spurred the building of river estates and suburban communities. It played a key role in deepening the north-south axis and unifying the worlds of nature and technology.

FERRIES

As the Hudson's transportation network matured, the patterns became more differentiated and an east-west link began to mature. Steamboat journeys up and down the river often drew passengers to river towns from cross-river villages utilizing smaller local lines frequently linked to railroad connections Local crossing points became hubs of economic and social activity and provided the local population with regular and intimate contact with the river. These journeys became the "warp and woof of the lives of millions of valley residents" (Adams 2006, 117). The scale of community ferries and their integration into the experiences of community life created a romantic aura and an enduring nostalgia which inspired numerous river stories.

Steam ferries appeared at Troy, originally called Ferry Hook, in the first half of the nineteenth century. While three steam ferries served Albany, modest local operations appeared in smaller towns south of the capital. The Hudson-Catskill ferry and the Kingston-Rhinecliff ferry were later displaced by bridges. Indeed, ferry crossings provided the blueprint for Hudson River bridges, which linked communities nurtured and matured by these older east-west bonds and served the automobile boom of the 1920s. The numerous Hudson River ferry-rail connections underscore just how networked and interdependent the transportation system had become by the late nineteenth century.

The Newburgh-Beacon ferry, with roots in the American Revolution, operated for two hundred years until the Newburgh-Beacon Bridge opened in

1963. Travelers often noted the scenic beauty of the crossing and were charmed by the Beacon ferry house and especially the Newburgh Terminal with its gables, towers, turrets, and vergerbord decoration. Shops, trolleys and buses, and the railroad/Dayliner connection made the Newburgh Terminal a busy place (Adams 2006, 126–27). Kings Ferry between Verplanck and Stony Point, with its Revolutionary War connections, reminds us again of the historic association of so many river crossings. Farther south, Nyack, a major shipbuilding port, initiated service to New York in 1827 and linked to Tarrytown in 1834, becoming a key juncture in the lower Hudson. The revolutionary association is reiterated by the Dobbs Ferry or Sneden's Landing Ferry connecting Westchester with Rockland County. In 1841 the Erie Railroad began an unsuccessful short-lived ferry service to Manhattan from Piermont (Adams 2006, 130–34).

The opening of the Palisades Interstate Park in 1909 drew thousands of New Yorkers up the river seeking respite from city life. They used the Edgewater-125th Street ferry, which connected to the New York subway. Ferries played a crucial role not only in providing an east-west connection between river towns but also offered access to parks, historic sights, religious meetings, and outdoor recreation.

The island of Manhattan had become the hub of the transportation network drawing the people and the goods of the valley and the hinterland to this metropolis. Success, which depended on access to the island and the North River, was under pressure from increasing commerce. An elaborate system of steam ferries crossing from New Jersey was improvised to meet the persistent growth of the city's commerce and population. In addition to the Edgewater-125th Street ferry, which would be replaced by the George Washington Bridge, the Weehawken and West 42nd Street ferry was introduced. In 1883 a five-slip, sixteen passenger track terminal was completed—an architectural witness to the importance of this crossing. By 1927 it served twenty-seven million passengers and was one of largest ferry operations on the Hudson (Adams 2006, 145–48). Control of this terminal, like many others on the Hudson River, changed often, reflecting the constant merging and reconfiguration of the railroad companies. The west shore ferries, fed by trolley lines, competed with railroad-operated smaller vessels—lighters, tugs, and carfloats adding to the congestion of the east-west crossings.

Hoboken New Jersey, with its historic link to the beginnings of steam ferries and the Stevens family, began ferry service in 1811 and connected to many points in lower Manhattan. The impressive Lackawanna Terminal completed in 1907 and the Union Ferry Terminal at 23rd Street in Manhattan along with other terminals, large and small, pockmarked the eastern and western shoreline of the Hudson providing visible evidence of New York City's dominant commercial position (Adams 2006, 150–53).

TUNNELS

The pressure, especially from the railroads, for more efficient and less expensive crossings, instigated a series of innovations in tunnel construction and bridge building. The extent of the crises was described in 1916 by Ellsworth Huntington, a professor of geography at Yale University who had an early interest in ecology, and served as president of the Ecological Society of America. Huntington's article entitled "The Water Barriers of New York City" asked whether New York in fact had conquered nature (Huntington 1916). He wondered whether "we diminished the influence of the water barriers?" How much did the ferries cost in money, time, health, and character? Indeed, what did New York City pay for its power and it river location? The compression of business activity in lower Manhattan, a by-product of the ferry system, raised real estate values and according to Huntington produced the skyscraper, which, he argued, was a direct result of the water barriers (Huntington 1916, 179). He believed these barriers cost the city hundreds of millions of dollars and that while "we may have conquered nature . . . in the struggle she has bound us as tightly as we have bound her" (Huntington 1916, 183). Thus, we can suggest that New York City paid dearly to serve as the hub city, the clearinghouse for the hinterland, and the link between river and ocean.

It is not surprising that engineers would again be challenged by this notion of the river as barrier. From the earliest post–Civil War musings of "Crazy Luke," who harangued New York financiers with his scheme for a one-mile tunnel under the Hudson, the city has been intrigued by plans to end its islandness (Cudahy 2002, 5). In 1874 DeWitt Clinton Haskin proposed a Hudson River tunnel, secured financing, and began digging in 1879. This first attempt was stopped by a blowout and the loss of twenty lives. In 1901 William Gibbs McAdoo proposed twin tube interurban railway tunnels. He obtained financing for the Hudson and Manhattan Railroad Company and on January 15, 1908, the completed tunnel to Hoboken transported the first passengers under the Hudson River. One year later the downtown tubes linking Exchange Place in Jersey City to Cortlandt Street in Manhattan were opened. Quickly, regular passenger service began. The Hudson Tubes were an engineering marvel—the first transportation tunnel under a major river.

Moving passengers was not the only challenge. The Pennsylvania Railroad slowly accepted the commercial ascendancy of New York City and organized its own ferry service from the flatlands of New Jersey to Manhattan. The New York Central, its chief competitor, could make the trip directly onto the island from the Hudson Valley while the Pennsylvania needed to cross the Hudson. In 1910 they competed a bold a $110 million project—including approaches, six separate tube tunnels, an elaborate terminal at 32nd Street, and a four-track subway connection (Cudahy 2002, 23–30). The *New York Times* judged the project

> the greatest thoroughfare to the westward out of Manhattan ever devised , making easily accessible to the undeveloped insular territory beyond the river the pulsing heart of the metropolis, America's chief center of life. (quoted in Cudahy 2002, 31)

Again, engineers had played a decisive role in modernizing the transportation network linking the city and the country. But increasingly in the twentieth century the Hudson River was seen as nature's obstacle to the easy movement of east-west commerce.

BRIDGES

Bridges offered another alternative. The success of the Brooklyn Bridge stimulated the possibility of Hudson River crossings. Railroads were eager to replace their fleets of ferries, lighters, and carfloats. As early as 1855, an engineer proposed that a railroad bridge be built across the Hudson River at Poughkeepsie. In 1886, the Manhattan Bridge Building Company was organized to finance the construction of a multispan cantilever and truss bridge connecting Poughkeepsie and Highland. Philadelphia investors hoped to use the bridge to carry Pennsylvania coal to New England and bypass New York City. The Poughkeepsie Railroad Bridge opened in 1889 and lasted until it fell victim to the railroad mergers of the 1960s when the Erie consolidated many freight routes on its own trackage. A fire closed the bridge in 1974. Recently a nonprofit volunteer organization called *Walkway over the Hudson* turned it into a pedestrian and cyclist roadway. The Walkway over the Hudson State Historic Park was opened the to the public on October 3, 2009, in time for the Hudson-Fulton-Champlain Quadricentennial. This remarkable effort provides not only a unique river experience but also serves as a monument to the transportation history of the region (Burke 2007, 11–24).

In 1910, the Palisades Interstate Park (PIP), organized to protect the Palisades along the Hudson's western shore, received a gift from Mrs. Mary Williamson Averill Harriman of ten thousand acres in memory of her husband. This provided the foundation for the fifty thousand-acre Bear Mountain and Harriman State Parks. In 1915, the Bear Mountain Inn was opened, including a restaurant, a cafeteria, overnight accommodations in a rustic setting, and a dock for steamboat landings. The inn and the park were also accessible by automobile via Route 9W. PIP's role as a safety valve for urban congestion was an informing idea of the park from its inception. It drew city dwellers in increasing numbers in search of recreation and the great outdoors.

Long and inconvenient steamboat and ferry journeys, winding roadways, and slow-moving traffic overwhelmed local hotels, at times transforming automobiles into improvised bedrooms (Burke 2007, 25–28). If ready access to the new parks was the objective, then travel needed to be more efficient and direct. Plans for a suspension bridge three and one-half miles north of Peekskill, linking Anthony's Nose on the east to Fort Montgomery on the west, were approved in 1922. The structure, which would be built of steel, drew the ire of critics for whom concrete, because of its massive look and reassuring density, was the only appropriate material. The aesthetic argument was grounded in the notion that the Hudson River crossings required special consideration and that a bridge was not just a roadway, a connector between two points, but needed to harmonize with its river setting. However, the steel Bear Mountain toll bridge, the longest suspension bridge at the time, opened in 1924. The bridge brought more urbanites into the valley and established the first bridge link between Westchester and Rockland counties (Panetta 2010, 133–35). In the subsequent debates about the Tappan Zee Bridge, the Bear Mountain Bridge would serve as a reference point for both sides, and when the George Washington Bridge was completed, it was fixed as the southern frame for the planners.

It is hard to imagine Manhattan as an island without a hard crossing of the Hudson River. Tunnels had a dark foreboding quality and lacked the capacity to meet the growing needs of the automobile culture. In 1887 the engineer Gustav Lindenthal proposed the North River Bridge to cross "the most important water highway in the United States" and "eliminate the Hudson river from the New York terminus problem" (Petroski 1996, 133–35)." Lindenthal was responding to growing public pressure for a bridge crossing. He submitted plans for a suspension bridge, which critics argued was too ambitious a span and not financially feasible. In the severe winter of 1918 the Hudson froze cutting New York off from food and coal supplies (Petroski 1996, 211). This crisis created a sense of urgency for an all-weather bridge crossing the Hudson. Othmar Ammann, Lindenthal's former assistant, submitted a design in 1923, construction began in 1927 and the bridge was dedicated on October 24, 1931. The initial designation as the "Hudson River Bridge," was dropped, and in keeping with the river's historic association it was renamed The George Washington Bridge, commemorated the fortified positions used by General Washington and his American forces during the battle for New York. The fourteen-lane structure, originally intended to be encased in concrete, was redesigned to expose its steel superstructure, which the architect Le Corbusier described as the most beautiful bridge in the world. The George Washington Bridge allowed truckers and motorists to bypass Manhattan, stimulated suburban development in New Jersey, and populated the southern Palisades and Fort Lee with a cluster of high-rise apartments. Each day three hundred thousand automobiles and trucks cross The George Washington Bridge, a testament to the automobile age. Access to the bridge required good roads and an extensive highway system, which had to be designed, constructed, maintained, and financed for almost a century. Bridging the Hudson had a dramatic impact on the riverscape and the character of the surrounding neighborhoods.

The automobile age exploded in the 1920s. Automobile registration in New York rose dramatically from one million to two million between 1922 and 1927, accelerating demand for more paved roads (Fein 2008, 79). Initially, local extensions of streets into the countryside and the linking of nearby cities with state roads seemed adequate (Thompson 1977, 185). Of course these new routes were overlays on the existing pre-automobile roads; many river village and towns continued their historic connections to the waterfront. Indeed the north-south Route 9, which incorporated parts of the Old Albany Post road, served as one the earliest highways.

But the local communities lacked the financial resources and the political will to lead the road revolution. The Federal Highway Act of 1916 strengthened the role of the states in highway construction and New York State progressively increased its power in both the planning and financing of new roads with the aid of federal dollars (Fein 2008, 73). One of the first construction projects of the new state highway commission was the Storm King Highway, which promised to shorten the existing route between New York and Newburgh by sixteen miles (Fein 2008, 87). The proposed highway would wind around the steep slopes of Storm King Mountain, 420 feet (130 m) above the water. The slope of the mountain, the location of the proposed road, and the narrow ledge on which the work would take place made this an engineering challenge. Work was limited to small machines and manual labor. The Tafts, one of the principal supporters of the road stressed the "importance of bringing the motorists to the river's edge." Several observation points encouraged drivers to stop to enjoy the panoramic view. Critics pointed to the scar that would ring the mountain, a indelible mark clearly visible from the river (Dunwell 1991, 177–79). When it was completed in 1922, travelers were uncertain whether engineers had marred Storm King or enhanced the view of the river.

The Palisades Interstate Park saved the Palisades from quarrymen mining the cliffs and provided new recreational opportunities for stressed urbanites. The park owned most of the cliff face and the shoreline but not the land atop the Palisades. Land was purchased and gifted to the Palisades Interstate Park Commission and a new road—The Palisades Interstate Parkway—was built between 1947 and 1961. From Fort Lee in New Jersey to the Bear Mountain Bridge in New York, five hundred feet above the Hudson River, the parkway offered open vistas of the Hudson—a self-conscious element of the original design. But it was also a highway to the new suburbs of northern New Jersey, now accessible as a result of the George Washington Bridge. Stretches of undeveloped land and old farmsteads historically insulated by the barrier of the Hudson River were now ripe for suburbanization—the city was about to invade the country (Binnewies 2001, 213–38).

No transportation development better captures the full extent of the breadth of issues associated with spanning the Hudson River than the Tappan Zee Bridge. It encapsulates the modern predicament represented by the Hudson River as a natural barrier and the response of the engineer to the automobile. It is laboratory for the study of the making of the modern Hudson Valley suburban landscape.

Rockland County, located twelve miles (19 km) north-northwest of New York City, had a commercial connection to the city as early as the seventeenth century. Initially, sloops, then steamboats and the railroad in the nineteenth century and automobiles and trucks in the twentieth, transported farm produce, bricks, and ice to Manhattan and returned with imports and consumer goods. The river was a barrier insulating the county from the pressures of modernization and urbanization. Artists and writers, utopian communities, and summer vacationers were drawn to this rural sanctuary. The bucolic landscape was especially hospitable to numerous county and state parks. But beneath this bucolic exterior the forces of change were gathering power. Some local politicians and planners thought that Rockland was not a sanctuary but a backwater county—behind the times. Beginning in the 1920s, Rocklanders looked enviously at their cross-river neighbor Tarrytown, jealously eyeing its direct connection to New York City, its burgeoning suburban development, and its easy exchange with an array of new north-south parkways. In the 1930s the possibility of New Deal public works funding stimulated plans for a bridge connection between Nyack and Tarrytown. County residents, divided between the modernizers and the preservationists, hotly debated the merits of such a crossing. These plans, which lacked the necessary highway infrastructure, failed to win the support of the federal and state governments.

The bridge idea remained dormant until New York State's postwar commitment to highway expansion. Plans for an autobahn-like super highway—The Thruway—from Buffalo to New York City gained rapid approval. Buried in the engineering plans was a Tappan Zee crossing. Indeed, state officials managed the less vocal opposition, condemnation procedures, home displacements and relocations, and unprepared county and local governments with relative ease. The Tappan Zee Bridge, which served both the east-west and north-south axes, opened in 1955. The rush across the bridge by New Yorkers in a desperate search for postwar housing made Rockland a hothouse suburb. Runaway development overwhelmed the outdated rural infrastructure—schools, shopping, highways, and even government had to play catch-up. Rockland County became a negative example of suburban development: the result of bridging the Hudson, which had served for so long as the protector of the rural haven. Today's caution may well be "cross at your own risk."[2]

It is one the ironies of history that we now once again face the issue of the Tappan Zee crossing. Wear and tear on the bridge structure, explosive out-migration from the city, the continued growth of western and northern suburbs, and the levels of traffic volume remind us of the centrality of the Hudson River as a transportation corridor. River travel has been a decisive element in forming the shape of the region and the character of its economy and culture. It merits far more attention and respect and a calls for a revision in the ways we have imagined the history of America's river.

NOTES

1. The prosperity of New York City was also connected to the cotton trade (Albion 1939).
2. For a fuller treatment of this issue see Panetta 2010.

REFERENCES CITED

Adams, A. 2006. *The Hudson through the years*, revised edition. New York: Fordham University Press.

Albion, R. G. 1939 *The rise of New York port 1815–1860.* New York: Scribners.

Anonymous. 1851. *The Hudson River and the Hudson River Railroad.* New York: Bradbury and Guild.

Baer, C. T., D. B. Klein, and J. Majewski. 1993. From trunk to branch: Toll roads in New York, 1800–1860. *Essays in Economic and Business History* XI: 191–209.

Bennett, D. C. 1998 . The quest for speed: An American virtue, 1825–1930. PhD Dissertation, Auburn University.

Binnewies, R. O. 2001. *Palisades: 100,000 acres in 100 years.* New York: Fordham University Press and Palisades Interstate Park Commission.

Bruegel, M. 2002. *Farm, shop, landing: The rise of a market society in the Hudson Valley, 1780–1860.* Durham: Duke University Press.

Buckman, D. L. 1907. *Old steamboat days on the Hudson River.* The New York: Grafton Press.

Burke, K. W. 2007. *Hudson River bridges* Charleston, SC: Arcadia Pub.

Cronon, W. 1991. *Nature's metropolis: Chicago and the Great West.* New York: W. W. Norton.

Cudahy, B. 2002. *Rails under the mighty Hudson: The story of the Hudson Tubes, the Pennsy Tunnels, and Manhattan Transfer.* New York: Fordham University Press.

Dunbar, S. 1915. *A history of travel in America. Showing the development of travel and transportation from the crude methods of the canoe and the dog-sled to the highly organized railway systems of the present. Together with a narrative of the human experiences.* Vol. 1. Indianapolis: Bobbs-Merrill.

Dunwell, F. F. 1991. *The Hudson River Highlands.* New York: Columbia University Press.

Fein, M. R. 2008. *Paving the way: New York road building and the American state, 1880–1956.* Lawrence: University Press of Kansas, 79.

Fontenoy, P. E. 1994. *The sloops of the Hudson River: A historical and design survey.* Mystic, CT: Mystic Seaport Museum in association with the Hudson River Maritime Museum.

Huntington, E. 1916. The water barriers of New York City. *The Geographical Review* II(3): 169–83.

Kasson, J. F. 1999. *Civilizing the machine: Technology and republican values in America, 1776–1900.* New York: Hill and Wang.

Marx, L. 1964. The *machine in the garden: Technology and the pastoral ideal in America.* New York: Oxford University Press.

Morgan, R. P. 1842. *Report to the Hudson River Railroad Committee.* Pougkeepsie: Robert B. Killey.

Panetta, R. 2010. *The Tappan Zee Bridge and the forging of the Rockland suburb.* New City: Historical Society of Rockland County.

Parkerson, D. H. 1995. *Agricultural transition in New York State.* Ames: Iowa State University Press.

Peluso, A. J. Jr. 1997. *The Bard brothers: Painting America under steam and sail.* New York: Harry M. Abrams Inc. in association with the Mariners' Museum.

Petroski, H. 1996. *Engineers of dreams : Great bridge builders and the spanning of America.* New York: Vintage Books.

Seelye, J. 1991. *Beautiful machine: Rivers and the republican plan 1755–1825.* New York: Oxford University Press.

Sheriff, C. 1996. *The artificial river: The Erie Canal and the paradox of progress, 1817–1862.* New York: Hill and Wang.

Thompson, J. H., ed. 1977. *Geography of New York State.* Syracuse; Syracuse University Press, 156–57.

White, R. 1996. *Organic machine: The remaking of the Columbia River.* New York: Farrar, Strauss, and Giroux.

CHAPTER 18

OUT OF THE FRAY

Scientific Legacy of Environmental Regulation of Electric Generating Stations in the Hudson River Valley

John R. Young and William P. Dey

ABSTRACT

Between 1949 and 1976, the electric generating industry increased its ability to withdraw cooling water from the Hudson estuary from 600 m³/min to 17,400 m³/min, with most of the increase occurring in the 1970s when six large generating units were completed. While these units were being constructed, §316(b) of the Clean Water Act was enacted in 1972 to regulate the impacts of entrainment—passage of small aquatic organisms into and through the cooling system—and impingement—entrapment of larger aquatic organisms on the screens in front of the intake structures. The law triggered an ecological study of unprecedented magnitude to understand and quantify the effects of cooling water withdrawals on the estuary's biota. After more than thirty-five years, the scientific contributions of the study are many, but consensus on the impact of cooling water withdrawals and appropriate means to reduce impacts has not been achieved. Even so, the use of the estuary's water for cooling purposes has declined substantially from the peak in the 1970s as new generating technology that uses little or no cooling water has replaced the output of older stations.

INTRODUCTION

The Hudson River estuary began serving as a source of cooling water for power generation in 1918 when the 59th Street Station began operation in Manhattan. Between 1949 and 1969, fourteen additional generating units were constructed along the shores of the Hudson River estuary, bringing the total withdrawal capacity to 5,488 m³/min (Hutchison 1988). All of these units used once-through cooling systems that pumped water from the estuary through steam condensers and discharged it back into the estuary at temperatures 8 to 15 °C warmer. Once-through systems are less expensive to build, and operate more efficiently, than closed cycle systems, which require additional land and have environmental impacts of their own. As a result, once-through cooling has been the standard for steam turbine generation in settings where sufficient water was available. Closed cycle systems, which employ cooling towers or large ponds, have only been common in locations where water quantity was insufficient for once-through.

In the early 1970s, a projected rapid increase in electrical demand in New York led to the construction of six additional large steam units alongside the estuary, consisting of two units each at Bowline Point, Indian Point, and Roseton with a combined generating capacity of approximately 4,400 MW. The increase in electrical generation in the Hudson Valley was accompanied by an increase in the quantity of water that could be withdrawn from the estuary (Fig. 18.1). By the mid-1970s, installed cooling water capacity totaled 17,400 m³/min.

This rapid increase in water withdrawal capacity for electrical generation, coupled with greater public environmental awareness, raised concerns about the possible detrimental effects of water

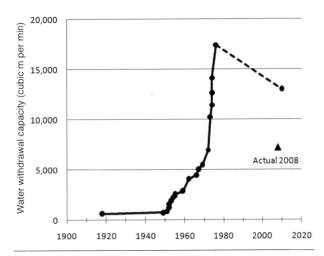

FIG. 18.1. Power generating industry cooling water withdrawal capacity from the Hudson River Estuary. Single value for 2008 represents actual withdrawal rate.

withdrawals on aquatic ecosystems, not only in the Hudson Valley but throughout the United States. The issues of concern for the cooling water withdrawals in the Hudson Valley primarily involved the capture of fish, generally less than about 10 cm in length, on the screens in front of the intakes (impingement), and the passage of fish eggs and larvae through the screens and into the pumps and cooling systems (entrainment). The thermal discharges, although not ignored, have been a lesser concern for the Hudson due to the large volume of the estuary and the rapid mixing of the bidirectional tidal flows.

Concerns about the potential effects of entrainment and impingement on aquatic ecosystems led Congress to include Section 316(b) in the Clean Water Act amendments of 1972, which required that the location, design, construction, and capacity of cooling water intake structures reflect the best technology available for minimizing adverse environmental impact (AEI). Neither the 316(b) statute nor the Environmental Protection Agency, which enforced it, provided an operational definition of adverse environmental impact, leading to a wide variety of approaches to measuring and minimizing it. On the Hudson, concern was primarily, but not exclusively, about the loss of anadromous fishes (e.g., striped bass, American shad, and river herring) passing by the intakes on their way into the estuary to spawn, and of their young during residence in the estuary and during emigration. For example, the Atomic Energy Commission predicted, with the minimal data then available, that entrainment and impingement losses from once-through cooling at Indian Point Unit 2 alone would reduce the striped bass population by 30 to 50 percent (USAEC 1972).

In response to the new statute, the owners of the new Hudson River generating stations initiated studies to examine both the local effects of cooling water withdrawals and discharges on the aquatic ecosystem (e.g., the benthic, phytoplankton, zooplankton, and ichthyoplankton communities), and riverwide programs to assess the cumulative effect of the facilities on the fish community of the estuary. This expanded program carried ecological sampling, principally on the fish fauna, to an unprecedented level.

As intended, the findings of the sampling program have been used in assessing the impacts of cooling water withdrawals. Barnthouse et al. (1988) summarized much of the information collected and analyzed prior to the Hudson River Settlement Agreement in 1980. The power generators analyzed information collected through 1997 in the 1999 Draft Environmental Impact Statement (Central Hudson et al. 1999), and the New York State Department of Environmental Conservation published its Final Environmental Impact Statement in 2003.

The environmental data from these studies, despite being what one New York regulator described as "the best data set on the planet," have not yet brought the cooling water issue to resolution. At present (March 2011), all six of the new units have draft permits that are awaiting adjudicatory hearings. Even so, the Hudson River environmental studies have advanced the science of impact assessment, and provided a wealth of additional basic information on the estuary and its aquatic communities. In addition, the existence of this background of knowledge and long-term continuation of the monitoring program has stimulated additional research. In this chapter we describe these study programs and present some of the key results.

ENTRAINMENT AND IMPINGEMENT MONITORING

Studies to estimate the number of fish entrained and impinged have been conducted at varying levels

TABLE 18.1. Estimated average annual entrainment and impingement at cooling water intakes on the Hudson River

Species	Entrainment[a]		Impingement[b]	
	Average	Range	Average	Range
American shad	1.47×10^7	$5.42 \times 10^5 – 7.08 \times 10^7$	4.13×10^4	$7.45 \times 10^3 – 2.78 \times 10^5$
River herring	8.37×10^8	$1.34 \times 10^8 – 2.70 \times 10^9$	4.29×10^5	$4.19 \times 10^4 – 1.94 \times 10^6$
Atlantic tomcod	– – – – Not Available – – – – – –		4.13×10^5	$1.72 \times 10^4 – 1.79 \times 10^6$
Bay anchovy	4.22×10^8	$1.21 \times 10^8 – 1.23 \times 10^9$	2.78×10^5	$8.49 \times 10^3 – 1.49 \times 10^6$
Spottail shiner	≈ 0	≈ 0	2.40×10^4	$1.23 \times 10^4 – 4.45 \times 10^4$
Striped bass	3.16×10^8	$1.21 \times 10^8 – 6.45 \times 10^8$	8.08×10^4	$1.44 \times 10^4 – 1.93 \times 10^5$
White perch	4.64×10^8	$2.68 \times 10^8 – 6.74 \times 10^8$	1.53×10^6	$6.66 \times 10^5 – 3.55 \times 10^6$
Total	2.05×10^9	$8.14 \times 10^8 – 4.83 \times 10^9$	3.08×10^6	$1.48 \times 10^6 – 7.52 \times 10^6$

Values are not adjusted for survival. Data from Central Hudson et al. 1999, EA 1991, LMS 1991a, LMS 1991b, NAI 1992.
a. Entrainment at Roseton, Indian Point, and Bowline Point stations from 1981, 1982–87. Total represents only the listed species.
b. Impingement at Roseton, Danskammer Point, Indian Point, Lovett, and Bowline Point stations, 1974–1990. Total represents all species impinged.

of intensity at the different cooling water intakes, ranging from weekly samples for one or two years at some intakes, to daily sampling over many years at others. Entrainment sampling methodology evolved substantially through research funded by the Empire State Electric Energy Research Corporation and the individual generating companies. Entrainment was sampled in the 1970s by suspending plankton nets in the intake and/or discharge flow, then in the 1980s automated sampling systems were developed (Occhiogrosso et al. 1981), which led to sampling over the entire twenty-four-hour period, seven days per week at some stations. Entrainment at most facilities was dominated by relatively few species that have pelagic eggs and larvae, principally bay anchovy, striped bass, white perch, and anadromous herrings (Central Hudson et al. 1999).

Early in the study of power plant impacts the combination of stresses experienced during passage was believed to be fatal to the vast majority of entrained organisms. However, new sampling gear and protocols developed on the Hudson (McGroddy and Wyman 1977; Muessig et al. 1988a) demonstrated that substantial survival is possible, depending principally upon species, size, and temperature of the discharge. For striped bass, one of the hardier species, up to 80 percent of the larvae that pass through the cooling system may survive (Young et al. 2009), but for more fragile species such as bay anchovy, entrainment survival is negligible (EA Engineering, Science, & Technology 1989).

The mean density (number per unit volume of water withdrawn) of fish eggs, larvae, and small juveniles from the abundance sampling, multiplied by the volume of cooling water withdrawn, provided an estimate of the number entrained. The total estimated annual entrainment for the Representative Important Species (RIS) at the six generating units completed in the 1970s averaged 2 billion and ranged from 0.8 to 4.8 billion (Table 18.1). The most commonly entrained was river herring (alewife and blueback herring combined) at 0.8 billion per year. White perch (0.5 billion), bay anchovy (0.4 billion), and striped bass (0.3 billion) were also major contributors to entrainment. However, adjustment for species-specific survival during passage through the cooling system produces average annual entrainment losses averaging 1.5 billion, with a range of 0.6 to 3.8 billion (Central Hudson 1999).

Impingement of fish, usually about 25 to 100 mm (1–4 in.) in length, was also monitored at various levels of intensity at the different stations, ranging from every few years at some stations to annually at others. A typical program would sample two to four days per month, but some programs sampled every day that the units operated. Research was also conducted on technologies to reduce impingement and to measure and improve impingement survival (Muessig et al. 1988b), such as bubble curtains (Alevras 1974), chains, lights, sound, angled screens, dual-flow screens (Fletcher 1994), barrier nets (Hutchison and Matousek 1988), fine mesh screens, and modifications to the traveling screens (Fletcher 1992).

Annual impingement of the RIS from 1974 through 1990 averaged 3.1 million, and ranged from 1.5 to 7.5 million (Table 18.1). The RIS comprised about 90 percent of total impingement numbers. White perch were the most commonly

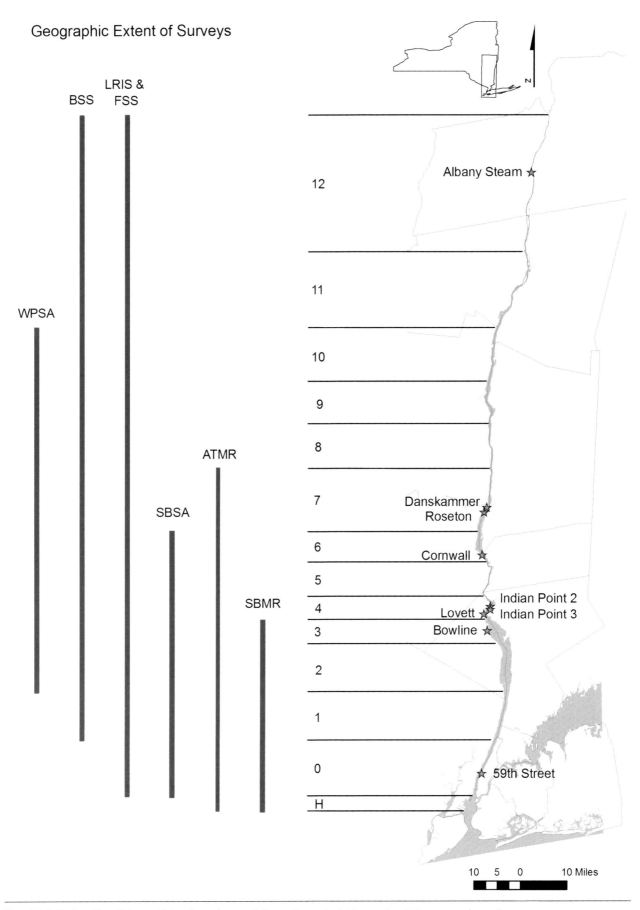

FIG. 18.2. Hudson River Estuary with locations of power station cooling water intakes and spatial coverage of sampling programs.

impinged species at 1.5 million per year, followed by river herring and Atlantic tomcod. Due to the measures that have been taken to improve survival of impinged fish, such as continuous screen rotation, Ristroph screen systems, fish return systems, actual impingement losses are far less than the numbers impinged.

RIVERWIDE SAMPLING PROGRAMS

The need to address cumulative impacts of cooling water withdrawals dictated that sampling be conducted throughout the entire estuary rather than just near the intake structures. In the mid 1970s the power plant owners initiated three sampling programs to describe the distribution and abundance of fish throughout the estuary between New York City and Troy, New York (Fig. 18.2).

The sampling focused on the life stages and species that would be susceptible to entrainment and impingement. Pelagic eggs and larvae, which are known collectively as ichthyoplankton, were sampled in the Long River Ichthyoplankton Survey (LRIS) during the spring and summer months in water more than three meters in depth. Once these early stages had transformed to juvenile fish, they were sampled both in the offshore habitats in the Fall Shoal Survey (FSS) and along the shoreline in beach habitats by the Beach Seine Survey (BSS). Although all fish species collected were counted and identified, the efforts focused on a set of Representative Important Species (RIS) that were selected by the regulatory agencies because they were common in the estuary, of particular recreational or commercial importance, or likely to be representative of other estuarine residents with respect to power plant impacts. The identified RIS for the studies were the anadromous striped bass, American shad, blueback herring and alewife, the marine bay anchovy, and the resident Atlantic tomcod, white perch, and spottail shiner. In addition to the biological data, water temperature, dissolved oxygen, and conductivity were measured with each survey. All three sampling programs, which continue to the present, have been conducted according to a set of detailed Standard Operating Procedures which are a key component of a quality assurance program that ensures high

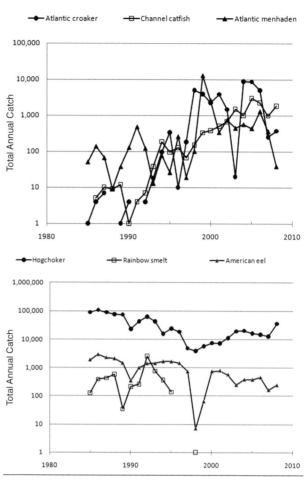

FIG. 18.3. Total annual catch of selected species in the Fall Juvenile Survey from 1985–2008. (Top) Species with increasing catch include Atlantic croaker, channel catfish, and Atlantic menhaden. (Bottom) Species with decreasing catch include hogchoker, rainbow smelt, and American eel.

quality data and comparability of results across the years (Young et al. 1992).

The distribution and abundance of many of the RIS and selected other species have been described from these sampling programs in a series of annual reports from 1974 through 2008 (e.g., ASA 2009), in the Hudson River DEIS (Central Hudson et al. 1999), and in efforts focused on individual species or families (e.g., Boreman and Klauda 1988; Englert and Sugarman 1988; Schmidt 1992; Klauda et al. 1988a; Heimbuch et al. 1992; Pace et al. 1993; Dey 1981; Limburg 1996a,b), and larval transport patterns (Dunning et al. 2006). These monitoring programs provide perhaps the only data available on the long-term abundance patterns of fish species that are not targeted by focused efforts. As would be expected, over the course of the studies annual catches of some species increased through time while catches of other species decreased. For example, total annual catch of Atlantic croaker, channel catfish, and Atlantic menhaden increased (Fig. 18.3 top), while catches of hogchoker, rainbow smelt, and American eel decreased in FJS sampling (Fig. 18.3 bottom). Indices of abundance were derived from the data sets by adjusting for the amount of sampling each year, and restricting the data to locations and time periods that match the life history of the particular species (Central Hudson et al. 1999). The intensity of the studies has made them useful for describing statistical patterns that could improve sampling survey design (Cyr et al. 1992; Wilson and Weisberg 1993).

Overall, the three programs have sampled much of the available fish habitat of the estuary over a thirty-five-plus-year period, providing information about the species and fish communities of the estuary (Beebe and Savidge 1988; Gladden et al. 1988; Strayer et al. 2004; Daniels et al. 2005) and the presence and abundance of individual species (e.g., Klauda et al. 1980; Barnthouse et al. 2003; Schmidt et al. 2003; Schmidt 2007).

ADULT STOCK STUDIES

In order to understand the effects of cooling water withdrawals on fish populations, rather than just on single age classes, it was necessary to know more about the life history characteristics of the populations. Information about the age and growth, maturity, fecundity, adult mortality rates, and migration patterns were also required. These life history aspects were studied intensively in the 1970s, and continued for some aspects up to the present, for striped bass, white perch, and Atlantic tomcod. Due to the large differences in life history and spatial and temporal distributions of these species, the sampling programs needed to be targeted to each species individually.

Striped bass stock assessment (SBSA) studies were conducted from 1976–1980 using large seines and gill nets in the middle estuary during their spring spawning run. In addition, a simulated commercial fishery was conducted to mimic certain aspects of the commercial striped bass harvest that was outlawed in 1976 due to PCB contamination. The program described the age, growth, sex ratios, age at maturity, and fecundity (Hoff et al. 1988), migration patterns (McLaren et al. 1981), food habits (Gardinier and Hoff 1982), and population sizes (SBMR). A separate study to determine the contribution of the Hudson River striped bass stock to the Atlantic coastal fishery was done in 1974 and 1975. The study used morphological characters to distinguish the Hudson, Chesapeake, and Roanoke stocks and determined that the Chesapeake stock was the dominant contributor to the coastal fishery between Maine and North Carolina (Berggren and Lieberman 1978). Due to the different migration patterns of the Hudson and Chesapeake stocks, and variations in the number of successful young produced each year class, the relative contribution of the Hudson varied from 0 percent to 79 percent for particular ages of fish (Van Winkle et al. 1988).

The resident white perch population was studied from 1971 to 1988, first from trawl samples taken near the Bowline Point station, and later from a wider range of directed sampling, and from the river-wide programs described above. White perch abundance trends have been described using data from these (WPSA) surveys (Wells et al. 1992). In addition, reproduction and growth (Holsapple and Foster 1975; Klauda et al. 1988b) and food habits (Bath and O'Connor 1985) have been described.

Atlantic tomcod, a northern species at the southern extreme of its range in the Hudson River, spawn in the middle part of the estuary in December and January (ATMR). The spawning stock was sampled with box traps set at fixed locations along the shoreline, from impingement collections, and since the late 1980s with bottom trawls near the mouth of the estuary. The Atlantic tomcod stock, composed primarily of fish one or two years of age (McLaren et al. 1988), has fluctuated between approximately 12 million and less than 0.5 million fish. Since about 1990, stock size generally has been lower than it was prior to 1990 (Central Hudson et al. 1999). In addition, these studies identified a high incidence of liver cancer in the stock (Smith et al. 1979; Klauda et al. 1981; Dey et al. 1993), which has led to substantial research into the chemical exposure and physiological response of this species to the organic pollutants in the estuary (e.g., Wirgin et al. 1989, 1990).

TABLE 18.2. Estimates of fractional reduction (conditional mortality rate) due to entrainment and impingement at all cooling water intakes for RIS species in the Hudson estuary.

Species	Entrainment		Impingement	
	Average	Range	Average	Range
American shad	0.239	0.031–0.520	<0.001	0.000–0.001
Alewife	0.232	0.077–0.407	0.014	0.011–0.019
Blueback herring	0.232	0.077–0.407	0.004	0.002–0.007
Atlantic tomcod	0.222	0.120–0.440	0.008	0.005–0.031
Bay anchovy	0.206	0.105–0.331	0.001	0.000–0.004
Spottail shiner	0.158	0.100–0.252	0.002	0.000–0.006
Striped bass	0.160	0.105–0.266	0.004	0.001–0.009
White perch	0.175	0.090–0.296	0.022	0.013–0.031

Data from Central Hudson et al. 1999.

IMPACT ASSESSMENT

What is AEI? Neither §316(b) of the CWA, nor New York state laws or regulations had defined AEI as it pertains to cooling water intake structures. However, a draft guidance document issued by USEPA in 1977 stated:

> Adverse aquatic environmental impacts occur whenever there will be entrainment or impingement damage as a result of operation of a specific cooling water intake structure. The critical question is the magnitude of any adverse impact. The exact point at which adverse aquatic impact occurs at any given plant site or water body segment is highly speculative and can only be estimated on a case-by-case basis by considering the species involved, magnitude of the losses, years of intake operation remaining, ability to reduce losses, etc. (USEPA 1977, 11)

> The study requirements necessary to evaluate losses of aquatic life at existing cooling water intakes can be considered in two separate steps. The first is assessment of the magnitude of the problem at each site through direct determination of the diel and seasonal variation in numbers, sizes and weights of organisms involved with operation of the intake. When losses appear to be serious, as a second step it may be necessary to conduct studies in the source water body if there is a need to evaluate such losses on a water-body-wide or local population basis. (USEPA 1977, 39)

Although this guidance was subsequently withdrawn, it established the basis for coupling in-plant sampling to estimate numbers entrained or impinged with field sampling programs to provide the context of population effects. Activities on the Hudson River led to development of calculation methods to express both entrainment losses and impingement losses as a conditional mortality rate, (defined loosely as the fraction of a year class or fish spawned within the year that is lost due to entrainment or impingement). The calculation procedures, the Empirical Transport Model (Boreman et al. 1978) for entrainment, and the Empirical Impingement Model (Barnthouse et al. 1979) were developed specifically to use the data available for the Hudson Estuary, but have subsequently been applied elsewhere. From 1974 through 1997 the average estimated conditional mortality rate due to all the Hudson River cooling water withdrawals for the RIS ranged from 0.158, or approximately 16 percent, (spottail shiner) to 0.239 (American shad) for entrainment and from <0.0001 (American shad) to 0.022 (white perch) for impingement (Table 18.2). These estimated reductions apply only to the ages of these species that are affected by the cooling water withdrawals. They are not indicators of the effect on the entire species population, nor do they apply to the many species whose life history characteristics do not make them susceptible to the withdrawals.

Because the monitoring data reflected what actually occurred with the existing Hudson River cooling water intakes in operation, questions about what might have occurred without the cooling water withdrawals could only be answered with impact modeling, and significant advances were also made in this area (e.g., Goodyear 1988; Lawler

1988; Savidge et al. 1988; Ginsberg and Ferson 1992). However, no consensus was achieved on the modeling results primarily due to disagreements over the strength of natural population regulatory processes (compensation) that would lessen the ultimate effect of cooling water withdrawals or other man-induced losses (Christensen and Goodyear 1988; Fletcher and Deriso 1988). For example, the 1999 DEIS (Central Hudson et al. 1999) concluded, based on modeling results, that total entrainment mortality at all Hudson River facilities combined reduced the size of the striped bass spawning stock of 652 thousand fish by only five thousand, and that elimination of all entrainment mortality would increase the harvest by less than 1 percent. Whether or not changes of this magnitude would be deemed adverse, they would likely be undetectable in the data.

When the NYSDEC completed the Hudson River FEIS in 2003 it abandoned the population effect paradigm and based its evaluation on the raw numbers of organisms involved with the intakes. Although NYSDEC expressed concern for aquatic populations, particularly those whose abundance was decreasing for whatever reason, it did not attempt to link power plant effects with population trends. The focus on numbers involved with the intakes, rather than populations, was consistent with the new cooling water intake regulations USEPA proposed in 2003 and finalized in 2004. USEPA again declined to define AEI, but instead proposed a compliance scheme requiring 60 percent to 90 percent reduction of entrainment and 80 percent to 95 percent reduction of impingement mortality at intakes located on estuaries. The reductions were calculated from a site-specific hypothetical baseline level of entrainment and impingement. The USEPA suspended the new regulations in 2007 after parts of the rule were remanded in the 2nd Circuit Court of Appeals. The U.S. Supreme Court overturned part of the remand in 2009, and as of 2010 USEPA had not reissued the rule.

Throughout the shifting federal regulatory field, NYSDEC continued to regulate cooling water intake structures under existing state regulations (6 NYCRR 704.5), requiring reductions in entrainment and impingement mortality approaching those that could be achieved with closed cycle cooling. The Danskammer Point station was required to minimize its cooling water flows, operate its screens in a manner to reduce impingement mortality, and install a sonic fish deterrent system. For the Bowline Point and Roseton stations, draft permits required entrainment and impingement reductions that were expected to be achieved through a combination of the reduced operation that was already occurring, careful management of cooling water use when operating, and other technology to exclude fish from the intakes or improve their survival. For Indian Point, the draft permit required continued use of the previously installed state-of-the-art screen systems with fish returns and multiple-speed pumps (which can pump less water when lower flows are sufficient). Closed cycle cooling was required if the units extended their nuclear operating licenses beyond the original forty-year terms, which end in 2013 (Unit 2) and 2015 (Unit 3). As of early 2011, the Bowline Point, Indian Point, and Roseton permits were in various stages of litigation. All three permits were challenged by environmental advocacy groups for being overly lenient, while the Indian Point permit was also challenged by the plant owners for being unnecessarily restrictive.

OUTGROWTH OF THE CONTROVERSY

The Hudson River monitoring program has provided an unsurpassed base of knowledge about the estuary, particularly the ichthyofauna. This base has supported a great deal of other ecological research on the estuary. Much of it has been funded by the Hudson River Foundation (HRF), which was established through the 1980 Hudson River Settlement Agreement (Sandler and Schoenbrod 1981) (see Butzel, ch. 19 in this volume) to support "scientific, ecological, and related public policy research on issues and matters of concern to the Hudson River," including "mitigating fishery impacts caused by power plants, providing information needed to manage fishery resources of the Hudson River, understanding the factors related to the abundance and structure of fish populations, and gaining knowledge of the Hudson River Ecosystem." Since 1983, HRF has provided grants totaling approximately $36 million, which have produced approximately 465 peer-reviewed publications. Efforts by the Hudson River Fisheries Unit of NYSDEC, the Marine Fisheries Unit of NYSDEC, Cornell University, SUNY Stony Brook, the Cary Institute of Ecosystem Studies, New York University, the University of Connecticut, the University of Massachusetts, Lamont-Doherty Earth Observatory, and others have contributed significantly to the ecological knowledge base.

CONCLUSION

The scientific issues associated with cooling water withdrawals were not unique to the Hudson River, but the intensity of effort to answer them was, and continues to be so. Today, almost forty years since §316(b) became law, we know a great deal more about the estuary's ecology than we did in the 1970s. Even though there has been no final decision on whether the six units constructed in the 1970s will employ cooling towers over some part of their remaining useful lives, the issue of cooling water withdrawal from the Hudson River may be following the wax and wane pattern of many of the other environmental issues described in this volume. Of the peak 17,400 m^3/min withdrawal capacity that existed in the 1970s, 13,000 m^3/min remains today (Fig. 18.1). As the older generating units were retired they have been replaced with units that did not have once-through cooling systems. In addition, much of the remaining withdrawal capacity is currently unused because some of the units operate much less frequently than in the past. In 2008, actual withdrawals of the Hudson River stations still operating averaged only 7,200 m^3/min (Nieder 2010). Whether the remaining units with once-through cooling ever convert to closed cycle cooling systems or they install alternative intake technology, the effects of cooling water withdrawals in the future will be much reduced from what they were over the last quarter of the twentieth century.

Permitting of water withdrawals for electric generation in the Hudson River valley over more than thirty-five years has resulted in expenditure of many millions of dollars on research and monitoring. The knowledge gained through these efforts has been disseminated through hundreds of reports and peer-reviewed scientific papers, but it has failed to achieve a consensus regarding adverse environmental impacts and appropriate cooling water intake technologies. Even so, the monitoring program continues today to provide data for more effective, science-based management of the Hudson River and its fish community.

REFERENCES CITED

Alevras, R. 1974. Status of air bubbler fish protection at Indian Point Station on the Hudson River. In Proceedings of the Second Entrainment and Intake Screening Workshop. Report No. 15: Entrainment and intake screening, ed. L. D. Jensen, 289–92. Electric Power Research Institute Publication No. 74-049-00-5.

ASA Analysis & Communication. 2010. 2008 year class report for the Hudson River Estuary Monitoring Program. Prepared on behalf of Dynegy Roseton, L.L.C., Entergy Nuclear Indian Point 2 L.L.C., Entergy Nuclear Indian Point 3 L.L.C., Mirant Bowline L.L.C.

Barnthouse, L. W., D. L. DeAngelis, and S. W. Christensen. 1979. An empirical model of impingement impact. Oak Ridge National Laboratory, ORNL/NUREG/TM-290. Oak Ridge, TN.

Barnthouse, L. W., R. J. Klauda, D. S. Vaughan, and R. L. Kendall, eds. 1988. Science, law, and Hudson River power plants—A case study in environmental impact assessment. American Fisheries Society Monograph 4, Bethesda, MD.

Barnthouse, L. W., D. Glaser, and J. Young. 2003. Effects of historic PCB exposures on reproductive success of the Hudson River striped bass population. *Environmental Science & Technology* 37:233–38.

Barnthouse, L. W., D. G. Heimbuch, W. Van Winkle, and J. Young. 2008. Entrainment and impingement at IP2 and IP3: A biological impact assessment.

Bath, D. W., and J. M. O'Connor. 1985. Food preferences of white perch in the Hudson River Estuary. *New York Fish and Game Journal* 32: 63–70.

Beebe, C. A., and I. R. Savidge. 1988. Historical perspective on fish species composition and distribution in the Hudson River Estuary. American Fisheries Society Monograph 4: 25–36.

Berggren, T. J., and J. T. Lieberman. 1978. Relative contribution of Hudson, Chesapeake, and Roanoke striped bass, *Morone saxatilis*, stocks to the Atlantic coast fishery. *Fishing Bulletin* 76: 335–45.

Boreman,, J., C. P. Goodyear, and S. W. Christensen. 1978. An empirical transport model for evaluating entrainment of aquatic organisms by power plants. U.S. Fish and Wildlife Service Biological Services Program FWS/OBS-78/90/.

Boreman, J., and R. J. Klauda. 1988. Distributions of early life stages of striped bass in the Hudson River Estuary, 1974–1979. American Fisheries Society Monograph 4: 53–58.

Central Hudson Gas & Electric Corp., Consolidated Edison Company of New York, Inc., New York Power Authority, and Southern Energy New York. 1999. Draft environmental impact statement for state pollutant discharge elimination system permits for Bowline Point 1 & 2, Indian Point 2 & 3, and Roseton 1 & 2 Steam Electric Generating Stations.

Christensen, S. W., and C. P. Goodyear. 1988. Testing the validity of stock-recruitment curve fits. American Fisheries Society Monograph 4: 219–31.

Cyr, H., J. A. Downing, S. LaLonde, S. B. Baines, and M. L. Pace. 1992. Sampling larval fish populations: Choice of sample number and size. *Transactions of the American Fisheries Society* 121: 356–68.

Daniels, R. A., K. E. Limburg, R. E. Schmidt, D. L. Strayer, and R. C. Chambers. 2005. Changes in fish assemblages in the tidal Hudson River, New York. In Historical changes in large river fish assemblages of the Americas, ed. J. N. Rinne, R. M. Hughes, and B. Calamusso, 471–503. American Fisheries Society Symposium 45, Bethesda, MD.

Dey, W. P. 1981. Mortality and growth of young-of-the-year striped bass in the Hudson River Estuary. *Transactions of the American Fisheries Society* 110: 151–57.

Dey, W. P., T. H. Peck, C. E. Smith, and G. L. Kreamer. 1993. Epizoology of hepatic lesions in Atlantic tomcod from the Hudson River estuary. *Canadian Journal of Fisheries and Aquatic Sciences* 50: 1897–1907.

Dunning, D. J., Q. E. Ross, A. F. Blumberg, and D. G. Heimbuch. 2006. Transport of striped bass larvae out of the lower Hudson River Estuary. In *American Fisheries Society Symposium* 51, ed. J. R. Waldman, K. E. Limburg, and D. A. Strayer, 273–86.

EA Engineering, Science & Technology. 1989. Final Indian Point generating station 1988 entrainment survival study. Prepared for Consolidated Edison Company of New York, Inc., and New York Power Authority.

———. 1991. Hudson River ecological study in the area of Indian Point 1990 Annual Report. Prepared for Consolidated Edison Company of New York, Inc., and New York Power Authority.

Englert, T. L., and D. Sugarman. 1988. Patterns of movement of striped bass and white perch larvae in the Hudson River estuary. In *Fisheries research in the Hudson River,* ed. C. L. Smith, 148–68. Albany: State University of New York Press.

Fletcher, R. I. 1992. The failure and rehabilitation of a fish-conserving device. *Transactions of the American Fisheries Society* 121: 678–79.

———. 1994 Flows and fish behavior: Large double-entry screening systems. *Transactions of the American Fisheries Society* 123: 866–85.

———, and R. B. Deriso. 1988. Fishing in dangerous waters: Remarks on a controversial appeal to spawner-recruit theory for long-term impact assessment. American Fisheries Society Monograph 4: 232–44.

Gardinier, M. N., and T. B. Hoff. 1982. Diet of striped bass in the Hudson River Estuary. *New York Fish and Game Journal* 29: 152–65.

Geoghegan, P., M. T. Mattson, and R. G. Keppel. 1992. Distribution of the shortnose sturgeon in the Hudson River Estuary, 1984–1988. In *Estuarine research in the 1980s*, ed. C. L. Smith, 217–27. Albany: State University of New York Press.

Ginzburg, L. and S. Ferson. 1992. Assessing the effect of compensation on the risk of population decline and extinction. In *Estuarine research in the 1980s*, ed. C. L. Smith, 392–403. Albany: State University of New York Press.

Gladden, J. B., F. R. Cantelmo, J. M. Croom, and R. Shapot. 1988. Evaluation of the Hudson River ecosystem in relation to the dynamics of fish populations. *American Fisheries Society Monograph* 4: 37–52.

Goodyear, C. P. 1988. Implications of power plant mortality for management of the Hudson River striped bass fishery. *American Fisheries Society Monograph* 4: 245–54.

Heimbuch, D. G., D. J. Dunning, and J. R. Young. 1992 Post yolk-sac larvae abundance as an index of year class strength of striped bass in the Hudson River. In *Estuarine research in the 1980s,* ed. C. L. Smith, 376–91. Albany: State University of New York Press.

Hoff, T. B., J. B. McLaren, and J. C. Cooper. 1988. Stock characteristics of Hudson River striped bass. *American Fisheries Society Monograph* 4: 59–68.

Holsapple, J. G., and L. E. Foster. 1975. Reproduction of white perch in the lower Hudson River. *New York Fish and Game Journal* 22: 122–27.

Hutchison, J. B. Jr. 1988. Technical descriptions of Hudson River electricity generating stations. *American Fisheries Society Monograph* 4: 113–20.

———, and J. A. Matousek. 1988. Evaluation of a barrier net used to mitigate fish impingement at a Hudson River power plant intake. *American Fisheries Society Monograph* 4: 280–85.

Klauda, R., W. P. Dey, T. P. Hoff, J. B. McLaren, and Q. E. Ross. 1980. Biology of juvenile Hudson River striped bass. *Mar. Rec. Fish* 5:101–23.

Klauda, R. J., T. H. Peck, and G. K. Rice, 1981. Accumulation of polychlorinated biphenyls in Atlantic tomcod (*Microgadus tomcod*) collected from the Hudson River estuary, New York. *Bulletin of Environmental Contamination and Toxicology* 27: 829–35.

Klauda, R. J., R. E. Moos, and R. E. Schmidt. 1988a. Life history of Atlantic tomcod, Microgadus tomcod, in the Hudson River Estuary, with emphasis on spatio-temporal distribution and movements. In *Fisheries research in the Hudson River,* ed. C. L. Smith, 219–51. Albany: State University of New York Press.

Klauda, R. J., J. B. McLaren, R. E. Schmidt, and W. P. Dey. 1988b. Life history of white perch in the Hudson River Estuary. *American Fisheries Society Monograph* 4: 69–88.

Lawler, J. P. 1988. Some considerations in applying stock-recruitment models to multiple-age spawning populations. *American Fisheries Society Monograph* 4 :204–18.

Lawler, Matusky, & Skelly Engineers. 1991a. 1990 Impingement study at the Bowline Point generating station. Prepared for Orange and Rockland Utilities, Inc.

———. 1991b. 1990 Impingement studies at the Lovett generating station. Prepared for Orange and Rockland Utilities, Inc.

Limburg, K. E. 1996a. Modelling the ecological constraints on growth and movement of juvenile American shad (*Alosa sapidissima*) in the Hudson River estuary. *Estuaries* 19(4), dedicated issue: The Hudson River Estuary (Dec. 1996): 794–813.

———. 1996b. Growth and migration of 0-year American shad (*Alosa sapidissima*) in the Hudson River estuary: Otolith microstructural analysis. *Canadian Journal of Fisheries and Aquatic Science* 53: 220–38.

McGroddy, R. M., and R.L. Wyman. 1977. Efficiency of nets and a new device for sampling live fish larvae. *Journal of the Fisheries Research Board of Canada* 34: 571–74.

McLaren, J. B., J. C. Cooper, T. B. Hoff, and V. Lander. 1981. Movements of Hudson River striped bass. *Transactions of the American Fisheries Society* 110: 158–67.

McLaren, J. B., T. H. Peck, W. P. Dey, and M. Gardinier. 1988. Biology of Atlantic tomcod in the Hudson River Estuary. *American Fisheries Society Monograph* 4: 102–12.

Muessig, P. H., J. R. Young, D. S. Vaughan, and B. A. Smith. 1988a. Advances in field and analytical methods for estimating entrainment mortality factors. *American Fisheries Society Monograph* 4: 124–32.

Muessig, P. H., J. B. Hutchinson, Jr., L. R. King, R. J. Ligotino, and M. Daley. 1988b. Survival of fishes after impingement on traveling screens at Hudson River power plants. *American Fisheries Society Monograph* 4: 170–81.

New York State Department of Environmental Conservation (NYSDEC). 2003. Final environmental impact statement concerning the application to renew New York State pollutant discharge elimination system permits for the Roseton 1 & 2, Bowline 1 & 2, and Indian

Point 2 & 3 steam electric generating stations, Orange, Rockland, and Westchester Counties.

Nieder, W. C. 2010. The relationship between cooling water capacity utilization, electric generating capacity utilization, and impingement and entrainment at New York State steam electric generating facilities. New York State Department of Environmental Conservation Technical Document. Albany, NY. July 2010.

Normandeau Associates Inc. 1992. Roseton and Danskammer Point generating stations impingement monitoring program 1990 annual progress report. Prepared for Central Hudson Gas & Electric Corporation.

Occhiogrosso, T., L. R. King, B. Muchmore, D. Belcher, B. Smith, and S. M. Jinks. 1981. A portable automated plankton abundance sampling system, In *Proceedings of the fifth national workshop on entrainment and impingement,* ed. L. D. Jensen, 277–80. Sparks, MD: EA Communications.

Pace, M.L., S. B. Baines, H. Cyr, J. A. Downing. 1993. Relationships among early life stages of *Morone americana* and *Morone saxatilis* from long-term monitoring of the Hudson River Estuary. *Canadian Journal of Fisheries and Aquatic Sciences* 50: 1976–85.

Sandler, R., and D. Shoenbrod, eds. 1981. The Hudson River power plant settlement. Materials prepared for a conference sponsored by New York University School of Law and the Natural Resources Defense Council, Inc., with support from the John A. Hartford Foundation, Inc.

Savidge, I. R., J. B. Gladden, K. P. Campbell, and J. S. Ziesenis. 1988. Development and sensitivity analysis of impact assessment equations based on stock-recruit theory. American Fisheries Society Monograph 4: 191–203.

Schmidt, R. E. 1992. Temporal and spatial distribution of bay anchovy eggs through adults in the Hudson River estuary. In *Estuarine research in the 1980s,* ed. C. L. Smith, 228–41. Albany: State University of New York Press.

———. 2007. Young striped searobins (*Triglidae: Prionotus evolans*) in the Hudson River. *Journal of Northwestern Atlantic Fishery Science* 38: 67–71.

———, B. M. Jessop, and J. E. Hightower. 2003. Status of river herring stocks in large rivers. *American Fisheries Society Symposium* 35: 171–82.

Smith, C. E., T. H. Peck, R. J. Klauda, and J. B. McLaren. 1979. Hepatomas in Atlantic tomcod *Microgadus tomcod* (Walbaum) collected in the Hudson River estuary in New York. *Journal of Fish Diseases* 2: 313–19.

Strayer, D. L., K. A. Hattala, and A. W. Kahnle 2004, Effects of an invasive bivalve (*Dreissena polymorpha*) on fish in the Hudson River estuary. *Canadian Journal of Fisheries and Aquatic Sciences* 61: 924–41

United States Atomic Energy Commission. 1972. Final environmental statement related to operation of Indian Point Nuclear Generating Plant Unit No. 2.

United States Environmental Protection Agency. 1977. Draft guidance for evaluating the adverse impact of cooling water intake structures on the aquatic environment: Section 316(b) P.L. 92-500.

———. 2004. National pollutant discharge elimination system: Final regulations to establish requirements for cooling water intake structures at phase II existing facilities. *Federal Register* 69: 131 (July 9, 2004): 41576–693.

———. 2007. National pollutant discharge elimination system—Suspension of regulations establishing requirements for cooling water intake structures at Phase II existing facilities. *Federal Register* 72: 130 (July 9, 2007): 37107–109.

Van Winkle, W., K. D. Kumar, and D. S. Vaughan. 1988 Relative contributions of Hudson River and Chesapeake Bay striped bass stocks to the Atlantic coastal population. American Fisheries Society Monograph 4: 255–66.

Wells, A. W., J. A. Matousek, and J. B. Hutchison. 1992. Abundance trends in the Hudson River white perch. In *Estuarine research in the 1980s,* ed. C. L. Smith, 242–64. Albany: State University of New York Press.

Wilson, H. T., and S. B. Weisberg. 1993. Design considerations for beach seine surveys of striped bass. *North American Journal of Fisheries Management* 13: 376–82.

Wirgin, I., D. Currie, and S. J. Garte. 1989. Activation of the K-ras oncogene in liver tumors of Hudson River tomcod. *Carcinogenesis* 10: 2311–2315.

Wirgin, I. I., M. D'Amore, C. Grunwald, A. Goldma, and S. J. Garte. 1990. Genetic diversity at an oncogene locus in mitochondrial NDA between populations of cancer-prone Atlantic tomcod. *Biochemical Genetics* 28: 459–75.

Young, J. R., R. G. Keppel, and R. J. Klauda. 1992. Quality assurance and quality control aspects of the Hudson River ecological study. *In Estuarine research in the 1980s,* ed. C. L. Smith, 303–22. Albany: State University of New York Press.

Young, J. R., W. P. Dey, S. M. Jinks, and D. T. Mosier. 2009. Survival of striped bass entrained into the cooling system of two Hudson River power plants. *North American Journal of Fisheries Management* 29: 1015–34.

PART IV

RIVER OF INSPIRATION

Robert E. Henshaw

He saw at a distance the lordly Hudson, far, far below him, moving on its silent but majestic course, with the reflection of a purple cloud, or the sail of a lagging bark, here and there sleeping on its glassy bosom, and at last losing itself in the blue highlands.
—Washington Irving (Rip Van Winkle awakening from a twenty-year sleep)

How I have walked . . . day after day, and all alone, to see if there was not something among the old things which was new!
—Thomas Cole (in Hood 1969)

> "Driver, what stream is it?" I asked, well knowing
> it was our lordly Hudson hardly flowing.
> "It is our lordly Hudson hardly flowing,"
> he said, "under the green-grown cliffs."
> Be still, heart! No one needs
> your passionate suffrage to select this glory,
> this is our lordly Hudson hardly flowing
> under the green-grown cliffs.
> "Driver, has this a peer in Europe or the East?"
> "No, no!" he said. Home! Home!
> Be quiet, heart! This is our lordly Hudson
> and has no peer in Europe or the east.
> This is our lordly Hudson hardly flowing
> under the green-grown cliffs
> and has no peer in Europe or the East.
> Be quiet, heart! Home! Home!
> — Paul Goodman, "The Lordly Hudson"

Cleaning up a river is a cause worth fighting for. . . . We had allowed some people to make good profit along the Hudson and then go somewhere else to enjoy clear water.
—Pete Seeger (1984)

INSPIRATION FROM UNIQUENESS

The Hudson River has often been referred to as America's Rhine for its utility, strategic importance, and beauty. Through the centuries the Hudson River region indeed was valued firstly for its utility, as conveyed in the preceding chapters. During colonial times human needs influenced acceptance, or at least tolerance, of associated impacts. When the impacts finally became intolerable, most people—feeling powerless to stop them—responded in the only way they could, by resignedly turning away from the river. But throughout its history the river also has inspired at least a few individuals to respond very differently.

The boisterous, but colorful, new populace, so lamented by Petrus Stuyvesant, spurred Washington Irving to take pen in hand and create an enduring body of New World fictional literature, becoming America's first internationally published writer. His reference to "the lordly Hudson" in many of his writings, for example, the 1819 short story "Rip Van Winkle," influenced generations, and finally was enlarged upon by Paul Goodman in 1962 (above). In chapter 21 Brackett discusses portrayal of the interplay of humans with the Hudson River in literature.

INSPIRATION FROM GROWTH AND DEVELOPMENT

In 1825, as development pushed up the river valley, encroaching on the vast scenic areas of the Hudson Highlands and the Shawungunk and Catskill Mountains, a then-unknown young artist, Thomas Cole, was moved to paint scenes which construction had not yet sullied, sometimes even excluding an existing factory or railroad track from the painted scene. So began the first school of art in America (Boyle 2009; Howat 1972), later known as the Hudson River School of landscape painting, as other artists followed Cole. Key among these artists was Asher B. Durand, a former engraver, whose work tightly captured scenic beauty. His painting, "Progress" (see Fig. Intro. 1) encapsulates the growing conflict in the mid-1800s of developing communities as railroads and commerce encroach. The longing for the original, pristine, and undeveloped—"the sublime"—expressed by the Hudson River School artists continues to inspire citizens today. Flad, in chapter 20, demonstrates how those early yearnings for the natural world provided the arguments used in citizen challenges of modern day environmental cases.

Commercial exploitation of the Hudson River Valley continued during the nineteenth and much of the twentieth centuries. Finally, by the beginning of the twentieth century toleration gave way to the first attempts to reverse environmental decline. Theodore Roosevelt, a Hudson River Valley resident in his youth and imbued with its history, aesthetics, and recognition of accumulating impacts, carried those influences to the nation's highest office. As president, he appointed his friend Gifford Pinchot to head a new National Forest Service, which he placed in the Department of Commerce (U.S. Forest Service 2010) to emphasize his interest in managing our forests. The American conservation movement can be traced to this singular action, and certainly to this Hudson River citizen (Almanac 2010). As New York City grew, a ready supply of trap rock to build the city's iconic brownstone houses lay just across the river. Roosevelt, responding to other (wealthy, Hudson River) friends, put an end to the blasting and destruction by buying the land and creating the Palisades Interstate Park, the first interstate park in the country.

INSPIRATION FROM ONE INDUSTRY

By the middle of the twentieth century the conservation ethic was ascendant (Library of Congress 2009), but the laws had not kept up and did not foresee growth of large industries and the utilities. In 1962 the Consolidated Edison Electric Company that supplies most of the electrical power to New York City proposed to construct a very large pumped storage power plant at Storm King Mountain. This mountain, of iconic importance to the Hudson River School artists, continued to be revered by the populace. ConEd's proposed massive alteration of the scenic cliff overlooking the Hudson spurred a small group of concerned Hudson River citizens into action. Rather than becoming dejected as so many had before, they organized as the "Scenic Hudson Preservation Conference" in 1966. Joined by the Hudson River Fishermen's Association, they filed case after case in federal court challenging ConEd's plans, as Butzel describes in chapter 19. They suffered repeated losses, but finally the Second Circuit Court of Appeals issued two key decisions: (1) Applications must include environmental impacts in addition to economic costs of projects (Scenic Hudson 1965); and (2) Citizens have legal standing to sue on behalf of the environment (Scenic Hudson 1971). These two points, codified into law, form the basis of the National Environmental Policy Act (NEPA) (Alm 1988). Butzel's chapter details how America's—and therefore the world's—environmental movement began in the Hudson River Valley at Storm King Mountain.

In addition to scenic impacts, the Storm King pumped storage plant would have withdrawn up to one-fourth of the river's flow each day. Because this site was in the "biological center" of the river (see "An Abbreviated Geography" in this volume) the entire river's fish community was feared to be in peril. Furthermore, fish were already being impacted by water withdrawals at large steam electric power plants as discussed by Young and Dey in the previous section. These very large generators produced vast amounts of waste heat and required a ready source of water for cooling. As cooling water was sucked into the power plants, fish eggs, larvae, and some adults were killed in great numbers, giving rise to a new term *fishkills* as the public became increasingly concerned. Fishkills, however, could be ameliorated by dissipating the waste heat into the air instead of the river by using cooling towers. Gargantuan natural draft cooling towers, however, could reduce aquatic impacts only at the expense of scenic quality and air quality. The U.S. Environmental Policy Administration, created by NEPA, sought to require construction of a 217 m (650 ft.) tall natural draft cooling tower at each of the six large power plants (Hudson River Fishermen 1974).

Because some power plants were coal fired, some oil fired, and some nuclear, they were regulated by different agencies and were involved in as many separate hearings. Eventually, all of these were consolidated into one hearing with four protagonists, the utilities, the federal agencies, the NYS Department of Environmental Conservation, and the citizen activist groups. Each party presented a mutually conflicting position and proposed resolution. When litigation failed to find a cure, parties entered into direct negotiations. (Three authors of the present volume participated in the settlement: Dey represented the utilities; Henshaw represented the state; and Butzel represented the citizens.) The resulting historic negotiated settlement required no gigantic cooling towers but rather protected fish populations by requiring that the power plants shut down operation during peak periods of fish spawning, as detailed by Butzel in chapter 19. Additionally, the settlement created the Hudson River Foundation (Hudson River Foundation 2009) with an initial endowment of $12 million (Sandler and Schoenbrod 1981, 273). Since 1981 HRF has supported more than $35 million of environmental research and educational projects (Suszkowski 2010). The published results of these studies benefit not only the Hudson River but also rivers everywhere. Thus, power generation introduced the modern era of sponsored research to support mitigation of impacts in the Hudson River Valley.

The consolidated power plant environmental impact case, in every respect, was a landmark battle with a landmark outcome. In the beginning, federal and state laws favored utilities proceeding to construct unfettered by regulatory oversight. Their applications mostly considered economic costs and paid scant attention to natural resources. Citizens had no legal standing to sue. All of these precedents were overturned. In the process, environmental law and environmental regulation effectively were born. The decision spurred creation of America's, and the world's, environmental movement. The settlement pioneered negotiation techniques and approaches that have since become commonplace. And the settlement firmly established the validity and value of science-based environmental management.

INSPIRING ACTION

The Hudson River has inspired a number of residents to create important prototypic environmental organizations. Frances S. Reese founded and led Scenic Hudson Preservation Conference. To many Hudson River preservationists she was known as the "grande dame of the Hudson Valley." Pete Seeger, nationally respected folk singer, was distressed by the debilitated condition of the Hudson River. In 1966 he conceived of recreating an eighteenth-century Hudson River freight Hauling sloop as a symbol to rally citizen interest. Launched in 1969, this 32 m (106 ft.) wooden boat took on the mission of preserving and protecting rivers through experiential environmental education. Seeger and *Clearwater* went to Washington, D.C., to lobby for passage of the Clean Water Act in 1972. Each day from April through October the sloop takes school groups and others for three-hour sails for hands-on experiences with all components of the aquatic ecosystem. Since 1969, about a half-million people have sailed on the

Clearwater and learned about sustainable uses of the Hudson River. This approach and mission have been emulated in many rivers around the world, building or adopting vessels and presenting *Clearwater* lookalike programs. Hudson River Sloop *Clearwater* created the first toxic discharge "pipe-watch" in the Hudson River Valley. Robert H. Boyle, a sportswriter, formed the Hudson River Fishermen's Association and joined Scenic Hudson in opposition to Storm King. Later joined by Robert F. Kennedy Jr., they renamed the organization Riverkeeper, and launched a small vessel that plies the river daily seeking out any illegal uses. Riverkeeper has spawned dozens of other Riverkeeper and Waterkeeper organizations throughout the world. Doug Reed created and for thirty years has led Hudson Basin River Watch, which sponsors monitoring and reporting of environmental conditions in the Hudson and its tributaries by sixty high school groups.

UNIQUE AMONG RIVERS

We can now see that the comparison of the Hudson to the Rhine River in Europe is not apt. The Rhine, like the Hudson River, has had a long history of increasing municipal and industrial impacts. But the responses in Europe were largely to *modify the Rhine itself* through damming, channeling, rip rapping, and rerouting so that it could better accommodate the uses of the river (Cioc 2002, 263). In the Hudson River, it was the other way around. With its unique history, superior scenic quality, and unusually spirited citizenry eventually empowered by law, as detailed in this volume, responses to impacts to the Hudson River mostly sought to *modify the impacts,* and to preserve the river. The roots of such reasoning surely were inspired by early and continuing appreciation of the unique qualities—"the sublime"—in the Hudson River region.

REFERENCES CITED

Alm, A. L. 1988. NEPA: Past, present, and future. (*EPA Journal,* Jan./Feb. 1988). http://www.epa.gov/history/topics/nepa/01.htm.

Almanac of Theodore Roosevelt—Forest Service Act. 1905. http://www.theodoreroosevelt.com/Forest Service1905.html; accessed Dec. 22, 2009.

Boyle, A. 2009. Thomas Cole: The dawn of the Hudson River School. http://hamiltonauctiongalleries.com/Cole.htm; accessed Dec. 22, 2009.

Cioc, M. 2002. The Rhine: An eco-biography. Seattle: University of Washington Press, 263.

Goodman, P. 1962. *Collected poems.* New York: Macmillan.

Hood, G. 1969. Thomas Cole's lost "Hagar." *American Art Journal* 1(2): 48.

Howat, J. K. 1972. *The Hudson River and its painters.* New York: Penguin, 29.

Hudson River Fishermen's Association v. Federal Power Commission. 1974. (2d. Cir. 1974), 498 F .2d 827.

Hudson River Foundation. 2009. About HRF. http://www.hudsonriver.org/about.htm.

Irving, W. 1809. Rip Van Winkle, from The sketch book of Geoffrey Crayon, Gent. In *Washington Irving: History, tales, and sketches,* ed. James W. Tuttleton, 774. New York: Library of America, 1983.

Library of Congress. 2009. The chronology of the conservation movement 1850–1920. http://rs6.loc.gov/ammem/amrvhtml/cnchron4.html; accessed Dec. 22, 2009.

Marist College Archives and Special Collections. 2006. The Cornwall pumped storage project collection 1953–1981. http://library.marist.edu/archives/cornwallCollection/cornwall.xml.

Sandler, R., and D. Schoenbrod. 1981. *The Hudson River power plant settlement.* New York: New York University School of Law, 31–48, 208–16.

Scenic Hudson Preservation Conference v. Federal Power Commission. 1965. (2d Cir. 1965) 354 F.2d 608.

———. 1971. (2d. Cir. 1971) 453 F.2d 463.

Seeger, P. 1984. The sloops of the Hudson. Presented at the 15th Anniversary of the Launching of the Hudson River Sloop Clearwater. Reprinted from *Clearwater Navigator.*

U.S. Forest Service. 2010. History: Gifford Pinchot (1865–1946). http://www.foresthistory.org/ASPNET/People/Pinchot/Pinchot.aspx; accessed Dec. 15, 2009).

CHAPTER 19

BIRTH OF THE ENVIRONMENTAL MOVEMENT IN THE HUDSON RIVER VALLEY

Albert K. Butzel

ABSTRACT

This chapter focuses on how the Storm King/Scenic Hudson lawsuit and its successful defense of the Hudson River Valley led to the birth of modern environmental law. It examines the background of the case, the legal strategy pursued by the newly formed citizen organizations Scenic Hudson and the Hudson River Fisherman's Association, and the central holdings of the courts that made this a landmark case. The chapter also references the Hudson River origins of the early environmental movement that culminated with Teddy Roosevelt's historic achievements—achievements that were perpetuated and expanded from 1965 onward, in part as a result of the Storm King decision.

BACKGROUND

In 1609, when Henry Hudson first sailed up the river that bears his name, the mark of man was barely to be seen. The Native American settlers lived largely in harmony with the land they occupied—a land filled with a diversity of flora and fauna, of sparkling streams and tracts of virgin woodland, of high hills rising starkly from the river and large expanses of marshy wetlands. "This is very good land to fall with and a pleasant land to see," wrote Robert Juet, Hudson's first mate on the *Half Moon*. The further narrative of the trip described a river rich in wildlife and overflowing with fish (Juet 1625).

These conditions did not last. Population followed once the Dutch established a foothold and grew even more quickly under the English and the Americans. The Hudson became a river of commerce supported by the towns along its shores and, with the opening of Erie Canal in 1830, the most important and heavily laden route to the West. With population and commerce also came degradation—small factories at first, then larger ones, municipal waste discharges, landfills to support the railroads and industry and widespread quarrying that despoiled some of the most beautiful stretches of the river. By 1880, the oyster beds of New York Harbor, once among the most productive in the world, were all but played out, and the river's sturgeon, which had once supported the export of caviar to Russia, were in sharp retreat, eventually to become endangered.

Little more could have been expected given the huge population center that became New York City and the extraordinary commerce in goods and wealth that the city generated. For many years, moreover, there was little concern for the damage being inflicted on the land and the river. The Hudson River School of Painting was, in many ways, a protest against encroaching industrialization, but the best that Cole, Church, and their colleagues could do was to excise the offending structures from the scenes they presented, substituting dead trees and other forms of natural camouflage in place of the belching smokestacks that actually rose in the viewsheds.

For all its grandeur and beauty, the Hudson suffered from being on the eastern seaboard. America was settled from east to west, and in the nineteenth century, when Americans began to recognize and seek to preserve some of its extraordinary natural treasures, the Hudson was already heavily settled with cities, towns, and villages. Industry was already widespread along its shores, which were themselves usurped by railroads, and the river was rapidly becoming a sink. The largely empty expanses that became Yellowstone and Yosemite and the Grand Canyon could be protected in virtually all their glory with little displacement of population or industry. The Hudson and its river valley, by contrast, were already home to millions of Americans, with the land and the river itself essential to the commerce that served those and many other millions. In 1966 hearings before the Federal Power Commission, Charles Callison, then executive vice president of the National Audubon Society, testified that if the Hudson and particularly its highlands were not already so marked by civilization, they would long ago have been dedicated as a National Park (Callison 1966). Timing did not allow that result.

Yet while the Hudson itself did not become a National Park, it was in the valley that the notion of conserving our greatest natural resources was, if not born, then nurtured and matured. This began toward the end of the nineteenth century when the phenomenon known as the conservation movement emerged. There was no such thing as environmentalism at the time, but much of the idea of saving the astonishing land forms of the west emanated from New York and the valley. It is no accident that Theodore Roosevelt, the greatest conservationist of all, was a New Yorker, who, while governor of New York, first staked his claim to that title by creating the Palisades Interstate Park in an effort to preserve the Hudson Highlands that were being despoiled by quarry operations (American National Biography 1999). Nor is it serendipity that J. P. Morgan, Colonel Rupert, the Rockefellers, the Harrimans, the Perkinses, and many other wealthy families built their summer homes in the highlands and supported the conservation movement with their money (Dunwell 1991, ch. 8). John Muir may have founded the movement, but it was the wealth of New York City, and the interest of its patricians, that fueled it.

Fast forward six decades to 1965—through two world wars and the Great Depression and the emergence of an increasingly affluent United States with increasing demands for housing and highways. The conservation movement had not died, but it was in eclipse; in the previous thirty years, only two new parks had been added to the National Park system and the suburbanization of America was gobbling up land without any sign of slowing and with little opposition.

However, there was an emerging malaise—a sense of concern over what was being lost. Nineteen sixty-five was three years after Rachel Carson published *Silent Spring*, which in many ways changed Americans' way of thinking about the environment. The concern over DDT had already resulted in one unsuccessful lawsuit. There were plans to dam the Colorado River in the Grand Canyon, and David Brower and the Sierra Club were already on that case (Brower 1990, 365–68). Increasing concern was being expressed about the new interstate highways being slashed through cherished landscapes, and more and more Americans found themselves impacted by the vast web of new transmission wires being woven across the country to meet the soaring demand for electricity. These concerns were also reflected in President Lyndon Johnson's emphasis on the "quality of life" with his Great Society programs, which included, as he declared in his 1965 State of the Union address, the preservation of "the beauty of America [that] has sustained our spirit and enlarged our vision." "We must act now," he continued "to protect this heritage," and make "the next decade a conservation milestone" (Johnson 1965).

THE STORM KING PROJECT

Now flash back to the Hudson. By 1965, the Palisades Interstate Park, which Teddy Roosevelt had helped create, had been considerably enlarged, protecting significant areas of forestlands and some of the more beautiful vistas in the Hudson Highlands, particularly on the west bank. But large gaps remained. Two of these were found at the dramatic northern gateway to the highlands—the lower slopes of Storm King Mountain on the western shore and a parallel and equally visible site at the base of Breakneck Ridge on the east. Both were now ticketed for development: the powerhouses of two large pumped storage hydroelectric projects were planned for the sites, and the land to support them was already owned by two major utilities.

The first and largest of the two was the Storm King Project. Proposed by the Consolidated Edison Company of New York, the project was to have a capacity of two million kilowatts and consist of three elements: a powerhouse located at the base of Storm King Mountain; a huge storage reservoir located one thousand feet above the river behind the mountain; and a two-mile-long tunnel some twenty feet in diameter connecting the two. During nighttime hours, when the company had excess capacity at its most efficient plants, it would use that cheap energy to pump water up from the Hudson into the storage reservoir. Then, during the day, when demand soared with air conditioning and worktime uses, the water would be released to course down the tunnel through the turbines (which, in reverse mode, had served as the pumps) and generate the electricity need to meet the rapidly increasing peak demand in the city. It would take three kilowatts to pump the water up to the reservoir, and only two kilowatts of that would be returned when the water was released. But the nighttime energy used for pumping was excess, while the electricity returned was both high value and essential to meet peak demand. In effect, what was being proposed was the creation of a huge storage battery at Storm King Mountain.

The concept was ingenious but the implications for the highlands were ominous. The plant would have carved into the mountain a powerhouse some eight hundred feet long and fifty feet high, with a bare wall rising another 150 feet above it. There would also have been eight large transformers on top of the powerhouse and a huge crane on tracks that was to be used to raise and lower screens intended to keep the fish out. When the project was announced by Con Edison early in 1963, the relatively thin conservation community had expressed shock. Later that year, when Con Edison published a rendering of what it planned to do (Fig. 19.1), the

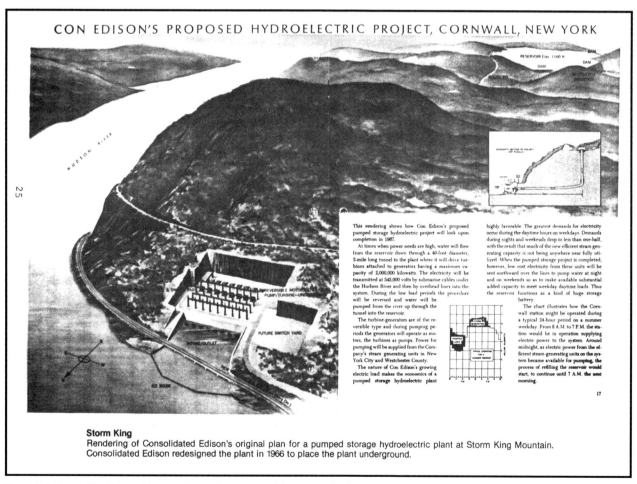

FIG. 19.1. Storm King Pumped Storage Power Plant proposed for construction in 1962 (photo from Con Edison Annual Report).

FIG. 19.2. Storm King Mountain viewed from Breakneck Ridge on the east side of the Hudson River.

shock turned to outrage. This illustration, later reproduced in *Popular Mechanics,* showed the side of the mountain cut away, leaving a gash the size of three football fields laid end to end, with a high cliff behind and electric accoutrement perched on the powerhouse roof (Con Edison 1963). It was this rendering, in particular, that roused lovers of the highlands to action.

It did not help—or at least it did not help Con Edison—that the utility selected one of the most beautiful and dramatic vistas on the Hudson (or any Eastern river) as the site for its pumped storage plant. The Hudson Highlands were, as *Life Magazine* described them in 1964, "one of the grandest passages of River scenery in the world." The great German traveler Baedeker had found the grandeur and beauty of the highlands as "finer and more inspiring than the Rhine," while *The New York Times,* editorializing against the project, described the area as "one of the most stunning regions in the Eastern United States" (*New York Times* 1963). And the northern portal to the gorge, with Storm King on the west and Breakneck Ridge on the east (Fig. 19.2) provided the most magnificent of all views.

SCENIC HUDSON

There was little precedent for citizen opposition in these circumstances, but a small group of individuals organized themselves as Scenic Hudson Preservation Conference and set out to defeat the project that threatened their beloved River (Talbot 1972, 91–96; Boyle 1979, 154–55; Dunwell 1991, 207–208). In many respects, they were heirs to the conservation movement, which itself had its origins along the Hudson. Their focus was the same landmass—the Hudson Highlands—that Teddy Roosevelt had sought to protect when the Palisades Interstate Park was established and that the magnates of industry and finance who had funded the movement had made their home. The Scenic Hudson group saw their opposition as a *conservation* battle—a struggle to protect a landscape they believed as worthy as Yosemite, much less some of the lesser National Parks. But they had no president in their corner to champion their cause, no men or women of great wealth to fund their campaign. They would need to use other tools; and it was those other tools that enabled the depleted conservation movement to evolve into a newly energized environmental movement.

Scenic Hudson's first approach was to Governor Nelson Rockefeller, whose response was that if they did not like the project, they should buy the mountain (Talbot 1972, 95). Short of the required funds and with Con Edison uninterested in selling, the founders of Scenic Hudson might well have given up; the general sense of public helplessness in the face of powerful industrial forces was rife and there was no tradition, and little sympathy, for citizens to stand up to "progress." But that is what they did, and in a unique way: they hired a public relations firm to get the story out. This worked: over a matter of months, the case had become a subject of nationwide attention. Then, learning that the project required a license from the Federal Power Commission, they hired a lawyer—Dale Doty, a former FPC commissioner—to represent them. This did not work as well. There were four days of hearings in 1964, in which the Scenic Hudson witnesses were ridiculed. They were described as lovers of dead trees and as effete birdwatchers who meant to stand in the way of the welfare of New Yorkers that only the Storm King project could protect. When the hearings were closed, the future looked grim (Talbot 1972, 96–106; Boyle 1979, 156–58; Dunwell 1991, 209–11).

But Scenic Hudson made the choice to fight on. The PR campaign was expanded. Among other things, a flotilla of several hundred boats sailed up the Hudson to Storm King to plant signs in response to Con Edison's then motto—"Dig We Must for a Growing New York." "Dig They Shall Not," the signs read, and the national media picked the story up. Stephen Currier, a philanthropist, took note of the story and decided he wanted to help. He was willing to give money if it could be used (Talbot 1972, 108–10; Boyle 1979, 156; Dunwell 1991, 212–13).

As to that, there was no doubt—Scenic Hudson intended to press on with the case. Alexander Lurkis, the recently retired chief engineer for the city's Department of Water Supply, Gas and Electricity, had written a letter to the editor of the *Times* identifying what he said was a superior alternative—a series of jet engine gas turbine generators that were new to the market. The PR firm reached out and hired him to develop the alternative in detail. Scenic Hudson then persuaded a state senator to call legislative hearings on the project in November 1964. Lurkis made a detailed presentation of the alternative, including cost comparisons showing his plan to be much less costly than the pumped storage plant (Talbot 1972, 111–12; Boyle 1979, 162; Dunwell, 1991, 213).

Also appearing was Bob Boyle, the Outdoors writer for *Sports Illustrated*, founder of the Hudson River Fishermen's Association and a Hudson River worshipper writ large. Boyle had discovered a report from ten years earlier that suggested the center of the spawning grounds for the recreationally—and commercially—important Hudson River striped bass was at Storm King Mountain. Since the project would suck in vast amounts of water—some eight million gallons a minute, equivalent to one-quarter of the total flow of the river—in which the eggs and fish larvae would be floating helplessly, the danger to the striped bass population was obvious. What made the disclosure all the more dramatic was that the study had been supervised by Con Edison's fisheries expert, who had testified in the FPC hearings that the plant posed no threat to aquatic or marine life (Boyle 1979, 158–62; Talbot 1972, 112–14).

Scenic Hudson promptly arranged to have the Lurkis and Boyle testimonies submitted to the FPC, with a request that the hearings be reopened to consider the new evidence. The FPC rejected the submissions as untimely. Then, in February 1965, the State Legislative Committee issued its report in which it found the Lurkis testimony compelling and the Boyle discovery disturbing. It urged the FPC to reopen the case, concluding that the scenic beauty of the area was unexcelled and recommending that the plant not be built if there were feasible alternatives, as appeared to be the case (New York State Legislature 1965; Talbot 1972, 114; Boyle 1979, 164–65).

All of this went unheeded. On March 9, 1965, the FPC granted the license application, finding, among other things, that the scenic beauty of Storm King Mountain would not be diminished by the plant but would actually be improved by the removal of a number of derelict structures. It also found that there was no feasible alternative and no danger to the fisheries (Federal Power Commission 1965).

THE COURT CASE

Two and a half weeks later, Stephen Currier, who had seen the flotilla article, agreed to finance a legal appeal on the condition that his attorney, Lloyd Garrison of the firm of Paul, Weiss, Rifkind, Wharton and Garrison, handle the case. The great-grandson of the abolitionist William Lloyd Garrison, a former dean of Wisconsin Law School, the first chairman of the National Labor Relations Board, and the erstwhile defender of such alleged subversives as Langston Hughes, Arthur Miller, and Robert Oppenheimer, Mr. Garrison was one of the country's preeminent attorneys, as well as a committed conservationist (*New York Times* 1991) But he was not an experienced litigator. So before he accepted the representation, he recruited his partner, Judge Simon Rifkind, one of the twentieth century's great trial and appellate lawyers, to work with him. Together, they came up with a unique strategy that, in the end, was to give birth to modern environmental law.

Their first coup was to give a new name to the proposed power plant. Con Edison had titled it the "Cornwall Project," after the village in which it was to be located, but that did nothing to identify the stakes involved. So Mr. Garrison and Judge Rifkind renamed it the Storm King Project. There was a certain irony in how central this seemingly minor adjustment became as the case progressed. Storm King was all very well. But this was a belated title. The early Dutch settlers had called the mountain Boterburg, which, translated into English, became—Butter Hill. Can anyone doubt that a battle over Butter Hill would have been considerably less impassioned—and very likely less successful—than a battle over Storm King? Happily, Nathaniel Willis, a romantic writer who lived in the highlands, felt Butter Hill was an indignity for such a grand geological feature. Remembering the clouds and lightning that raged around the mountain in the summer, he renamed it Storm King and in doing so significantly raised the stakes in the drama that was soon to play out (Talbot 1972, 9–10; Dunwell 1991, 63–64).

The case was a difficult one. Never in its history had a license granted by the FPC been annulled by a court on the grounds that the commission had misjudged the impacts of a project or improperly weighed the pros and cons. The law that governed the judicial review of expert agency decisions—the so-called substantial evidence rule—was that if there was evidence in the record to support the decision, no matter how much evidence there might be on the other side, the agency decision was to be upheld. The central issue that Scenic Hudson had brought before the FPC—and which had been the subject of many pages of testimony to both sides—was the preservation of natural beauty and historic areas. Under the substantial evidence rule, to try to attack the FPC decision because it had misjudged the impact of the plant on scenic beauty was a formula for defeat.

On the other hand, there were other areas of the record—particularly the consideration of alternatives and the potential impact on striped bass—where the FPC had had little meaningful evidence to rely on; and the commission had made matters worse for itself by refusing to consider the Lurkis report on the gas turbine alternative and Bob Boyle's testimony on the threat to striped bass. These became the focus of the briefs that Lloyd Garrison crafted. But he did not leave out scenic beauty. He used it as the background against which the FPC failure to look seriously at alternatives stood out. In these circumstances, where the project was to be located in an area of great and dramatic natural beauty and the raw cut into the side of Storm King Mountain could not be disguised, the commission should have bent over backward to find an option.

Instead, it closed its eyes to the Lurkis report and thereby blinded itself to a promising way to avoid the damage. The public interest deserved better (Garrison 1965).

Predictably, the FPC, in its response to Mr. Garrison's brief, took an opposite view with regard to alternatives and emphasized the commission's finding, fully supported by Con Edison's testimony, that the huge project would improve the scenery of the highlands. However, there must have been some uncertainty among the commissioners in taking this position. So as a first line of defense, the FPC argued that Scenic Hudson did not have the right to challenge the licensing decision. Neither it nor any of the groups allied with it had an economic interest in Storm King Mountain and thus in the outcome of the case and, so the commission asserted, under constitutional principles Scenic Hudson did not have "standing" to bring the lawsuit (Federal Power Commission 1965a). This contention, more than any other, once it was finally decided by the court, laid the groundwork for modern environmental law and the environmental movement that accompanied it.

The case was argued before the United States Court of Appeals for the Second Circuit in late October 1965. Mr. Garrison was eloquent and persuasive. Con Edison's counsel was less so. And the panel found it incredible when the attorney for the FPC argued that the project would improve the scenery of the highlands. The handwriting was on the wall; and two months later, on December 27, 1965, the Court of Appeals came down with its landmark Scenic Hudson decision setting aside the license for the Storm King plant.

THE 1965 SCENIC HUDSON DECISION

The court began by addressing the FPC position that Scenic Hudson had no right to bring a lawsuit challenging the commission's licensing order. The court reviewed the history and language of the Federal Power Act, noting that among the factors that had to be considered under the law was the impact of a project on recreation and, further, that in earlier decisions the FPC had itself defined recreation as encompassing the preservation of natural beauty. The court went on to observe that if only parties with an economic interest could bring a lawsuit, recreational interests would never be protected. As a result, the court concluded that economic injury was not necessary to maintain a lawsuit and that Scenic Hudson, because of its demonstrated interest in protecting the natural beauty of the Hudson and its highlands, had standing to bring the case. In effect, the court held that Scenic Hudson could sue on behalf of Storm King Mountain—that citizens could sue to protect the environment—even though they had no economic stake in the outcome (U.S. Court of Appeals for the Second Circuit 1965, 615–16). It was this holding that opened the courts to citizen suits on behalf of the environment.

But the Second Circuit Court of Appeals did not stop at that. In the balance of the opinion, it found that the FPC had improperly failed to consider the Lurkis report and the Boyle evidence that the striped bass fishery would be seriously threatened by the operation of the project. Indeed, the court held that in the circumstances of the case, where the impact on the Storm King and the Highlands could not be hidden, the commission was duty-bound to seek out potential alternatives on its own initiative, rather than simply acting like an umpire calling balls and strikes as the parties battled it out. Finally, the court concluded with the ringing mandate the ushered in a new era of environmental protections: "In its renewed proceedings, the Commission must take into account the preservation of natural beauty and national historic shrines as a basic concern, keeping in mind that in our affluent society, the cost of a project is only one of several factors to be considered" (U.S. Court of Appeals for the Second Circuit 1965, 624–25).

The impact of the Court of Appeals' decision cannot be understated. Among other things, it gave judicial sanction to the emerging environmental movement and provided it with the tool that, over time, allowed concerned citizens and organizations to defend cherished landscapes and other natural resources across the United States. The earlier conservation movement had relied largely on political lobbying, the leadership of Teddy Roosevelt, and the wealth of the magnates of industry who themselves saw value in the cause. But there had been no successful lawsuits—and very few at all—at the time that the National Parks came into being. The environmental movement, by contrast, was propelled, particularly at the outset, by court challenges. In time, politics and lobbying led to the creation of new laws—the Clean Air Act in 1970, the Clean Water Act of 1972, and others that followed—which became the primary basis for the movement's successes in protecting and cleaning up the environment. But litigation remained (and remains) an essential tool in enforcing environmental laws and achieving their stated goals. In a very real sense, all of this grew out of the Scenic Hudson decision (Houck 2002; 2010; Tarlock 2002; Bonine 2007).

Equally important, the decision led directly to the National Environmental Policy Act, which was signed into law on January 1, 1970. This was manifest in the dual requirements that federal agencies evaluate the impacts of actions they were proposing or asked to approve in environmental impact statements and the correlative obligation to identify reasonable alternatives to those actions. NEPA, in turn, has led to most states adopting similar laws governing proposed actions affecting the environment at the state and local levels, extending the reach of the decision far beyond the Federal Power Commission. And where did all this begin? On the Hudson River at Storm King Mountain.

THE NEXT ROUND: ENTER THE STRIPED BASS

The Storm King case did not, however, end with the Second Circuit's 1965 decision. The court did not conclude that the Storm King Plant could not be built. Its decision was limited to finding that the FPC had violated the law in granting the 1965 license. But it was free to reconsider the case, cure its errors and make a new licensing decision. In the words of the court, the case was "remanded"—or returned—to the commission for "renewed" proceedings.

These further proceedings continued before an FPC hearing officer for more than three years. They had begun with the Con Edison's announcement that it would revise its plans to place the powerhouse underground. In fact, the new plans still left plenty to be seen—a cut at the base of Storm King Mountain that was still the size of two football fields—but nonetheless, with the powerhouse itself relocated underground and the 150-foot sheer wall behind it reduced by two-thirds, the visual impacts were significantly reduced. In addition, the site on Breakneck Ridge that had been identified for a second pumped storage plant was bought up by the state as a part of the newly created Hudson Highlands State Park on the east bank of the river.

This did not dissuade Scenic Hudson, now joined by the Sierra Club and some forty other groups, from presenting testimony on the scenic impacts from a who's who of eminent conservation leaders, including David Brower, Charles Callison of the National Audubon Society, Anthony Wayne Smith of the National Parks Association, and Yale professor Vincent Scully, whose eloquent testimony was a lyrical poem centered on the mountain and the river. Still, the fact that the main structures of the plant would be out of sight reduced the passion that had characterized the initial set of hearings.

Over time, the impact of the project on fish, and particularly the striped bass, became the central issue. It was possible to relocate the powerhouse underground, but there was no way the project could function without drawing in and later discharging immense amounts of water—up to eight million gallons per minute—that was basic to the pumped storage concept. And with that water would come an immense number of eggs and larvae and very young fish that could fit through the plant's protective screens. Should these organisms be fortunate enough to survive their journey through the pumps, they would be subjected to a very rapid one thousand foot pressure change on the both the ascending and descending trips, something few, if any could be expected to survive. If Bob Boyle was correct that Storm King was at the center of the spawning grounds for the Hudson River striped bass, the consequences could be severe indeed. The striped bass were the preeminent commercial and game fish in the river and they contributed significantly to the North Atlantic stock that migrated from Maine to North Carolina before returning to their spawning grounds. Any threat to the viability of the stock was a serious concern not only to New York, but also to New Jersey, Rhode Island, Massachusetts, and other states along the Atlantic seaboard; and there were already signs that the Chesapeake Bay stock, which usually contributed the greatest percentage of stripers along the coast, was in decline.

Faced with this problem, in 1966 Con Edison sponsored a three-year study to help determine the potential impacts on striped bass. This was carried out during the extended second round of the FPC proceedings. However, the resulting report, entitled "Hudson River Fisheries Investigation 1965–68," was not issued until after the hearings ended and thus was not subject to review, much less cross-examination, by Scenic Hudson and, more importantly, the Hudson River Fishermen's Association, which had become a party to the proceedings in 1966. When the report was finally issued late in 1969, its conclusion was that the project would not have a significant impact on the striped bass population, resulting in only four percent of the river-wide eggs and larvae being exposed to intake into the plant (Boyle 1979, 294–95). By this time, the FPC was mulling its licensing decision, and it relied on the report, and adopted its findings, as the answer to the fisheries concerns. On this basis, and finding that (1) the scenic beauty of the highlands would not be marred by the largely underground powerhouse and (2) there were no reasonable alternatives, in September 1970 the FPC licensed the project for a second time (Federal Power Commission, 1970).

Scenic Hudson and seven other organizations, including New York City, which feared that its Catskill Aqueduct was threatened by the blasting for the underground powerhouse, appealed to the Second Circuit. This time, they lost in a two-to-one split decision. In its opinion, the Court of Appeals emphasized, in part, that the Con Edison–financed fisheries study amply supported the FPC's conclusion that the project would not have a significant impact on striped bass (U.S. Court of Appeals for the Second Circuit 1971).

At this juncture, it seemed that all was lost: Con Edison had its license and nothing seemed to prevent it from starting construction. That was not the case. A recent amendment of the Clean Water Act required the project to secure a state water quality permit from the Department of Environmental Conservation; and while this was granted after four days of hearings, the DEC decision was overturned by a lower court, only to be reinstated on appeal (State Cases 1972–73). But by the time the New York courts issued a definitive decision in 1973, information had come to light that largely undercut the conclusions of the Hudson River Fisheries Investigation Report. This information had been developed by the Hudson River Fishermen's Association in the course of licensing hearings on Con Edison's Indian Point 2 Nuclear Plant, and it was fundamental. The report had failed to take account of the tidal nature of the Hudson, treating it instead as a river that only flowed downstream rather than one reversing directions four times a day. As a result, the potential impact of the plant on the striped bass population had been grossly understated. Rather than 4 percent of the first-year striped bass being sucked into the plant, the corrected number was 40 percent—a much greater threat to the fisheries than had been previously assumed (Boyle 1979, 294–95).

The discovery resulted in yet another appeal to the Second Circuit Court of Appeals. In a 1974 decision, the court concluded that the FPC was legally obligated to reassess the impact of the project on the striped bass population (U.S. Court of Appeals for the Second Circuit 1974). In the meantime, Con Edison had begun to excavate the power tunnel at Storm King, but with the court's decision, it called a halt to the work. As it turned out, work never resumed. Over the next several years, the company, Scenic Hudson, and the Fishermen's Association, joined by the New York State Attorney General and the New York State Department of Environmental Conservation, battled over the potential damage the project would inflict on the fisheries, until Con Edison asked for more time to conduct more studies. The Court of Appeals agreed, but also enjoined any further construction unless and until it lifted the injunction.

This, however, did not end the battle over the striped bass. Under the Clean Water Act Amendments, power plants that drew their cooling water from a navigable river were required to use the best available technology to limit the impacts of those withdrawals on the environment. Applying this standard, the U.S. Environmental Protection Agency had ordered Con Edison and several other utilities to install closed cycle cooling at the Indian Point, Bowline, and Roseton plants, all of which used the Hudson for their cooling water. Because of the high costs that would be required to build and operate cooling towers, the utilities protested, and this required EPA to hold hearings to sustain its position. These hearings began in 1976. EPA took the lead in developing the case in support of cooling towers, and they were joined by NRDC, representing the Hudson River Fisherman's Association, and the New York State Attorney General, representing the NYS Department of Environmental Conservation and other state interests. Con Edison came forward with studies and models that the company claimed demonstrated the limited impact of the water withdrawals on the River fisheries (see Young and Dey 2010, ch. 18 in this volume). EPA assembled a team of experts that came to the opposite conclusion. The hearings ground on for more than two years—and twenty thousand pages of transcript—and the government had not yet begun to present its affirmative case. The issues became increasingly arcane, while the prospect of resolution appeared to be far in the future.

THE SETTLEMENT: FISH PROTECTION AND THE END OF THE PROJECT

It was in the context of the Court of Appeals' injunction and the ongoing and potentially endless EPA hearings that in 1979 the parties concluded it might be worthwhile to sit down and talk; perhaps there was a way the controversy could be resolved through negotiation. By this time, the involved parties included, in addition to Scenic Hudson, the Hudson River Fishermen's Association, and Con Edison, EPA, NRDC, which was representing the fishermen, the State Attorney General, the State DEC, and six other utilities. Moreover, by reason of its regulatory role and the potential cumulative effects of the power plant water withdrawals, EPA had become a major voice in the discussions, the state DEC was also a critical actor, and six other utilities, including the New York Power Authority, had a major stake in the outcome.

Russell Train, the first chairman of the Council on Environmental Quality and later EPA Administrator, had been recruited to act as an independent mediator and conciliator. Over the year and a half that the settlement negations continued, he had to be both. It was a given from the outset that there would be no resolution unless Con Edison abandoned its plans for the Storm King plant, and the signal that it was willing to do so came early on.

Several other issues, including the Hudson River Fishermen's demand that the utilities fund a scientific foundation to develop independent information on the river environment and man's impact on it, also seemed capable of resolution. The sticking point was EPA's insistence that each of the major base load plants, including the nuclear plants at Indian Point, be equipped with cooling towers. While none of these plants, individually, was as voracious as the pumped storage project would have been, their cumulative impacts in terms of killing striped bass eggs and larvae was equal, if not greater, than the Storm King installation. But Con Edison and the New York Power Authority, the utilities that owned the units, were equally adamant that there could no resolution if there was *any* requirement for cooling towers (Talbot 1983, 1–24; Sandler and Schoenbrod 1981).

For some months, this standoff threatened to sink the entire negotiations. In time, however, the discussions began to focus on how, if at all, the operations of the base load plants might be limited during the critical spawning and larval season as means of mitigating the impact of those plants. The EPA representatives took the position that operational limitations could never be enough and, even if there was some merit in the approach, the offer the utilities had come up with was short of the critical number. The tacit goal was to reduce the intake of water (and thus eggs and larvae) by 50 percent from late March through May, but the utilities saw no way that they could achieve the number without significantly compromising system reliability. Finally, the parties recognized that if there was to be a solution, it would depend on being able to evaluate more accurately the benefits of operational limitations and that, in turn, depended on complex issues that required input from the technical experts. This led to the creation of, and delegation of the problem to, an eight-member technical committee, consisting of representatives from the state and federal agencies, the utilities, and the public intervenors.

The initial discussions in the technical committee focused on reconciling the competing mathematical models in an effort to reach agreement on the level of impact the power plants were having on the striped bass fishery. This was a very complex task, but in time, the biologists for all parties were able to agree on parameters that brought the outcomes of the models close together. However, while this established a common basis for assessing impacts, it did not provide an answer to mitigating those impacts.

The breakthrough came when the suggestion was made to use the common model to anticipate the impacts of a series of mitigation scenarios that would involve closing down the various base load plants at different times during the critical spawning and early growth seasons. Depending on the time when a particular plant was shut down, it would result in greater or lesser degrees of mitigation, because there would be greater or lesser numbers of eggs, larvae, and early juveniles drawn into it. Suddenly, the technical committee had a basis upon which they could work with different plant closing scenarios and determine their effect, namely, by giving "credit" for fish *not* killed due to the suspended operation of any plant at a particular time or for a particular period. At this point, the biologists representing all parties effectively coalesced into one team seeking outcomes that would come as close as possible to the 50 percent mitigation goal.

But there was yet another problem—due to their obligation to provide uninterrupted electric service, the utilities could not guarantee that any particular plant could be shut down at a particular time. This constraint was overcome when the committee agreed that all the plants could be considered together, with the evaluation made on the basis of their *combined* impact to the fishery. These two concepts—credits and sharing—were the keys to unlocking the impasse and permitted the negotiation of a final settlement.

In the end, the resolution was to require that in the aggregate the plants be shut down sufficiently during the critical season to reach the overall mitigation goal. Thus, for example, if the largest of the predators, Indian Point, were closed down for six weeks during the season (as it could be for scheduled maintenance), the resulting mitigation value might totally fulfill the combined utilities' obligations for the year, even though the units at Roseton and Bowline continued to operate throughout the period, and these credits could be successive rather than contemporaneous six weeks. In addition, the idea of using excess credits from one year to offset a failure to reach the credit goal in a later year added additional flexibility to the operational scenarios. Eventually, after several meetings and continuous consultation with the expert EPA, AEC, DEC, and utility modelers, the technical committee arrived at a scenario that was as far as the utilities were willing to go. The result approached, but did not reach, a 50 percent reduction in impact.

The technical committee brought the results back to the full negotiating group, where, in time, they were accepted by all parties, including EPA, which grudgingly gave up its demand that the Indian Point units be equipped with natural draft cooling towers. Once that happened, the negotiations became a matter of filling in the details.

Finally, in December 1980, the parties, under the wary eye of Russell Train, signed the settlement agreement, which, the next day in *The New York Times*, was hailed as a peace treaty for the Hudson (Talbot 1983; Sandler and Schoenbrod 1981). Thus, the saga of Storm King came to an end.

CONCLUSION

The Storm King battle was truly a landmark. Before the Court of Appeals ruled that the Federal Power Commission needed to reconsider its licensing of the pumped storage project taking into account the impacts on scenic beauty and Hudson River fisheries population, the law had favored industry and accepted what was regarded as the necessity of material progress. The court's decision changed that balance, injecting environmental values into the judicial review process, paralleling the waking environmental consciousness within the country. At the same time, the court's decision for the first time allowed citizens to sue on behalf of those values—a ruling that, while cut back in recent years, continues to be the base upon which citizens are able to bring their grievances before the courts. In addition, the decision laid the groundwork for the National Environmental Policy Act, which incorporated the court's focus on environmental impacts and the urgency of considering alternatives. By the time the case ended in 1980—fifteen years after the Court of Appeals had issued its opinion—the judicial landscape had undergone a sea change, with hundreds of cases brought to court in efforts to protect our landscapes, our air and water, and our biota.

Of equal significance, the settlement process that brought the battle to a conclusion not only recognized the importance of protecting important wild resources, but also pioneered negotiation techniques and approaches that have since been repeated many times. In addition, the settlement established the validity and value of science-based environmental management. And finally, it demonstrated that while utilities, regulators, and citizens may come at issues from very different orientations and with very different biases, they have the capacity to collaborate in allowing necessary development to proceed while protecting natural resources to the greatest extent possible.

REFERENCES CITED

American National Biography. 1999. Volume 18, 829–35 (Teddy Roosevelt) New York: Oxford University Press.

Bonine, J. 2007. Private public interest environmental law: History, hard work and hope. Thirteenth Annual Lloyd K. Garrison Lecture on Environmental Law, found in 26 *Pace Environmental Law Review*, 465.

Boyle, R. 1979. *The Hudson River: A natural and unnatural history*. New York: W. W. Norton.

Brower, D. 1990. *For Earth's sake: The life and times of David Brower*. Salt Lake City: Peregrine Smith.

Callison, C. 1966. Testimony before the Federal Power Commission on Project No. 2338 (Consolidated Edison Company of New York, Inc. Cornwall Project).

Consolidated Edison Company of New York, Inc. 1963. Annual report for 1962.

Dunwell, F. 1991. *The Hudson River Highlands*. New York: Columbia University Press.

Federal Power Commission. 1965. Opinion No. 452, Project No. 2338. Opinion and order issuing license and reopening and remanding proceeding for additional evidence on the location of the primary lines and design of fish protective facilities (March 9).

———. 1965a. Brief on behalf of Federal Power Commission to U.S. Court of Appeals for the Second Circuit. *Scenic Hudson Preservation*

Conference v. Federal Power Commission, Docket No. 29853.

Garrison, L. K., et al. 1965. Brief on behalf of Scenic Hudson Preservation Conference to U.S. Court of Appeals for the Second Circuit. *Scenic Hudson Preservation Conference v. Federal Power Commission,* Docket No. 29853.

Houck, O. 2001. Environmental law and the general welfare. Fourth Annual Lloyd K. Garrison Lecture on Environmental Law, found in 19 *Pace Environmental Law Review,* 675.

———. 2010. *Taking back Eden: Eight environmental cases that changed the world.* Washington, DC: Island Press.

Johnson, President L. B. 1965. State of the Union Address, January 4, 1965. http://www.presidency.ucsb.edu/ws/index.php?pid=26907.

Juet, R. 1625. Journal of Hudson's 1609 voyage, from 1625 Edition of *PurchasHis Pilgrimes.* http://www.halfmoon.mus.ny.us/Juets-modified.pdf.

Life Magazine. 1964. Editorial, July 31.

New York State Legislature. 1965. Preliminary report of the Joint Legislative Committee on Natural Resources on the Hudson River Valley and the Consolidated Edison Company Storm King Mountain project, R. Watson Pomeroy, Chair, Feb. 16.

New York Times. 1963. Editorial, May 22.

———. 1991. Obituary of Lloyd K. Garrison, October 3.

Sandler, R., and D. Schoenbrod. 1981. *The Hudson River power plant settlement.* New York: New York University School of Law.

State Cases. 1972–73. *Matter of de Rham v. Diamond,* 69 Misc.2d 1 (Sup. Ct. Albany Co., 1972), *reversed* 39 A.D.2d 302 (3d Dept 1972), *affirmed* 32 N.Y.2d 34 (1973).

Talbot, A. 1972. *Power along the Hudson: The Storm King case and the birth of environmentalism.* New York: Dutton.

———. 1983. *Settling things: Six case studies in environmental mediation.* Washington, DC: The Conservation Foundation.

Tarlock, A. D. 2002. The future of environmental rule of law litigation. Sixth Annual Lloyd K. Garrison Lecture on Environmental Law, found in 19 *Pace Environmental Law Review,* 575.

United States Court of Appeals for the Second Circuit. 1965. Opinion in *Scenic Hudson Preservation Conference v. Federal Power Commission*, 354 F.2d 608 (1965).

———. 1971. Opinion in *Scenic Hudson Preservation Conference v. Federal Power Commission*, 453 F.2d 463 (1971).

———. 1974. Opinion in *Hudson River Fisherman's Association v. Federal Power Commission*, 498 F.2d 827 (1974).

Young, J., and W. Dey. 2011. Out of the fray: Scientific legacy of environmental regulation of electric generating stations in the Hundson River Valley. In *Environmental history of the Hudson River,* ed. Robert E. Henshaw. Albany: State University of New York Press.

CHAPTER 20

THE INFLUENCE OF THE HUDSON RIVER SCHOOL OF ART IN THE PRESERVATION OF THE RIVER, ITS NATURAL AND CULTURAL LANDSCAPE, AND THE EVOLUTION OF ENVIRONMENTAL LAW

Harvey K. Flad

ABSTRACT

The seeds of American conservation were sown by many artists of the Hudson River School during the nineteenth century. Progress toward scenic and landscape preservation began in the nineteenth century. In this chapter I consider several twentieth and twenty-first century examples where Hudson River School paintings of the natural and cultural landscape of the river and the valley established a landscape aesthetic that aided local citizens in their efforts to preserve the "Landscape that Defines America." All of these efforts integrated a concern for scenic views and a sense of place. Views of the natural and cultural landscapes of the Hudson River Valley region forged a national identity, and, along with other forms of cultural discourse, offered a humanistic framework to the development of national environmental law and policy.

INTRODUCTION

The Hudson River regional landscape has been nationally recognized for its beauty and historical significance since the early nineteenth century. Its natural environment—rivers, waterfalls, forests, mountains, and geology—offered elements of a cultural landscape where the presentation of a philosophical ideology of nature through the arts became instrumental in forming a national identity. The region's role in formulating American culture became manifest by the close of the twentieth century when the Hudson was declared one of the first National Heritage Rivers and in 1996 the valley a National Heritage Area; subsequently, the U.S. Congress declared it to be the "Landscape that Defines America" (National Park Service 1996, 32; U.S. Congress 2009). As described in the management plan for the National Heritage area, it is the "landscape that defined America . . . an exceptionally scenic landscape that has provided the setting and inspiration for new currents of American thought, art, and history . . . [and] the fountainhead of a truly American identity" (Hudson River Valley National Heritage Area 2000, 4).

Over the course of the last two centuries, paintings of the Hudson River School of artists developed a national interest in the natural landscape that would be crucial in the development of a conservation ethic. In this chapter a short introduction to the art and artists of the Hudson River School introduces the argument that changing attitudes toward nature and progress in the nineteenth century, generated initially from sketching and painting landscapes of the Hudson River and the Catskill Mountains, created both regional and national policies, projects, and legal frameworks that emerged to strengthen conservation efforts in the twentieth century. Nineteenth-century landscape art, artists, and architects offered crucial support to the creation of urban, state, and national parks and scenic and wilderness preservation, and led to twentieth and twenty-first century efforts to preserve places significant to American cultural history and identity and the evolution of environmental law.

In several twentieth and twenty-first century examples, paintings of the natural and cultural landscape of the Hudson River Valley became important images used by local citizens in their preservation efforts. All of these efforts integrated a concern for scenic views and a sense of place. Preserving the Palisades led to the creation of Palisades Interstate Park, the first interstate open space compact in America. Protecting Storm King Mountain from a proposed hydroelectric power plant led to the National Environmental Policy Act (NEPA) under which all subsequent federal environmental legislation would be framed. Defeating the proposed Greene County Nuclear Power Plant established Visual Impact Analysis as a requirement in New York's State Environmental Quality Review Act (SEQRA). Deterring the construction of the St. Lawrence Cement Plant underlined the importance of aesthetics in the construction and definition of "community character" in environmental law.

NATURE, ART, AND THE ORIGINS OF A NATIONAL CULTURE

Thomas Cole (1801–1848), considered the founder of the Hudson River School of Art, was also an essayist and an early advocate of conservation. In his "Essay on American Scenery" (Cole 1836) he not only presented the natural elements that are essential for a landscape painting (such as forests, waterfalls, and sky) and the emotions they symbolized, he also argued that the dialectic between the American wilderness and the cultivated landscape was being challenged by "improvements" such that "the sublimity of the wilderness should pass away" (1836, 5). He lamented that the "ravages of the axe are daily increasing—the most noble scenes are made desolate, and oftentimes with a wantonness and barbarism scarcely credible in a civilized nation" (1836, 12). Furthermore, he offered one of the earliest arguments for conservation, where "spots, now rife with beauty" should be preserved for future generations: "The way-side is becoming shadeless, and another generation will behold spots, now rife with beauty, desecrated by what is called improvement; which, as yet, generally destroys Nature's beauty without substituting that of Art" (1836, 12; see also Schuyler 2005, 32–33).

He was greatly concerned, although somewhat ambivalent (Robinson 1993, 81). Of his walk on July 31, 1836, Cole wrote, "I took a walk, last evening up the valley of the Catskill, where they are now constructing the railroad. This was once my favourite walk; but now the charm of solitude and quietness is gone. It is, however, still lovely: man cannot remove its craggy hills, nor well destroy its rock-rooted trees: the rapid stream will also have its course" (Noble 1856, 221; see also Stilgoe 1993, 17; Bruegel 2002, 87).

Although he questioned the "improvements" that were changing the landscape, he was not without hope. As he looked toward the future, he declared, "We are still in Eden; the wall that shuts us out of the garden is our own ignorance and folly" (1836, 12). Cole admonished the emerging middle classes to develop a conservation ethic: "It would be well to cultivate the oasis that yet remains to us, and thus preserve the germs of a future and a purer system" (1836, 3). The landscape he sought to preserve was the wilderness that was fast disappearing beneath the axe, since "the most distinctive, and perhaps the most impressive, characteristic of American scenery is its wildness" (1836, 5).

Nineteenth-century writers similarly extolled Nature's virtues. In the essay "Nature," published the same year as Cole's "Essay on American Scenery," Ralph Waldo Emerson declared, "In the woods we return to reason and faith" ([1836] 1985, 39), while his student Henry David Thoreau later exclaimed, "In Wildness is the preservation of the world" (1862, 665). Thoreau "visited museums and galleries in New York and Boston in which paintings by Cole and Durand were exhibited . . . was influenced by the same landscape aesthetics that affected them, and he conceived of himself as a landscape painter" (Smithson 2000, 93). By the end of the century, efforts were made to preserve significant aspects of the national natural and cultural landscape (Nash 2001), and it would be to paintings by members of the Hudson River School that environmental conservationists and historic preservationists would turn.

PLACES: HUDSON RIVER VALLEY AND CATSKILL MOUNTAINS

Cole's early wilderness paintings, such as *Kaaterskill Falls* (1826) and *Sunny Morning on the Hudson River* (1827), inspired the artists that came after him as well as public figures, poets, and essayists in early conservation efforts. The poet and journalist William Cullen Bryant (1794–1878), for example, published poems on the spiritual power of Nature and wrote newspaper editorials to bring nature into New York City through the construction of Central Park (Bryant 1894; Schuyler 1986). The artist Asher B. Durand (1796–1886) memorialized both Cole and Bryant upon Cole's death in 1848 in his 1849 painting *Kindred Spirits* (Foshay and Novak 2000; Ferber 2007).

Yet, as the nineteenth century progressed, the natural landscape continued to be domesticated; the wildness of the land, as depicted in the paintings of a vanishing wilderness, was being lost. In the Catskills the forest was under siege. "Across the hillsides travelers saw great barren tracts where once hemlocks had stood," harvested for their bark to tan "animal hides into leather for boots and shoes, belts and gloves, and saddles and harnesses. . . . Tanning wrought havoc upon the landscape" (Lewis 2005, 216).

Railroad lines penetrated the forests and traversed the valleys. The "machine" had entered the "garden," to the concern of artists (Marx 1964); some attempted to accommodate the railroad and industrialization while others removed the images from their compositions. Cole offered two views of his "beautiful valley" in the Catskills, a bucolic pastoral scene *View of the Catskills—Early Autumn* (1837), and the same scene six years later after the coming of the railroad *River in the Catskills* (1843), with a small locomotive and train nestled in the middle distance, while a foreground "with its tangled array of logs and stumps, a grove reduced to 'fragments,' has the appearance of a tree massacre, and, in accord with the conventions of the prospective view, the man with the axe . . . looks to the future" (Wallach 2002, 344). Four years later Cole wrote of the railroad as a "Machine which is merciless & tyrannical" (cited in Wallach 2002, 334).

Cole and other artists remained ambivalent about "progress" and industrialization; for example, the railroad train in *The Lackawanna Valley* (c. 1856) by George Inness (1825–1894) may seem to be harmoniously integrated into the landscape to please his railroad company patron, although the many tree stumps in the foreground and the dreary landscape of Scranton's factories, chimney smoke, and straight dirt streets offers an ambiguous perspective on the scene (National Gallery of Art 2010). Nevertheless, artists would occasionally seek out the picturesque in the emerging industrial landscape (Maddox 1983).

Many of the artists of the Hudson River School, such as Durand, focused on the details of the natural landscape in their quest for a truthful telling of place, even while often exclaiming of its beauty in Romantic prose (Ferber 2009). For example, in Durand's *Where the Streamlet Sings in Rural Joy* (ca. 1850) the rocks, trees, and lichen are all drawn with felicitous concern for detail in a painting with a title that resonates with Bryant's poetry (Fig. 20.1). Other artistic works, such as *Woodland Interior, Shawangunk Mountains* (1850) by Sanford R. Gifford (1823–1880), examine the geology (the blocks of durable conglomerate), lichens, and wild forest that constitute the ridge's ecology.

Artists followed in Cole's wake to search out Nature in the mountains, glens, and forests of the Hudson Valley region (Novak 1969; Miller 1993; Novak 1995). In their search for "wilderness," however, they stayed at rather elegant establishments, such as the Catskill Mountain House, built in 1824 at the Pine Orchard on a prospect overlooking the Hudson River Valley (Van Zandt 1966; O'Toole 2005, 124). Many other boarding houses and mountain houses were built in the Catskills to accommodate the growing number of tourists and artists (Myers 1988). Mohonk Mountain House began operations in 1869 nearby on the Shawangunk ridge (Partington [1911] 1970; Burgess 1980) where panoramic views of the surrounding mountains and valleys were juxtaposed with more intimate scenes of Lake Mohonk, craggy cliffs, and talus slopes. A number of Hudson River School artists, such as Gifford, Worthington Whittredge (1820–1910) and Jervis McEntee (1828–1890) painted in the 'Gunks (Wilton and Barringer 2002, 163; Avery and Kelly 2003, 163–66). The landscapes surrounding the mountain houses were constructed with carriage roads and specific places for

FIG. 20.1. Asher B. Durand, *Where the Streamlet Sings in Rural Joy*, ca. 1850; 24 1/8" x 18 1/4", oil on canvas, courtesy of The Frances Lehman Loeb Art Center, Vassar College, Poughkeepsie, New York, Gift of Matthew Vassar, 1864.

scenic views to encourage visitors to engage the natural environment (Flad 2009): to "walk in the woods" as Emerson advised, and as McEntee and others sketched.

NINETEENTH-CENTURY CONSERVATION EFFORTS

Although most Hudson River School artists lived and sketched in the region (Phillips and Weintraub, 1988), they also ventured farther afield to seek sublime and picturesque landscapes. They traveled throughout the northeast to the White Mountains of New Hampshire and the Adirondacks in the upper reaches of the Hudson River, west to the Rocky Mountains, north toward the arctic, south to the Andes, and most went to the British Isles, Italy, and the rest of Europe. For example, in his search for nature's sublime, Frederic E. Church (1826–1900), Cole's only student, traveled to South America in 1853 on the route of the 1803–05 explorations of the great naturalist and geographer Alexander von Humboldt (Bunkse 1981), to "confirm Humboldt's sense of 'what an inexhaustible treasure remains still unopened by the landscape painter between the tropics'" (Sachs 2006, 96). Church's South American paintings of tropical vegetation and volcanic eruptions were shown throughout the east. Thoreau saw *The Andes of Ecuador* (1855) on July 4, 1855, at the Athenaeum gallery in Boston (Walls 1995, 125). Seeking the sublime, Church's paintings grew larger, and following success in painting Niagara Falls (*Niagara* [1857]), Church enjoyed further success with the composition *Heart of the Andes* (1859). Later, Church traveled north to the arctic to paint the northern light and seascape (*The Icebergs* [1861]).

Places: The American West

Artists traveled across the prairies to the western mountains and returned with sketches of the relatively unexplored landscapes, often alongside federal expeditions; for example, Whittredge traveled on an "inspection" tour in 1866 (Janson 1989, 111). His paintings of Indians on the plains camped on the Platte River became "a simile of the red man's disappearance" (Janson 1989, 115), much as the poetry of Bryant and the wilderness paintings of Cole and others decried the loss of the natural landscape and its wildness. Four years later, he returned to the west with fellow artists John F. Kensett (1816–1872) and Gifford, by the recently constructed transcontinental railroad. In Denver, Gifford joined the 1870 survey party to the Wyoming Territory of F. V. Hayden, director of the U.S. Geological Survey (Anderson 1991; Avery and Kelly 2003, 196–98), while Kensett and Whittredge toured the Colorado Rockies.

Albert Bierstadt (1830–1902) also sought the sublime in the western United States. His paintings, such as *The Rocky Mountains-Landers Peak* (1863) and *The Domes of the Yosemite* (1867), introduced the majestic western landscapes to the eastern public (Anderson and Ferber 1990). Yosemite Valley had already become a California state park in 1864, by an act signed by President Lincoln, through the efforts of Frederick Law Olmsted (1822–1903), chairman of the first Yosemite Commission (Ranney 1995, vii). Olmsted, as a founder of landscape architecture in America and co-designer of Central Park in New York City in 1855–57 with Calvert Vaux (1824–1895), was instrumental in creating a language to establish public parks throughout America. Bierstadt's paintings were purchased by patrons with railroad interests and promoted tourism, which began to leave its mark on the landscape. Bierstadt, like Cole and others, erased such changes; these more sublime landscape paintings would help persuade the federal government to create Yosemite as a national park in 1890.

Meanwhile, paintings by Thomas Moran (1837–1926), such as *Grand Canyon of the Yellowstone* (1872), purchased by the U.S. Congress (Wilton and Barringer 2002, 250), were instrumental in Yellowstone's becoming the world's first national park in 1872. As historian Stephen J. Pyne noted about Moran's role, "For his catalytic work at Yellowstone, the National Park Service proclaimed him the 'father of the park system.' . . . What Moran had fixed in paint was now fixed in law" (Pyne 1999, 93).

Places: New York State—Niagara Falls

Even before the rise of the Hudson River School of Art, artists and tourists traveled up the Hudson and west along the Mohawk to Niagara Falls; for example, Thomas Davies's *Niagara Falls (from above)* (c. 1766) and Louisa Davis Minot's *Niagara Falls* (1818). Cole visited the falls in 1847 and many of his followers made their early reputations with views of Niagara, such as Church in 1857. The falls and its immediate area became so overrun with tourists, hotels, and facilities for their amusement, as well as projects to harness the energy for industrial purposes, that New York State established a park to preserve its scenic views (Dow 1914). In an 1879 letter supporting preservation of the falls and surrounding landscape, Olmsted declared, "My attention was first called for the rapidly approaching ruin of its characteristic scenery by Mr. F. E. Church, about ten years ago" (cited in Dow 1914, 11). Olmsted and Vaux were engaged to develop the plans for the Niagara Reservation (Irwin 1996, 72; Kowsky 1998, 303), and through efforts in 1883–86 it "was the beginning of scenic preservation by the State" (American Scenic and Historic Preservation Society 1916, 46).

Places: New York State—Adirondacks

Aspects of the sublime and picturesque were also sought in the mountains, forests, and lakes of the Adirondacks. Cole sketched at Schroon Lake in 1837, McEntee sketched around Raquette Lake in 1851, and Kensett often painted Lake George starting in 1853 (Mandel 1990, 44–45, 78–79, 88). Gifford considered his 1864 *A Twilight in the Adirondacks* to be one of his "chief pictures" and chose it to exhibit in the United States Centennial Exhibition in 1876 (Mandel 1990, 58–60; Avery and Kelly 2003, 167–69). However, the scenic views, forests, and watersheds were "fast disappearing beneath the axe" by lumbering and forest fires in the 1880s and 1890s.

Early conservation efforts in New York State included state legislation to create the Adirondack Forest Preserve in 1885, Adirondack Park in 1892, and a state constitutional provision in 1894 as Article VII, Section 7 to guarantee that the publicly owned lands of the Forest Preserve would "be forever kept as wild forest lands" (Graham Jr. 1978, 131; Terrie 1997, 102). Artists' "images of a ravaged landscape," such as Julian Rix's engravings of *A Feeder of the Hudson—As It Was* and *As It Is,* published in *Harper's Weekly* in 1885, "promoted public sentiment for protecting the Adirondacks" (Terrie 1997, 95). Article VII, Section 7, protecting the Adirondack "forest lands" as "forever wild," remained in the 1938 state constitution as Article XIV, Section 1 (Robinson 2007), and would be continually defended throughout the twentieth century (McMartin 2002).

New York State's unique legislative protection of the Adirondacks influenced the writing of the 1964 Wilderness Act and federal enactment of the Wilderness Preservation System. Howard Zahniser, executive director of the Wilderness Society and author of the Wilderness Act, had been introduced to the Adirondacks by conservationist Paul Schaefer in 1946 (Schaefer 1989; Zahniser 1992; Harvey 2005). New York State's wilderness landscape, sought after by Hudson River School artists, nineteenth-century tourists, and twentieth-century outdoors enthusiasts, became the foundation for the legal protection of wild lands and the nation's natural heritage throughout America (Scott 2004).

Places: Palisades

To paint and hike in the Catskills and Adirondacks, study the New World's natural ecology and geology, or visit Niagara Falls and other scenic and historic sites, artists, scientists, and tourists traveled by steamboat up the Hudson River, past the Palisades and through the Hudson Highlands. In the early nineteenth century, in the same decades that Cole wrote his essay on scenery and the art critic John Ruskin advised artists to study the emerging sciences of geology and botany, Sir Charles Lyell, one of the founders of geology as a natural science, on a tour of the New World in 1841, accurately described the river as "an arm of the sea or estuary," and the Palisades as "a lofty precipice of columnar basalt" (Lyell 1845, 12).

Lyell was similarly impressed with the Palisades sill as scenery as he described the rocky cliffs as "extremely picturesque," using the Romantic rhetoric of the era (Flad 2002, 46). Hudson River School

FIG. 20.2. Charles Herbert Moore, *The Upper Palisades*, 1860; 12" x 20 1/8", oil on canvas, courtesy of The Frances Lehman Loeb Art Center, Vassar College, Poughkeepsie, New York, Gift of Matthew Vassar, 1864.

painters agreed and often painted views of the Palisades from the east bank or the river itself. For example, in *The Upper Palisades* painted by Charles Herbert Moore (1840–1930) in 1860 (Fig. 20.2), the Palisades seem to control the background with geologic majesty, while the foreground and midground reveal a picturesque scene of a sailboat and rowboat upon a placid river.

By the end of the nineteenth century the Palisades were being destroyed by rampant quarrying of the diabase igneous sill for "trap rock" (O'Brien 1981, 241–43; Roseberry 1982, 249–56; Binnewies 2001, 1–4). Trap rock was used for paving stones and building foundations in New York City and crushed for railroad track ballast or to make macadam for paving roads. The thirty-mile-long cliff face, ranging to a height of 550 feet above the Hudson River, had long been viewed as an imposing sight, from early descriptions by Robert Juet in 1609 and Giovanni da Verrazano in 1524 to paintings by Hudson River School artists such as Gifford's *Sunset on the Hudson* (1876) and Jasper Cropsey (1823–1900), who painted many views of the Palisades from his home and studio in Hastings-on-Hudson, including *Winter on the Hudson* (1887) and *Sunset on the Palisades, Hastings* (1890) (Brennecke 1987, 126–31). For estate owners on the east bank, travelers on steamboats or on the New York Central Railroad, the formerly inspiring view was being erased.

The American Scenic and Historic Preservation Society was chartered by the state of New York in 1895. Among its concerns for preserving the natural wonders and historical landmarks in America was the destruction of the Palisades (American Scenic and Historic Preservation Society 1906, 91–212). With the additional effort of the New Jersey State Federation of Women's Clubs and the political force of Theodore Roosevelt as governor of New York, the Palisades Interstate Park Commission was formed in 1900. Roosevelt "considered the Palisades Park between New York and New Jersey a landscape masterpiece" (Brinkley 2009, 352). Its success spurred him as president to formulate his "grand preservationist accomplishment—the Antiquities Act of 1906" (Brinkley 2009, 414).

At the official opening of the park in 1909 during the Hudson-Fulton Tri-centennial Celebration, New York State governor Charles Evans Hughes elaborated on the preservation efforts: "The preservation of the scenery of the Hudson is the highest duty with respect to this river imposed upon those who are the trustees of its manifold benefits. It is fortunate that means have already been taken to protect this escarpment, which is one of its finest features. The two States have joined in measures for this purpose." And, in a visionary comment toward the end of the century he added, "The entire watershed which lies to the north should be conserved, and a policy should be instituted for such joint control as would secure adequate protection" (Johnson 1910, 638).

The Palisades and the lands on the ridge were consolidated as a park through purchase and donation by property owners on both the east and west banks of the river. The numerous land and park gifts of wealthy river families, many of whom had large art collections, "left a mark on the conservation movement in the valley . . . by acquiring unspoiled tracts of land, such as the Palisades, for the purpose of protecting them from tawdry commercial interests" (Talbot 1972, 54). The Harriman family donated ten thousand acres to create Harriman State Park in 1910, while the Palisades Interstate Parkway became possible when the Rockefeller family, in the 1930s, donated the parcels for the project (Binnewies 2001). It was the first conservation project to preserve a landscape between two states. For the property owners the goal was to preserve the natural scenery that artists portrayed and prevent further defacement of the cliffs, and hence, the scenic view (O'Brien 1981, 265). This citizen action was among the first preservation efforts that engaged landscape appreciation as a core value, a moral value articulated decades earlier in essays and paintings by artists of the Hudson River School.

Places and Cases: Storm King

In 1861 Moore painted another Hudson River scene, *Down the Hudson to West Point* (Fig. 20.3). In this remarkably interesting scene, the mountainous topography of the Hudson Highlands, with the mountains Storm King (previously known as Butter) Mountain and Dunderberg on the west bank

FIG. 20.3. Charles Herbert Moore, *Down the Hudson to West Point*, 1861; 19 ½" x 29 ¾", oil on canvas, courtesy of The Frances Lehman Loeb Art Center, Vassar College, Poughkeepsie, New York, Gift of Matthew Vassar, 1864.

and Breakneck Ridge on the east bank, frames the Hudson River in the center of the painting. As with Moore's painting of the Upper Palisades, the Hudson River is relatively calm, and the sloops and sailing vessels are slowly making their way up and down river. The foreground, however, gives one pause (Flad 2000, 86). The artist is situated on the east bank, presumably on the edge of a dirt wagon road. This road, moreover, extends south across a causeway that would have separated one of the numerous small bays along the river shore and through a tunnel blasted out of Breakneck Ridge. A tunnel was indeed there in 1861; it had been built for the railroad that had already reached Poughkeepsie and Albany a decade earlier. However, in this painting, there are no signs of the existing rail line. It seems that Moore, as did many of the Hudson River School, engaged in "artistic license" by removing the railroad tracks from the scene as a way to advance the Romantic past against the industrial "improvements" of the Civil War era. Similarly, he did not include any steamboats on the river.

But it is to the mountains I wish to return. They are presented with a bold and powerful sensitivity. It is not so much the wilderness of the forests that engages one's attention but rather the tectonic strength of the topographic features. They form the gateway to the Hudson Highlands and create a symbolic relationship between the natural and the cultural landscape. Thrust into view is the iconic form of Storm King Mountain (Fig. 19.2). A century after Moore's painting, Storm King would become the site of the most important battle in recent environmental history—a legal case that began the environmental movement of the twentieth century (Dunwell 1991; also see Young and Dey, ch. 18 and Butzel, ch. 19 in this volume).

The birth of the modern American environmental movement may be placed in the Storm King legal case (*Scenic Hudson Preservation Conference v. Federal Power Commission and Consolidated Edison Company of New York, Inc.*) (Talbot 1972; Sandler and Schoenbrod 1981). In the early 1960s Con Ed proposed to build the world's largest hydroelectric pump storage facility on the top of Storm King Mountain (Fig. 19.1). The environmental preservation group Scenic Hudson was formed in 1963 to contest Con Ed's plans (Talbot 1972, 91–116; Binnewies 2001, 245–68; Dunwell 2008, 279–304). Their members viewed the proposed power plant as a major industrial defilement of one of the Hudson Valley's most visible natural features and as an attack on the nation's cultural identity. An editorial in *The New York Times* titled "Defacing the Hudson" quoted the nineteenth-century traveler Baedecker as finding the Hudson's scenery "grander and more inspiring" than the Rhine's and declared that the proposed plants "would desecrate great areas that are part of the natural and historic heritage of our country, are still largely unspoiled and should remain that way" (Binnewies 2001, 252; Dunwell 2008, 284).

Two key rulings by the United States Court of Appeals for the Second Circuit decision in 1965 allowed Scenic Hudson to continue its opposition to the construction of the power plant; the delays eventually wore down Con Ed and led to the defeat of their proposal. The decision expanded the role of the courts in environmental and land use law by addressing the role of aesthetics. Testimony by professor of architectural history Vincent Scully was "most strikingly eloquent" and "effective" according to the counsel for the plaintiffs David Sive, and persuaded the court that "the aesthetic qualities of Storm King were so great that any diminishing of these qualities would leave society without these values" (Smardon 1979, 684).

Two of the court's rulings would further strengthen environmental law. The court held that environmental concern extended to natural and scenic beauty and the historical fabric, not only to the economic cost of a project. The court also held that citizen groups had the legal right, or "standing," even if they did not have a direct economic interest, to challenge the potential environmental impacts of proposed construction, and that alternatives must be presented (Talbot 1972, 192–98; Sandler and Schoenbrod 1981, 55–67; Binnewies 2001, 266–67; Lewis 2005, 264–68). These features were then incorporated into the nation's most important environmental legislation, the National Environmental Policy Act of 1969 (Smardon 1979, 682; Flad 2002, 51; Dunwell 2008, 290). Described by legal scholar David Sampson as the "cornerstone of all subsequent federal environmental laws," he concluded, the "statute directly led to states developing their own 'little' NEPAs, such as New York's State Environmental Quality Review Act" (Sampson 2004, 220).

Cases: Lloyd Nuclear Power Plant

Over the next several years utilities continued to press for construction of new power plants in the Hudson Valley using Hudson River water for cooling. In the early 1970s the Atomic and Space Development Authority (ASDA) focused on nuclear power. In 1975 the New York state legislature replaced ASDA with the New York State Energy Research and Development Agency (NYSERDA). However, according to a statement in opposition to a proposed nuclear plant in 1976, the new agency was "created . . . to replace ASDA and preclude undue emphasis on nuclear development," but "ironically . . . has instead taken up ASDA's last cause" (Scenic Hudson Preservation Conference 1976, B32). In 1976 NYSERDA engaged consultants to draw up a master plan for a site in the town of Lloyd in Ulster County, on the west bank of the river across from the city and town of Poughkeepsie. Opponents to the $1 million plan that proposed the Lloyd-Esopus site as "suitable for up to four, 1,000 megawatt nuclear reactors (complete with four 500-foot cooling towers)" declared the plan to be "truly remarkable and disturbing" (Brown and Egemeier 1976, i).

Following the precedent of Scenic Hudson's intervention in the Storm King case, including the opportunity under New York State's recently enacted mini-NEPA, the State Environmental Quality Review Act (SEQRA) (Salkin 2001), citizens and environmental groups critiqued the consultants' master plan. They raised general concerns about the project's cost, future energy need, the safety of nuclear power, radioactivity, and problems of waste disposal, as well as deficiencies in the consultants' work (Konigsberg 1976). A review of potential biologic impact on the local soils and biota also noted that the consultants' discussion of "aesthetic effects" of the cooling towers was mentioned only in a single "subparagraph" and did not even mention the visual impact of transmission lines over the river or of the potential aesthetic effect of the plumes (Barnett 1976, 16A).

A review of the towers under the heading "visual pollution" was equally deficient according to a review of economic and social impacts: "ERDA does not mention that, at 500 feet tall, the towers would be the highest structures in south-eastern New York outside of Manhattan, and that they would be as high as any mountain peak from New Paltz to the river, if they are built as ERDA suggests. . . . [ERDA] does not mention the visibility of the plume from these places, nor does it stress that the towers would be visible from Poughkeepsie, Lake Mohonk, and elsewhere" (Stillman 1976, 10A). Moreover, a thorough visual impact study prepared for a field trip for participants at the 1976 annual meeting of the New York State Geological Association distinguished numerous specific locations from which the 137.3 m (450 ft.) cooling towers, sited on deep muck soils and peat bogs at 320 feet and rising to 770 feet in altitude, could be observed (Flad 1976, B-9-18). Also, the potential plume would be seen from urban and rural areas and historic sites within the viewshed, including the Roosevelt and Vanderbilt national parks.

Three years after the proposal to build a nuclear power complex in Ulster County, a second attempt was made a few miles north along the banks of the Hudson River in Greene County. Although an analysis of the aesthetic impact of the Lloyd proposal had been of limited value in the final negative decision to proceed with construction, visual and aesthetic impact would become the most significant argument against the new proposal with the addition of scenic views painted by Hudson River School artists.

Cases: Greene County Nuclear Power Plant

The Storm King precedents became significant in a subsequent major battle over a proposed electric power plant in the Hudson River Valley. A nuclear power plant was proposed in the late 1970s in Greene County on the west bank of the Hudson within the viewshed of "Olana," the historic home and studio of the Hudson River School painter Frederic Church. The name "Olana" evoked the eclectic multicolored Persian-style design of the building, which recalled architectural designs that Church and his wife had seen during their travels in the Near East. From atop "Church's Hill" south of Hudson, Church could see his mentor's home and studio "Cedar Grove" across the Hudson River in Catskill. Views from Olana became the significant factors in opposition to the proposed nuclear power plant.

After Cole's death, Church became acknowledged as the leading landscape painter of the mid-nineteenth century (Huntington 1966; Carr 2000; Wilton and Barringer 2002). His large exhibit pictures *Niagara* (1857), *Heart of the Andes* (1859), and *Icebergs* (1861) were considered masterpieces in their time and earned the artist both fame and fortune. Over the course of the Civil War, Church's paintings *Twilight in the Wilderness* (1860), *Cotopaxi* (1862), and *Rainy Season in the Tropics* (1866) offered an aesthetic response to the progress of the Union and the course of the war.

A citizens' coalition joined together to challenge the Greene County power plant proposed by the utilities, the New York State Power Authority and the U.S. Nuclear Regulatory Commission (NRC). As in the Storm King and Lloyd hearings, information was presented on the potential environmental impact of the 1,200 mw facility on the river ecology and the atmosphere, along with consideration for the immense projected costs of more than three billion dollars and the economic need for the plant. However, the crucial, and ultimately compelling testimony focused on the aesthetic and visual impact of the 450-foot cooling towers and the associated steam plumes on the cultural and historic landscape (Flad 1979; Petrich 1979; U.S. Nuclear Regulatory Commission 1979).

The proposed site for the nuclear power plant was at Cementon on the west bank of the Hudson six miles downriver from Olana. In testimony opposing the proposal, art historian Barbara Novak argued that the natural and cultural landscape encompassed by views from Olana were nationally significant. As Novak declared, "The Cementon-Athens area is the hea[r]t of American nineteenth century culture . . . one area of landscape that the nation as a whole and the federal government specifically should designate as a national landmark." (U.S. Nuclear Regulatory Commission 1979, 5–70). Cementon is easily seen in a view south-southwest, from the porch at Olana (Fig. 20.4).

FIG. 20.4. View from Olana south-southwest of Hudson River and west bank of proposed Greene County Nuclear Power Plant site at Cementon, ca. 1979, courtesy of the author.

SCENIC VIEWS

In testimony in opposition to the construction of the plant, drawings, including a computer mock-up, clearly showed that the cooling towers and associated plumes would be visible from Olana. The massive tower and the six-mile steam plume extended above the ridgeline of the Shawangunks to the south and the Catskills to the southwest. They became focal points of the view from Church's studio. Written and oral testimony by nationally renowned artists and art historians concluded that such an industrial intrusion on the cultural and natural landscape would severely diminish the emotional power of the scene (Gussow 1979; Huntington 1979; U.S. Nuclear Regulatory Commission 1979). Church's paintings that focused on that immediate viewscape, of the view southwest from Olana, such as *The Hudson Valley in Winter from Olana* (ca. 1866–1872) (Trebilcock and Balint 2009, 46) were particularly effective (Fig. 20.5).

Views west from the carriage roads that traverse Olana's historic landscape of the alternative site for the proposed power plant in Athens were equally compelling in their aesthetic impact on the historic village. Church designed Olana's landscape following the principles laid out by Andrew Jackson Downing (1815–1852), known as the "arbiter of taste" for his efforts at domestic improvement of country estates, gardens, and homes in the antebellum period (Tatum and MacDougall 1989; Schuyler 1996). Church constructed carriage roads and planted trees in the picturesque manner of Downing and his partner Calvert Vaux (1824–1895). After Downing's death in 1852, Vaux in collaboration with Olmsted won the competition to design and construct New York City's Central Park. Central Park brought nature into the urban environment and helped frame the role of landscape aesthetics in an emerging American culture (Schuyler 1986; Beveridge and Rocheleau 1998; Schuyler 2008). Later, Vaux assisted Church in his architectural designs for Olana and Olmsted and Vaux arranged for Church to be appointed a park commissioner in 1871 as he was working on Olana's grounds (Rybczynski 2000, 309–11).

FIG. 20.5. Frederic Edwin Church, *The Hudson Valley in Winter from Olana*, ca. 1866–72; 11 ¾ x 18 ¼ in., oil on academy board, courtesy of Olana State Historic Site, New York State Office of Parks, Recreation and Historic Preservation.

LANDSCAPE DESIGN AND VIEWSCAPES

At Olana, Church laid out more than seven and one-half miles of carriage roads and trails. These trails were designed to give the traveler an aesthetic experience, with concern shown for differences of intimate and expansive views, light and shade, and topography and water. That these were a measure of Church's artistic sensibilities is attested to by a quotation from his letter to Erastus Dow Palmer, a friend and fellow Hudson Valley artist, in 1884:

> I have made about one and three-quarters miles of roads this season, opening entirely new and beautiful views—I can make more and better landscapes in this way than by tampering with canvas and paint in the Studio." (Church 1884; see also Huntington 1966, 116; Flad 1979, 13; Ryan 1989, 147; Toole 2004, 52; Trebilcock and Balint 2009, 73)

Views from the carriage roads became important in the GCNPP case in constructing the argument of adverse impact on the visual environment.

Views from the city of Hudson, New York, were also significant as examples of aesthetic impact on a cultural landscape, such as the historic parade ground on the bluff overlooking the Hudson River in the foreground, the village of Athens in the middle ground, and the Catskills in the background (Flad 1979, 27–30). The parade ground had been sited during the settlement of Hudson in 1790 for its scenic view as one of the first urban design features to incorporate an appreciation of a prospect view from a city's core in America. A mock-up of a view of the cooling tower located west of the village indicated that its immense scale would have completely dominated the village's nineteenth-century architecture and streetscape (Fig. 20.6).

In an examination of the full impact of the proposal on the region's scenic quality, especially as viewed from Olana, the NRC concluded:

FIG. 20.6. View west across Hudson River of Village of Athens from Parade Hill, City of Hudson, with proposed Greene County Nuclear Power Plant cooling tower superimposed, 1979, courtesy of the author.

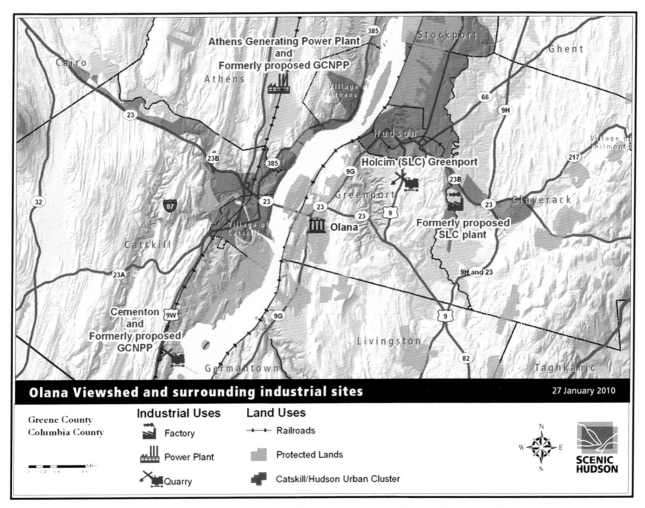

FIG. 20.7. Locations of proposed Greene County Nuclear Power Plant at Cementon and Athens, Athens Power Plant and St. Lawrence Cement Plant, courtesy of Scenic Hudson, 2010.

From the total perspective of the analysis of what is there to be seen in this stretch of the mid-Hudson Valley and how this area might be affected by the construction and operation of the proposed power plant, the GCNPP at the Cementon location is seen to be quite disruptive to the existing scenic ambience. The analysis of the individual photographs from visually sensitive and intensively used areas points to this conclusion. The relatively high number of scenic features in the Cementon area underscores the uncommon ambience of the area. Although the hilly terrain surrounding the Cementon site diminishes its visibility, it also enhances its scenic quality. The construction and operation of the power plant would seriously affect an area of high scenic quality for the Hudson Valley, one of unique rural and small village atmosphere. (U.S. Nuclear Regulatory Commission 1979, 5–68)

The NRC analysis cited the importance of Church's paintings in their decision: "A 10-mile radius around the proposed power plant would take in literally dozens of the scenic views and picturesque areas that were eventually transferred to canvases now hanging in the country's major museums and art galleries" (U.S. Nuclear Regulatory Commission 1979, 5–68). For the NRC, it was specifically the southwesterly view from Olana, "painted by Church at least 35 times," that was a crucial determinant in the final negative assessment (1979, 5–71).

The NRC denied the license, citing its potential irreversible negative impact on the cultural and historic landscape. The decision was unprecedented. As noted soon after at a national conference "on applied techniques for analysis and management of the visual resource," the decision was "the first impact statement issued by the NRC ever recommending the denial of a license to construct a nuclear power plant. That this recommendation is primarily for aesthetic reasons documents the progress in credibility and defendability [*sic*] visual analysis has made" (Petrich 1979, 483). The methodologies (U.S. Nuclear Regulatory Commission 1979, Appendix M and N; Flad 1979, Appendix A) included the introduction of industrial images, such as cooling towers and smoke plumes, to photographs of the same views as painted by Church; they may be seen as an ironic twist to the "artistic license" taken by Hudson River School artists such as Moore (Fig. 20.3) who often removed "improvements." The decision by the NRC was a significant test for the role of the visual impact of potential development and soon led New York State to add Visual Impact Analysis as a separate part of environmental impact analysis, and of equal legal standing to other natural and social environmental impacts (Smardon 1986, 151–55; Smardon and Karp 1993, 196–99).

Cases: St. Lawrence Cement Plant

Views from Olana were instrumental in another, equally significant case that has important roles for aesthetics, community character, and the cultural landscape. In 1999, St. Lawrence Cement Co., LLC (SLC), a subsidiary of the Swiss-owned Holcim Group, applied to the New York State Department of Environmental Conservation for permits to construct a new 2.6 million ton dry-process cement manufacturing facility on its property in the town of Greenport and city of Hudson (Fig. 20.7). The quarry and proposed plant would lie in the Olana viewshed, north toward Hudson and the Becraft Hills.

Opponents analyzed the environmental and social impacts of the proposal as a result of the enormous scale of the plant, its stack and associated plume, and docking and loading facilities (Silverman 2006). Environmental impacts on the Hudson River were of special concern as most municipalities had recently developed plans for their waterfronts by enacting Local Waterfront Revitalization Plans (LWRPs) under the aegis of the Department of State. Developments along the Hudson's shoreline would have to be consistent with the policies enacted in the local LWRP and in the municipality's economic development plans. Scenic Areas of Statewide Significance (SASS) had also been mapped during the late twentieth century and would require impact assessment of any proposed industrial development.

Two decades after the Greene County Nuclear Power Plant (GCNPP) case, aesthetic impact analysis had expanded to include local and regional land use focused on an elusive concept termed "community character" (Duerksen and Goebel 1999, 146; Ghilain 2009). In the SLC case, potential social and economic impacts were deemed to be significant, especially to the emerging economy tied to heritage tourism and the historic sense of identity associated with the cultural landscape. In an attempt to define the area's local and regional identity and sense of place, elements that constituted the social construction of the cultural landscape were presented as contributing to the community character of the local region (Flad 2005, 7).

Landscape and Community Character

A cultural landscape approach to five aesthetic elements (scenic views, artistic images, historic architecture, landscape architecture, and site and town planning) contributed to, but was separate from, objective visual analysis, as it also required the expression of aesthetic and community values (Flad 2005; Ghilain 2006; Ghilain 2009). Paintings by Church and Gifford and views from Cole's home Cedar Grove and Olana were used to document the historic and aesthetic character of the landscape that would be visually impacted by the proposal.

As with Church's views south from his studio of the Hudson River and the flanks of the Catskills, Cole's early view of Mount Merino on the east bank from his prospect in Catskill, *Point Merino* (ca. 1837) documented his interest in the scene. Many landscape artists visited Cole's home and painted it and views from its grounds. In a painting by Thomas C. Farrer, *Buckwheat Field on Thomas Cole's Farm* (1863), the view of the east bank, Mount Merino, and the city of Hudson in 1863 is focused on the location of the proposed SLC plant. Both paintings

FIG. 20.8. Photosimulation of East Bank, Hudson, N.Y., and Mount Merino, of proposed St. Lawrence Cement Plant and Docking Facilities, courtesy, T. DeWan, 2004.

from the property of the founder of the Hudson River School were significant in an examination of the visual impact of the proposed plant and its associated docking facilities, as they documented both an aesthetic and historic scenic view. Other eighteenth and nineteenth-century-paintings of Mount Merino and Hudson added important data to the visual impact analysis (Piwonka 1978). A photosimulation of the visual impact offered a powerful analysis of the proposed development on the Hudson shoreline (Fig. 20.8).

An aesthetic and community character analysis determined that views to the north and northeast from the North circular drive and from Cosy Cottage, a small cottage built on the original 126 acre farm as Church's first home, indicated significant visual impact of the proposed plant on Church's property (Sampson 2004, 226). Church's perspective from Olana encompassed 360 degrees. His paintings of views south and west of the Hudson River and the Catskill Mountains had been used in the GCNPP proceedings while several sketches of the Becraft Hills in the northeastern viewshed such as *Be Craft Mountain from Church's Farm* (1863) and *Blue Hill from Cosy Cottage* (c. 1869–1872), indicated that these views were also of interest to Church (Trebilcock and Balint 2009, 40).

A full cultural landscape and community character analysis documented the potential aesthetic impact of the proposed plant and its riverfront facilities (Flad 2005). Views from the city of Hudson and from the surrounding historic rural landscape formed an overall narrative of the settlement history of the area that had historically included a regional identity and community character in the cultural landscape. Testimony on the GCNPP had developed a methodology for examining visual impact of a proposal under the guidelines of SEQRA, which was further refined during proceedings related to a proposal to construct the Athens Generating Plant in 2000. In addition, impacts associated with the power plants' plumes and the relationship of the proposed projects to local economic plans, LWRPs and SASSes in the regional viewshed were assessed. New York State's Department of State addressed all of these issues with respect to the "consistency" of the SLC proposal to the state's Coastal Management Program (Palmer 2005).

Heritage Tourism and Scenic Views

In 2005 the secretary of state for New York, Randy Daniels, denied the application of St. Lawrence Cement to construct its plant and docking facilities as proposed (Daniels 2005). His ruling incorporated opponents' concerns about potential adverse impacts on the local economy and river ecology as well as the historic, cultural, and aesthetic landscape and its community character. In a review of the proposal's potential impact on "actions" that "should improve adjacent and upland views of the water, and, at a minimum, not affect these views in an insensitive manner," Daniels wrote: "The Hudson River viewshed in this area is important. The Hudson Valley was the setting for the Hudson River School of artists and the geographic center of the American Romantic Movement, a cultural movement that took place during the first half of the 19th Century" (Daniels 2005, 9). Referring to riverfront park development efforts by both the village of Athens and the city of Hudson, Daniels noted, "The region is also a significant resource of tourism and recreation." After a review of SLC's plans, he concluded, "The current outstanding scenic views of the water from the adjacent and surrounding areas, however, will be profoundly changed. . . . The impact on visual quality will be to impair, not to improve adjacent and upland views of the water." In addition, he specifically noted the "discordant" features of the massive plant and plume that would severely disrupt views from Olana (Daniels 2005, 18–19). The opinion noted that the region was entering a postindustrial economy of growth focused on heritage tourism, an era in which community character and landscape aesthetics were significant resources (Shapley 2005; Bonopartis 2005). St. Lawrence Cement subsequently dropped its plans.

CONCLUSION

The art and discourse of Hudson River School artists in the nineteenth century continues to engage conservation efforts into the twenty-first century. After a review of several ways in which nineteenth-century artists influenced national perspectives on the natural landscape, a number of examples where scenic and cultural interests underlay the creation of public parklands and preservation policies are presented. In the twentieth and twenty-first centuries several regional examples of attempts to preserve scenic views that engaged the nineteenth-century artists, architects, and landscape designers further represent the evolution of landscape preservation law and the continuing efforts by environmentalists and local citizens to preserve the Hudson River and its valley.

ACKNOWLEDGMENTS

I thank historians J. Winthrop Aldrich and Ruth Piwonka and legal counsels Warren P. Reiss, Marc S. Gerstman, and the late Robert C. Stover for assistance and commentary related to the author's testimony on two of the examples cited, an anonymous reviewer, and Robert E. Henshaw and Lucille L. Johnson of the Hudson River Environmental Society for their interest and advice.

REFERENCES CITED

American Scenic and Historic Preservation Society. 1906. *Eleventh annual report*. Albany: State of New York Assembly.

———. 1908. *Thirteenth annual report*. Albany: State of New York Assembly.

Anderson, N. K. 1991. "The kiss of enterprise": The western landscape as symbol and resource. In *The west as America: Reinterpreting images of the frontier, 1820–1920*, ed. William H. Truettner, 237–83. Washington, DC: Smithsonian Institution Press.

———, and L. S. Ferber. 1990. *Albert Bierstadt: Art and enterprise*. Brooklyn: Brooklyn Museum.

Avery, K. J., and F. Kelly. 2003. *Hudson River School visions: The landscapes of Sanford R. Gifford*. New York: The Metropolitan Museum of Art.

Barnett, R. W. 1976. Summary critique on "Lloyd Site Master Development Plan Concept Phase Report" proposed by the New York State Energy Research and Development Authority, Nov. '75—Biologic impact. In *Nuclear Power in the Hudson Valley*, ed. Peter D. G. Brown and Steven G. Egemeier, 14A–20A. Highland, NY: Mid-Hudson Nuclear Opponents.

Beveridge, C. E., and P. Rocheleau. 1998. *Frederick Law Olmsted: Designing the American landscape.* New York: Universe.

Binnewies, R. O. 2001. *Palisades: 100,000 acres in 100 years.* New York: Fordham University Press.

Bonopartis, N. 2005. Cement firm abandons plant bid. *Poughkeepsie Journal*, April 25, 1-2A.

Brennecke, M. 1987. Catalog of the paintings. In *Jasper F. Cropsey: Artist and architect.* New York: The New York Historical Society, 31–133.

Brinkley, D. 2009. *The wilderness warrior: Theodore Roosevelt and the crusade for America.* New York: HarperCollins.

Brown, P. D. G., and S. J. Egemeier, eds. 1976. *Nuclear power in the Hudson Valley: Its impact on you.* Highland, NY: Mid-Hudson Nuclear Opponents.

Bruegel, M. 2002. *Farm, shop, landing: The rise of a market society in the Hudson Valley, 1780–1860.* Durham: Duke University Press.

Bunkse, E. V. 1981. Humboldt and an aesthetic tradition in geography. *Geographical Review* 71(2): 127–46.

Burgess, L. E. 1980. *Mohonk: Its people and spirit, a history of one hundred years of growth and service.* New Paltz, NY: Smiley Brothers.

Carr, G. L. 2000. *In search of the promised land: Paintings by Frederic Edwin Church.* New York: Berry-Hill.

Church, F. E. 1884. Letter to Erastus Dow Palmer. 18 October. Albany: McKinney Library, Albany Institute of History and Art.

Cole, T. 1836. Essay on American scenery. *American Monthly Magazine* n.s. 1: 1–12.

Daniels, R. A. 2005. Opposition to consistency certification, Re: F-2004-0863, Army Corps of Engineers New York District Permit Application #2000-00943-YN, April 19.

Dow, C. M. 1914. *The state reservation at Niagara: A history.* Albany: J. B. Lyon Company.

Duerksen, C. J., and R. M. Goebel. 1999. *Aesthetics, community character, and the law.* Chicago: American Planning Association.

Dunwell, F. F. 1991. *The Hudson River Highlands.* New York: Columbia University Press.

———. 2008. *The Hudson: America's river.* New York: Columbia University Press.

Emerson, R. W. [1836] 1985. *Nature.* Boston: Beacon Press.

Ferber, L. S., ed. 2007. *Kindred spirits: Asher B. Durand and the American landscape.* Brooklyn: Brooklyn Museum.

———. 2009. *The Hudson River School: Nature and the American vision.* New York: Rizzolli.

Flad, H. K. 1976. Visual pollution of the proposed nuclear reactor site in the town of Lloyd, Ulster County, New York, Trip B-9. In *Guidebook to Field Excursions at the 48th Annual Meeting of the New York State Geological Association, October 15–17, 1976*, ed. John H. Johnsen. Poughkeepsie, NY: Vassar College.

———.1979. *Prepared testimony on aesthetic impact of Greene County nuclear power plant*, submitted on behalf of Columbia County Historical society, et al., to Power Authority State of New York and U.S. Nuclear Regulatory Commission, Case 80006, NRC Docket 50-549 (1979).

———. 2000. Following "the pleasant paths of Taste": The traveler's eye and New World landscapes. In *Humanizing landscapes: Geography, culture, and the Magoon Collection*, ed. Sheila Schwartz, 69–102. Poughkeepsie, NY: Frances Lehman Loeb Art Center, Vassar College.

———. 2002. The Hudson River Valley and the geographical imagination. *Watershed Journal* 1: 45–55.

———. 2005. Community character testimony re: Coastal consistency determination, St. Lawrence Greenport project. Poughkeepsie, NY: Scenic Hudson.

———. 2009. The parlor in the wilderness: Domesticating an iconic landscape. *Geographical Review* 99: 356–76.

Foshay, E. M., and B. Novak. 2000. *Intimate friends: Thomas Cole, Asher B. Durand, William Cullen Bryant.* New York: New-York Historical Society.

Ghilain, K. 2006. Cultural landscape preservation in the Hudson Valley: St. Lawrence Cement's place in the evolving legal protection of "place." BA thesis, Vassar College.

———. 2009. Improving community character analysis in the SEQRA environmental impact review process: A cultural landscape approach to defining the elusive "community character." *N.Y.U. Environmental Law Journal* 17: 1194–1242.

Graham, F. Jr. 1978. *The Adirondack Park: A political history.* NewYork: Alfred A. Knopf.

Gussow, A. 1979. Prepared testimony on aesthetic impact of Greene County nuclear power plant, submitted as addendum to Flad, 1979.

Harvey, M. 2005. *Wilderness forever: Howard Zahniser and the path to the Wilderness Act.* Seattle: University of Washington Press.

Hudson River Valley National Heritage Area. 2000. *Hudson River National Heritage Area draft management plan.* Albany: Hudson River Valley Greenway.

Huntington, D. C. 1966. *The landscapes of Frederic Edwin Church: Vision of an American era.* New York: George Braziller.

———. 1979. Prepared testimony on aesthetic impact of Greene County nuclear power plant, submitted as addendum to Flad, 1979.

Irwin, W. 1996. *The new Niagara: Tourism, technology, and the landscape of Niagara Falls, 1776–1917.* University Park: Pennsylvania State University Press.

Johnson, R. U. 1910. The neglect of beauty in the conservation movement. *The Century Illustrated Monthly Magazine* 79: 637–38.

Kealy, J. 2002. The Hudson River Valley: A natural resource threatened by sprawl. *Albany Law Environmental Outlook Journal* 7: 154–56.

Kelly, F. 1989. *Frederic Edwin Church.* Washington, DC: National Gallery of Art.

Konigsberg, A. S. 1976. Environmental effects of cooling towers: A critique of the Dames & Moore meteorological studies of the Lloyd, New York site. In *Nuclear power in the Hudson Valley*, ed. Peter D.G. Brown and Steven J. Egemeier, 26A–32A. Highland, NY: Mid-Hudson Nuclear Opponents.

Kowsky, F. R. 1998. *Country, park, and city: The architecture and life of Calvert Vaux.* New York: Oxford University Press.

Lewis, T. 2005. *The Hudson: A history.* New Haven: Yale University Press.

Lyell, C. 1845. *Travels in North America, in the years 1841–2.* New York: Wiley and Putnam.

Maddox, K. W. 1993. *In search of the picturesque: Nineteenth century images of industry along the Hudson River Valley.* Annandale-on-Hudson, NY: Edith C. Blum Art Institute, Bard College.

Mandel, P. C. F. 1990. *Fair wilderness: American paintings in the collection of The Adirondack Museum.* Blue Mountain, NY: The Adirondack Museum.

McMartin, B. 2002. *Perspectives on the Adirondacks: A thirty-year struggle by people protecting their treasure.* Syracuse: Syracuse University Press.

Miller, A. 1993. *The empire of the eye: Landscape representation and American cultural politics, 1825–1875.* Ithaca: Cornell University Press.

Myers, K. J. 1988. *The Catskills: Painters, writers, and tourists in the mountains, 1820–1895.* Yonkers: Hudson River Museum of Westchester.

Nash, R. F. 2001. *Wilderness and the American mind*, 4th ed. New Haven: Yale University Press.

National Gallery of Art. 2010. George Inness (artist), *The Lackawanna Valley* (c. 1856). http://www.nga.gov/fegi-bin/tinfo_f?object=30776; accessed Feb. 1, 2010.

National Park Service. 1996. *Hudson River Valley Special resource study report.* Boston: New England System Support Service.

Noble, L. L. 1856. *The life and works of Thomas Cole*, 3rd ed. New York: Sheldon, Blakeman.

Novak, B. 1969. *American painting of the nineteenth century: Realism, idealism, and the American experience.* New York: Praeger.

———. 1980. *Nature and culture: American landscape and painting, 1825–1975.* New York: Oxford University Press.

O'Brien, R. J. 1981. *American sublime: Landscape and scenery of the Lower Hudson Valley.* New York: Columbia University Press.

O'Toole, J. H. 2005. *Different views in Hudson River School painting.* New York: Columbia University Press.

Palmer, J. F. 2005. Coastal consistency determination review of St. Lawrence Cement Company's Greenport project on behalf of the Hudson Valley Preservation Coalition and Friends of Hudson.

Partington, F. E. [1911] 1970. *The story of Mohonk.* New Paltz, NY: Smiley Bros.

Petrich, C. H. 1979. Aesthetic impact of a proposed power plant on an historic wilderness landscape. In *Proceedings of our national landscape*, ed. Gary H. Elsner and Richard C. Smardon, 477–84. Berkeley: Pacific Southwest Forest and Range Experiment Station.

Phillips, S. S., and L. Weintraub, eds. 1988. *Charmed places: Hudson River artists and their houses, studios, and vistas*. New York: Harry N. Abrams.

Piwonka, R. 1978. *Mount Merino: Views of Mount Merino, South Bay, and the city of Hudson painted by Henry Ary and his contemporaries*. Kinderhook, NY: Columbia County Historical Society.

Pyne, S. J. 1999. *How the Canyon became Grand: A short history*. New York: Penguin.

Ranney, V. P. 1995. Introduction. In Frederick Law Olmsted, *Yosemite and the Mariposa Grove: A preliminary report, 1865,* vii–xx. Yosemite National Park, CA: Yosemite Association.

Robinson, C. T. 1993. Thomas Cole: Drawn to nature. In *Thomas Cole: Drawn to nature*, 47–88. Albany: Albany Institute of History and Art.

Robinson, N. A. 2007. "Forever wild": New York's constitutional mandates to *enhance* the Forest Preserve. Paper presented for the Arthur M. Crocker Lecture of the Association for the Protection of the Adirondacks, February 15, 2007, at the Center for the Forest Preserve, Niskayuna, NY.

Roseberry, C. R. 1982. *From Niagara to Montauk: The scenic pleasures of New York State*. Albany: State University of New York Press.

Ryan, J. A. 1989. Frederic Church's Olana: Architecture and landscape as art. In *Frederic Edwin Church*, ed. Franklin Kelly, 126–56. Washington, DC: National Gallery of Art.

Rybczynski, W. 2000. *A clearing in the distance: Frederick Law Olmsted and America in the 19th century*. New York: Simon and Schuster (Touchstone).

Sachs, A. 2006. *The Humboldt Current: Nineteenth-century exploration and the roots of American environmentalism*. New York: Viking.

Salkin, P. E. 2001. SEQRA's silver anniversary: Reviewing the past, considering the present, and charting the Future. *Albany Law Review* 65: 577–86.

Sampson, D. S. 2004. Maintaining the cultural landscape of the Hudson River Valley: What grade would the Hudson River School give us today? *Albany Law Environmental Outlook Journal* 8: 213–35.

Sandler, R., and D. Schoenbrod, eds. 1981. *The Hudson River power plant settlement conference materials*. New York: New York University School of Law.

Scenic Hudson Preservation Conference. 1976. Remarks. In *Nuclear power in the Hudson Valley*, ed. Peter D. G. Brown and Steven G. Egemeier, B30–B33. Highland, NY: Mid-Hudson Nuclear Opponents.

Schaefer, P. 1989. *Defending the wilderness: The Adirondack writings of Paul Schaefer*. Syracuse: Syracuse University Press.

Schuyler, D. 1986. *The new urban landscape: The redefinition of urban form in nineteenth-century America*. Baltimore: Johns Hopkins University Press.

———. 1996. *Apostle of taste: Andrew Jackson Downing, 1815–1852*. Baltimore: Johns Hopkins University Press.

———. 2005. The mid-Hudson Valley as iconic landscape: Tourism, economic development, and the beginnings of preservationist impulse. In *Within the landscape: Essays on nineteenth-century American art and culture*, ed. Phillip Earenfight and Nancy Siegel, 11–41. Carlisle, PA: The Trout Gallery, Dickenson College.

———. 2008. Central Park at 150: Celebrating Olmsted and Vaux's greensward plan. *Hudson River Valley Review* 24(2): 1–21.

Scott, D. 2004. *The enduring wilderness: Protecting our natural heritage through the Wilderness Act*. Golden, CO: Fulcrum Publishing.

Shapley, D. 2005. Cement plant halt applauded. *Poughkeepsie Journal*, April 21, 1B.

Silverman, M. D. 2006. *Stopping the plant: The St. Lawrence Cement controversy and the battle for quality of life in the Hudson Valley*. Albany: State University of New York Press.

Smardon, R. C. 1979. The interface of legal and esthetic considerations. In *Proceedings of our national landscape*, ed. Gary H. Elsner and Richard C. Smardon, 676–85. Berkeley: Pacific Southwest Forest and Range Experiment Station.

———. 1986. Review of agency methodology for visual project analysis. In *Foundations for visual project analysis*, ed. Richard C. Smardon, James F. Palmer, and John P. Felleman, 141–66. New York: John Wiley and Sons.

———, and J. P. Karp. 1993. *The legal landscape: Guidelines for regulating environmental and aes-*

thetic quality. New York: Van Nostrand Reinhold.

Smithson, I. 2000. Thoreau, Thomas Cole, and Asher Durand: Composing the American landscape. In *Thoreau's sense of place: Essays in American environmental writing*, ed. Richard J. Schneider, 93–114. Iowa City: University of Iowa Press.

Stebbins, T. E. Jr. 1978. *Close observation: Selected oil sketches by Frederic E. Church*. Washington, DC: Smithsonian Institution.

Stilgoe, J. R. 1993. Walking seer: Cole as pedestrian spectator. In *Thomas Cole: Drawn to nature*, 17–25. Albany: Albany Institute of History and Art.

Stillman, P. G. 1976. Lloyd power plants: Economic and social impact. In *Nuclear power in the Hudson Valley*, ed. Peter D. G. Brown and Steven G. Egemeier, 5A–13A. Highland, NY: Mid-Hudson Nuclear Opponents.

Talbot, A. R. 1972. *Power along the Hudson: The Storm King case and the birth of environmentalism*. New York: E. P. Dutton.

Tatum, G. B., and E. B. MacDougall, eds. 1989. *Prophet with honor: The career of Andrew Jackson Downing, 1815–1852*. Washington, DC: Dumbarton Oaks.

Terrie, P. G. 1997. *Contested terrain: A new history of nature and people in the Adirondacks*. Syracuse: Syracuse University Press.

Thoreau, H. D. 1862. Walking. *Atlantic Monthly* 9: 657–74.

Toole, R. M. 2004. "The art of the landscape gardener": Frederic Church at Olana. *The Hudson River Valley Review* 21: 39–63.

Trebilcock, E. D., and V. A. Balint. 2009. *Glories of the Hudson: Frederic Edwin Church's views from Olana*. Ithaca: Cornell University Press.

U.S. Congress. 2009. *Hudson River Valley Special Resource Study Act*, H.R.4003. Washington, DC: 11th Congress, 1st Session.

U.S. Nuclear Regulatory Commission. 1979. *Final environmental statement related to construction of Greene County nuclear power plant, Power Authority of the State of New York*. Docket No. 50-549. NUREG-0512.

Van Zandt, R. 1966. *The Catskill Mountain House*. New Brunswick: Rutgers University Press.

Wallach, A. 2002. Thomas Cole's *River in the Catskills* as antipastoral. *The Art Bulletin* 84: 334–50.

Walls, L. D. 1995. *Seeing new worlds: Henry David Thoreau and nineteenth-century natural science*. Madison: University of Wisconsin Press.

Wilton, A., and T. Barringer. 2002. *American sublime: Landscape painting in the United States 1820–1880*. Princeton: Princeton University Press.

Zahniser, H., ed. 1992. *Where wilderness preservation began: Adirondack writings of Howard Zahniser*. Utica: North Country Books.

CHAPTER 21

"Thy Fate and Mine Are Not Repose"

The Hudson and Its Influence

Geoffrey L. Brackett

ABSTRACT

This chapter examines the cycle of influence of the Hudson River and its inhabitants by analyzing the literary and historical works that take up the subject of the river across several centuries. These works function as a kind of feedback loop: the writing about the river serves as a record of the way communities of a certain time thought about and treated the Hudson; it crystallizes and illuminates the competing interpretations of the river; it influences future interpretations; it both archives the past and forecasts a future vision.

The changing relationship between its human inhabitants and the Hudson River Valley is nothing if not complex, and like many feedback loops the cycle of influence does not flow in one direction. The attitudes of the original Native American inhabitants of the valley are exceedingly difficult to map out, though clearly different from the early European settlers, whose first contact narratives illustrated a wide range of expectation up to arrival as well as a rapidly changing set of assumptions after their arrival. Subsequent eras were influenced by the river and responded to it with changing priorities, some driven by the river itself; from its early recorded human history the Hudson Valley was the site of some of the most wrenching social, political, and economic change on the continent, and that state of flux is even reflected in the current writing about the Hudson. The title of this chapter comes from William Cullen Bryant's "Scene on the Banks of the Hudson" (1827), *and is indicative of the complexity of the relationship between the river and the human being on its banks since it posits a combined—or at least parallel—fate between the human speaker and the river that is based on flux, not repose. Through all of this "lack of repose," as Bryant might say, the Hudson River is driving change while being driven by it.*

FROM THE FIRST written record of the Hudson we see a region in a state of flux through cultural, economic, and social forces that are interconnected with its own ecology. The writing reflects this, and like the river itself, the cycle of influence does not flow in one direction. What we can extricate from the records of original inhabitants, while exceedingly difficult to map out, shows a different relationship to those who followed them in the river's region. The early European explorers had expectations before arriving that rapidly changed upon arrival, and that change continued for each generation. From the point of first contact the Hudson River Valley was the site of some of the most wrenching social, political, and economic change on the continent. Through all of these cultural and literary periods, the river was driving change and being driven by it.

Through each era, the human relationship to the river changes for a variety of reasons, driven by culture, social or political change, technological and economic development, or even individual inspiration. The Hudson is either a driver or a recipient of that change, but intimately connected with it. In the several centuries of written record, we see the cycle of interaction and the main characteristics that determine each era and serve as the foundation for future eras. In each part of the cycle there are several key elements of the human/river relationship, the main determinant of which is whether the river is viewed as a "source" or a "resource." The former assumes a kind of superior relationship of the river to the human world: as a source it determines the worldview, provides living circumstances, and wields its influence over the human world. In eras where the river is viewed as a resource it assumes a subordinate relationship to the human world, providing navigation, natural resources, and real property whose effects are owned and determined by the human interest rather than the estuarine ecology. The historical eras are related to each other in a continuum of succession, and the chart of such a relationship would look something like this:

The original relationship of the pre-Columbian peoples to the land, and even to those who were here during and immediately prior to Hudson's arrival, is almost entirely an archeological record. We have no evidence in writing to help us understand that relationship in detail other than the European record of first contact, the secondhand stories of the native populations by early travelers, and the existing traditions of the descendents of those first people. The record that does exist illustrates an important distinction in the relationship of the Hudson to its populations. The critical distinction

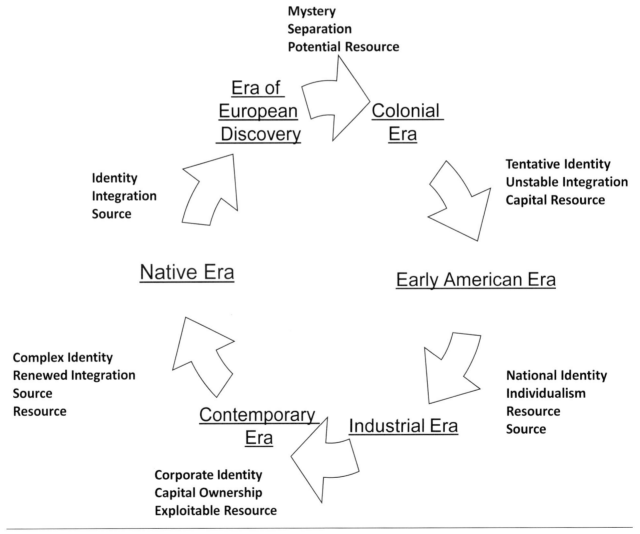

FIG. 21.1. This cycle illustrates the dynamic influence the Hudson has had on successive eras. Inhabitants have viewed the Hudson's influence as superior (as a source) or subordinate (as a resource) in different ages and with different effects on the society living by the Hudson.

here begins with the differences in the names of region recorded from the languages spoken before contact and after European discovery.

The characteristic of names in the native Lenape (or Unami) language serves as a window into the relationship to the Hudson River region because the dominant characteristic in naming is so different between the native and the European cultures. The names we have in the pre-contact period emphasize a different concept, marking the land more as a "source" than "resource" in what records we have. The Lenape name for the Hudson, "Mahicanituk," for "Constantly moving waters" (or, sometimes "River that flows two ways"), is perhaps the most famous example of this dynamic. In the specifics of the region we see even more detailed examples: the northern section of Manhattan and corresponding shore in New Jersey was called "Shorakapkok" or "the sitting-down place"; the University Heights area of the Bronx was known as "Keskeskick," the term for "sharp sedge grass"; the Harlem River was known as "Muscooten," which translates as "the place where rushes grow on the banks" (Pritchard 2002, 96). The language here indicates a tightly knit association between the land and its usefulness in such a way as is sometimes determined by the land (e.g., "sharp sedge grass") and other times determined by human will (e.g., "sitting down"). What this illustrates is that the Hudson was mapped out and known by its individual attributes and usefulness as a source of materials and experiences. It indicates a very different relationship than the directional or honorific titles with which the region came to be dominated.

The identity of a given Lenape group was often directly determined by the geography of the Hudson's region, as Mark Kurlansky points out:

> There were three major groups of Lenape and numerous subdivisions within those. They had few unifying institutions except language, and even that broke down into dialects. One of the groups, the Munsey, which means "mountaineers," controlled the mountains near the headwaters of the Delaware. They also maintained hunting grounds in what is now the New York City area.

The Lenape determination of the land's functionality is also thought to be the source behind the most famous island in the region:

> It is the Munsey language that gave Manhattan and many other New York places their names. It is uncertain from which Munsey word the name Manhattan is derived. One theory is that it comes from the word manahactanienk, which means "place of inebriation," but another is that it comes from manahatouh, meaning "a place where wood is available for making bows and arrows." (Kurlansky 2006, 10–11)

The language of reference for the European explorers and early settlers indicates a very different dynamic, driven in large part by the focused purpose of the enterprise of discovery. The Dutch first named the river the North River (in relation to the "South," later Delaware, River), then the Mauritius (perhaps for Prince Maurice of Nassau), and later the English named it Hudson. The region was established as the colony of "New Amsterdam" and then "New York," names that illustrate ownership of state. The notion of direction and named ownership are illustrative of purpose: the land is to be traveled through in a certain direction, or labeled in celebration of a royal patron, the recognition of an explorer, or to identify it as a subservient political and economic entity. Each of these names references something "other" than the physical geography of the river, indicating usefulness or ownership, implying the region is a resource to some external entity. The dynamic of European discovery of the Hudson Valley is built around this concept of the region as a resource, inherently separating the explorer from identity with the land. As David Quinn points out, in navigating and naming rivers, assessing cultivatable land, discerning vegetation, observing Native American customs, early explorers were inherently alienated from the landscape: they "were concerned with the thing-in-itself, with the objective situation in which they themselves acted as detached observers" (Quinn 1977, 72).

The detailed record of successive visitations illustrates the gradual revelation of the abundant possibilities of the Hudson Valley. Early explorers begin in the realm of mystery and then move quickly to an analysis of the region's economic resources. They typically remark about the clothing and offerings of the native populations with attention to the details of minerals and resources. Europeans saw native use of these elements as fundamentally different: the explorers and early colonists were viewing the resources as marketable commodities rather than functional ones: whereas the native population used copper for pipes, the explorers recognized it for its value in European markets; furs were used as functional clothing for the populations in the area, but their ability to be traded into those same European markets—specifically the beaver pelt—launched the Dutch economic interest in the region (Axtell 1988, 161).

The first European to record his entrance into the bay was Giuseppe Verrazano in 1524. He uses language that is descriptive and focused on the geographical particulars: the words he uses in describing the discovery alternate between a measured wonder at the characteristics of the native population—as well as their sheer numbers—and a calculated assessment of the physical resources promised in by the river valley:

> After a hundred leagues we found a very agreeable place between two small but prominent hills; between them a very wide river, deep at its mouth, flows out into the sea; and with the help of the tide, which rises eight feet, any laden ship could have passed from the sea into the river estuary. Once we were anchored off the coast and well sheltered, we did not want to run any risks without knowing anything about the river mouth. So we took the small boat up this river to land which we found densely populated. The people were almost the same as the others, dressed in birds' feathers of various color and they came toward us joyfully, uttering loud cries of wonderment, and showing us safest place to beach the boat. We went up this river for about half a league, where we saw that it formed a beautiful lake, about three leagues in circumference. About 30 small boats ran to and from across the lake with innumerable people aboard who were crossing from one side to the other to see us.

He concludes the visit because of a sudden turn of the weather, which reads as if he is only marginally interested in the bay and uses the weather as a convenient reason to leave (surely he could have waited out a storm in the bay). He ends with a passing comment about the possible mineral wealth that almost reads like an afterthought:

> Suddenly, as often happens in sailing, a violent unfavorable wind blew in from the sea, and we were forced to return to the ship, leaving the land with much regret on account of its favorable conditions and beauty; we think was not without some properties of value, since all the hills showed signs of minerals.[1]

The question of the river's value was at the heart of the next expedition to enter the Hudson, since the *Half Moon* was charged with finding the shortest route to the trading ports in the Far East. Where Verrazano's was a cursory assessment of the region's worth, Hudson's trip was a calculated assessment of the river's navigable potential. Hudson's first mate, Robert Juet, begins with language very similar to Verrazano's, except the focus is less for an external audience than an internal one: the language is largely prosaic and lists a series of events characteristic of a ship's log. On September 4, 1609, Juet describes the Hudson estuary, its fish, and its physical characteristics:

> The fourth, in the morning as soone as the day was light, wee saw that it was good riding farther vp. So we sent our Boate to sound, and found that it was a very good Harbour; and foure and fiue fathoms, two Cables length from shoare. Then we weighed and went in with our ship. Then our Boate went on Land with our Net to Fish, and caught ten great Mullets, of a foot and a halfe long a peece, and a Ray as great as foure men could hale into the ship. So we trimmed our Boate and rode still all day. At night the wind blew hard at the North-west,

and our Anchor came home, and we droue on shore, but took no hurt, thanked bee God, for the ground is soft sand and Oze.

Juet continues with a matter-of-fact observation about the population along the Hudson River, but his language is even less excitable than Verrazano's. Later in his entry for September 4th he makes a series of quick and unqualified observations of the native inhabitants:

> This day the people of the Countrey came aboard of vs, seeming very glad of our comming, and brought greene Tabacco, and gaue us of it for Kniues and Beads. They goe in Deere skins loose, well dressed. They have yellow Copper. They desire Cloathes, and are very ciuill. They have great store of Maiz or Indian Wheate, whereof they made good Bread. The Countrey is full of great and tall Oakes.

On September 5, 1609, Juet simply continues the narrative interspersed with nautical details for reference. He explains the exploration of the land and the interaction with its people. This description is remarkable—perhaps the most remarkable in his narrative—for the way he ends it. What starts as a description of a fruitful and peaceful exchange of goods ends with the comment that undercuts the whole preceding narrative: "[we] durst not trust them."[2]

What Juet's narrative accomplishes, of course, is a brief log of the most famous trip in the valley, an assessment of its geographical resources, a record of the first meaningful interaction—communication, trade, and murder—between the European and Native American populations. It also, however, illustrates the shifting priorities in the human relationship to the land: he gives us the image of a native people aware of the Hudson Valley as a source for clothing, food, accoutrements such as copper pipes, etc., but the language he uses illustrates that he views it as a resource. Like Verrazano, his perspective is inherently foreign to the population as he describes them: he looks to the shore to see wood, minerals like copper, as raw materials rather than elements of a living landscape. This narrative of discovery illustrates a critical difference in the perspective of the river by juxtaposing the European explorers' ethic against the native population's relationship to the land.

The acuity with which Juet evaluates the natural resources suggests the overarching incentive for the Dutch colonization of the Hudson Valley. The observations in his journal about the region's potential resources, in addition to the warnings about the native population, set the stage for further exploration and colonization. European settlers in the Hudson Valley arrived here primarily to plumb the resources for economic gain, marking a different ethos than the Puritan one in New England, which was imbued with much more religious fervor.[3] Indeed, the written record of early colonization illustrates that Juet's brief assessment of the Hudson's economic potential was underestimated, if anything. More often than not the bounty of the Hudson is wondered at, celebrated, and occasionally even lamented. Jacob Steendam's poem "The Complaint of New Amsterdam" (1659) reads as an apex of delight and distress about those resources first spied by Verrazano and Juet. The primary message of his poem is that he lacks the resources to be able to control the bounty of the Hudson:

> For, I venture to proclaim,
> No one can a maiden name
> Who with richer land is blessed
> Than th' estate by me possessed.
> See: two streams my garden bind.
> From the East and North they wind, —
> Rivers pouring in the sea.
> Rich in fish, beyond degree.
> Milk and butter; fruits to eat
> No one can enumerate;
> Ev'ry vegetable known;
> Grain the best that e'er was grown.
> All the blessings man e'er knew,
> Here does our Great Giver strew
> (And a climate ne'er more pure),
> But for me,—yet immature,
> Fraught with danger, for the swine
> Trample down these crops of mine;
> Up-root, too, my choicest land;
> Still and dumb, the while, I stand,
> In the hope, my mother's arm

Will protect me from the harm.
She can succor my distress.
Now my wish, my sole request,—
Is for men to till my land;
So I'll not in silence stand.
I have lab'rors almost none;
Let my household large become;
I'll my mother's kitchen furnish
With my knick-knacks, with my surplus;
With tobacco, furs and grain;
So that Prussia she'll disdain.
             (Murphy 1969, 41–43)

The narrative is so rich in detail that it almost oozes off the page. As a citizen working to control this cornucopia, he feels that the distant burghers in the homeland need more than anything else to abet his work by sending more people; without it he stands to lose control of the land's bounty. The mystery the explorers felt as they encountered the region is no longer invested in the land itself—the mystery is how one overworked economic colonist will be able to manage it. The "resource" overwhelms the exploiter in conflict with the land and its bounty, which he endeavors to cultivate, control, and ultimately, profit from. The irony in Steendam's poetic plea is that the "resource" of the land in his poem creates the need for additional "resources" from the "mother's arm" to control it. Indeed, this is one of the conundrums of capital: one inevitably needs more capital to manage capital's continued growth.

The conception of the land as capital resource is at the heart of the colonial transactions between the settlers and the Indian populations throughout the Hudson Valley (and elsewhere of course). We all know the famous anecdote of the Peter Minuit transaction for Manhattan, but other examples abound. One example is the Sopes Indian treaty of 1665, negotiated by Richard Nicholls in what is modern day Esopus, New York. In the first paragraph the agreement outlines the mutual pledge to do no harm to each other, but that is quickly followed by an injunction against committing crime against property—with a general phrase to "seek satisfaction for the same." The governing impulse here is the preservation of capital, built upon the principle of land ownership and the physical capital Europeans invested in the planting, harvesting, and, importantly, enclosing of that land. The Europeans believed their right to land was secure against Indian claims both because of the legal treaties arranged like this one but also because they saw no evidence of the capital investment in the land by the Native Americans. In the words of Nicholls's fellow English settler John Winthrop, the tribes they encountered "inclose noe Land, neither have any settled habitation, nor any tame Cattle to improve the Land by," leaving them with little or no legal claim to the land (Axtell 1981, 51).

In the Sopes Treaty, Nicholls arranged for there to be a "Convenient House" built for Indians to visit, leave their arms in order that they may "come without molestation to Sell or Buy" whatever they want from the settlers. The references here to the land are descriptive of the boundaries of ownership rather than their functional utilization (as we saw in the earlier native era). The attributes of the land are to become markers and signifiers of the new boundaries: in some places creeks perform that function; in other places trees shall be planted to mark them. Two sticks are traded for three red coats to mark the transaction, and Nichols puts in the provision that every year the young Indians will be brought to the site to "mark forever in their memory" the transaction, since written record exists only on the English side.[4] Nicholls establishes a ritual attached to the land for the native inhabitants to preserve the memory of its sale—in effect, creating a function for the land that is not inherent in its landscape but the legal arrangement he has effected. In this way, this treaty and the countless others like it serve as markers of a dramatic shift in consciousness about the Hudson, dividing its regions into tracts of land measured as capital—acknowledging them as resources for their owners.

After the colonial settlers, the next people to claim ownership of the land were the American Revolutionaries. In the one hundred and fifty years between Nicholls and Washington Irving, the concept of colonial ownership had been erased and the population of settlers had lived in the place long enough to assume a "native" perspective and ownership principle. Irving writes after the Revolution and documents the transition from English to American in his short story "Rip Van Winkle," whose narrative structure leapfrogs the political change from English to American when Rip falls asleep under King George's rule and awakens during the first term of George Washington, a transition cleverly illustrated by the change in the name of the tavern in Rip's home town. The Revolution was in part fuelled by colonists reversing Steendam's complaint one hundred years later: instead of not providing enough capital to its colony, the motherland was accused of extracting a disproportionate amount in the form of tax revenues and tariffs. Where Steendam had cried to Amsterdam for more men to manage the land, the colonists' bounty was the victim of England's protectionism.

Irving's focus is not, however, on the details of the political argument. Instead, he uses the political change simply as a structure to parody the precursor to the English colonial population: the Dutch, in the form of Diedrich Knickerbocker. Irving's parody shows that the Dutch are now considered, like the Native Americans before them, erstwhile owners. They once controlled the region but have become historical footnotes; now they are associated with magic and mysticism in much the way that the Indians had been. Irving highlights that mystical tradition sardonically in his own historical footnote to the tale:

> Note
> The foregoing Tale, one would suspect, had been suggested to Mr. Knickerbocker by a little German superstition about the Emperor Frederick der Rothbart, and the Kyffhäuser mountain: the subjoined note, however, which he had appended to the tale, shows that it is an absolute fact, narrated with his usual fidelity:
> "The story of Rip Van Winkle may seem incredible to many, but nevertheless I give it my full belief, for I know the vicinity of our old Dutch settlements to have been very subject to marvellous events and appearances. Indeed, I have heard many stranger stories than this, in the villages along the Hudson; all of which were too well authenticated to admit of a doubt. I have even talked with Rip Van Winkle myself who, when last I saw him, was a very venerable old man, and so perfectly rational and consistent on every other point, that I think no conscientious person could refuse to take this into the bargain; nay, I have seen a certificate on the subject taken before a country justice and signed with a cross, in the justice's own handwriting. The story, therefore, is beyond the possibility of doubt.
> D.K." (Irving 1961, 53-54)

Americans in the nineteenth century were well aware of the mystical properties of the land, and these had more uses than simply a backward-looking nostalgia (as humorous as it might be). They also became intimately connected with the sense of self: in broad strokes, if the Dutch felt the land's bounty was a natural economic opportunity that would be shameful to let lie fallow, and the English felt they were doing the king's and God's will in bringing a kind of ordered industry to the land, then nineteenth-century Americans looked at the Hudson Valley as a bounty that provided a new, critical element to their world: legitimacy. While the colonial populations looked at the land as a resource to be managed and carefully exploited in concert with the mother countries, the first few generations of politically independent Americans looked at the land as an identifying asset against the cultural imperative of their former colonial masters. This idea has been well documented in the growth of American literature and in the famous Hudson River School of painting, but for this investigation it illuminates the process by which the inhabitants of the land redefine it as a "source." Like the Native American populations who utilized the land to provide identity and a way of life, American citizens saw the region as intimately connected with their own sense of identity. For American citizens involved in their cultural self-identification, the Hudson was a source of mythology and artistic vitality: it was the source of poetic and mythological inspiration, of artistic and cultural legitimacy.

Joseph Rodman Drake provides one example of this stylized romanticism. In "The Culprit Fay" (1835) Drake narrates the tale of a misbegotten fairy ("fay") who falls in love with a human girl. The poetry is somewhat stilted and Drake's poem never ascended the heights of the canon, but the architecture of the myth makes it clear that the landscape of the Hudson is at least as mystical as the storied Rhine river of Germany which gave rise to the myth of the "Rheingold."

XI

Soft and pale is the moony beam,
Moveless still the glassy stream,
The wave is clear, the beach is bright
With snowy shells and sparkling stones;
The shore-surge comes in ripples light,
In murmurings faint and distant moans;
And ever afar in the silence deep
Is heard the splash of the sturgeon's leap,
And the bend of his graceful bow is seen—
A glittering arch of silver sheen,
Spanning the wave of burnished blue,
And dripping with gems of the river dew.

XII

The elfin cast a glance around,
As he lighted down from his courser toad,
Then round his breast his wings he wound,
And close to the river's brink he strode;
He sprang on a rock, he breathed a prayer,
Above his head his arms he threw,
Then tossed a tiny curve in air,
And headlong plunged in the waters blue.
                    (Adams 1980, 137)

Drake's language is florid and ornamental in order to emphasize the mystical depth of the region he is describing, and his poetry is an earnest complement to Irving's more sardonic approach to the mystical power of the Hudson as a source for American literary mythology. Where Irving caricatures the Dutch fondness for legend (indeed, Irving did much to create this as a stereotype), Drake works hard in "The Culprit Fay" to legitimize the region's mystical properties in a way that the Hudson River School's work legitimized the region for Europeans familiar with the works of Turner and the other English and European Romantic painters then in vogue.

William Cullen Bryant's poem "A Scene on the Banks of the Hudson" captures the solemnity and intensity of the experience felt by those early-nineteenth-century artists who looked to the Hudson as a source of inspiration. Byrant recognized, like other artists of his generation, that the Hudson was more than a byway or a land rich in resources for economic gain. It was also a magnificent identifier of the American essence, a spiritual guide whose very motions mirrored the actions of those who lived on its banks. The Hudson anchors Bryant as the valley a few miles above Tintern Abbey anchored Wordsworth as a spot for pure contemplation of those moments when we see into the life of things.

Cool shades and dews are round my way,
And silence of the early day;
'Mid the dark rocks that watch his bed,
Glitters the mighty Hudson spread,
Unrippled, save by drops that fall
From shrubs that fringe his mountain wall;
And o'er the clear still water swells
The music of the Sabbath bells.
All, save this little nook of land
Circled with trees, on which I stand;
All, save that line of hills which lie
Suspended in the mimic sky—
Seems a blue void, above, below,
Through which the white clouds come
  and go;
And from the green world's farthest steep
I gaze into the airy deep.
Loveliest of lovely things are they,
On earth, that soonest pass away.
The rose that lives its little hour
Is prized beyond the sculptured flower.
Even love, long tried and cherished long,
Becomes more tender and more strong,
At thought of that insatiate grave
From which its yearnings cannot save.
River! in this still hour thou hast
Too much of heaven on earth to last;
Nor long may thy still waters lie,
An image of the glorious sky.
Thy fate and mine are not repose,
And ere another evening close,
Thou to thy tides shalt turn again,
And I to seek the crowd of men.
                    (Bryant 1854, 191–92)

At the same time that Bryant and others were trying to define the spiritual and artistic importance of the Hudson, the industrial period on the river began in earnest. As a means of transportation the river was dramatically altered by Fulton's invention of the steamboat, since it drastically reduced the time needed for travel up and down the river. Despite its inherent risks (with such disasters as the one that killed Andrew Jackson Downing occurring on a regular basis), the steamboat removed the serendipitous travel of sail and allowed reasonable transport time against the tide-borne waters of the Hudson. The changing landscape of transport had a dramatic effect on the relationship of the locals to the Hudson, making it possible for more business to occur with regularity. It also changed the role of leisure on the river: as New York City matured into a center of business predicated on dependable timetables, steamboats provided a regularized access to the river itself and the points along its shore. The culture of the city, then, carried itself onto the river, and James Kirke Paulding's narratives are a striking example of how society and its mores quickly adapted to riding the waves of the Hudson. In his sarcastic opinion piece on how to properly ride the steamboat published in his *New Mirror for Travelers* (1828), Paulding opens a door into the early-nineteenth-century New York society that seems shockingly modern:

> The following hints will be found serviceable to all travelers in steam boats.
>
> In the miscellaneous mélange usually found in these machines, the first duty of a man is to take care of himself—to get the best seat at table, the best location on deck; and when these are obtained to keep resolute possession in spite of all the significant looks of the ladies.
>
> If your heart yearns for a particularly comfortable seat which is occupied by a lady, all you have to do is keep your eye steadily upon it, and the moment she gets up, don't wait to see if she is going to return, but take possession without a moment's delay. If she comes back again, be sure not to see her.
>
> Keep a sharp outlook for meals. An experienced traveler can always tell when these amiable conveniences are about being served up, by a mysterious movement on the part of the ladies, and a mysterious agitation among the male species, who may be seen gradually approximating towards the cabin doors. Whenever you observe these symptoms, it is time to exert yourself by pushing through the crowd to the place of flagons. Never mind the sour looks, but elbow your way with resolution and perseverance, remembering that a man can eat but so many meals in his life, and that the loss of one can never be retrieved. (Adams 1980, 12–13)

There are many other examples of how the Hudson and its valley became a center of cultural struggle, from the narratives of New York Society in Edith Wharton to the explorations and personal travails of Herman Melville. But one hundred years on from Paulding, the poet Hart Crane looked to the Hudson River—and specifically the East River—for an image that would suffice for an epic anchor to tell the story of what it meant to be an American. In the "Proem" that opens his epic *The Bridge* (1930) he envisions the Brooklyn Bridge as a manifestation of the spirit of the age: its geographical importance and mechanical majesty combine to locate the heart of America in the Lower Hudson Valley.

> How many dawns, chill from his rippling rest
> The seagull's wings shall dip and pivot him,
> Shedding white rings of tumult, building high
> Over the chained bay waters Liberty—
> Then, with inviolate curve, forsake our eyes
> As apparitional as sails that cross
> Some page of figures to be filed away;
> —Till elevators drop us from our day . . .
> I think of cinemas, panoramic sleights
> With multitudes bent toward some flashing scene
> Never disclosed, but hastened to again,
> Foretold to other eyes on the same screen;
> And Thee, across the harbor, silver-paced
> As though the sun took step of thee, yet left
> Some motion ever unspent in thy stride,—
> Implicitly thy freedom staying thee!
> (Frank 1946, 3)

The change with Crane's work is substantial, because while he locates the imaginative heart and soul of America in the lower Hudson Valley, the secondary and tertiary attributes of the valley are the focus of wonder: the human population that gathered in New York, and the structures that it has created—namely the Brooklyn Bridge. The topic is the engineered structure rather than the landscape's natural bounty. For Crane, writing in the twenties, the Brooklyn Bridge is old enough (having been completed in 1883) to anchor a myth of America, and while its gothic design is sympathetic with that mystical role, its use marks another significant shift in perception of the Hudson River Valley: the accomplishments of its inhabitants outrank the attributes of the landscape itself. It is worth noting that the bridge is a structure that physically bypasses the river completely through engineering. A bridge eliminates contact with the river—and the need for a ferry boat. In this way, Crane's apostrophe to the Brooklyn Bridge celebrates the modern separation from the river. This is a significant shift, of course, and here held as a marker of progress.

If Hart Crane's apostrophe to the spirit of America's industrial imagination in *The Bridge* represents the modern inspirational high point of the Hudson River Valley, then Robert Lowell's poem "The Mouth of the Hudson" (1961) captures the nadir of consciousness about the river. Thirty-one years on from Crane's poem, the Hudson had been taken over in large part as an industrial river whose natural purpose had been subverted by the swollen needs of manufacturing and trade. Lowell's poem illustrates the negative effects of being separated from the landscape, and represents a complete reversal of Steendam's theme in that the river is now choked by mankind's industriousness rather than providing too much bounty to control.

> A single man stands like a bird-watcher,
> and scuffles the pepper and salt snow
> from a discarded, gray
> Westinghouse Electric cable drum.
> He cannot discover America by counting
> the chains of condemned freight-trains
> from thirty states. They jolt and jar
> and junk in the siding below him.
> He has trouble with his balance.
> His eyes drop,
> and he drifts with the wild ice
> ticking seaward down the Hudson,
> like the blank sides of a jig-saw puzzle.
> The ice ticks seaward like a clock.
> A negro toasts
> wheat-seeds over the coke-fumes
> of a punctured barrel.
> Chemical air
> sweeps in from New Jersey,
> and smells of coffee.
> Across the river,
> ledges of suburban factories tan
> in the sulphur-yellow sun
> of the unforgivable landscape.
>   (Lowell 2007, 42)

The imagery of nature is only referenced as an ironic counterpoint: the man stands "like a bird-watcher" while the poem calls into question whether any birds exist in this landscape. The ice is disconnected like a jigsaw puzzle, and "ticks" like a mechanical clock toward the sea. The man in the text is unable to balance and is unable to discover America—an ironic nod to the early explorers like Amerigo Vespucci and Verrazano (whose eponymous bridge defines the mouth of the Hudson and was under construction from 1959–1964). The river's function, alluded to in the title, is much like involuntary emesis since it vomits out the region's waste to the sea. Here we see the hallmarks of the industrial phase of the Hudson's history as an unrelenting separation of the human from the landscape: the river performs a function of business by distributing waste, is owned by corporate rather than human interests, and has no other measurable influence on the lives of its inhabitants.

In our own time the Hudson has become a touchstone for renewed consciousness about the human relationship to one's environment. One might say we have been engaged in climbing down from the majestic parapets of our bridges to feel the water once more. With the increased understanding of our biological interdependence with the river's ecosystem as symbiotic elements in the biosphere, we have a greater appreciation of the Hudson as source as well as resource. It is still one of the busiest waterways in the nation, as easily plied as Henry Hudson had hoped. It is also, of course, the birthplace of the environmental movement, and intimately involved in the birth of American literary and artistic consciousness. It still functions as the contemplative source for many poets, artists, and working commuters on a daily basis, the latter of whom ride its shores every day going to and coming from New York City.

Poet Billy Collins describes that journey in his poem "Albany" and illustrates the parallax that occurs when the individual locks into a kind of consciousness with the river. Here the focus is so intently developed that one loses track of whether the river or the train is in motion. The effect produces "wonderment" in the observer and simultaneously illustrates the differences in the relationship with the river we have in the modern world. Collins, speeding along faster than Robert Juet could have imagined, is nevertheless intimately involved in the human contemplation of place: here the Hudson is a source of inspiration and wonderment for both its beauty and bounty.

> As I sat on the sunny side of train #241
> looking out the window at the Hudson River,
> topped with a riot of ice,
> it appeared to the untrained eye
> that the train was whizzing north along the rails
> that link New York City and Niagara Falls.
>
> But as the winter light glared
> off the white river and the snowy fields,
> I knew that I was as motionless as a man on a couch
> and that the things I was gazing at—
> with affection, I should add—
> were really the ones that were doing the moving,
> running as fast as they could
> on their invisible legs
> in the opposite direction of the train.
> The rocky ledges and trees,
> blue oil drums and duck blinds,
> water towers and flashing puddles
> were dashing forever from my view,
> launching themselves from the twigs
> of the moment into the open sky of the past.
> How unfair of them, it struck me,
> as they persisted in their flight—
> evergreens and electrical towers,
> the swing set, a slanted fence,
> a tractor abandoned in a field—
> how unkind of them to flee from me,
> to forsake an admirer such as myself,
> a devotee of things—
> their biggest fan, you might say.
> Had I not stopped enough times along the way
> to stare diligently
> into the eye of a roadside flower?
> Still, as I sat there between stations
> on the absolutely stationary train
> somewhere below Albany,
> I was unable to hide my wonderment
> at the uniformity of their purpose,
> at the kangaroo-like sprightliness of their exits.
> I pressed my face against the glass
> as if I were leaning on the window
> of a vast store devoted to the purveyance of speed.
> The club car would open in fifteen minutes,
> came the announcement
> just as a trestle bridge went flying by.
> (Collins 2003, 55–57)

Collins ends the poem with a counterpoint that illustrates the conflict in the modern world between such profound recognition of place against the speed and distractions of our time: the moment of profundity is cut off by the announcement that the bar car has opened. It is in this modern conundrum that most of us now exist: torn between the items offered by the Hudson for our contemplation, drawn to the blue oil barrel and the eye of a roadside flower, often as the riverscape speeds by.

The arc of human relationship with the Hudson as revealed in the literary record is understandably complex. Its dynamic is determined by many disparate factors that include the writer's individual temperament, the intent and the interpretation of the written record, and the genre in which the record takes place. But in its wider perspective, the human record of the Hudson reveals the oscillating dynamic its inhabitants have shared with its role in their lives. It has been a source and determinant of culture and living conditions; a newly discovered land of wonder; a gathering of resources to be profitably exploited; an impossibly challenging landscape to manage; a political lifeline; a source of national inspiration; an industrial resource and liability; and a landscape for the reflection of the complexity of life. The Hudson has mirrored human aspirations and simultaneously influenced them, making life on its banks a significant model of how an environment and its human population actively and extensively interact.

NOTES

1. This is Susan Tarrow's translation in Lawrence C. Wroth, ed., *The Voyages of Giovanni da Verrazzano, 1524–1528* (New Haven: Published for The Pierpont Morgan Library by Yale University Press, 1970). It is also reprinted in Pritchard 2002, 356–58.
2. *Henry Hudson's Voyages from Purchas His Pilgrimes by Samuel Purchas* (Ann Arbor: University Microfims, 1966), 591–92. A transcription by Brea Barthel is available on the New Netherland Museum's Web site: http://www.halfmoon.mus.ny.us/Juets-journal.pdf.
3. See Russell Shorto, *Island at the Center of the World* (New York: Vintage, 2007), 301–307 for a distinction between the English and Dutch sentiments in Massachusetts and New Amsterdam during colonization. James Axtell makes a similar point: "Nearly all the colonial charters granted by the French and English monarchs in the sixteenth and seventeenth centuries assign the wish to extend the Christian Church and save savage souls as a principal, if not the principal, motive for colonization." The Dutch are distinctly different in that regard (1981, 43).
4. Selections quoted from Richard Nicholls, *Esopus Indian Treaty 1665: an Agreement made between Richard Nicholls Esq. Governor under his Royal Highness the Duke of Yorke and the Sachems and people called the Sopes Indyans* (New York: Ulster County Clerk Records Management Program Archives Division, 2002).

REFERENCES CITED

Adams, A., ed. 1980. *The Hudson River in literature.* New York: Fordham University Press.

Axtell, J. 1981. *The European and the Indian: Essays in the ethnohistory of colonial North America.* Oxford: Oxford University Press.

———. 1988. *After Columbus: Essays in the ethnohistory of colonial North America.* Oxford: Oxford University Press.

Bryant, W. C. 1854. *Poems.* Philadelphia.

Collins, B. 2003. *Nine horses: Poems.* New York: Random House.

Frank, W., ed. 1946. *The collected poems of Hart Crane.* New York: Liveright.

Irving, W. 1961. *The legend of Sleepy Hollow and other stories in the Sketch Book.* New York: Signet.

Kurlansky, M. 2006. *The big oyster.* New York: Ballantine Books.

Lowell, R. 2007. *Selected poems.* New York: Farrar, Strauss, and Giroux.

Murphy, H. 1969. *Anthology of New Netherland; or, translations from the early Dutch poets of New York, with memoirs of their lives.* Port Washington, NY: Friedman.

Pritchard, E. T. 2002. *Native New Yorkers: The legacy of the Algonquin People of New York.* San Francisco: Oak Council Books.

Quinn, D. B. 1977. *North America from earliest discovery to first settlements: The Norse voyages to 1612.* New York: Harper and Row.

CHAPTER 22

THE PAST AS GUIDE TO A SUCCESSFUL FUTURE

Robert E. Henshaw

IN 1978 A HISTORIC battle raged. Utilities had constructed power plants along the shores of the Lower Hudson River during the preceding several decades relying on the river as a heat sink. At the time they were built, the river was heavily polluted throughout its length, there was little public interest in the river, and little was known about the distribution of fish. "Fishkills" were common at the cooling water intakes. As regulatory oversight evolved in the 1970s these losses of fish began to concern regulatory agencies and particularly citizens, as described by Young and Dey. All of the nine large power plants were facing renewal of their operating licenses. Armed with NEPA, the EPA consolidated all of the agencies' cases into one litigation, and eventually negotiation, as described by Butzel and myself. EPA sought to require closed cycle cooling at six of the power plants, which would reduce their water withdrawals by ca. 95 percent. At that time, state-of-the-art for closed cycle was 200 m (650 ft.) tall natural draft cooling towers. If built, these would have been the tallest man-made structures between New York City and Montreal. The environmental organizations, Hudson River Fishermen's Association and Scenic Hudson Preservation Conference, intervened in support of the federal government case seeking protection of the fish populations. In contrast, New York State's DEC and Office of Parks, Recreation, and Historical Preservation argued that while protecting fish populations should be done, construction of gigantic towers would unacceptably degrade the scenic quality of the natural forested mountainous setting. The state agencies acknowledged the conundrum: protecting one environmental resource came at the expense of another.

Of course, it could hardly have been otherwise in the Hudson River Valley. This river has served multiple interests since well before the arrival of European immigrants. Throughout this volume our authors have considered the obligatory close feedback relation of human uses of the resources that changed (i.e., impacted) those resources, and the changes in the resources that forced, or "drove," subsequent modifications of use of those resources (recall Fig. 2.1). As Findlay pointed out in chapter 2, many societal forces determine the uses of resources. These may stress the resources, requiring management of either the resources or the uses, or both. If demands conflict, so too may the regulatory responses. It would be difficult to find a river system with more striking resources and mutually conflicting demands than the Hudson River. If the ultimate goal is to live sustainably with these resources in and on this river, then past history must be a guide to that future.

HUMAN USES THAT CHANGED THE ECOLOGY

Forests

Hudson's crew noted the abundant forests along the shore in their log. Resident Native Americans relied on them for nuts, medicines, and access to forest dwelling animals (Lindner, ch. 7). Arriving immigrants relied on them to build their houses, but cleared them to create their pastures and croplands (as Grossman describes in chapter 8). During the seventeenth century the forests in the Hudson River Valley were progressively removed as new homesteads required more land for pastures and agriculture (Henshaw, ch. 1 and Part II; Vispo and Knab-Vispo, ch. 12; Litten, ch. 11). During the eighteenth and early nineteenth centuries forests were virtually cleared for tanning bark, charcoal to sustain a growing iron industry, and agriculture (Peteet et al., ch. 9; Thompson and Huth, ch. 10; Vispo and Knab-Vispo, ch. 12). By the middle of the nineteenth century, New York State was largely deforested. Accounts at the time indicated it was recognized that animal populations and distributions also were heavily impacted. Daniels, in chapter 4, and Breisch, in chapter 5, describe population declines in fish and in amphibians and reptiles. Declines in forest-dwelling birds and mammals were known. During the late nineteenth and early twentieth century, most of the forests regenerated from seeds and sprouts, albeit not necessarily with the same species mix as the primordial forests (as noted by Thompson and Huth, ch. 10). Today, New York State is more than 60 percent forested again. As these forests age, their future is uncertain. In the Hudson River Valley urban sprawl, real estate development, and conversion to orchards and vineyards frequently trump forest preservation. Old growth forests probably will not return to the Hudson Region except in isolated pockets in the Hudson Highlands.

Agriculture

Colonial homesteads, crop fields, and pastures were carved from the surrounding forests. As farm soils became exhausted, farmers moved on and cleared new tracts until virtually all tillable lands were in cultivation or various stages of abandonment (Litten, ch. 11). Much of the reforestation that has occurred is on fallow farmland. Because farmlands mostly occupy gentle terrain they increasingly became targets for commercial and residential development. By the mid-twentieth century, loss of prime agricultural land, particularly in the northern portion of the Hudson River Valley, had become alarming. In response, the NYS Department of Agriculture and Markets created Agricultural Districts in 1971. This first-in-the-nation program sought to forestall the conversion of prime farmlands to nonagricultural uses through a combination of landowner incentives and protections. Landowners who agreed to keep their land in production received favorable taxation rates and protection from overly restrictive local laws or government acquisition of the land. Development pressure throughout the Hudson River Valley is unrelenting, however, as urbanization spreads northward from New York City.

Today, and hopefully tomorrow, there is rekindled interest in eating locally produced foods. Farmers markets throughout the valley and in New York City during the growing season are providing access to local fresh fruits, vegetables, and meats, and at many, musical entertainment. One may hope that the present new interest in "eating healthy" will persist and further stimulate preservation of sustainable agriculture in the valley.

Suburban Communities

Suburban development means replacement of forests and agricultural lands with lawns and horticultural, often exotic, species (as Teale discussed in chapter 13). Typically, these lawns demand increased use of chemical fertilizers, pesticides, and defoliants. Simultaneously, increasing amounts of impervious surfaces, such as parking lots and roadways, provide quick direct runoff of residues.

Tributaries to the Hudson River

Threading through communities, the tributaries to the Hudson are most directly impacted by land uses and maintenance. Non-point sources of pollution from agriculture, impervious surfaces, and suburban yards most directly affect the tributaries. Robert Boyle, the originator of the Hudson River Fishermen's Association, recently passionately argued that regulatory attention is inadequate on these headwater streams and must be enhanced in the future (Boyle, RH 2010, speaking at HRES Annual Meeting, Staatsburg, NY). Concern for the main river channel is inadequate, he said, without protection and management of the tributaries.

Water Quality

For the most part, industries that were heavy polluters of the Hudson River are gone or their impacts ameliorated, and the remaining companies minimize effluents or eliminate releases of hazardous wastes. One example of a company that released heavy metals, caused biological impacts, was closed and removed, and whose site was remediated was discussed by Levinton in chapter 16. The car assembly plant that colored the river with paint wastes, as well as many other industrial sites, are being redeveloped for riverside housing or river-benign commerce. Chemical releases today are regulated and reduced to low levels under permits from NYS DEC. One particularly problematic pollutant, polychlorinated biphenols (PCBs), once released under DEC permit into the Upper Hudson from a General Electric capacitor manufacturing plant, now so taint fish in the Lower Hudson that no fish caught in the river is considered safe to eat. The PCBs are being dredged from "hotspots" in the Upper Hudson River bottom in the largest such cleanup effort in history. During dredging some of the PCBs are resuspended in the water column and move downriver. GE and the EPA are debating whether the amount of resuspension is likely to be environmentally important.

Old-timers remember swimming in the Hudson River using a sweeping stroke to clear flotsam out of the way. Today, sanitary wastes are much reduced. Primary treatment of sewage mandated during the 1970s and secondary treatment during the 1980s and 1990s now largely control sewage contamination (as discussed by Tonjes et al., ch. 15). The oxygen block south of Albany that precluded fish passage is gone. The entire Lower Hudson River is swimmable from just south of Albany to New York City during dry periods. However, during heavy rains the river is temporarily contaminated. All communities use combined sanitary and storm sewers because of the cost of separate systems. These urban systems may be overwhelmed during large rain events, and raw sewage is washed directly into the river. Under pressure from the state and federal governments, communities are searching for affordable ways to manage sewage during high rains. Solving sanitary waste contamination looks easy however, compared to the newly recognized threats from discarded and excreted pharmacological residues and metabolic byproducts. Sanitary treatment systems simply were not designed to manage such materials.

Shorelines

Former rip rapped shorelines and wooden bulkheads are aging and eroding, returning lost riparian habitat to the river. Embayments that were created by deposition of dredged materials during shipping channel maintenance (see Nitsche et al., ch. 6), and by railroad construction on trestles and raised gravel beds are valued. Several are managed today by the Hudson River National Estuarine Research Reserve, a federal/state research, education, and management program administered by DEC.

Aquatic Ecosystems

We have come a long way since 1966 when the Hudson River Valley Commission published a report on the biological resources of the Lower Hudson River in which there was only a list of possibly interesting locations. The fish populations have been more studied in the Hudson than any other river—by 2010 well in excess of $110 million between the power companies and the Hudson River Foundation alone—and more than all other components of the aquatic ecosystem. Recent estimates indicate that the populations of most or all key fish species are currently reduced as stated by Daniels et al. in chapter 4. The causes are not readily apparent. Human predation is now minimal. Power plants and communities withdrawing drinking water from the river are now the only major water withdrawals. Power plant predation continues, but as discussed by Young and Dey in chapter 18, studies sponsored by the utilities lead them to conclude that the power plant–related impacts to fish populations may be sustainable. Any "fishkills" however, remain contentious to the public, and DEC currently is seeking to require closed cycle cooling at all major power plants as their operating permits are renewed. The 1970s gigantic natural draft cooling towers are no longer the state-of-the-art; convective/radiant "dry" cooling is now feasible, though expensive. If constructed, this would reduce utility predation on fish by ca. 95 percent. Carefully sited to be mostly obscured by forest cover, the dry cooling towers, although large, would impose on the scenic quality less than natural draft cooling towers would have. An in-between technology, called hybrid cooling, combines a dry tower with a fan-assisted evaporative tower. Their steam plumes, however, when emitted, would impact the scenic quality by conveying an industrial character, and could cause fogging and icing at times.

In addition to water withdrawals, other stresses continue. Impacts to fish populations are mixed and in most cases uncertain. Since 1993, zebra mussels, arriving from Asia, probably in a ship's ballast water, have caused major decreases in algae and detritus particulates, thus decreasing the amount of energy available to mainstream species of the food web (Strayer et al. 1999). The river is now exhibiting increased annual temperature probably due to global warming. Non–point source pollutants continue to increase. Many organic pollutants are hormone disruptors or have subtle or long-term metabolic effects in fish and other organisms. Some bioaccumulate in the food web. Reduced reproduction has been shown in shoreline populations of predatory birds due to PCBs. Direct effects on fish populations have not been established. One may surmise, however, that greatest impact of these pollutants in fish might be manifested in the species at the top of the food web, that is, in several of the species of concern at the power plants.

Demography

While the population of the Upper Hudson River region remains quite static, the population of the Lower Hudson River region is growing more rapidly than in any other part of the state. Urban sprawl is sweeping northward from New York City. With the entire Lower Hudson now swimmable and without unpleasant odors, communities are repopulating the shores with boat accesses and riverside parks where access under or over the railroad tracks permits. Homeowners are peppering the shores with interesting, quality new homes wherever accessible to the shoreline and on lands overlooking the river. Multiple-family housing areas are being constructed on larger parcels, especially in the southern half of the river. Second homes, and even primary dwellings in the upper half of the estuary, are increasingly attractive options for commuters relying on reliable mass transit to New York City. Vigilance will be required in the new century to assure that the river will not be "loved to death."

RESPONSES TO CHANGED ECOLOGY

Regulation, Planning, and Environmental Oversight

Early in the twentieth century, the Hudson River was largely abandoned by the public. Agencies lacked regulatory oversight and even enthusiasm for such responsibility. Today, many federal and state agencies have regulatory responsibilities (see list of web addresses for agencies at the end of this volume). Most of these enjoy support among citizens and, with caveats, also among industries. Federal agencies oversee major industrial sites and power plants. By state statutes, DEC's environmental quality divisions have broad responsibilities for virtually all water, air, land, and wastes. Permits for most federal regulation of siting, water withdrawals, and effluent discharges are administered by DEC, with the federal agencies in an oversight role. Additionally, the Division of Coastal Resources of the NYS Department of State has administrative control of major siting along the Lower Hudson River because the river is tidal up to Troy (see "An Abbreviated Geography" in this volume). Coastal Resources seeks to protect scenic quality in declared Scenic Areas of Statewide Significance (SASS 2010).

Large proposed construction, then, is well managed by agency oversight. However, most development is from many smaller projects. Because Home Rule controls, local communities more than state agencies determine whether, where, and in what form much siting occurs. Many smaller communities cannot afford a highly trained environmental review staff. Many pet projects are declared, based on little analysis, to have no important environmental impacts. That is, they receive a Negative Declaration, or a "Neg Dec." Although state agencies do have the opportunity to reserve out important cases for their administration, this happens too infrequently because the state agencies also are strapped for funding. Currently limited coordination occurs among communities. The result also means uneven application of environmental criteria.

Fiscal problems present the greatest hurdles to adequate environmental oversight at both the state and local levels. DEC, strapped for funds, has increasingly routinized environmental analysis and permit review. Much environmental assessment and field analysis is relegated to simply completing forms to aid quick agency oversight. SEQR, much like NEPA, requires that the applicant for a permit to construct a proposed project must complete an Environmental Assessment Form (the "EAF"). Too many proposed projects all but automatically receive a "Neg Dec" by in-office review.

Through the years, attempts at regional planning with and without controlling regulation have been tried and disbanded. In the 1960s and 1970s the Hudson River Valley Commission sought to create region-wide siting criteria, including design and aesthetics. Developers and industries were to meet these criteria for approval of their applications. HRVC was terminated when communities and industries objected to imposition of siting criteria. HRVC was followed by the Hudson River Valley Greenway that emphasized a voluntary buying-in by communities to siting guidelines. This too was deemed too much central control, and its mission was restructured into a trails and advocacy agency (Greenway 2010).

A different tack was taken by Congressman Maurice D. Hinchey, who hailed from the Hudson River Valley. He devised the concept of a Hudson River Valley National Heritage Area to recognize the national importance of the estuary. The HRVNHA was designated by Congress in 1996 and became the first of forty-nine federally recognized National Heritage Areas administered by the National Park Service (HRVNHA 2010).

In 1987, the Hudson River Estuary Program (HREP) was established as an office of the commissioner of DEC. HREP maintains research, education, and support programs that bridge many other parts of DEC. Its Web site shows an Action Agenda of: (1) ensure clean water including swimmable water quality from Troy to New York City; (2) protect and restore fish, wildlife, and their habitats, including wetlands and tributaries, and operating the joint federal/state Hudson River National Estuarine Research Reserve; (3) provide recreation and river access and trails; (4) adapt to climate change including mapping flood-prone areas; and (5) conserve the world-famous scenery including helping communities to adopt scenic and open space plans and improve land use plans (HREP 2010). It likely is this latter approach of enabling communities rather than seeming to limit them that has saved HREP from going the way of its predecessor agencies. Superbly administrated, it works to facilitate communication among communities, and encourages appreciation for regional perspectives and cooperation. Defying the stereotype of regulatory agencies, HREP is respected, supported and applauded, and is a sought-after partner by communities and developers alike. We have to hope its future in these days of fiscal austerity is secure.

Recent trends to preserve, enhance and develop the historic, agricultural, scenic, natural, and recreational resources have involved groups of communities in the region south of the Adirondack Park banding together. For the Upper Hudson River, state legislation in 2006 created the "Historic Saratoga-Washington on the Hudson Partnership" (Partnership 2010). It acknowledged the tradition of independent community home rule, and then provided enhanced power through regional collaboration. River communities can opt into the Partnership at any time simply by local resolution. The Partnership facilitates collaboration among not-for-profit organizations, local governments, and private groups to develop projects that enhance agricultural and open space protection, economic and tourism development, and the protection and interpretation of our natural and cultural heritage. As of 2010 the Partnership claims strong support and cooperation among fifteen cooperating communities and fifteen institutional and NGO Partners.

The most important process mechanism for protecting ecosystems, which may be traced back to the principles laid down in the Storm King controversy, is the concept of self-disclosure. Anyone who would use environmental resources may be presumed to have at least some effect on those resources or others. NEPA adopted environmental impact assessment, that requires the applicant to analyze and describe, that is, to self-disclose how its proposed activity might impact the environment in an Environmental Impact Statement (the "EIS"). The public and its agencies then may decide whether the likely impacts will be acceptable. The State Environmental Equality Review Act (SEQRA) also adopted the self-disclosure approach. At a minimum, this means filing the Environmental Assessment Form; at most an EIS (SEQR 2010). State regulations then require the state to take a "hard look," that is, to examine all potential direct, indirect, and cumulative impacts of a proposed action.

Ideally, siting today occurs with more forethought and required protections.

Litigation, Mediation, and Negotiation

Both NEPA and SEQRA presume all disputes will be resolved in adjudicatory hearings. In the 1970s many siting cases went to protracted hearings where decisions often turned on skillful argument or entrapment of an expert witness. That decade saw the birth and growth of environmental law. By the 1980 settlement of the cooling tower case, participants of all parties had come to recognize that it is impossible to find perfect truth through litigation. They learned at the deepest level that face-to-face respectful negotiation among parties with competing legitimate points of view should most effectively and cheaply produce an outcome that all parties can live with. (The present reader should ponder every single word in this sentence.) It is, in fact, possible for each party to walk away satisfied that it got more than it would have hoped to get had it litigated. New York's State Environmental Quality Review Act, written by, and revised by, that generation of litigation-phobes now encourages all parties to seek accommodation by direct negotiation in the hope of avoiding costly litigation. During the 1980s and 1990s public dispute resolution services came to be provided by counties and towns and are today available for settlement of all sorts of disputes from environmental to property to domestic.

In the new century, all of the power utilities along the Hudson River are owned by national or international corporations. Their leadership can no longer be assumed to be citizens of the area; no longer can the community expect either pride of ownership, or embarrassment over malfunctions, which we might hope for from a resident. However, recent events suggest parties may have been guided by the earlier cases that occurred here in the Hudson River Valley. All of the large fossil fueled steam generating plants were successfully relicensed by 2011 with modified operating standards that are expected to be protective of the environment. The relicensing of Indian Point nuclear power plants was "progressing." Representatives of both the power companies and the state claim (pers. comm.) they are cognizant of the earlier battles and the importance of the 1980 settlement. They characterize discussions as substantive, respectful, and collaborative. One may be optimistic that history will NOT repeat itself. It was particularly for this reason that the present volume emphasizes the environmental history associated with Hudson River power generation.

Looking ahead, the fiscal restraints on regulatory agencies at local, regional, state, and national levels can be expected to force changes in environmental surveillance and management. This need not require reduced regulation in the name of enabling development and being business-friendly. Nor should environmental analysis be reduced because of inadequate staff. Quite the opposite. Forcing society to accept potentially permanent environmental losses or modifications that indeed might further one business but impact widely the natural environment is shortsighted. Rather, what is needed is for applicants and regulators to come to accept they have interlocking responsibilities to future societies and that this requires collaboration and not adversity, and certainly not reduced overall surveillance and regulation. All parties that wish to begin development of any kind should avail themselves of SEQR and NEPA environmental analysis processes. They should voluntarily assure full disclosure of their plans and the possible environmental impacts from the first conceptual stage onward. If they demonstrate complete openness throughout the scoping process, they stand to obtain more acceptable decisions more amicably and cheaply, and most expeditiously. At the same time, regulators must dispense with the presumption that every regulatory process should be adversarial. When the applicant has demonstrated complete openness, and the public needs are accommodated to the maximum extent, then the regulatory agencies can, and should, become collaborators. They can *facilitate* timely approvals.

The 1970s New York State power plant siting law, Article VIII, was designed to be adversarial and to require onerous amounts of studies that did not facilitate decision making. Of all power plants proposed, only one was sited and built. All others became entangled in endless squabbling that could not be resolved. The one plant that was approved and constructed, Somerset, was my responsibility as an environmental analyst within DEC. Looking back now, my greatest professional satisfaction was to develop a decision-making process that enabled approval at the secondary site. DEC then supported the applicant, New York State Electric and Gas Company, and discussions evolved into finding optimum ways to meet the many environmental challenges.

RESEARCH

Utilities continue today to sponsor applied research on potential impacts of power plant operations, especially to fish populations. These studies build on the over $70 million of studies they had sponsored before the consolidated power plant case settlement in 1980. The Hudson River Foundation, created in that settlement, sponsors applied and basic studies throughout the river. As of 2010 it has awarded more than $35 million in competitive research and environmental education grants since 1980 (Suszkowski, D.J. 2010, pers. comm.). Many institutions maintain active research programs. Scientists at Cary Institute for Ecosystem Studies conduct studies of nutrient availability, ecological productivity, and other fundamental processes in embayments and the river channel. Lamont Doherty Earth Observatory of Columbia University and the State University of New York at Stony Brook, as well as several environmental consulting firms, deploy research vessels throughout the Lower Hudson. In 2010 the Beacon Institute for Rivers and Estuaries was developing teaching and research facilities at Rensselaer Polytechnic Institute in Troy, in Beacon, and in New York City. A new program, the Hudson River Environmental Conditions Observing System (HRECOS), by Stevens Institute of Technology and others deployed permanently placed as well as free-floating monitoring devices continuously reporting water quality parameters via satellite (HRECOS 2010; Hudson, 2010). It was truthfully said in 1978 that the Hudson was the most studied river in the world, and that surely remains true in the new century.

Education

Virtually all of the research programs except for the utility-sponsored studies are linked with environmental education programs. Many colleges and universities maintain graduate and undergraduate research programs, field courses, and summer institutes. The Cary Institute of Ecosystem Studies supports undergraduate, graduate, and postdoctoral student research. In 2004, the Environmental Consortium of Hudson River Colleges and Universities was formed to facilitate communication and cooperation among the more than forty academic institutions with field and classroom, and semester and summer, education programs.

Environmental educational opportunities for K-12 students abound in the Lower Hudson River today. The most prominent and respected is the Hudson River Sloop Clearwater. Since 1969 it has provided twice-daily three-hour sails from the beginning of April through the end of October during which students directly experience all components of aquatic ecosystems at five on-deck teaching stations and hear discussions and songs about ecology and good management practices. Clearwater also provides classroom support especially to urban schools with limited financial resources. Often referred to as the "Flagship of the Environment," Clearwater's floating classroom has educated more than one-half million people, and has been replicated in many rivers worldwide.

In the 1980s, Hudson Basin River Watch was established to provide education through in-field monitoring studies in the Hudson River and its tributaries for more than two hundred high school students, teachers, and volunteers each year. Their annual day-long conference for students to report their findings are of such quality that it is difficult to realize one is not in a meeting of professional scientists. The program has been acclaimed as a valuable approach to improving the water quality of the Hudson River Basin through research-based education. We have come a long way since the early 1970s when Boyce Thompson Botanical Institute, then located in Yonkers, was roundly criticized for deploying presumed-to-be-incompetent high school students to study Haverstaw Bay!

Advocacy

A very important adaptive response to impacts to the river's ecosystems has been improved public involvement. Citizens were empowered by the decision of the Second Circuit Court of Appeals in the Storm King pumped storage power plant case that enabled them to sue on behalf of the environment. This power has been somewhat reduced through subsequent litigation, so that citizens must demonstrate direct harm. Citizen participation was codified in NEPA, and then in New York's SEQRA.

In the Hudson River Valley, the residents are well educated, attentive to corporate and private proposals that might harm their valley, and most importantly they are willing to support a plethora of topical grassroots not-for-profit organizations to express impassioned and energetic positions on environmental concerns. It is common for these groups to find motivation in the Hudson River School of landscape painting (Flad, ch. 20). Hudson River Sloop Clearwater devotes one-third of its mission to environmental advocacy. Riverkeeper deploys a surveillance vessel daily to watch out for illegal discharges and other potential problems. Scenic Hudson Preservation Conference, which was created by citizens to oppose the Storm King pumped storage power plant, and which may be directly credited with initiating the environmental movement (Butzel, ch. 19), continues actively and has metamorphosed into a major land conservancy organization channeling scenically, historically, and environmentally important properties into public protection.

Liaison

Keeping abreast of so many research and education programs within the Hudson River Valley is a challenge. In 1969, the Hudson River Environmental Society was established to provide for non-advocacy nonadversarial information exchange among decision makers, researchers, and educators, as well as interested public. It does so through conferences, books (of which this is one), monographs, and newsletters. HRES believes that sound public policy must be science-grounded, although it takes no position on specific technical issues.

Very recently, the Hudson River Watershed Alliance was created to integrate all environmental aspects, users, and advocates. It was nurtured into existence within Clearwater Inc. before fledging to independence in 2010. One can hope that one comprehensive liaison organization such as the HRWA will succeed in providing umbrella coordination among all not-for-profit organizations, and support for all users and organizations on the river. At the end of this volume we provide a list of Web addresses of many Hudson River organizations.

There is emerging coordination among research, education, and advocacy programs within the Hudson River Valley. The same cannot yet be said for coordination between the Hudson River and other riversheds. One organization in 2010, the River Network, is hoping to become the "alliance of alliances" to provide coordination among riversheds. It provides extensive how-to training for not-for-profit organizations, nearly all of which suffer similar leadership, management, membership maintenance, and funding problems.

Industries that would seek to impact this river or its surroundings should anticipate active, informed, and supported citizens. This does not mean that applications will not receive a fair evaluation, or that they will be summarily rejected by the administrative agency. With due diligence, and especially with openness, most applications will be approved after considering all options, minimizing most impacts, and often accommodating interests other than the applicant's at the same time.

What the Hudson River has derived from its distant and recent past surely will prepare it for its future:

- awareness of "the Sublime" that makes this river a national treasure
- knowledge of its resources through continuing surveillance and research
- multiple sources of relevant education enabling its informed public
- enforcement of protective science-based regulations by agencies at all levels
- access to public and private funding by interested citizens of the river region
- active, responsible, and empowered citizen participation

To these we must hope will be added in the near future:

- increasing intercommunity cooperation leading to collaboration
- more regional planning

To be true to my generation of professional environmental analysts, I must include a strong personal plea for:

- environmental assessment and full self-disclosure of proposed actions based on thorough required environmental analyses

followed by:

- complete review by trained professionals using field observations to confirm the facts in the EIS; and throughout the process from initial "Scoping" to final determination
- friendly cooperation and respect among all participants.

It appears that today's uses of the Hudson River, responsive to the many drivers, seem to be leading to a very rosy future for the Hudson River.

REFERENCES CITED

Greenway. 2010. Hudson River Valley Greenway. http://www.hudsongreenway.state.ny.us/; accessed December 2010.

Henshaw, R. E. 1978. Introduction, NYS DEC testimony in USEPA consolidated power plant adjudicatory hearing.

Hudson. 2010. http://www.hudson.dl.stevens-tech.edu/maritimeforecast/mobile; accessed December 2010.

Hudson River Environmental Conditions Observing System (HRECOS). 2010. http://www.hrecos.org/joomla/; accessed December 2010.

Hudson River Estuarine Program (HREP). 2010. Hudson River estuary action agenda. http://www.dec.ny.gov/lands/5104.html; accessed October 2010.

Hudson River Valley National Heritage Area (HRVNHA). 2010. http://www.hudsonrivervalley.com/; accessed December 2010.

Lifset, R. D. 2005. *Storm King Mountain and the emergence of modern American environmentalism, 1962–1980*; PhD diss, Columbia University.

Partnership. 2010. The Historic Saratoga-Washington on the Hudson Partnership. http://www.upperhudsonpartnership.org; accessed November 2010.

Significant Areas of Statewide Significance (SASS). 2010. http://www.nyswaterfronts.org/SASS/SASS1/AppendixC.htm; accessed December 2010.

State Environmental Quality Review Act (SEQR). 2010. The SEQR Handbook, 3rd edition. http://www.dec.ny.gov/docs/permits_ej_operations_pdf/seqrhandbook.pdf; accessed December 2010.

Strayer, D. L., N. F. Caraco, J. J. Cole, S. Findlay, and M. L. Pace. 1999. Transformation of freshwater ecosystems by bivalves a case study of zebra mussels in the Hudson River. *BioScience* 49: 19–27.

Afterword

Robert E. Henshaw

WE HAVE SOUGHT to demonstrate that the Hudson River—*Muh-he-kun-ne-tuk*—is unique. Its history, like its prehistory, weaves together its special resources—the Sublime—and its spirited entrepreneurial residents. For thirteen millennia we used the resources and changed them little. During the recent four centuries, we used the resources much more intensively, changed the ecosystems, and repeatedly have been forced to rethink what we were doing. Now as we begin another four centuries, or (wishfully) thirteen millennia, we must learn from our history.

The Hudson River has been uniquely important to New York's culture, history, commerce, and tourism. It provided identification, water, food, livelihood, transportation, strategic protection, recreation, and scenic splendor for its inhabitants and for countless travelers. It connected interior to coast, producers to consumers, rural to urban, and to the access westward into an expanding nation. It carried raw materials from the interior to New Amsterdam, now New York City, spurring growth of the largest port and metropolis in North America. Cultures and commerce have flourished, declined, or been removed throughout the river's length, each contributing long-term changes to the Hudson River. But despite its heavy use, and the increasing settlement of upriver communities, the characteristic landscape today remains one of scenic grandeur; the Upper Hudson a tumbling mountain stream; the Lower Hudson a fjord penetrating the Hudson Highlands. The Hudson River, then, is many resources, connecting New York's past to its future; and the River is the Bridge.

Contributors

Omkar Aphale: Dept. of Technology and Society, Stony Brook University, Stony Brook, NY; oaphale@stonybrook.edu. Environmental issues in and around the New York metropolitan area in solid waste and in environmental management; impacts of sewage treatment on the environment.

Robin Bell: Lamont-Doherty Earth Observatory of Columbia University, Palisades, NY; robinb@ldeo.columbia.edu. Marine and polar geophysics.

Geoffrey L. Brackett: Vice President, Marist College; Geoffrey.Brackett@Marist.edu. Romantic literature; literature and the environment.

Alvin R. Breisch: arbreisch@yahoo.com. Amphibian and Reptile Specialist for the NYS Department of Environmental Conservation's Endangered Species Unit (ret.); project director for the New York Amphibian and Reptile Atlas Project; contributor to the *Conservationist* and co-author of two books: *The Amphibians and Reptiles of New York State: Identification, Life History and Conservation* and *Habitat Management Guidelines for Amphibians and Reptiles of the Northeastern United States*.

Albert K. Butzel: Albert K. Butzel Law Offices; albutzel@nyc.rr.com. Attorney for Scenic Hudson and the Hudson River Fishermen's Association throughout the Storm King case. Currently practices law in New York City.

Suzanne Carbotte: Lamont-Doherty Earth Observatory of Columbia University, Palisades, NY; carbotte@ldeo.columbia.edu. Marine geology and geophysics.

Robert A. Daniels: Curator of Ichthyology, New York State Museum, Albany, NY; rdaniels@mail.nysed.gov. Research fisheries biologist; long-term monitoring of fish species in the Hudson River; community structure, abundance, macrodistribution, and life history of key species; structure in near-shore fish assemblages. Curates the largest collection of Hudson River fishes.

William P. Dey: Senior Scientist, ASA Analysis & Communication, Inc.; wdey@asaac.com. Hudson River environmental program since 1974; the effects of cooling water intakes and thermal discharges at stations across the country. Pioneered the use of risk assessment techniques for assessing adverse environmental impacts.

Stuart Findlay: Senior Ecologist, Cary Institute of Ecosystem Studies, Millbrook, NY 12545; findlays@caryinstitute.org. Nutrient cycling in tidal wetlands of the Hudson; microbial carbon decomposition and the microbial ecology of streams. Hudson River Estuary Management Committee for DEC.

Harvey K. Flad: Emeritus Prof. Geography, Vassar College, Poughkeepsie, NY; flad@vassar.edu. Cultural and historic landscapes; conservation and urban history; and environmental and urban planning in America. Visual impact analysis; nineteenth-century landscape design theory and practice; environmental and aesthetic perception of landscape, including artists of the Hudson River School and contemporary photographers; tourism; urban waterfront and main street revitalization. Co-author of *Main Street to Mainframes: Landscape and Social Change in Poughkeepsie.*

Roger D. Flood: Stony Brook University, Stony Brook, NY; rflood@notes.cc.sunysb.edu. Marine geology and geophysics.

Joel W. Grossman: Geospatial Archaeology; jwg@GeospatialArchaeology.com. Prehistoric and historic archaeology of the northeast; environmental and economic history; archaeology of toxic and hazardous sites.

Wendy E. Harris: gullyroad@aol.com. Archaeological consultant, partner in Cragsmoor Consultants, Ulster County, New York. Historian for the Shawangunk Mountains Scenic Byway. Formerly Sr. Archaeologist, New York District Army Corps of Engineers. Co-author of *Yama Farms: A Most Interesting Catskills Resort.*

Robert E. Henshaw: New York State Department of Environmental Conservation, Division of Fish and Wildlife (ret.). rhenshaw@nycap.rr.com. NYS DEC Aquatic and Terrestrial Ecologist and Environmental Analyst (ret.); Adjunct Professor, SUNY Albany and Bard College (ret.); NYS technical representative to negotiations in historic Consolidated Hudson River Power Plant Case.

Paul C. Huth: Director of Research, Daniel Smiley Research Center Mohonk Preserve, New Paltz, NY; www.mohonkpreserve.org. Botanist and naturalist; Shawangunks natural history and long-term records (for the last thirty-seven years); climate change; and land management practices in the Shawangunks and greater Hudson Valley.

Lucille Lewis Johnson: Professor of Anthropology, Vassar College, Poughkeepsie, NY; johnsonl@vassar.edu. Prehistory of North America; technology and ecology; environmental studies.

Timothy C. Kenna: Lamont-Doherty Earth Observatory of Columbia University, Palisades, NY; tkenna@ldeo.columbia.edu. Marine geochemistry and sediment processes.

Claudia Knab-Vispo: Hawthorne Valley Farmscape Ecology Program; claudia@hawthornevalleyfarm.org. Interaction of plants and human land uses; use of historical flora of the county to locate botanically special areas of the county and document recent vegetation change.

Jeffrey S. Levinton: Department of Ecology and Evolution, Stony Brook University, Stony Brook, NY; levinton@life.bio.sunysb.edu. Marine benthic ecology; evolutionary biology; effect of metal pollution in the Hudson on evolutionary changes in aquatic populations. Coeditor of *The Hudson River Estuary*, published by Cambridge University Press.

Karin E. Limburg: Professor, SUNY College of Environmental Science and Forestry, Syracuse, NY; klimburg@esf.edu. Research fisheries biologist; long-term monitoring of fish species in the Hudson River; community structure, abundance, macrodistribution, and life history characteristics of key species especially herring, moronids, and eels. Co-author of *The Hudson River Ecosystem.*

Christopher R. Lindner: Archaeologist in Residence, Bard College, Annandale, NY; lindnerarch@gmail.com. Archaeology and history of Hudson Valley landscapes, natural resource uses, and social lives of early Native, African, German, and British Americans.

Simon Litten: New York State Department of Environmental Conservation, Division of Water (ret.); slitten@nycap.rr.com. Relation between land use and water quality; intersection of technology and history; toxic chemicals in water; agriculture.

Elizabeth Markgraf: Lamont-Doherty Earth Observatory of Columbia University, Palisades, NY; esm2003@ldeo.columbia.edu. Paleoecology; wetland environments; secondary school education.

Frank O. Nitsche: Lamont-Doherty Earth Observatory of Columbia University, Palisades, NY; fnitsche@ldeo.columbia.edu. Marine geology and geophysics.

Christine A. O'Connell: Reduction and Management Institute, School of Marine and Atmospheric Sciences, Stony Brook University, Stony Brook, NY; caoconne@ic.sunysb.edu. Water quality surveys in the Hudson River, New York Harbor, and New York Bight; development of research agenda for New York Ocean and Great Lakes Ecosystem Conservation Council; ecosystem management and marine zoning in New York metropolitan region.

Roger Panetta: Lecturer in History, Fordham University; panetta@fordham.edu. Urban and industrial historian; Adjunct Curator at Hudson River Museum; author or co-author of several books on the Hudson River, including *The Hudson: An Illustrated Guide to the Living River;* other Hudson River books.

Dee C. Pederson: Soil Scientist, US Dept. of Agriculture, Greensboro, GA; dee.pederson@ga.usda.gov. Soils; Hudson marshes; wetland environments.

Dorothy M. Peteet: NASA/Goddard Institute for Space Studies, New York, NY 10045; Lamont-Doherty Earth Observatory of Columbia University, Palisades, NY; peteet@ldeo.columbia.edu. Paleoecology; paleoclimate of northeastern United States; wetland environments; Hudson River marshes; Alaskan wetlands; carbon storage.

Arnold Pickman: 150 East 56th St., New York, NY; apickman@aol.com. Archaeological consultant and partner in Cragsmoor Consultants, Ulster County, New York. Principal investigator for numerous archaeological surveys and excavations in New York City, the surrounding metropolitan area, and throughout the Hudson River Valley Region.

William B. F. Ryan: Lamont-Doherty Earth Observatory of Columbia University, Palisades, NY; billr@ldeo.columbia.edu. Marine geology and geophysics.

Robert E. Schmidt: Professor, Bard College at Simon's Rock, Great Barrington, MA; schmidt@simons-rock.edu. Research fisheries biologist; long-term monitoring of fish species in the Hudson River; community structure, abundance, macrodistribution, and life history characteristics of key species. Recent work with American eel and the invasive Chinese mitten crab.

Angela L. Slagle: Lamont-Doherty Earth Observatory of Columbia University, Palisades, NY; aslagle@ldeo.columbia.edu. Marine geology and geophysics.

Sanpisa Sritrairat: Lamont-Doherty Earth Observatory of Columbia University, Palisades, NY; sanpisa@ldeo.columbia.edu. Hudson marshes; paleoecology; paleoclimate; human impact; trace metals.

R. L. Swanson: Waste Reduction and Management Institute, School of Marine and Atmospheric Sciences, Stony Brook University, Stony Brook, NY; robert.swanson@stonybrook.edu. Management and determination of impacts associated with sewage, sewage treatment, and sewage sludge generated in New York City and the metropolitan region; water quality in the Hudson River, New York Harbor, and New York Bight. Co-author, the *Hudson River Estuary Program Monitoring Plan.*

Chelsea Teale: Geography, Pennsylvania State University, University Park, PA; clt198@psu.edu. Environmental change; paleoecology; agriculture.

John E. Thompson: Natural Resources Specialist, Daniel Smiley Research Center, Mohonk Preserve, New Paltz, NY; www.mohonkpreserve.org. Baseline monitoring of ecology of the Shawangunk Mountains; land management planning to protect and adaptively steward the 6,700 acre Mohonk Preserve.

David J. Tonjes: Dept. of Technology and Society, Stony Brook University, Stony Brook; david.tonjes@stonybrook.edu. Impacts of New York City sewage treatment on the Hudson River; Harbor Estuary Program-New York Bight Restoration Plan; co-author of the *Hudson River Estuary Program Monitoring Plan*.

Conrad Vispo: Hawthorne Valley Farmscape Ecology Program, 327 Route 21C, Ghent, NY; conrad@hawthornevalleyfarm.org. Applied animal ecology; role of agriculture in providing habitat for native species and the importance of native (or at least wild) species in providing ecological services to farms.

John R. Young: Senior Scientist, ASA Analysis & Communication, Inc.; jyoung@asaac.com. Hudson River environmental studies since 1976; assists power generating industry in compliance with §316 of the Clean Water Act both as a consultant and while employed within the generating industry at power stations in New York and throughout the country.

Web Addresses of Cited and Key Agencies, Not-For-Profit Organizations, and Academic Institutions in the Hudson River Basin

FEDERAL AGENCIES

US Army Corps of Engineers, New York District	http://www.nan.usace.army.mil/index.php
Federal Energy Regulatory Commission	www.ferc.gov/
Hudson River Valley National Heritage Area	http://www.hudsonrivervalley.com/Home.aspx
US Environmental Policy Administration, Region 2	http://www.epa.gov/region02/
US Geological Survey, National Water Quality Survey	http://ny.water.usgs.gov/projects/hdsn/fctsht/su.html
Water Quality in the Hudson River Basin	http://pubs.usgs.gov/circ/circ1165/

STATE AGENCIES

NYS Department of Agriculture and Markets, Agriculture Districts	http://www.agmkt.state.ny.us/AP/agservices/agdistricts.html
NYS Department of Environmental Conservation	http://www.dec.ny.gov/
Hudson River Estuary Program	http://www.dec.ny.gov/lands/4920.html
Hudson River National Estuarine Research Reserve	http://www.dec.ny.gov/lands/4915.html
State Environmental Quality Review	http://www.dec.ny.gov/permits/357.html
NYS Dept. of State, Division of Coastal Resources	http://www.nyswaterfronts.com/index.asp
Scenic Areas of Statewide Significance	http://www.nyswaterfronts.com/SASS/SASS1/HR_Scenic_Map.htm>
NYS Museum	http://www.nysm.nysed.gov
NYS Office of Parks, Recreation, and Historical Preservation	http://nysparks.state.ny.us/
Regional Plan Association: NY-NJ-CT	http://www.rpa.org/welcome.html

NOT-FOR-PROFIT ORGANIZATIONS (NGOS)

American Museum of Natural History	http://www.amnh.org
American Rivers	http://www.amrivers.org/
Beacon Institute for Rivers and Estuaries	http://www.thebeaconinstitute.org/home/
Catskill Center for Conservation and Development	http://www.catskillcenter.org/
Croton Watershed Clean Water Coalition	http://www.newyorkwater.org/
Directory of Environmental Organizations	http://www.swimfortheriver.com/pdf/directory.pdf
Environmental Consortium of Hudson River Colleges and Active	http://www.environmentalconsortium.org/
Hudson Highlands Nature Museum	http://www.museumhudsonhighlands.org/
Hudson Basin River Watch	http://www.hudsonbasin.org/
Hudson River Fishermen's Association New York	(see Riverkeeper)
Hudson River Fishermen's Association New Jersey	http://www.hrfanj.org/

Hudson River Foundation	http://www.hudsonriver.org/
Hudson River Sloop Clearwater (or "Clearwater")	http://www.clearwater.org/category/latest-news/
Hudson River Watershed Alliance (list of participating organizations)	http://www.hudsonwatershed.org/
Hudson River Valley Greenway Council	http://www.hudsongreenway.state.ny.us/home.aspx
Hudson River Valley Institute	http://www.hudsonrivervalley.org/
Hudson Valley Smart Growth Alliance	http://www.sustainhv.org/hvsga
Hudsonia, Ltd.	http://www.hudsonia.org/
Natural Resource Defense Council	http://www.nrdc.org
Newtown Creek Alliance	http://www.newtowncreekalliance.org/
New York League of Conservation Voters	http://www.nylcv.org/
New York Public Interest Research Group	http://www.nypirg.org/
New York State Association of Conservation Councils	http://www.nysaccny.org/
New York State Association of Environmental Management Councils	http://nysaemc.org/
Our Hudson	http://www.ourhudson.org/
River Network	http://www.rivernetwork.org/
Riverkeeper	http://www.riverkeeper.org/?gclid=CMbgr7fJ_aMCFZJ95QodgVmSLA
Scenic Hudson Preservation Inc.	http://www.scenichudson.org/
Sustainable Hudson Valley	http://www.sustainhv.org/
The Nature Conservancy, Eastern Chapter	http://www.nature.org/wherewework/northamerica/states/newyork
The Nature Conservancy—NYC	http://www.nature.org/
Environmental Defense	http://www.environmentaldefense.org/home.cfm
Center for Community and Environmental Development	http://www.prattcenter.net/
Adopt-a-Waterway	http://dbserv.pace.edu/execute/page.cfm?doc_id=12540
Hudson River Maritime Museum	http://www.ulster.net/~hrmm/
Hudson River Heritage (HRH)	http://www.hudsonriverheritage.org/about.html

ACTIVE COLLEGES AND UNIVERSITIES

Cary Institute for Ecosystem Studies	http://www.ecostudies.org/
Cornell Cooperative Extension (offices in each county)	http://cce.cornell.edu/LEARNABOUT/Pages/Local_Offices.aspx
Cornell University	http://cornell.edu/
LaMont Doherty Earth Observatory of Columbia University	http://www.ldeo.columbia.edu/
Marist Hudson River Collection	http://library.marist.edu/archives/hrcs/conservationsociety.xml
Pace Institute for Environmental and Regional Studies	http://www.pace.edu/pace/dyson/research-and-resource-centers/academic-centers-and-institutes/piers/
Rensselaer Polytechnic Institute	http://rpi.edu/
State University of New York at Stony Brook	http://www.stonybrook.edu/
Stevens Institute of Technology	http://www.stevens.edu/sit/

INDEX

aal, 29*table*, 30. *See also* eel (*Anguilla rostrata*)
abandonment of farmland, 174–177, 326
aboriginal adaptation, 65–76
abortions, 107
Abramis brama (bream), 28, 29*table*
acacia (*Acacia* spp.), 88–89*table*, 91–93*table*
Acalypha spp. (copperleaf), 88–89*table*, 91–93*table*
accelerator mass spectrometry (AMS), 70, 72, 124
Acer spp. (maples), 124–125, 126–128*figures*, 137, 138, 146*figure*, 148, 168*figure*, 169, 186, 190; *A. platanoides* (Norway maple), 186, 190; *A. pseudoplatanus* (sycamore maple), 186; *A. rubrum* (red maple), 137, 138, 141, 146*figure*, 148, 170, 174; *A. saccharinum* (silver maple), 170; *A. saccharum* (sugar maple), 124–125, 138, 146*figure*, 148, 190; (silver), 170; *A. spicatum* (mountain maple), 169
acid precipitation, 46
Acipenser spp. (sturgeon), 9, 29*table*, 32; *brevirostrum* (shortnose sturgeon), xxv, 32–33; *A. oxyrinchus* (Atlantic sturgeon), 32–33
Ackerly, S., 36
acorns (*Quercus* spp.), 68, 69, 70, 71, 72, 73, 149, 172
Acorus americanus (sweetflag), 171
Acris crepitans (Eastern cricket frog), 48
Adams, Denise, 186, 188
adaptations, aboriginal, 65–76
adder's tongue fern (*Ophioglossum* spp.), 171
Adirondack Forest Preserve, 296
Adirondacks, xxi, xxii, 23, 46, 196, 295, 296, 329
Advanced Geoservices Corporation (AGC), 238
adverse environmental impact (AEI), 266–267, 268
advocacy, 332. *See also* public attitudes and awareness
aesthetics: bridges and, 258; effects on humans, 17; power plants and, 277, 282, 283–285, 289, 292, 299–307; preservation and, 296; regulation and, 329; tourism and, 307; transportation and, 251, 255, 256, 259; waste and, 229; West Point and, 233. *See also* art; scenery
aesthetic services, 10*figure*

AGC (Advanced Geoservices Corporation), 238
Agelaius phoeniceus (red-winged blackbird), 173, 190
Agkistrodon contortrix (copperhead), 48, 49
Agricultural Districts, 326
agriculture: pre-contact period, 67–73, 113; colonial period, 5–6, 153–156, 173, 177*figure*, 326; 19th century, 155–159, 167, 169; 20th century, xxi, 155–161, 175; 21st century, 326; canals and, 253; cities and, 10; climate and, 13, 97; Columbia County, 165–182; forests and, 24, 125, 153, 326; marshes and, 131; Native Americans and, 24, 96; sediment and, 13, 60; Shawangunks and, 135, 139, 142; wetlands and, 129; women and, 111. *See also* cultivation; farms; horticulture; soil; tractors
Ailanthus altissima (tree of heaven), 186
air quality, 222, 277, 286
Akebia quinata (Asian chocolate vine), 188
Albany: colonial period, 72; 19th & 20th centuries, 196; basics, xxv; canals and, 254; ferries and, 255; generating stations and, 264*figure*; herpetofauna, 44; invasive vines and, 189; railroads and, 254; shoreline data, 55; transportation and, 250; whales and, x. *See also* other *Albany locales*
"Albany" (Collins), 323
Albany beef (smoked sturgeon), xxv, 32–33
Albany County, 46
Albany Pine Bush, 44, 47
Alces alces (moose), 177
alewife (*A. pseudoharengus*), 31, 33, 263, 265, 267*table*
Alexander, Mrs., 111
alfalfa (*Medicago sativa*), 173
algae, xxiv, xxv, 131
Algonquin tribe, 73, 107, 108, 110
alkali, 154
Allaire, James, 208
Allegheny Mountains, xvi*map*, xxiii*map*
Allegheny Plateau, xxii, 67
Allium tricoccum (wild leek), 177

allochthonous material, xxi
Alnus spp., 126–128*figures*
Alosa spp., 29*table*; *A. aestivalis* (blueback herring), 31; *A. pseudoharengus* (alewife), 31, 33, 263, 265, 267*table*; *A. sapidissima* (American shad), 30, 31*figure*, 32*figure*, 33, 34, 36, 37, 262, 263*table*, 265, 267
Alpine Swamp, 123, 124, 129
altitudes, xxi, 58
amaranth (*Amaranthus* spp.), 88–89*table*, 91–93*table*, 95, 96, 103, 104–105
Ambloplites rupestris (rock bass), 30*table*
Ambrosia artemisiifolia (common ragweed), 125, 126–128*figures*, 129, 169, 176
Ambulacrum, 94
Ambystoma tigrinum (tiger salamanders), 44
ambystomid salamanders, 47
Ameiurus catus (white catfish), 29, 30*table*
Amelanchier arborea (shadbush), 138, 172
American ash (*Fraxinus americana*), 190
American chestnut (*Castanea dentata*), 137, 139, 141, 142, 145*figure*, 146, 148
American cockroach (*Periplaneta americana*), 29*table*
American eel (*Anguilla rostrata*), xxv, 29, 30, 42, 70, 265*figure*
American elm (*Ulmus americana*), 138
American halibut (*Hippoglossus hippoglossus*), 32, 33
American Indians. *See* Native Americans
American nightshade (*Solanaceae*), 97, 108
American pennyroyal (*Hedeoma pulegioides*), 171
American Rhine, xv, 275, 278, 282, 299, 319
American Scenic and Historic Preservation Society, 297
American shad (*Alosa sapidissima*), 30, 31–32, 31*figure*, 32*figure*, 33, 34, 36, 37, 262, 263*table*, 265, 267
American wormseed (*Chenopodium ambrosioides*), 105
America's Rhine, xv, 275, 278, 299
amino acids, 95
Ammann, Othmar, 258
Ammodramus savannarum (grasshopper sparrow), 173
ammonia, anhydrous, 158
ammunition, xvi*map*
Ampelopsis brevipeduculata (porcelain berry), 189
amphibians, 5, 17, 24, 41–52, 43*table*, 265*figure*. *See also specific amphibians*
Amphicarpaea bracteata (hog-peanut), 138, 168
AMS (accelerator mass spectrometry), 70, 72, 124
Amur honeysuckle (*Lonicera mackii*), 187
anadromous fishes, xxv, 27, 28, 36, 38, 67, 262. *See also individual species*
Anchoa mitchilli (bay anchovy), 263, 265, 267*table*
anchovies, 263, 265, 267*table*
ancient humans. *See* eras: pre-contact (native)
Andes, 95, 295
Andes of Ecuador, The (Church), 295
Anguilla rostrata (American eel), xxv, 29, 30, 42, 70, 265*figure*

anhydrous ammonia, 158
animals: woodland, 172. *See* fauna; *specific animals*
Anoplophora glabripennis (Asian long-horned beetle), 186
Antennaria sp. (pussytoe), 171
anthelmintic treatments, 103, 105
Anthony's Nose, xix*figure*, 43, 257
Antiquities Act of 1906, 297
Apalone spinifera (soft-shell turtle), 44
Appalachian brown butterfly (*Satyrodes appalachia*), 171
Appalachian Mountains, 58, 124
Appalachian oak hickory, 148
apple (*Malus* spp.), 155, 157–158, 161, 167*figure*
aquatic life, xxiv–xxv, 233–246, 261. *See also specific species*
Aralia nudicaulis (wild sarsaparilla), 177
Arbutus, trailing, 169
ArcGIS (Geographic Information System), 55
archaeology: Dutch West India Company (WIC) and, 78*map*; ethnobotany and, 77–115; flora and, 95; herpetofauna and, 41–42; Lower Hudson and, 79, 115; Native Americans and, xx, 24; New Amsterdam and, 77–121; regulation of, 65; seeds and, 96; sites, 66*map*. *See also* artifacts
architecture, 183, 184, 251, 291, 295, 299, 300, 302, 303, 305, 307
Arisaema dracontium (green dragon), 170
Arisaema triphyllum (Jack-in-the-pulpit), 177
Arlington High School, 47
arrowheads, 72. *See also* projectile points
arrowhead violet (*Viola sagittata*), 171
arrow-leaved tear-thumb (*Polygonum sagittatum*), 108
arrowwood (*Viburnum dentatum*), 71, 177
art, x, 250, 251, 253. *See also* aesthetics; Hudson School of landscape painting
Arthur Kill, 160
artifacts, 79, 82–85*table*. *See also specific artifacts*
Asarum canadense (wild ginger), 177
"A Scene on the Banks of the Hudson" (Bryant), 320
Asclepias exaltata (tall milkweed), 169
Asclepias quadrifolia (whorled milkweed), 171
ASDA (Atomic and Space Development Authority), 300
ash. *See Fraxinus* spp.
asheries, 155
Ashokan Catskill rockshelters, 73
Asian autumn olive (*Eleagnus umbellata*), 187
Asian chocolate vine (*Akebia quinata*), 188
Asian goldfish (*Carassius auratus*), 34, 189
Asian long-horned beetle (*Anoplophora glabripennis*), 186
asparagus, 100
aspen (*Populus*), 190
asters, 168, 169
astringents, 103, 104, 107
Athens, 209–210*map*, 250, 301, 302, 303, 304*map*
Athens Generating Plant, 302, 306, 307

Atlantic croaker (*Micropogonias undulatus*), 265
Atlantic menhaden (*Brevoortia tyrannus*), 265*figure*, 266
Atlantic salmon (*Salmo salar*), 28, 29*table*
Atlantic sturgeon (*Acipenser oxyrinchus*), 32–33
Atlantic tomcod (*Microgadus tomcod*), 29*table*, 30, 263, 265, 266, 267*table*
Atomic and Space Development Authority (ASDA), 300
atomic bomb fallout, 161
Atomic Energy Commission, US, 262
Atrazine, 159
atringents, 107
Audubon Field Guide Series, 45
Augustine's Warehouse, 86
Augustus (emperor), 108
automobile plants, 327
automobiles, 59, 198, 255, 257, 258, 259. *See also* roads
autumn olive (*Eleagnus umbellata*), 187
Awosting Lake, 144

backcountry, 65, 67, 69, 73
backs, sore, 107
bacteria, xxiv
Baedeker, Karl, 282, 299
Bagg, 174
Baird, Spener, 33–34
bald eagle (*Haliaeetus leucocephalus*), xviii*figure*
baleage, 173
Baltimore checkerspot (*Euphydryas phaeton*), 171
Baptisia tinctoria (yellow wild indigo), 171–172
barbed wire, 172
barbil [*sic*], 28, 29*table*
Bard, James and John, 251–252
Bard College, 47, 68, 71
barges, xxv, 34, 201, 207–208, 209, 211, 213*figure*, 253
barley (*Hordeum vulgare*), 253
barrels, 141–142
barrens, pine, 135, 138, 139, 148, 149
barrens, sand, 171
Barton site, 70
Bartram, John, 102
Bartramia longicauda (upland plovers), 173
bases (fish), 28, 29*table*
basses. *See Morone* spp. (temperate basses) (Moronidae)
bathymetry, 17, 54, 55, 55*map*, 57*map*, 58, 61*figure*
Battery, the, 203
Battery Park City, 58, 229
battery plant, 233–242
bay anchovy (*Anchoa mitchilli*), 263, 265, 267*table*
Bay Ridge, 31
beaches, 223. *See also* coastal areas
Beach Seine Survey (BSS), 264*figure*, 265
Beacon Institute for Rivers and Estuaries, 331
bead-lily (*Clintonia borealis*), 169
beads, 70, 72, 110
beaked hazel (*Corylus cornuta*), 169

beans (*Phaseolus* spp.), 23, 72, 95, 96, 153; *P. vulgaris*, 153
bear, black (*Ursus americana*), 69, 177
Bear Mountain area, xix*figure*, 73
Bear Mountain Bridge, xix*figure*, 59, 257–258, 259
Bear Mountain Inn, 257
Bear Mountain State Park, xvi*map*, xix*figure*, 257
bears, 72
beauty. *See* aesthetics; scenery
beaver (*Castor canadensis*), 4–6, 17, 23, 171, 176–177
beaver meadows, 174
Be Craft Mountain from Church's Farm (Church), 306
bedstraw (cheese rennet) (*Galium* spp.), 88–89*table*, 90*figure*, 91–93*table*, 95, 96, 106–107
beech (*Fagus grandifolia*), 125, 137, 138, 148, 169, 190
Beers, F. W., 211
bees (*Apis* spp.), 153
beetle, ground (Carabidae), 170
beets (*Beta vulgaris*), 100*table*
Behler, Deborah, 45
Behler, John, 45
Belgium, 101
Bender, S., 28, 71
benthic mapping, 55*map*, 58*map*
benthic species, 233, 236–243, 262
bentonite caps, 238
Berberis thunbergii (Japanese barberry), 187
Berkshire County MA, 170, 175
Bernstein, D. J., 69
berries, 88–89*table*, 88–94, 91–93*table*, 101. *See also specific berries*
Beta vulgaris (beets), 100*table*
betony (heal-all/woundwort) (*Stachys* spp.), 88–89*table*, 91–93*table*, 108, 113
Betula spp. (birch), 126–128*figures*, 141, 142, 169, 176, 190; *B. alleghaniensis* (yellow birch), 138; *B. lenta* (black birch), 138, 146*figure*, 148, 190; *B. lutea* (yellow birch), 125; *B. papyrifera* (paper birch), 169; *B. papyrifera* (white birch), 142; *B. populifolia* (gray birch), 140, 141*figure*
Bidens frondosa (devil's beggar-ticks), 169
Bierstadt, Albert, 295
biodiversity: colonial period, 87–94, 101; 19th century, xvii*figure*, 167; basics, xxiv, 3, 131; beaver trapping and, 5; cadmium and, 236; clearing and, xvi*map*; deer browsing and, 148; ecosystem services and, 11; fisheries and, 37; Foundry Cove and, 242; Hackensack Meadowlands and, 129; herpetofauna, 47*map*, 49; ice harvesting and, 212–213; invasive vines and, 189; naturalized species and, 185; plant, 87–94, 101, 113, 114; Shawangunks and, 136–137; soil and, 159
biological center, xxii, xxii*map*, xxv, xxvi
biological oxygen demand (BOD), 224
biologist, role of, xviii, 1–6, 13–21, 288
biology, xxiv–xxv
biomass, aquatic, xxi
birch. *See Betula* spp.

birds: pre-contact period, 71; abandoned farmland and, 176; dredging and, 243; grassland, 173, 177*figure*; hay fields and, 172, 173–174; invasive vines and, 188; Laurentide Ice Sheet and, 41; marshes and, 131; midwestern, 173; Native Americans and, 23; oak and, 149; PCBs and, 328; prairie, 165, 169–170, 172–173, 176; resources, 46; sheep and, 169–170; shrubland, 177*figure*. *See also specific birds*
Birket, James, 248
Bishop, Sherman C., 46
bistort (knotgrass/knotweed) (*Polygonum* spp.), 88–89*table*, 91–93*table*, 95, 97, 103
bites, 107
bitternut hickory (*Carya cordiformis*), 170
black ash (*Fraxinus nigra*), 170
black bear (*Ursus americana*), 69, 177
blackberry (*Rubus* spp.), 68, 72, 73, 88–89*table*, 91–93*table*, 94, 103, 113, 176; *R. ideus*, 138
black birch (*Betula lenta*), 138, 146*figure*, 148, 190
black crappie (*Pomoxis nigromaculatus*), 30*table*
black dash (*Euphyes conspicua*), 171
black-eyed Susan (*Rudbeckia hirta*), 173
black gum (*Nyssa sylvatica*), 146*figure*, 148
black huckleberry (*Gaylussacia baccata*), 72, 137, 138
black locust (*Robinia pseudoacacia*), 186
black oak, 139, 146*figure*, 148, 168*figure*
Black Rock Forest, 123, 125
black salts, 155
black swallowtail (*Papilio polyxenes*), 169
black swallow-wort (*Cynanchum louiseae*), 188
bladder maladies, 105, 106
Blanding's turtle (*Emydoidea blandingii*), 47, 48
bleeding, 107
Bleeker, Elizabeth, 203
Block, Adriaen, 4, 195
blood milkwort (*Polygala sanguinea*), 171
blood purifiers, 105
bloodroot (*Sanguinaria canadensis*), 177
blueback herring (*Alosa aestivalis*), 31, 263, 265, 267*table*
blueberry ("huckleberry"), 142–143
blueberry (*Vaccinium* spp.), 88–89*table*, 91–93*table*, 103, 142–143, 176
blue cohosh (*Caulophylum thalictroides*), 177
blue crab (*Callinectes sapidus*), 233, 238, 240, 242
blue-eyed grass (*Sisyrinchium* spp.), 171
bluefish (*Pomatomus saltatrix*), 37
bluegrass, 153
Blue Hill from Cosy Cottage (Church), 306
Blue Mountain PA, 136
blue-stem goldenrod (*Solidago caesia*), 168
blue toadflax (*Linaria canadensis*), 173
boats, 109, 196–197, 248. *See also* barges; freight hauling; sailing vessels; steamboats
bobcat (*Lynx rufus*), 177
bobolink (*Dolichonyx oryzivorus*), 173, 174, 176
bobwhite quail (*Colinus virginianus*), 170, 172

BOD (biological oxygen demand), 224
bog turtle (Muhlenburgh tortoise) (*Glyptemys muhlenbergii*), 44, 45, 48, 171, 190
bone tools, 72
Bontius, Jacobus, 112
bood remedies, 106
Boone, Daniel, 153
Bos primigenius (cow), 157. *See also* livestock
Boston MA, 184, 187, 188, 254
Boterburg, 284
bounties, 48
bovine growth hormone, 157
Bowline Point, 261, 264*figure*, 266, 268, 287, 288
box turtle, Eastern (*Terrapene carolina*), 42, 48
Boyle, Robert H., 36, 228, 278, 283, 284, 285, 326
brambles (raspberry/blackberry) (*Rubus* spp.), 68, 72, 73, 88–89*table*, 91–93*table*, 94, 103, 113, 176; *R. ideus*, 138
Brannon's Gardens, 203
Branta canadensis (Canada geese), 240
Brassica/Cruciferae spp. (cabbage/mustard Family), 88–89*table*, 91–93*table*, 94, 97–98, 100–101, 100*table*, 108, 113, 114, 169
Braun, E. L., 168
Brazil, 101, 106
Breakneck Ridge, 280–281, 282, 286, 299
bream (*Abramis brama*), 28, 29*table*
breasts, sore, 105, 109
Brevoortia tyrannus (Atlantic menhaden), 265*figure*, 266
bricks, 15, 19, 196
Bridge, The (Crane), 321–322
bridges, xxiv, 19, 32, 59, 62, 255, 257–259, 322. *See also specific bridges*
brikken, 29*table*, 30
British Isles, 295
Broad Street Canal, 221
Broad Street, 78*figure*, 80, 81, 82–85*table*, 86, 87, 95, 109
broccoli, 98, 100*table*, 101
Bromus latiglumis (Canada brome), 170
Bronx, 315
Bronx Zoo, 44, 45
bronze copper butterfly (*Lycaena hyllus*), 171
Brooklyn Botanic Garden, 185
Brooklyn Bridge, 257, 321–322
Brooklyn Promenade, 229
brook trout (*Salvelinus fontinalis*), 29*table*, 30*table*
Broussonettia papyrifera (paper (birch) mulberry), 187–188
Brower, David, 280, 286
brownstone, 196
brown swift (Eastern fence lizard) (*Sceloporus undulatus*), 44
brown thrasher (*Toxostoma rufum*), 176
bruises, 105, 106
Brumbach, H. J., 28, 71
Brussel sprouts, 98, 100*table*, 101

Bryant, William Cullen, 293, 295, 313, 320
BSS (Beach Seine Survey), 264*figure*, 265
BT (builder's trenches), 86
bubble curtains, 263
buckthorn (*Rhamnus cathartica*), 176
buckwheat, 253
Buckwheat Field on Thomas Cole's Farm (Farrer), 305–306
buffalo, xviii, 170
Buffalo NY, 259
builder's trenches (BT), 86
Bull Brook site (MA), 68
bullfrogs, 48
bunchberry (*Cornus canadensis*), 169
bunions, 105
bur-cucumber (*Sicyos angulatus*), 169
burials, 70, 71
burning bush (*Euonymus alatus*), 187
burns, controlled, 19, 148
Burns, D., 13
burns, salves for, 105, 106, 111
Burr, Aaron, 220
Burroughs, John, 211
butter, 157, 253
buttercup (*Ranunculus* spp.), 72
butterflies, 169, 171, 173, 174, 176, 188–189
Butter Hill, 284
butternut (*Juglans cinerea*), 69, 71, 72
buttonbush (*Cephalanthus occidentalis*), 170, 174

cabbage/mustard Family (*Brassica/Cruciferae* spp.), 88–89*table*, 91–93*table*, 94, 97–98, 100–101, 100*table*, 108, 113, 114, 169
cadmium, 9–10, 233, 235–242
Calhoun, John C., 252
California, 157*table*
Callinectes sapidus (blue crab), 238, 240, 242
Callison, Charles, 280, 286
Camp Smith, xix*figure*
Canada, 72, 97, 100, 101, 103
Canada brome (*Bromus latiglumis*), 170
Canada geese (*Branta canadensis*), 240
Canada lily (*Lilium canadense*), 171
Canada mayflower (*Maianthemum canadense*), 177
Canadian onion, 100
Canadian salamanders, 46
canals, xvi*map*, 154, 252–254. *See also specific canals*
Canoe Place, 109
canoes, 19, 248
cantaloupe, 105
capacitor plant, 198, 327
Cape Saint Thomas (Brazil), 106
capitalism, 212, 251, 318
Carabidae (ground beetle), 171
Carassius auratus (Asian goldfish), 34, 189
carbonates, 154–155
carbon dioxide, 9, 159

carcinogens, xxv
Cardamine hirsuta (hairy bittercress/scurvy grass), 98
Carex spp. (true sedges), 94, 129, 138, 169, 170, 171, 174, 315; *C. pensylvanica* (Pennsylvania sedge), 137, 177; *C. platyphylla* (tufted sedge), 138
caribou, 65, 67, 68
carpetweed (*Mollugo* spp.), 88–89*table*, 91–93*table*, 94, 95, 113
Carpinus/Ostrya spp., 126*figure*
carps. *See Cyprinus* spp.
carriage roads, 144, 293, 300–301, 302, 303
carrots, 100, 169
Carson, Rachel, 161, 280
Carya spp. (hickory): pre-contact, 31, 69, 71, 72, 126*figure*, 128*figure*, 168; contact period and after, 68, 125, 126*figure*, 128*figure*; abandoned fields and, 148; Appalachian oak hickory, 148; Columbia County, 169; *C. cordiformis* (bitternut hickory), 170; pollen depth, 127*figure*; rock walls and, 172; Shawangunk, 137, 138, 141, 146*figure*
Cary Institute of Ecosystem Studies, 268, 331
Cassedy, D., 71
Castanea spp. (chestnut), 67, 125, 127*figure*, 128*figure*, 168*figure*, 169; *C. dentata* (American chestnut), 126*figure*, 137, 139, 141, 142, 145*figure*, 146, 148
Castleton, 32
Castor canadensis (beaver), 4–6, 17, 23, 171, 176–177
catadromous fishes, xxv
catbird (*Dumetella carolinensis*), 170, 176
caterpillars, 169, 173, 174
cat-nip, 106
Catostomidae (suckers), 29*table*, 30, 34–36, 35; *Catostomus commersonii* (white sucker), 30, 31, 34
Catskill Examiner, 212
Catskill Mountain House, 293, 295
Catskill NY, 209, 250
Catskills: colonial period, 72; 19th century, 293; art and, 291, 292, 305–306; basics, 136*map*; deer and, 143–144; drinking water and, xxv; ice harvesting and, 209–210*map*, 214; Mohawk River and, xxii
cattails (*Typha* spp.), xvi*map*, 123, 125, 129, 130, 189–190, 240
cattle, 172, 253, 318. *See also* livestock
cauliflower, 98, 100, 100*table*, 101
Caulophylum thalictroides (blue cohosh), 177
causeways, xvi*map*, 59
caviar, 33, 279
Cedar Grove, 300, 305
Celastrus orbiculatus (oriental bittersweet), 188
Celastrus scandens (native bittersweet), 188
cement barrels, 141–142
Cementon, 301, 304
cement plants, 292. *See also specific projects*
censuses, 166, 167*figure*, 185
Census of Agriculture (2007), 173
Centennial Exhibition (1876), 296
Central Hudson et al., 262, 265, 268

Central Park, 293, 295, 302
Centrarchidae (sunfishes), 28
Cephalanthus occidentalis (buttonbush), 170, 174
ceramics (pottery), 69–71, 72, 80–81, 82–85*table*, 102, 109
cereal rust, 187
ceremonies, 69, 70, 71, 72, 73, 318
chain pickerel (*Esox niger*), 30*table*
chains (across Hudson), xix*figure*, 233, 263
Chamaecrista nictitans (wild sensitive plant), 172
Champlain, Samuel de, 97
Champlain Canal, 253
chancre sores, 106
change, environmental, 77–121, 313
Channa argus (snakehead), 30*table*
channel catfish (*Ictalurus punctatus*), 30*table*, 265, 265*figure*
channels, xxiv, 9, 24, 32, 55–56, 62, 327, 331
Chant, R., 54
Charadrius vociferus (killdeer), 169
charcoal ("chark"), 140, 141*figure*, 144*figure*
charcoal analysis, 123, 125, 127*figure*, 128*figure*, 130, 131, 142, 143
chark (charcoal), 140, 141*figure*, 144*figure*
Charlotte Creek, 71
cheese, 106, 155*table*, 157, 253
cheese rennet (bedstraw) (*Galim* spp.), 88–89*table*, 91–93*table*
chelonians, 44
Chelydra serpentina (common snapping turtle), 42, 43, 48
chelydra serpentina (snapping turtle), 42, 43, 48
Cheney, A. N., 31
Chenopodiaceae spp. (goosefoots), 104–105
Chenopodium Oil, 105
Chenopodium spp. (various), 88–89*table*, 91–93*table*, 103, 104–105, 126–128*figures*, 129; *C. ambrosioides* (American wormseed), 105; *C. quinoa* (wild spinach), 68, 70, 95, 96–97, 100*table*, 101, 105
Chenopods, 95, 96, 104–106
Cherokee tribe, 90, 106, 109
cherry (*Prunus* spp.), 88–89*table*, 90, 91–93*table*, 94, 172
Cherry Valley Turnpike, 250
chert flakes, 72
Chesapeake Bay area, 80, 266, 286
chestnut. See *Castanea* spp.
chestnut blight, 146
chestnut oak (*Quercus montana*), 124, 137, 139, 142, 144*figure*, 146*figure*, 148
chestnut-sided warbler (*Dendroica pensylvanica*), 176
Chicago, 249
chills, 107
China, 201
chipmunks, 172
Chippewa tribe, 106, 108
chironomids, 236, 240
chloride, 236, 242
chlorpropham, 160

chocolate vine, Asian (*Akebia quinata*), 188
cholera, 107, 228
"cholera infantism," 106
Christiaensen, Hendrick, 4, 195
chrysemys picta (painted turtle), 43*table*, 44, 46*figure*
Church, Frederic E., 295, 296, 300, 302, 303, 304, 305, 306
churches, xvi*map*
cider, 155*table*, 157, 158
Circus cyaneus (harriers/day hawks), 171
Cistothorus palustris (marsh wren), 190
cities, xxi, xxii*map*, 10, 131, 201–218, 249, 255, 259. *See also specific cities; specific features and locations*
citizens, 9, 289
citrus (*Citrus* spp.), 88–89*table*, 90–91, 91–93*table*
City Common Council, 221
City of Hudson, 168
civilization (progress), xvi*figure*, xix, 10, 259. *See also* development
Civil War, 233, 301
Cladium spp. (sawgrass), 129
clammy cuphea (*Cuphea viscosissima*), 171, 172
clams, 69
Claude glass, 184
claw muscle, 240
clay, 9, 15
clay pipes, 72, 81
clay tiling, 174
Clean Water Act (CWA) (1972): AEI and, 267; environmental movement and, 285; industry and, 235, 261, 262, 286, 287; Pete Seeger and, 277; sewage and, 61, 225, 228
Clearwater (sloop), x–xi, 277–278, 331, 332
Clearwater, Inc., 332
Clemmys guttata (spotted turtles), 44, 45, 171
Clermont, 170, 189*figure*
Cliff House, 144
cliffs, 137*map*, 138
climate, 10*figure*, 13, 70, 97, 185, 317
climate change, 46, 77, 97, 123–134, 329
Clinton, DeWitt, 252
Clintonia borealis (bead-lily), 169
clothing manufacture, 196
clover (*Trifolium* spp.), 88–89*table*, 91–93*table*, 108–109, 153, 173
clubmosses, 169
Clusisus (de L'Ecluse/Calolinus/Carolus), 98, 100, 103, 111, 112, 114
CO_2, 9, 159
coal, 42, 140, 154
coastal areas, 33, 69, 97, 125, 185, 223, 227, 328
cobweb skipper butterfly (*Hesperia metea*), 171
Coeymans, 209–210*map*, 214
Cohoes Falls, 29
cohosh, blue (*Caulophyllum thalictroides*), 177
coke, 24
cold remedies, 108, 109

Cold Spring, 234, 235
Cole, Thomas, xvi*map*, 276, 292, 293, 296, 301, 305
colic, 106, 108
coliform, fecal, 226, 228
Colinus virginianus (bobwhite quail), 170, 172
collaboration, 329, 330–331
Collect Pond, 206
Collins, Billy, 323
colonial period. *See eras*: colonial period
Colorado River, 280
Columbia County, 19, 156, 165–178, 184, 189
Columbia Garden, 203, 204
Columbia River, 253
Columbia University, 331
commerce (trade): 21st century, 330; art and, 276; basics, 195–199, 247; industry and, 233; medicinal plants and, 102–103; transportation and, 248, 249–250; turnpikes and, 252; unity and, 250. *See also* transportation; *specific commercial enterprises*
common carp (*Cyprinus carpio*), 31, 33–36
Common Council, 222, 229
common five-lined skink (*Plestiodon faciatus*), 43
common lousewort (*Pedicularis canadensis*), 173
common monkeyflower (*Mimulus ringens*), 171
common ragweed (*Ambrosia artemisiifolia*), 125, 126–128*figures*, 129, 169, 176
common reed (*Phragmites*), 123, 125, 129, 130
common snapping turtle (*Chelydra serpentina*), 42, 43, 48
common vervain (*Verbena hastata*), 171
"community character," xxv, 305–306
"The Complaint of New Amsterdam" (Steendam), 317
Compositae, 129
Comptonia peregrina (sweet fern), 171
concrete, 155, 258
Con Ed (Consolidated Edison Company of New York), 17, 161, 276–277, 283, 286. *See also* Storm King Project
confectioners, 202–204
Connecticut State, xxi, 68, 73, 178, 250. *See also specific locations*
Connecticut Valley, 73, 174
Conrad, Joseph, 10
conservation: 19th century, 145, 295; Adirondacks and, 296; art and, 291–301, 295, 307; exotic flora and, 184; farmland and, 178; feedback loops and, 9; herpetofauna and, 47; Hughes on, 298; industry and, 198, 233, 243, 280–287; invasive plants and, 191, 192; Shawangunks, 135–136; water and, 227. *See also* conservation movement; environmental movement; the future; management; preservation; restoration; *specific government agencies & legislation*
conservation movement, x, 48, 280–287, 298. *See also* conservation; restoration
conservation tillage, 159
Consolidated Edison Company of New York (Con Ed), 17, 161, 276–277, 283, 286. *See also* Storm King Project

Constitution Island, 234*figure*, 235
Constitution Marsh, 234, 238, 240, 243
Contact Period. *See eras*: Contact Period
contact period. *See eras*: contact period
"Contacts between Iroquois Herbalism and Colonial Medicine" (Fenton), 102
Contoit, John H., 203
"Convenient House," 318
cooling systems, power plant, xxvi, 10, 198, 261–269. *See also specific projects*
Cooper, H. P., et al., 171
Cooper, William, 155
cooperage, 141–142
cooperation, 332, 333
copper, 316, 317
copperhead (*Agkistrodon contortrix*), 48, 49
copperleaf (*Acalypha* spp.), 88–89*table*, 91–93*table*
cordgrass (*Spartina patens*), 125
corn (*Zea mays*), 23, 70–71, 72, 95, 96, 97, 153, 156, 160–161, 167*figure*, 169, 317
Cornell University, 45, 98, 174, 191, 268, 269, 342
Corning, Erastus, 197
Cornus canadensis (bunchberry), 169
Cornus florida (flowering dogwood), 138
Cornus spp., 126*figure*
Cornus spp. (dogwood), 176, 177
Cornwall, 264*figure*
Cornwall Project, 283. *See also* pumped storage; Storm King Project
Corre, Joseph, 202, 203–204, 206–207
Cortlandt Street (NYC), 257
Corylus spp. (hazelnuts), 68, 72, 126*figure*; *C. cornut* (beaked hazel), 169
Cosy Cottage, 306
Cottage Residences: A Series of Designs for Rural Cottages and Cottage Villas (Downing), 184
cotton, 196
cottonwood (*Populus deltoides*), 170
court cases, 9, 48, 276, 299. *See also specific projects*
cover, terrestrial, 24, 67, 135, 138, 165–178, 328
cow (*Bos primigenius*), 157. *See also* livestock
cow-wheat (*Melampyrum lineare*), 138
Coxingkill, 139
Coxsackie, 209–210*map*
crabs, 42, 233, 238, 240, 242, 243
Crane, Hart, 321–322
Crataegus spp. (hawthorns), 68, 176
crayfishes (*Orconectes* spp.), 239
"Crazy Luke," 256–257, 261–262
credits, 288–289
creek chubsucker (*Erimyzon oblongus*), 30
creeks, 58–59
crimson-spotted triton (*Triton puctatus*), 44
CRM (cultural resource managment), 65, 67
crocodilians, 45
Cronon, W., 5, 249
cropland, 168–170, 172. *See also* farmland; *specific crops*

Cropsey, Jasper, 297
Crotalaria sagittalis (rattlebox), 172
Crotalinae spp. (rattlesnakes), 42–44, 48
Crotalus horridus (timber rattlesnake), 42, 43, 48
Croton Aqueduct, 154, 220, 221, 255
Cruciferae/Brassica spp. (mustard/cabbage Family), 88–89*table*, 91–93*table*, 94, 97–98, 100–101, 100*table*, 108, 113, 114, 169
cucumbers, 105
Cucurbita spp. (squash/pumpkin): pre-contact period, 71, 96; contact period, 72, 101; colonial period, 88–89*table*, 88–94, 90*figure*, 91–93*table*, 113; floodplains and, 95; Lenape and, 23; medicinal uses, 103; as medicine, 105; Native Americans and, 153; 17th century, 113
Culpeper, Nicholas, 104, 106, 107, 109
"The Culprit Fay" (Drake), 319–320
cultivation, 70, 71–74, 78*figure*, 94–100, 95–96, 97, 109, 135, 156, 159, 185, 208. *See also* agriculture; horticulture
cultural landscape, 137*maps*, 305, 306, 307
cultural resource management (CRM), 65, 67
culture, x, 10*figure*, 11, 252
cumulative effects, 330
Cuphea viscosissima (clammy cuphea), 171, 172
Currier, Stephen, 283
Custer, J. F., 70
Cygnus olor (mute swan), xix*figure*, 189
Cynanchum spp. (swallow-worts), 188–189; *C. louiseae* (black swallow-wort), 188; *C. rossicum* (pale swallow-wort), 188
Cyperaceae spp., 126–128*figures*
Cyperaceae spp. (sedges): *Carex* spp. (true sedges), 94, 129, 138, 169, 170, 171, 174, 315; *Carex pensylvanica* (Pennsylvania sedge), 137, 177; *Carex platyphylla* (tufted sedge), 138; *Cladium* spp. (sawgrass), 129; *Cyperus* spp. (nut sedges/nut grasses), 71, 88–89*table*, 90*figure*, 91–93*table*, 94, 129, 169; *Eleocharis* spp. (spike grasses), 129; *Scirpus* spp. (bulrushes), 129
Cyprinus spp. (carps), 28, 29*table*, 30, 35, 36; *C. carpio* (common carp), 31, 33–36

dairy, 155, 157, 178, 253
dams: basics, xxi, 61*map*; beaver, 5; fisheries and, 32; nature and, 253; PCBs and, 242; sediment and, 60–61, 62, 129, 130; Troy, at, xxv; wetlands and, 17
damselflies, 170
Danaus plexippus (monarch butterfly), 188–189
D & H (Delaware and Hudson Canal), 139–140, 154
Daniels, Randy, 307
Daniels, Robert A., et al., 30*table*, 36
Danskammer Point, 264*figure*, 268
Danthonia spicata (poverty oatgrass), 171
da Verrazano, Giovanni, 297
David, Elizabeth, 203
Davies, Thomas, 296
Davis, George, 142

Day, Gordon, 111
day hawks (harriers) (*Circus cyaneus*), 171
DDT, 160, 280
Death in Venice (Mann), 10
DEC. *See* NYS Department of Environmental Conservation
deer, white-tailed (*Odocoileus virginianus*): pre-contact period, 71, 72; 19th century, 143–144; habitat disturbance and, 177; invasive plant species and, 186, 187; naturalized plants and, 190; overbrowsing, 131, 135, 168; populations, 69; Shawangunk, 145, 146, 148; skins, 317; wet meadows and, 171. *See also individual species*
deforestation, xv, 129, 154, 326. *See also* forests; land clearing
DeGarmo Tannery, 140
DEIS (Draft Environmental Impact Statement) (Central Hudson et al.), 262, 265, 268
DeKay, James E., 33, 44
Delacroix, Jacques Madeline, 203
Delaware and Hudson Canal (D&H), 139–140, 154
Delaware River, 36, 68, 70, 72, 73, 315
Delaware tribes, 73, 105, 108
Delaware Water Gap (NJ), 68, 136
Delcourt, H. R., 96
Delcourt, P. A., 96
Delft tiles, 80, 83*table*, 84*table*
demographic center, xxii*map*
demography. *See* population
Dendroica spp. (warblers): Canada, 169; *D. discolor* (prairie warbler), 176; *D. pensylvanica* (chestnut-sided warbler), 176; *D. petechia* (yellow warbler), 170, 176
Department of Environmental Conservation (DEC). *See* NYS Department of Environmental Conservation
Depression, Great, 224
de Saussure, Nicholas-Théodore, 158
A Description of the New Netherlands (van der Donck), xvi*map*, 42, 103, 110
Desmodium spp. (tick-trefoil), 168
desserts, 203
detritus (detris), xxiv, xxv, 328
development: 21st century, 328–329; abandoned farms and, 175; art and, 276, 299, 305; basics, xix; Cole on, 292; forest decline and, 178; LWRPs and, 305; waste and, 229, 230; water and, 220; wetlands and, 175. *See also* progress; suburbs; *specific projects*
devil's beggar-ticks (*Bidens frondosa*), 169
Devoe, Thomas F., 208
De Vries, David Pieterszoon, 29, 30
dewberry (*Rubus flagellaris*), 171
DeWitt, Benjamin, 250
Diamond, J. P., 72
diarrhea, 104, 105, 106, 107
dichloropropane, 160
dichotomies, 249, 251, 255
Dickau, R., 72
dicksissel (*Spiza americana*), 173

digestion, 107
dikes, 55, 239
dirtienen, 29*table*, 30
diseases, 10, 70, 72, 202, 213–214, 222, 227, 229. *See also specific diseases*
dissolved oxygen. *See* oxygen
disturbances, habitat: deer and, 177; invasive plant species and, 186
Ditmars, Raymond L., 44
DiTomaso, J. M., 98
diuretics, 103, 106
diversity. *See* biodiversity
Dobbs Ferry, 256
dock (*Rumex*), 125, 129
Doellingeria umbellata (whorled aster), 169
Dogan Point shell midden, 69
dogs, 65
dogwood (*Cornus* spp.), 176, 177
Dolichonyx oryzivorus (bobolink), 173, 174, 176
Domes of Yosemite, The (Bierstadt), 295
domestication, plant, 96
Doodletown, xvi*map*, xix*figure*
Doty, Dale, 282
Downing, Andrew Jackson, 184–185, 186, 302, 320–321
Down the Hudson to West Point (Moore), 298–299
downy trailing Lespedeza (*Lespedeza procumbens*), 172
Draft Environmental Impact Statement (DEIS) (Central Hudson et al.), 262, 265
dragonflies, 170
Dragon's Blood, 103
drainage, 174
Drake, Joseph Rodman, 319–320
dredging and filling: basics, 9, 56–57, 62; bathymetry and, 17; fishes and, 15, 32, 37; Foundry Cove and, 233, 235, 238, 242–243; ice harvesting and, 209; invasive species and, 240; New York Harbor and, 61; PCBs and, 327; transportation and, 19
Dreissena polymorpha (zebra mussels), xxiv–xxv, 37, 56, 328
drill bits, 70
drinking water, xxv, 10, 19, 218, 327. *See also* Clean Water Act (CWA) (1972)
droughts, 97, 125, 129, 131, 148, 188
dry hillsides, 171
Duco, Don, 81
Dumetella carolinensis (catbird), 170, 176
Dunderberg, 298–299
Dunwell, Frances, x–xi
Durand, Asher B., xvi*map*, xix*figure*, 276, 292, 293, 294*figure*
Dutch colonizers, 4, 5–6; agriculture and, 153; artifacts, 78–79; basics, 3, 23, 195; drinking water regulations, 218; fishing, commercial and, 24; fish reports, 29, 30; fur and, 316; herpetofauna and, 41; Irving on, 319; medicinal plants and, 102, 103; names and, 315; Native Americans and, 4, 17, 79, 97, 109–113; resources and, 317; sloops and, 248. *See also* Europeans; New Amsterdam; New Netherland; *individual colonizers*; *specific organizations*
Dutch East India Company, 23, 94, 111–112, 114, 195
Dutch elm disease (*Ophiostoma* spp.), 161
Dutchess County, 45, 47, 48, 68, 69, 71, 160, 184, 189, 250
Dutch Gardener, The (Van der Groen), 98
Dutch scholarship, 80–81, 98, 112
Dutch West India Company (WIC): archaeological sites, 78*map*; basics, 81, 195; deforestation and, 24; directors, 86, 108; ethnobotanical data, 77–115; Manhattan land grants, 79, 81, 86; medicinal plants and, 102, 111–112, 114; Native Americans and, 109; orangeries and, 94; plant data, 96. *See also* Kierstede, Hans
dwarf pine, 137*maps*, 138
dyes, 95, 97, 106. *See also specific plants*
dysentery, 104, 106, 107

eagle, bald (*Haliaeetus leucocephalus*), xviii*figure*
ear disorders, 107
early American era: influence of Hudson on, 314
early Woodland garden complex, 95
Earth Day, x
Eastern Agricultural Complex, 94–97
Eastern box turtle (*Terrapene carolina*), 42, 48
Eastern cricket frog (*Acris crepitans*), 48
Eastern fence lizard (brown swift) (*Sceloporus undulatus*), 44, 48
Eastern hognose snake (*Heterodon platirhinos*), 45*map*
Eastern hop-hornbeam (*Ostrya virginiana*), 138
Eastern milksnake (*Lampropeltis triangulum*), 43*table*
Eastern mud turtle (*Kinosternon subrubrum*), 48
Eastern red-backed salamander (*Plethodon cinereus*), 43*table*
East India Company, 23, 94, 111–112, 114, 195
East River, 224, 229
de L'Ecluse, Charles (Calolinus (Carolus) Clusius), 98, 100
ecological analogies, 165, 169–170, 170–171, 172, 176, 177, 249
ecology, 7–8, 10, 100, 114
economic factors: colonization and, 317; exotic ornamental plants and, 185; explorers and, 316; ice harvesting and, 212; names and, 315; power plants and, 306; Shawangunks, 135; SLC and, 305; sloops and, 249; transportation and, 247. *See also specific factors*
ecosystem services, 10–11
Edgewater-125th Street ferry, 256
Edison, Thomas, 208
education, 10*figure*, 11, 191, 331–332, 333. *See also* public attitudes and awareness
eel (*Anguilla rostrata*), xxv, 29, 30, 42, 70, 265*figure*
Eights, James, 44
EISs (Environmental Impact Statements), 268, 276, 285, 330, 333
Elbe River, 57
elderberry, red-berried (*Sambucus pubens*), 169

Eleagnus umbellata (autumn olive), 187
electric power generation, x, 259–274, 277, 280. *See also* capacitor plant; Storm King Project
Eleocharis spp. (spike grasses), 129
elft, 29*table*, 30
elk, 69
Ellenville area, 139, 140, 142
Elliot, 174
Ellis, F., 166
Ellsworh, J. M., 59
elm disease (*Ophiostoma* spp.), 161
Elmendorf, Case, 142
elms (*Ulmus* spp.), 126*figure*, 138, 155
embayments, 327, 331
Emerson, Ralph Waldo, 253, 292
Emperor Augustus, 108
Empire State Electric Energy Research Corporation, 263
Empirical Impingement Model, 267
Empirical Transport Model, 267
employment, 135, 140, 142, 144–145, 175, 196, 211–213
Emydoidea blandingii (Blanding's turtle), 47, 48
Endangered Species Act, 9
endangered/threatened species, 48, 136, 174, 191, 198, 279
Enderly Mill, 139
endocrine disruptors, 228. *See also* hormone mimetics
energy, 253; canals and, 154; Niagara Falls and, 296. *See also specific types of energy*
Energy Research and Development Agency (ERDA), NYS, 300
engineers, x
English colonization: artifacts, 78–79; Dutch *versus*, 195; ethnobotany and, 77–115; exotic ornamentals and, 183; Irving and, 319; medicinal plants and, 102, 103; Native Americans and, 73; water gardens and, 189. *See also* Europeans
English explorers, 4, 23. *See also* Europeans; *individual explorers*
English grasses, 173
engravings, 276
entrainment, 262–265, 267, 268
environmental change, 77–115; public attitudes toward, 198
Environmental Consortium of Hudson River Colleges and Universities, 331
environmental history, 77, 249
Environmental Impact Statements (EISs), 268, 276, 285, 330, 333
environmental law, 277, 291–307, 293–314, 330. *See also specific cases; specific laws*
environmental movement: art and, 276; basics, xi, xix–xx, 198, 277; birth of, 279–289; Storm King Mountain Project and, 234, 332; water quality and, 219. *See also specific organizations*
Environmental Protection Agency, US (USEPA), 238, 240, 262, 268, 287, 289, 325
environmental quality, xxv
Environmental Quality Review Act, 299

EPA (Environmental Protection Agency), US, 238, 240, 262, 268, 287, 289, 325
epidemiology, 70
Epigaea repens (trailing Arbutus), 169
epilepsy, 106, 107
Equus asinus x E. caballus (mule), 159, 173
Equus ferus caballus (horse), 158–159, 172, 173, 222–223. *See also* livestock
eras: pre-contact (native): agriculture, 65–76, 113; basics, 314; ecological change, 79, 96; indigenous plants, 90, 94–97, 112; prairie biota, 173. *See also* Eastern Agricultural Complex
eras: contact period, 28, 42, 78, 79, 95–98, 100, 101, 107–113*passim*, 153, 313–315, 325–326. *See also* explorers; Native Americans; *specific nationalities*
eras: colonial period: agriculture, 5–6, 153–156, 173, 177*figure*, 326; Albany and Catskills, 72; attitudes and, 313; basics, xviii–xix, 3–4, 135–152, 154; drinking water and, 218; economic factors, 317; fire and, 5; fish and, 31; flora, 73, 77–114, 184; floral diversity, 79, 87–94, 90*figure*, 114; forests, 123, 326; fur trade, 3, 4–5, 11, 23, 195, 316; herpetofauna and, 41; ice harvesting and, 201–207; influence of Hudson on, 314, 317–318; names and, 315; Naturalism and, 43–44; paintings, 306; pollen analysis, 126*figure*; pollen analysis and, 125; populations, 27; potherbs, 113; resources and, 23–24, 316, 317; sediments and, 53; Shawangunks and, 135, 138–139; transportation, 196–197, 248, 249–250; waste and, 220–221. *See also specific colonies & nationalities*
eras:19th century: abandoned farms and, 175–176; agriculture, 155–159, 167, 169; agriculture and, 177*figure*; art and, 292, 302, 306; artifacts, 85*table*; Columbia County, 166; fishing, commerical, 31–36, 32; flora, 17; forests and, 326; hay fields, 172, 173; herpetofauna and, 42; ice harvesting and, 201, 207–211; improved acreage, 176*map*; industry, xxv–xxvi; influences of the Hudson on, 314, 319–321; landscaping trends, 186–190; Naturalism and, 43–44; Shawangunks and, 139–145; transportation, 197, 247, 249–250, 252–253, 253–254, 255, 256; vegetables, 101; waste and, 221–223, 226–227, 228, 229, 230; writings, 292. *See also* exotic ornamental plants; Hudson River School of landscape painting; Naturalist Period
eras: 20th century: agriculture, xxi, 155–161, 175; art and, 292; artifacts, 84*table*, 85*table*; commercial fishing, 35; electricity and, 261–269; farmland change and, 177*figure*; forests, 125, 326; hayfields, 173; herpetofauna and, 44–49, 48; ice harvesting and, 213–214; improved acreage, 176*map*; industry, xxv–xxvi, 233–242; influence of Hudson on, 314, 321–323; power industry, 198; regulation, 328; science and, 45–48; shad, 32; Shawangunks, 145–149; transportation, 197, 255–256; waste and, 223–230, 229; wetland flora, 130
eras: 21st century, 149, 177*figure*, 219, 292, 326
ERDA (Energy Research and Development Agency), NYS, 300

Ericales spp., 126*figure*
Erie Canal, xxiii, 154, 155, 251, 252–254
Erie Railroad, 256, 257
Erigeron spp. (fleabanes), 173
Erimyzon oblongus (creek chubsucker), 30
erosion: 21st century, 59; basics, 155; Battery Park City and, 58; Columbia County and, 177; dredging and, 57, 62; farming and, 159, 161; railroads and, 58. *See also* runoff
Esopus, 300, 318
Esopus Creek, 72
Esox spp. (pickerel/pike), 29*table*; *E. Esox americanus* (redfin pickerel), 30*table*; *E. lucius* (northern pike), 30*table*; *E. niger* (chain pickerel), 30*table*
"Essay on American Scenery" (Cole), 292
Essex County, 48, 160
estates, xix*figure*, 174, 187*figure*, 189, 198, 208, 255, 302, 317
ethanol, 156
ethnobotanical exchange, 86–115
Euonymus alatus (burning bush), 187
Eupatorium sessilifolium (upland boneset), 171
Euphydryas phaeton (Baltimore chekerspot), 171
Euphyes conspicua (black dash), 171
Eurasian gypsy moth (*Lymantria dispar*), 148, 160
Eurasian swallow-wort (*Cynanchum rossicum*), 188
Europeans. *See* eras: colonial period; *specific nationalities*
eutrophication, 158, 227
Evans, John, 138
evening primrose (*Oenothera biennis*), 173
evolution, cadmium and, 233, 237, 241, 242
E. Waters & Sons, 157
exotic fauna, xix*figure*. *See also specific exotic fauna*
exotic fishes, 30, 32, 37. *See also individual species*
exotic flora, 88–94, 91–93*table*, 94, 183–194, 198
exotics. *See also* invasive species
expectorants, 105
explorers, 23, 314, 315–317. *See also individual explorers*
extinctions, 9, 34, 36, 68, 185
eyed butterfly, 171

Fagus grandifolia (beech), 125, 126–128*figures*, 137, 138
fallfish (*Semotilus corporalis*), 29
Fall Juvenile Survey (FJS), 265
false mermaid weed (*Floerkea proserpinacoides*), 170
false Solomon's seal (*Maianthemum racemosum*), 168
farmland: abandonment of, 175–177, 326; Columbia County, 19, 165–178; New York City and, 208–209. *See also* cropland; ecological analogies; *specific crops*
farm ponds, 17. *See also* ponds and fens
farms, 159–160, 161, 165, 175, 178, 253. *See also* agriculture
Farrer, Thomas C., 305
fat, 95
fatty acids, 70, 106
fauna: pre-contact period, 69, 73; beaver trapping and, 5; cadmium and, 236–238; colonial period, 6; Foundry Cove, 237*figure*; Foundry Cove and, 239; historical perspectives and, 19; human impacts, 24; Laurentide Ice Sheet and, 41; mega-, 68; Shawangunk domestic, 139; waste and, 222. *See also* aquatic life; game; *specific fauna*
fecal coliform, 226, 228
federal agencies, 277, 285
Federal Highway Act of 1916, 258
Federal Power Commission (FPC), 280, 282–287*passim*
Federal Rivers and Harbors Act (1948), 225
feedback loops, 1–2, 7–11, 8*figure*, 15–16*figure*, 18*figure*, 20*figure*, 25, 242–243, 313
feet, sore, 106, 107
FEIS (Final Environmental Impact Statement) (NYS Department of Environmental Conservation), 268
Feldspar Brook, xxi
fences, 48, 172
fens and ponds, 123
Fenton, William A., 102, 106, 111
ferns, 169
ferries, 19, 59, 255–256, 257
Ferry Hook, 255
fertilizers, 155, 156, 158, 161, 221, 326
fevers, 104, 109
fiber, 10*figure*
Fiedel, , 70
field pennycress (*Thlaspi arvense*), 98
field sparrow (*Spizella pusilla*), 170, 176
59th Street Station (electricity generation), 261
Final Environmental Impact Statement FEIS (NYS Department of Environmental Conservation), 268
financial industry, 196
Finger Lakes, 70
Fire Island, 36
firepits/hearths/ovens, 67–68, 69, 70, 71–72, 73
fires: pre-contact era, 5, 72, 78*map*; colonial period, 5, 154; berry pickers and, 143; bridges and, 257; deer and, 177; erosion and, 155; invasive species and, 185; rattlesnakes and, 44; Shawangunks and, 135, 140, 142, 144*figure*, 149; suppression, 131, 135, 148; water supply and, 220; wetlands and, 129. *See also* burns, controlled
Fish Creek, 71
fisher (*Martes pennanti*), 177
fisheries: biodiversity and, 37; dams and, 32; management and, 36, 37; mitigation, 36; overfishing, 24, 32; PCBs and, 24, 37, 242, 266, 327; pollution and, 32, 36–37, 266; power plants and, 268, 286, 287; restoration of, 38; stocking of, 32, 33, 36; studies, 265
fishes: pre-contact period, 69–70, 71; basics, xxv, 28; canals and, xvi*map*; colonial period, 24; ecology and, 8–9; ecosystem services and, 11; exotic ornamentals, 189; hatcheries, 32, 34; human impacts, 131; Juet on, 316; marine, 30, 265; monitoring studies, 262–268; native, 30*table*, 31, 37; New York City purchases, 34, 36; oxygen and, 223; pollution and, 328; power plants and, 36, 198, 262–268, 328; pre-contact era, 41, 68, 69, 70,

fishes (*continued*)
  71; predatory, 37; resources, 46; shape of the river and, 15; studies, 331; Tappen Zee and, xxiii; 21st century, 30–31, 327; van der Donck on, 42; waste and, xvi*map*, xxv–xxvi, 228, 229. *See also* fisheries; fishing; Foundry Cove; larvae, aquatic; spawning; *specific families and species*; *specific kinds of fish*
fishing, commercial: basics, 27–31, 37–38; history, 8–9; industry and, 19; Native Americans and, 3, 27; 19th century, 31–36; pollution and, xxvi; pre-contact era, 71, 72; productivity and, 24; 17th and 18th centuries, 24, 27, 28–31; shape of the river and, 15; stocking programs, 32, 33, 36; 20th century, 36–37; 21st century, 31, 327. *See also specific fishes*
fishing, personal use, 34
fishing, sport, xxv–xxvi, 9, 28, 36, 37
Fishkill NY, 44, 48
fishkills, 198, 277, 325, 327
Fitch, John, 250
FJS (Fall Juvenile Survey), 265
flax (toadflax) (*Linum* spp.), 88–89*table*, 91–93*table*, 94, 103, 108, 109, 113, 155*table*, 173
fleabanes (*Erigeron* spp.), 173
flies, 236
Flinn, K. M. et al., 175
flint, 67, 68, 70, 71
Floerkea proserpinacoides (false mermaid weed), 170
flooding: agriculture and, 153; Columbia County and, 174; deer and, 177; ecosystem services and, 11; hay meadows and, 170; historical, 70, 153, 177; mapping projects, 329. *See also* tides
floodplains, xvii*figure*, 67, 68, 71, 95, 153, 174
flora: pre-contact period, 70–73; art and, 98, 103; flooding and, 174; Foundry Cove and, 239; herbaceous, 107, 148, 168, 173, 175, 176, 188 (*see also specific plants*); highland, 124–125; historical, 68, 94; indigenous/introduced, 77–115, 88–89*table*, 88–97, 91–93*table*, 94, 95, 98, 101, 103 (*see also specific plants*); Laurentide Ice Sheet and, 41; native, 90*figure*, 101; non-food, 91–93*table*, 94, 96 (*see also specific plants*); origins, 104*table*; 17th and 18th century Lower Manhattan, 78*map*–79; Shawangunk, 17, 19, 135–149; stratigraphy and, 130; vascular, 131. *See also* agriculture; aquatic life; ethnobotany; exotic ornamental plants; gardens; horticulture; medicinal plants; soil; weeds; *specific flora*; *specific types of flora*
flotation technique, 71–72, 73, 87
flounder (*Trinectes maculatus*), 29*table*, 30
flour, 253
flowering dogwood (*Cornus florida*), 138
flow of the Hudson River. *See* Hudson River, flow of
flushing rate, xxiv
fly-honeysuckle (*Lonicera canadensis*), 169
Fontenoy, Paul, 249
food chain, 237–238, 242. *See also* food webs
foods: pre-contact period, 69–73, 113; colonial period, 17, 113; 17th century, 95; 21st century, 326; consumption patterns, 161; ice and, 202, 207, 208; pollution and, 228; supplies, 8, 9, 10*figure*, 32, 154; transportation and, 254; waste disposal and, 219. *See also specific foods*; *specific nutritional components*
food webs, xxiv, 24, 198, 227, 233, 236, 242, 328. *See also* food chain
forams (Foraminifera), 127*figure*, 128*figure*, 130
forbs, 170. *See also specific forbs*
"For Comfort and Affluence" (Leighton), 186
Ford site, 69
Fordson tractor, 159
Forest, Fish and Game Law (1905), 48
forests: pre-contact period, 3, 4, 68–71, 73, 325–326; contact period, 24; 19th–21st centuries, 125, 129, 326; abandoned farmland and, 174, 176; agriculture and, 24, 153, 159; art and, 293; basics, xxi, xxii*map*; beaver trapping and, 5; butterflies and, 176; canopies, 148, 149; Columbia County, 168, 175*figure*; decline of, 123, 177–178; ecosystem services, 11; fire and, 5; herpetofauna and, 41; human impacts, xv, 129, 130, 154, 155, 326; industry and, 139–140, 197; Lenape and, 23; lowland, 67; old growth, 138; pollen analysis and, 24; restoration of, 177; sediment and, 60–61; Shawangunk, 19, 135, 136–137, 137*maps*, 138–149, 140–149. *See also* forests; land clearing; pollen analysis; reforestation; trees; woodland
Fort Clinton, xix*figure*
Fort Edward, xix–xx
Fortin 2 site, 71
Fort Lee NJ, 258, 259
Fort Montgomery, xix*figure*, 257
Fort Nassau, 72
Fort Orange, 72, 80
forward purchases, 158
fossil fuels, 155, 330
Foundry Cove, 233–242, 233–246
Fox Creek, 71
Fox Meadows site, 72
FPC (Federal Power Commission), 282–287*passim*
fragaria spp. (strawberry), 88–89*table*, 91–93*table*, 94, 105–106, 113
fragmentation, 178
Frances Skiddy (steamboat), 251
Fraxinus spp. (ash), 125, 126*figure*, 170; *F. americana* (American/white ash), 125, 138, 146*figure*, 190; mountain ash, 141; *F. nigra* (black ash), 170; *F. pennsylvanica* (green ash), 170
Freeder of the Hudson, A (Rix), 296
freight hauling, xvi*map*, xix*figure*, xxiii, xxiv, 168, 197, 222–223, 327. *See also* transportation
Frenchmen, 4, 79, 102, 103, 112, 202, 203. *See also* Europeans
French Pox, 104
freshwater, xxiv
Fresh Water Pond, 206
frogs, 41, 43, 45–46, 48, 171. *See also specific frogs*
frontier, 249

frostfish (*Microgadus tomcod*), 29*table*, 30
frostfish (tomcod) (*Microgadus tomcod*), 29*table*, 30, 263, 265, 266, 267*table*
fruits, 88–94, 101, 142, 178. *See also specific fruits*
FSS surveys, 264*figure*
Fuchs, L., 97, 105, 106, 107, 109
fuel, 10*figure*, 11, 135, 139, 145, 330. *See also specific fuels*
Fulton, Robert, 197, 250–251, 320–321
Fulton Fish Market, 34
functional ecosystem analysis, 5–6
Fundulus spp. (killifishes), 239
fungi, xxiv, 156, 159, 189
fungicides, 160
Funk, Robert E., 67, 68, 69, 70, 71
fur trade, 3, 4–5, 11, 23, 195, 316
the future, 77, 313, 325–336, 333
fyke nets, 31

Gabry, Carel & Pieter, 86
Galium spp. (bedstraw/cheese rennet), 88–89*table*, 90*figure*, 91–93*table*, 95, 96, 106–107
Gallatin, Albert, 252
gall midge (*Mayetiola destructor*), 156
game animals, 23, 24, 67, 69, 72. *See also specific prey*
gardens: canals and, 253; exotic plants and, 183, 184; Kierstede's, 110; layouts, 100; pleasure, 203, 204; restoration of, 186; 17th and 18th century, 113, 114; waste and, 222; water, 189–190
Garrison, Lloyd, 283, 284
gastropods, 240
Gaylussacia baccata (black huckleberry), 72, 137, 138
GCNPP (Greene County Nuclear Power Plant), 292, 300–307
Gehring, Charles T., 29
General Electric, 327
Genesee River, 71
geographic center, xxii*map*, 307
Geographic Information System (ArcGIS), 55
geography, xxi–xxvii, xxvi, 315, 316
geology, 136–137*maps*, 137
George Washington Bridge, 59, 61, 62*figure*, 256, 258
Georgia State, 45
geotextile caps, 238
Geranium maculatum (wild geranium), 168
Gerard, John, 104, 109
German Sarsaparilla, 94
Germantown, 170
Geum aleppicum (yellow avens), 171
Gibbons, W., 43
Gibbons v. Ogden (1824), 251
Gifford, Sanford R., 293, 295, 296, 297
Gill, T., 31
gill nets, 71
ginseng (*Panax quinquefolius*), 111
GIS (geographic information system), 55
glacial lake basin mosaic model, 68

glacial lakes, 73, 206
glaciers, xxiii, 124, 129, 137
glandular swelling, 105
glass beads, 72
glassworks, 142
Gleditsch, N. P., 77
Glens Falls, 254
Glenville, 43
global warming. *See* climate change
Glyptemys spp. (tortoises/turtles): *G. insculpta* (wood turtle), 42, 45, 48, 170; *G. muhlenbergii* (bog turtle/Muhlenburgh tortoise), 44, 45, 48, 171, 190
Goat Island rockshelter, 71
Goedhuys, D. W., 103
golden ragwort (*Packera* [formerly *Senecio*] *aureus*), 171
goldenrods. *See Solidago* spp.
golden shiner (*Notemigonus crysoleucas*), 28
goldfish (*Carassius auratus*), 34, 189
Goldkrest site, 72
gonorrhea, 106, 107, 109
Goodman, Paul, 276
goose, 69
goosefoots (Chenopodiaceae spp.), 104–105
Gorge, the, xxiii, xxiii*map*
Gould, Jay, xvi*map*
gourds, 95
Government House, 204
grain, 155, 167*figure*, 168, 249. *See also specific grains*
Gramineae (grass), 125, 129
Gramly, R. Michael, 68, 72
Grand Banks, 32
Grand Canyon, 280
Grand Canyon of the Yellowstone (Moran), 295
Grand Street market, 208
grapes, 89, 90*figure*, 91–93*table*
grasses (*Gramineae*), 70, 125, 126–127*figures*, 129, 153, 169, 171. *See also* pollen analysis; *specific grasses*
grasshopper sparrow (*Ammodramus savannarum*), 173
grassland birds, 173, 177*figure*
grasslands, 169–170, 178
gravel, 55*map*, 56
gray birch (*Betula populifolia*), 140, 141*figure*
gray goldenrod (*Solidago nemoralis*), 171
gray treefrog (*Hyla versicolor*), 43*table*
grazing. *See* pasture
Great Fire of 1835, 19
Great Lakes, 34, 36, 111, 252
Great Lakes Salmon, 34
Great Society program, 280
green ash (*Fraxinus pennsylvanica*), 170
green dragon (*Arisaema dracontium*), 170
Greene County ice industry, 211
Greene County Nuclear Power Plant (GCNPP), 292, 300–307
Greenfield, Haskell, 114
green frogs, 48

green-headed coneflower (*Rudbeckia laciniata*), 171
Greenhouse Consultants, 114
Green Island Dam, 60
Greenport, 305
Grieve, Mrs. M., 97, 104, 105, 106, 107
Grieves, 108
Gross, M. G., 61
ground beetle (Carabidae), 170
ground cover, 138, 169
ground water, 19
grouse, 69
Grouse Bluff site, 71, 72
guano, 158
guidebooks, 251, 255
Gulf Coast, 90
"Gunks" (Shawangunk Mountains), 123, 133–152
gypsy moth (*Lymantria dispar*), 148, 160
gypsy moth (*lymantria dispar*), 148, 160

habitat: Columbia County loss, 168–169, 170; disturbed, 113; mesic, 124–125, 137, 138. *See also* ecological analogy; *specific habitats*
Hackensack Marsh and Meadowlands, 123, 124, 129, 130, 160
Haie, Jacob, 83*table*
hair loss, 108
hairy bittercress (scurvy grass) (*Cardamine hirsuta*), 98
Haiti, 101
Half Moon, 316
Haliaeetus leucocephalus (bald eagle), xviii*figure*
halibut, American (*Hippoglossus hippoglossus*), 32, 33
halocline, xxiv
Hamamelis viiniana (witch hazel), 138
"Handbook of Salamanders" (Bishop), 46
hangings, 206
Harber-Bosch process, 158
harbors, ix
hardwoods, 41, 67, 138, 146, 148–149. *See also specific hardwoods*
Harlem River, 224, 229, 315
harriers (day hawks) (*Circus cyaneus*), 171
Harriman family and State Park, 257, 280, 298
Harriman State Park, 257
Harris, Wendy E., 19
Harrisburg Peneplain, 168
Hart, John P., et al., 70, 71, 72
Harvard Forest, 178
harvesting of crops, 158
Haskin, DeWitt Clinton, 257
hatcheries, 32, 34
Haverstraw Bay, 57, 240
Hawkins, Richard, 106
hawthorn (*Crataegus* spp.), 176
hawthorn plum (*Crataegus* spp.), 68
Hayden, F. V., 295
hayfields, 167*figure*, 171, 172–175, 176

hazelnut, beaked (*Corylus cornut*), 169
hazelnuts (*Corylus* spp.), 68, 72
headaches, 108
heal-all (betony/woundwort) (*Stachys* spp.), 88–89*table*, 91–93*table*, 94, 108, 113
heal-all (betany-) (woundwort) (*Stachys* spp.), 88–89*table*, 91–93*table*
health, human, 11, 222, 224, 228, 230, 238, 242, 243
hearths/firepits/ovens, 67–68, 69, 70, 71–72, 73
heart-leaved tear-thumb (*Polygonum arifolium*), 108
heart medicines, 107
Heart of Darkness (Conrad), 10
Heart of the Andes (Church), 295, 301
heat. *See* temperature
heaths, 124, 138
Hedeoma pulegioides (American pennyroyal), 171
hedge mustard, 98
hedgerows, 172
hedges, 186–188
Hedrik, U. P., 98, 154
Heerman, Agustijn, 86
Helderberg Escarpment, 46, 67
hematomas, 108
hemlock. *See Tsuga canadensis*
hemlock woolly adelgid (*Adelges tsugae*), 149
hemorrhages, 106, 107
hemorrhoids, 107
hepatopancreas, 238, 240
herbaceous plants, 107, 148, 168, 173, 175, 176, 188. *See also specific plants*
herbalism, 79
herbicides, 160–161
herbs, 126*figure*
heritage gardens, 183
herpetofauna, 5, 17, 24, 41–52, 265*figure*. *See also specific herpetofauna*
herring, 30, 31, 32, 34, 37, 38, 262, 263, 265, 267*table*
Hesperia leonardus (Leonard's skipper butterfly), 171
Hesperia metea (cobweb skipper butterfly), 171
Hesperia sassacus (Indian skipper butterfly), 171
Hessian fly (*Mayetiola destructor*), 156
Heterodon platirhinos (Eastern hognose snake), 45*map*
hickory. *See Carya* spp.
high fructose corn syrup, 156, 161
Highland, 257
highlands. *See* Hudson Highlands
highway system, 258, 280
Hinchey, Maurice D., 329
Hippoglossus hippoglossus (American halibut), 32, 33
Hiscock site (NY), 68
historical eras. *See* eras . . .
Historic Horizon, 79, 114
"Historic Saratoga-Washington on the hudson Partnership," 329
history: archaeology and, 79–86; biology and, xviii, 1–6, 13–21; climate change and, 123; feedback loops and,

9, 17; human adaptations, 65–76; management and, 131. *See also* communities; human impacts
HNERR (Hudson National Estuarine Research Reserve), 123
hobble bush (*Viburnum lantanoides*), 169
Hoboken NJ, 256, 257
Hoffman, C., 69
hogchoker (*Trinectes maculatus*), 29, 34, 265
hognosed snakes, 48
hogpeanut (*Amphicarpaea bracteata*), 138, 168
Holcim Group, 305
Holland, 97, 101
Holman, J., 41
Holocene epoch, 125
homesteading, 24
honeybees (*Apis* spp.), 153
honeysuckle. *See Lonicera* spp.
hookworms, 105
hoop poles, 141–142, 143, 146
Hordeum vulgare (barley), 253
hormonally active substances, xxv, 160, 228, 326, 328
Hornaday, W. T., 36
horse (*Equus caballus*), 158–159, 172, 173, 222–223. *See also* livestock
horticulture, 70–71, 72; pre-contact period, 71, 73; exotic ornamental plants and, 184. *See also* agriculture
Hortus Botanicus (Leiden), 94, 98, 100, 108, 111, 112
Hosack, David, 184
Houghtaling Island, 211
hounding, 145
Housatonic River, 68, 73
houses, 138, 139*figure*, 327
HRBMP (Hudson River Benthic Mapping Project), 54, 62
HRC (Hudson River Valley Commission), 327, 329
HRECOS (Hudson River Environmental Conditions Observing Systems), 331
HRES (Hudson River Environmental Society), ix, xi, xv, 13, 332
HRF (Hudson River Foundation), 62, 268–269, 277, 331
HRWA (Hudson River Watershed Alliance), 332
huckleberry (*Gaylussacia* spp.), 72, 137, 138
"huckleberry" (blueberry), 142–143
Hudson, city of, 250, 305–306
Hudson, Henry, xv, 3–4, 23, 41, 195
Hudson, The (Dunwell), xi
Hudson and Manhattan Railroad Company, 257
Hudson-Catskill ferry, 255
Hudson-Champlain lowlands, 67
Hudson-Fulton-Champlain Quadricentennial, 257
Hudson-Fulton Tricentennial Celebration (1909), 297
Hudson Highlands: pre-contact period peoples and, 65; art and, xvii*figure*, 275–276, 280, 296, 298, 299; basics, 19, 136*map*, 233, 335; depths of Hudson in, 54; forests and, 124–125, 326; frogs, 46; Native Americans, 4; pollution and, 234–235; power plants and, 281–282, 284, 285, 286; Roosevelt and, 280; snakes, 42, 43, 44; topography, 67; transportation and, 19. *See also specific locations*
Hudsonia, Ltd., 47
Hudson Lowlands, 19, 67, 68, 137
Hudson National Estuarine Research Reserve (HNERR), 123
Hudson River: basics, ix, xv, xvi*map*, 59, 129; depths of, xxiii, 54–55, 56, 57; historical eras and, 314; land use and, 178; as model, xv; morphology, 15–16 (*see also* shorelines); names of, xv, 73, 315; population, 154, 317; tributaries, xxii*map*, 24, 32, 60, 70, 326–327; watershed, xxi–xxii. *See also* Hudson River, flow of; Hudson River Estuary; Lower Hudson; Upper Hudson
Hudson River, flow of: basics, xvi*map*; Battery Park City and, 58; bridges and, 59; cadmium and, 240, 242; channelization and, 56; dredging and, 62; human effects, 9; Lower Hudson, xxiii–xxiv; Mohawk River, xxii; power plants and, 277; railroads and, xix*figure*, 59; seasonal, xxii, xxiii–xxiv, 54; transportation and, 17; tributaries and, 19; near Troy, xxii; Upper Hudson, xxi
Hudson River Benthic Mapping Project (HRBMP), 54, 62
Hudson River Environmental Conditions Observing Systems (HRECOS), 331
Hudson River Environmental Society (HRES), ix, xi, xv, 13, 332
Hudson River Estuary, 54–63; pre-contact period, 65–74, 124, 316; basics, ix, xviii*figure*, xxi, 53–63; dredging and, 37; fishes and, xxv, 266, 267*table*, 268; flora, 153; flow and tides of, xxiv; industry and, 19, 233–243, 261–269; marshes and, 124; pollution of, xxv, 233, 236, 238, 240, 266; power plants and, 261, 262, 264*map*, 265, 270; sediment and, 13; transportation and, 197
Hudson River Estuary Program (DEC), 178, 329
"Hudson River Fisheries Investigation 1965-68" (ConEd), 286
Hudson River Fishermen's Association, 276, 278, 286, 287, 288, 325, 326
Hudson River Foundation (HRF), 62, 268–269, 277, 331
Hudson River labor strife monopolies origins physical extent of ruins of technology for harvesting and storing workforce working conditions National Estuarine Research Reserve Power Plant Settlement School of Art sloop(s) striped bass Valley Watershed
Hudson River Line Rail Road, xvi*map*, xix*figure*
Hudson River National Estuarine Research Reserve, 62, 327
Hudson River Railroad, 254–255
Hudson River School of landscape painting, xvi*map*, 183, 276, 279, 291–307, 320, 332. *See also individual painters*
Hudson River Settlement Agreement (1980), 262, 268
Hudson River Sloop, 196–197
Hudson River valley, xxiii, 17
Hudson River Valley Commission (HRVC), 327, 329

Hudson River Valley Greenway, 329
Hudson River Valley National Heritage Area (HRVNHA), 329
"The Hudson River Watershed Alliance," xxvi
Hudson River Watershed Alliance (HRWA), 332
Hudson Tubes, 257
Hudson Valley in Winter, The (Church), 302
Huey, Paul, 80
Hughes, Charles Evans, 297–298
Hughes, Langston, x, 283
human adaptations, ancient, 65–76
human impacts: basics, ix–xi, xv, 1, 7–12, 24; on fishes, 36, 37; on forests, 130; good, bad and subtle, 9–10; on land, 17; on marshes, 123–132; poetry and, 322; productivity and, xxiv; sediment and, 53–64, 61; on Shawangunks, 19, 135; on shoreline, 24. *See also* development; feedback relationships; history; *specific activities, impacts & industries*
von Humboldt, Alexander, 295
Hummer, C. C., 69
huneysuckles (*Lonicera* spp.), 187, 188
Hunter's Home site, 70
hunting, 24, 71, 145, 177. *See also* game; projectile points; *specific prey*
Huntington, Ellsworth, 256
Huth, Paul C., 47
hybrid plants, 160, 328
Hyde Park, 33, 184
hydrology, xxiii–xxiv, 129, 189
Hyla versicolor (gray treefrog), 43*table*
hysteria, 106

ice, 196, 254
ice ages, xvi*map*, 97, 130
Icebergs (Church), 301
ice cave talus, 137*map*, 138
ice clearing, 19
ice cream, 202, 203
ice harvesting, 9, 15, 17, 19, 201–214
Ice House Garden, 204
ice houses, 204, 205*figure*, 207, 209–214, 213*figure*
ice-scouring, 177
ichthyoplankton, 262, 265
Iconography of Manhattan Island (Stokes), 81
Ictalurus punctatus (channel catfish), 30*table*, 265, 265*figure*
Idaho State, 155–156, 157
identity, 248, 291–292, 299, 305, 306, 315, 319, 320
Ilex spp., 126*figure*
Ilex verticillata (winterberry), 170
Illinois River, 34
Illinois State, 71
immigrants, x. *See also specific immigrant nationalities*
Impatiens (jewelweed), 129
impervious surfaces, 10, 130, 221, 297, 326
impingement, 262–265, 267
improved acreage, 167, 169, 172, 175*figure*, 176*map*

improvement, 292. *See also* development
Incas, 95
Indian Point, 261, 262, 263*table*, 264*figure*, 268, 287, 288, 330
indians. *See* Native Americans; *specific tribes*
Indian skipper butterfly (*Hesperia sassacus*), 171
indigenous flora, 88–94; characterizations, 94–95; weeds and, 94–95
Indonesia, 103
Industrial Revolution, 8; fishing and, 31
industry: 20th century, 54; 21st century, 332; aquatic life and, 233–246; art and, 252; basics, xix, xxv–xxvi; canals and, 254; the Collect and, 206; extractive, 5*figure*, 11, 24; fishing, commercial, and, 24; history of, 196, 197; ice harvesting and, 208–210, 211; influence of Hudson on, 314, 320–321; marshes and, 131; pollution and, xvi*map*; railroads and, 17; sediment and, 61; sewers and, 222; Shawangunks and, 135, 139–144, 142; too much, 322; upland, 123; waste and, 225. *See also specific industries & locations*
inflammation, 107
inlets, 58–59
Innes, J. H., 81, 86
Inness, George, 293, 295
insect bites, 106
insecticides, 160
insects, 131, 156, 189
Instructions to Apothecaries and Surgeons who will Board the Fleet to the East Indies in the Year 1602 (Clusius & Pauw), 112
integrated pest management (IPM), 161
International Panel on Climate Change (IPCC), 77
interregional exchange, 102–103
Interstate Route 88, 67
Interstate Sanitation Commission (ISC), 224, 225, 229
intestinal maladies, 105
introduced species, 9, 49, 108. *See also individual species*
Invasive Plant Council of New York State, 185, 191
invasive species: pre-contact period, 95, 113; contact period and after, 125, 126; agriculture and, 159; information, 191; Lower Hudson, 129; naturalized, 185–191; sediments and, 56; upland, 123, 129, 131; wetland, 129–131, 240. *See also individual species*
invertebrates, 36, 42, 131, 233, 236, 240. *See also specific invertebrates*
Iona Island and Marsh, xvi*map*, 123, 124, 127*figure*, 129, 130
IPCC (International Panel on Climate Change), 77
IPM (integrated pest management), 161
iris (*Iris versicolor*), 171
Iris pseudacorus (yellow flag), 189–190
iron mining, 142
Iroquois Pipeline site, 230-31, 72
Iroquois tribes, xix, 96, 97, 105, 106, 107, 108
irrigation, 156, 161
Irving, Washington, 275, 276, 318–319
ISC (Interstate Sanitation Commission), 224, 225, 229

Isherwood and Grieg (confectioners), 202–203
Italian businessmen, 202
Italy, 295

Jack-in-the-pulpit (*Arisaema triphyllum*), 177
jack-lights, 145
Jacobs, Jaap, 77
Jacoby, R. M., 70
Jamaica Bay, 57, 123, 124, 130, 227
Jameson, J. F., 108
Jamestown, VA, 80, 82*table*, 100, 103, 105, 201
Jans, Anneke, 110
Japanese barberry (*Berberis thunbergii*), 187
Japanese beetle (*Popillia japonica*), 189
Japanese honeysuckle (*Lonicera japonica*), 188
Jarvis, P. J., 184
jasper, 68
jaundice, 106
Jersey City NJ, 257
Jervis, John B., 254–255
Jesuits, 103, 112
Jet, Robert, 23
jewelweed (*Impatiens*), 129
John Milner Associates Inc., 204*figure*
Johnson, Lyndon, 280
joints, 105
Juet, Robert, 24, 27, 28–29*table*, 279, 297, 316–317
Juglans spp., 126*figure*; *J. cinerea* (butternut), 71; *J. nigra* (walnuts), 72
Juniperus spp., 126*figure*
Juniperus virginiana (red cedar), 176

Kaaterskill Falls (Cole), 293
kale, 98, 100*table*, 101
Kalm, Peter, 102, 248
Kalmia angustifolia (sheep laurel), 138
Kalmia latifolia (mountain laurel), 137, 138, 143
Kane, Michael, 240
Kaulfield, Carl, 44–45
Kennedy, M., 96
Kennedy Jr., Robert F., 278
Kensett, John F., 295, 296
Kent, Edwin C., 173
Kentucky State, 153
Keskeskick, 315
kidney disorders, 106, 107
Kieft, Director, 110
Kierstede, Hans, 83*table*, 85*table*, 86, 102, 109, 110, 114
Kierstede, Sara Roelofs (Madame, Mrs. Hans), 86, 102, 109, 110–113, 114
killdeer (*Charadrius vociferus*), 169
killifishes (*Fundulus* spp.), 239
Kindred Spirits (Durand), 293
kingbird (*Tyrannus tyrannus*), 170
Kings Ferry, 256
Kingston, 56, 72, 139, 143, 250, 256

Kingston-Rhinecliff ferry, 255
Kinosternon subrubrum (Eastern mud turtle), 48
Kittattiny Mountain NJ, 136
Kiviat, Erik, 47
Klerks, P. L., 236
Knickerbocker Ice Company, 207
knotgrass (knotweed) (bistort) (*Polygonum* spp.), 88–89*table*, 91–93*table*, 95, 97, 103, 107–108
K'oeh, Hans Skola, 106
kohlrabi, 100*table*
Kraft, H. C., 70

labor, 253, 318, 331
Labrador, A. M., 71
labsquarter/pilewort (smearwort), 104
Lackawanna Terminal, 256
Lackawanna Valley, The (Inness), 293
de Laet, Johan, 107–108
Lafitau, Joseph-François, 111
Lake, T. R., 28
Lake Albany, 68
Lake Fort Ann stage, 68
Lake George, 44, 296
Lake House, 145*figure*
Lake Minnewaska, 135
Lake Mohonk, 135, 144, 300
lakes, 73, 123, 125, 129, 130, 144, 206. *See also specific lakes*
Lake Tear of the Clouds, xxi
lake to marsh records, 122–134
lambsquarter (pilewort/smearwort) (*Chenopodium* spp.), 88–89*table*, 91–93*table*, 96–97, 103, 104–105
Lamont-Doherty Earth Observatory, 268, 331
lamp oil, 33
lamprey, 29, 30
Lampropeltis triangulum (Eastern milksnake), 43*table*
land clearing, 154–155, 169. *See also* deforestation
landgrants, 138
landscapes, xvi*figure*, 19, 183–193, 305, 307, 318, 322. *See also* Hudson River School of landscape painting; scenery; *specific landscapes*
land trusts, 136, 146
land use, 5–6, 13, 17, 36, 48, 62, 79, 131. *See also specific locations*; *specific uses*
larch (*Larix laricina*), 126*figure*
largemouth bass (*Micropterus salmoides*), 30*table*, 36
Largy, Tonya, 72
Larix laricina (larch), 126*figure*
larvae, aquatic, xxiv, xxv, xxvi, 236, 262, 263, 265, 277, 283, 286, 288
Late Woodland artifacts, 110
Lathyrus palustris (marsh pea), 170
Laurentide Ice Sheet, 41, 124
law of the minimum, 158
lead arsenate, 160
leaded glass, 81
leatherwood (*Dirca palustris*), 170

L'Ecluse, Charles de (Clusius/Calolinus/Carolus), 98, 100, 103, 111, 112, 114
Le Corbusier, 258
leg disorders, 107
legislation: Adirondacks and, 296; fishing and, 33, 36; invasive plants and, 191; power generation and, 19, 285; scenery and, xxi; shad and, 31; Upper Hudson and, 329. *See also specific legislation*
legumes, 70. *See also specific legumes*
Leiden, Germany, 94, 98, 100, 107
Leiden Medical Garden, 107
Leiden University, 111–112, 114
Leighton, Ann, 186
Lemon, J. T., 167, 169
Lenapes (Lenai Lenapi/Lenni Lenape), 3, 4, 5, 23, 73, 248, 315
Lenzi, Philip, 202, 203
Leonard's skipper butterfly (*Hesperia leonardus*), 171
leopard frog (spring frog) (*Rana pipiens*), 48, 171
Lepomis auritus (redbreast sunfish), 31
Lepomis gibbosus (sunfish), 28, 29*table*, 30
Lerner, Gerda, 21
Lespedeza procumbens (downy trailing Lespedeza), 172
Libri Picturati (Clusius/de L'Ecluse), 98, 100*table*, 101
von Liebig, Justus, 158
Life Magazine, 282
Lighthouse Cove site, 71
lighthouses, xvi*map*
Ligquidambar, 128*figure*
Liguliforae spp., 126*figure*
Ligustrum spp. (privets), 188
Lilium canadensis (Canada lily), 171
Lilium philadelphicum (wood lily), 169
lime industry Napanoch, 142
limes, 94
limestone, 137, 166, 196
Limnodrilus hoffmeisteri (oligochaete), 237, 239, 240, 241, 242
Linaria canadensis (blue toadflax), 173
Lincoln, Abraham, 295
Lindenthal, Gustav, 258
Linnaeus, Carl (Linné), 102, 112
Linum spp. (flax/toadflax), 88–89*table*, 91–93*table*, 94, 103, 108, 109, 113, 155*table*, 173
Linum virginianum (Virginia yellow flax), 172
lipfis, 30
Lipomis gibbosus (pumpkinseed), 30
Liquidambar spp., 126*figure*
literary historical eras, xxvi, 313–324
Lithobates spp. (frogs), 41, 43, 46; *L. septentrionalis* (mink frog), 46; *L. sylvaticus* (wood frog), 41, 43*table*
litigation, xxvi, 277, 280, 285, 289, 325. *See also negotiation; specific cases*
little bluestem (*Schizachyrium scoparium*), 171
Little Falls rockshelter, 72, 73
Little Ice Age, 97, 130
little sundrops (*Oenothera perennis*), 172

Live Carp (barge), 34
liver, 105
liver cancer, 266
livestock, 5, 24, 172. *See also pastures; specific livestock*
Livingston, Robert, 197, 250–251
lizards, 43, 44
Lloyd Nuclear Power Plant, 300
Lobelia spicata (spiked lobelia), 173
lobster, 42
Local Waterfront Revitalization Plans (LWRPs), 305, 306
logging. *See lumbering/timber*
LOI (loss-on-ignition), 123, 125, 128*figure*
long-billed marsh wren (*Cistothorus palustris*), 190
Long Island, 69
Long Island Sound, 227
Long River Ichthyoplankton Survey (LRIS), 264*figure*, 265
Long-Term Ecological Research sites, 8
Lonicera spp. (honeysuckles): *L. canadensis* (fly-honeysuckle), 169; *L. japonica* (Japanese honeysuckle), 188; *L. mackii* (amur honeysuckle), 187; *L. morrowii* (Morrow's honeysuckle), 187; *L. tatarica* (tartarian honeysuckle), 187
Loop, A. S., 229
loosestrife (*Lythrum*), 129
Lopuch 3 site, 70
Lossing, B. J., 32
loss-on-ignition (LOI), 123, 125, 128*figure*
Lothrop, Jonathan C., 72
lousewort, common (*Pedicularis canadensis*), 173
love medicines, 105, 107
Lovett NY, 264*figure*
lowbush blueberry (*Vaccinium* spp.), 137, 138
Low Countries, 98
Lowell, Robert, 322
Lower Hudson: 17th century, xvi*map*, 4; agriculture, 123, 155*table*, 156, 158*table*, 160*tables*; archaeology and, 79, 115; basics, xix*figure*, xxi, xxiii–xxvi, 153–154, 335; fishes and, 15, 28, 31, 327; forests, 5, 149; ice houses, 209; industry and, 233, 238, 240, 242, 325; invasive species and, 129, 185, 191; morphology/sediment, 53–63; PCBs and, 198, 242, 327; poetry and, xvi*figure*, 321–322; population, 154, 328; power plants and, 19; research and, 331; transportation and, 256; turtles, 42; wetlands, 129. *See also specific locations*
LRIS (Long River Ichthyoplankton Survey), 264*figure*, 265
lumbering/timber, 24, 123, 135, 139, 141–142, 143, 155, 172, 197, 254, 296
Lurkis, Alexander, 282–283, 285
LWRPs (Local Waterfront Revitalization Plans), 305, 306
Lycaena hyllus (bronze copper butterfly), 171
Lycopodium lucidulum (shining clubmoss), 125, 126*figure*, 130
Lyell, Charles, 296–297
Lymantria dispar (Eurasian gypsy moth), 148, 160

lymantria dispar (gypsy moths), 148
Lynx rufus (bobcat), 177
lysine, 95
Lythrum spp. (loosestrife), 129; *salicaria* (purple loosestrife), 189–190, 240

Macauley, James, 43–44
machine, the, 253–254
Machine in the Garden, The (Marx), 255
macrofossil analysis, 123, 125, 129, 130, 131
Mahicanituk (*Muhheakunnuk/Muh-he-kun-ne-yuk*) (River that Flows Both Ways), xv, 73, 315
Mahicans (Mohicans/Mohegans), 3, 4, 5, 28, 72, 73, 97, 105, 108, 248
Maianthemum canadense (Canada mayflower), 177
Maianthemum racemosum (false Solomon's seal), 168
maize (*Zea mays*), 23, 70–71, 72, 95, 96, 97, 153, 156, 160–161, 167*figure*, 169, 317
Malaclemmys terrapin (Northern Diamond-backed Terrapin), 42
malaria, 104
Malthus, Thomas Robert, 153–154
Malus spp. (apple), 155, 157–158, 161, 167*figure*
mammals, 23, 41, 46, 70, 149. *See also specific mammals*
management: ecosystem services and, 11; fisheries, 36, 37; resources, 49; sediment and, 131; Shawangunk forests, 148, 149; soils, 159, 161. *See also* conservation; preservation; stocking programs, fish
Manhattan: contact period, 4, 318; drinking water and, 218; environmental data, 78; ethnobotany, 77–115; ferries and, 256; names for, 4, 24, 195, 315; population of, 220*figure*, 223*figure*, 224; rye and, 156; shoreline and, 15, 58; waste and, 224. *See also* Battery Park City; New Amsterdam; *specific bridges*; *specific locations*
Manhattan Bridge Building Company, 257
Manhattan Company, 220
Manifest Destiny, xix
Mann, Thomas, 10
Manna site (PA), 72
maple-leaf viburnum (*Viburnum acerifolium*), 138
maples. *See Acer* spp. (maples)
Maratanza Lake, 144
Marathon Battery Company, 235
marine bay anchovy (*Anchoa mitchilli*), 263, 265, 267*table*
marine biota, 17, 29, 30, 37, 219, 227, 228, 229, 236. *See also individual species*
Marmota monax (woodchuck), 69, 169
Marone chrysops (white bass), 30*table*
marshes: cadmium and, 233; climate change and, 122–134; development and, 69; flora, 125; pollen analyses, 175; railroads and, 197–198; restoration of, 131; sedimentation rates, 130; upland, 123. *See also specific marshes*
marsh pea (*Lathyrus palustris*), 170
marshpepper (*Polygonum hydropiper*), 107

marsh wren (*Cistothorus palustris*), 190
Martes pennanti (fisher), 177
Martinsburg Formation, 137
Marx, Leo, 255
Massachusetts Bay Colony, 90
Massachusetts State, xxi, 68, 170, 172, 175, 178, 286. *See also* Boston MA
mass spectrometry, 69, 70
mast foods/ forests, 68–69, 70, 72
mastodons, 68
Materia Medica, 112
Matteuccia struthiopteris (ostrich fern), 170
Mattice 2 site, 69
Mauritius River, xv, 315
Mayetiola destructor (gall midge/Hessian fly), 156
McAdoo, William Gibbs, 257
McDonnell, M. J., 9
McEntee, Jervis, 293, 296
McKinstry Tannery, 140
McMahon, Bernard, 184
McShea, W. J., et al., 149
McVaugh, R., 174, 177
MDS (multidimensional-scaling) plots, 237*figure*
meadowlark (*Sturnella magna*), 173, 174, 176
meadows, 170–171, 173. *See also* hayfields
meadowsweet (*Spiraea alba*), 171, 177
Mearns, E. A., 31, 46
meat, 70
mediation, 330, 332
Medicago sativa (alfalfa), 173
medicinal fish, 33
medicinal plants: colonial period and after, 88–89*tables*; basics, 17, 88–97, 101–102; collection procedures, 112; European, 102–104, 106–109; Native American, 94, 96, 102–113 (*see also specific tribes*); University of Leiden and, 98, 100. *See also specific plants*
Medieval Warm Period (MWP), 123, 130
megafauna, 68
Megapolensis, Johannes, 29, 111
Melampyrum lineare (cow-wheat), 138
Meleagris gallopavo (wild turkey), 177
melons, 105
Melospiza melodia (song sparrow), 170
Melville, Herman, 321
memory disorders, 108
Memphis TN, 227
Menidia spp. (silversides), 29*table*
menstrual disorders, 105–106, 107, 108
mercury, 242
mesic habitats, 124–125, 137, 138, 148
Messner, T., 72
metals, 228, 235–236, 327. *See also specific metals*
metam-sodium, 160
Metropolitan Sewerage Commission, 223, 224, 225, 229
Mexico, 95
Microgadus tomcod (Atlantic tomcod/frostfish), 29*table*, 30, 263, 265, 266, 267*table*

Micropogonias undulatus (Atlantic croaker), 265, 265*figure*
Micropterus spp. (freshwater basses): *M. dolomieu* (smallmouth bass), 30*table*, 36; *M. salmoides* (largemouth bass), 30*table*, 36
Midwestern US, 96
midwives, 110
Mifflinville sites (PA), 70
migrations, 68, 69, 71, 73
milk, 155*table*, 157, 167*figure*
milk flow treatments, 105
milk purslane (*Portulaca oleracea*), 169
Millennium Assessment, 10
Miller & Whitbeck Ice House, 213*figure*
Mimulus ringens (common monkeyflower), 171
Mimus polyglottos (mockingbird), 176
minerals, 316
mining, 15, 48, 142, 155
mink frog (*Lithobates septentrionalis*), 46
Minnewaska Lake, 135, 144
Minnewaska State Park Preserve, 136, 149
minnow, 29*table*, 30
Minot, Louisa Davis, 296
mint, mountain- (*Pycnanthemum* spp.), 171
Minuit, Peter, 318
miscarriages, 107
missionaries, 103
Mississippian Moundbuilders, 105
Mississippi River, 34, 36
Mitchell, Samuel, 253
mockingbird (*Mimus polyglottos*), 176
Moeller, Rober, 72–73
Mohawk and Hudson Railroad, 254
Mohawk River: pre-contact period, 70; 17th century fisheries, 29; 21st century, 30; agriculture and, 13, 155*table*, 156, 158*table*, 160*table*; altitutude, 58; basics, xxi, xxii–xxiii, 153–154
Mohawk tribe, 23, 102, 111
Mohegans (Mohicans). *See* Mahicans
Mohonk Lake, 135, 144
Mohonk Lake Cooperative Weather Station, 124
Mohonk Mountain House, 140, 142, 144, 145, 146, 293, 295
Mohonk Preserve, 46, 49, 136, 140, 142, 144*figure*
Mohonk Ridge, 126*figure*
Mohonk Trust, 146, 148
moisture, 130
Mollugo spp. (carpetweed), 88–89*table*, 91–93*table*, 94, 95, 113
molluscs, 69
monarch butterfly (*Danaus plexippus*), 188–189
Monardes, Nicolás, 112–113
monkeyflower, common, 171
monocultures, 160, 186
monoplies, 195, 197, 210, 213, 250–251, 253
Monsanto, 157
Monticello, 204

Moore, Charles Herbert, 298–299
moose (*Alces alces*), 177
Moran, Thomas, 295
Morgan, J. P., 280
Morgan, R. P., 254, 255
Morone spp. (temperate basses) (Moronidae), 28, 29*table*, 30; *M. americana* (white perch), 28–29, 29*table*, 263, 265, 266, 267. *See also Morone saxatilis* (striped bass)
Morone saxatilis (striped bass): contact period, 29; colonial period, 30; 19th and 20th centuries, 31, 34, 36, 228; 21st century, 37; generating stations and, 262; mortality rates, 267*table*; PCBs and, 24; predation by, 37; regulation and, xxvi, 27; Storm King mountain project and, 283, 284, 285–287; studies, 263, 266, 268
Morris, Robert, 204*figure*
Morrow's honeysuckle (*Lonicera Morrowii*), 187
Morse, Samuel F. B., xvi*map*
Morus alba (white mulberry), 186
Mossy Glen, 140
moths, 176
Moundbuilders, 105
mountain ash, 141
mountain laurel (*Kalmia latifolia*), 137, 138, 143
mountain lion, 145
mountain maple (*Acer spicatum*), 169
mountain-mint (*Pycnanthemum* spp.), 171
Mount Merino, 305–306
Mount Vernon, 204
mouth disorders, 107
"The Mouth of the Hudson" (Lowell), 322
Moxostoma macrolepidotum (shorthead redhorse), 30
Mt. Marcy, xxi
mud, 56
Muddy Brook Rockshelter, 69
Mugil cephalus (striped mullet), 28
Muh-he-kun-ne-yuk (Mahicanituk/Muhheakunnuk) (River that Flows Both Ways), xv, 73, 315
Muhlenburgh tortoise (bog turtle) (*Glyptemys muhlenbergii*), 44, 45, 48, 171, 190
Muir, John, 280
mulberry wing (*Poanes massasoit*), 171
mule (*Equus asinus x E. caballus*), 159, 173
mullets, 28, 29, 316
multidimensional-scaling (MDS) plots, 237*figure*
multiflora rose (*Rosa multiflora*), 176
munitions, xvi*map*
Munsee tribe, 73
Munsey tribes, 315
Muscooten, 315
Musell, J., 33
muskrat (*Ondatra zibethicus*), 190
mustard/cabbage Family (*Cruciferae/Brassica* spp.), 88–89*table*, 91–93*table*, 94, 97–98, 100–101, 100*table*, 108, 113, 114, 169
mute swan (*Cygnus olor*), xix*figure*, 189
MWP (Medieval Warm Period), 123, 130
mysticism, 319–320, 322

nannyberry (*Viburnum lentago*), 177
Narragansett Bay, 69
Natchez tribe, 105
National Audubon Society, 280, 286
National Environmental Policy Act (NEPA) (1969), 276, 285, 292, 299, 325, 330, 332
National Estuarine Research Reserve System, xvi*map*
National Forests, 9
National Heritage Areas, 291, 329
National Invasive Species Council, 185
National Ocean Services, 54
National Parks, 9, 280, 285, 291
National Parks Association, 286
National Priorities LIst, 238
National Resources Defense Council (NRDC), 287
"National Security and the Threat of Climate Change: Report from the Panel of Retired Senior US Military Officers" (Military Advisory Board 2007), 77
Native Americans: agreements with, xix–xx; agriculture and, 153; artifacts, 78–79; attitudes of, 313; basics, xv, 3; Dutch and, 79, 97, 109–113, 318; ecosystems and, 23; explorers and, 315–317; fish and, 27, 29, 31, 34; foods, 97; forests and, 325; fur trade and, 4; herpetofauna and, 41, 42–43; paintings of, xvi*map*, 295; plants and, 78*map*, 88–94, 96, 101–103, 105, 107, 109–110; populations, 28, 317; reservations, 73; resources and, 23–24; sustainability and, xviii; women, 109–113, 114. *See also* ancient humans; *specific tribes*
native bittersweet (*Celastrus scandens*), 188
Native Plant Center, The, 191
native species, 49, 165–178, 184, 190, 191, 192
native whites, 169
Naturalist Period, 41, 43–44
naturalization, 185, 189, 190, 191, 192
Natural Resources Defense Council, 227
nature, x, 249, 251, 252, 253, 255, 322
Nature Conservancy, The, 47, 136
Nature's Metropolis: Chicago and The Great West (Cronon), 249
navigation, 61, 250. *See also* channels
Neal, J. C., 98
negotiation, 289, 330
nematodes, 160
NEPA (National Environmental Policy Act) (1969), 276, 285, 292, 299, 330, 332
Nerodia sipedon (Northern watersnake), 43*table*
nerves, 106, 108
nests, 173
Netherlands, The, 80
nets, fishing, 33, 70, 72
New Amsterdam, xvi*map*, 77–121, 78*figure*, 195–196, 220–221, 315
Newark Bay, 61
New Baltimore, 209–210*map*, 211–212, 214
Newburgh, 184, 250, 258
Newburgh Bay, 54
Newburgh-Beacon ferry and bridge, 255–256

Newburgh & Cochecton Turnpike, 250
New Deal, 259
New England, 3, 5, 31, 100, 109, 187, 188, 250, 317
New England cottontail (*Sylvilagus transitionalis*), 174
New France, 3
New Hampshire State, 295
New Herbal (Fuchs), 106, 107
New Jersey State, xxi, 58, 68, 111, 136, 224, 251, 256, 286, 315. *See also specific bodies of water; specific locations; specific tribes*
New Jersey State Federation of Women's Clubs, 297
New Lebanon, 170
New Mexico State, 157*table*
New Mirror for Travelers (Paulding), 321
New Netherlands, xvi*map*, 86, 97
New Netherlands Company, 4, 5–6, 23, 195
New Netherlands project, 29
New Paltz, 123
New Spafford, 170
newspapers, 157, 202, 212–213. *See also specific newspapers*
newt, red-spotted (*Notophthalmus viridescens*), 44
Newtown Creek, 225
New Vauxhall Gardens, 204
New York and Albany Post Road, 252
New York Bay, 61
New York Central, 257
New York City: 21st century, 328; basics, ix–x, 196; boroughs of, xxv; canals and, 253–254; countryside and, 249; demographics, xxv; development and, 276; disturbed habitats, 113–114; exotic plants and, 183, 186, 187; fish purchases, 34, 36; ice harvesting and, 201, 204, 208–209; invasive vines, 188; names of, 315; Native Americans and, 315; population changes, 154, 206; shoreline, 19, 58; sloops and, 249; Storm King Mountain project and, 283, 286; transportation and, 250, 256, 259, 321; trap rock and, 297; waste and, xxv, 219, 229; water barriers and, 256. *See also* New York Harbor; *specific locations*
New York City Department of Health, 224
New York City Landmarks Commission, 78
New York Environmental Quality Review Act, 299
New York Harbor, 61, 223–224, 225, 227, 228, 230
New York Herald (newspaper), 222
New York Horticultural Society, 184
New York Ontario and Western Railroad (O&W), 139–140
New York State: censuses, 166–167; courts & Storm King project, 286–287; farms, 159–160; Flora Atlas, 185; land trusts, 136; milk and, 157*table*; transportation and, 254, 258; watershed, xxi. *See also entries starting* NYS . . . ; *specific bodies of water, legislation & locations*
New York State Electric and Gas Company, 331
New York Thruway, 259
New York Times (newspaper), 214, 257, 282, 283, 289, 299

New York University, 268
Niagara (Church), 301
Niagara Falls, 295, 296
Niagara Falls (Davies), 296
Niagara Falls (Minot), 296
Nicholas, George Peter, 68
Nicholls, Richard, 318
nickel, 235
Nicotiana spp. (tobacco), 88–89*table*, 90*figure*, 91–93*table*, 95, 96, 153
Nieuwenhof, Evert, xvi*figure*
nightshade (*Solanaceae*), 97, 108
Nike missiles, 235
nitrogen, 9, 11, 227
NOAA, 54, 55
nodding lady's tresses (*Spiranthes cernua*), 171
Nordås, R., 77
North Carolina State, 36
Northern Diamond-backed Terrapin (*Malaclemmys terrapin*), 42
northern pike (*Esox lucius*), 30*table*
Northern watersnake (*Nerodia sipedon*), 43*table*
North River, 225, 226, 256, 315
North River (steamboat), xv, 250, 251
North River Bridge, 258
Norway maple (*Acer platanoides*), 186, 190
nose bleeds, 103
Notemigonus crysoleucas (golden shiner), 28
no-till, 159
Notophthalmus viridescens (red-spotted newt), 44
Novak, Barbara, 301
NRDC (National Resources Defense Council), 287
nuclear power, 28, 235, 277, 292, 300–307
Nuclear Regulator Commission U.S. (NRC), 301, 303–305
nurseries, 184, 186, 190, 191
nut grasses/nut sedges (*Cyperus* spp.), 71, 88–89*table*, 90*figure*, 91–93*table*, 94, 129, 169
nutrients, 10*figure*, 123, 158, 173, 227, 229, 331
nuts, 67, 69, 70, 72, 73. *See also specific nuts*
Nutton Hook, 65, 67
Nyack, 256, 259
NYC Landmarks Commission, 78
NYS Coastal Management Program (DEC), 306
NYS Conservation Department, 48
NYS Department of Agriculture and Markets, 326
NYS Department of Environmental Conservation (DEC), 48–49, 55, 178, 191, 240, 306, 325, 329; power plants and, 262, 268, 277, 286, 325, 327–328, 331
NYS Department of Transportation, 191
NYS Division of Coastal Resources (Department of State), 328
NYS Energy Research and Development Agency (ERDA), 300
NYS Environmental Quality Review Act (SEQRA), 65, 292, 300, 306, 330, 332
NYS Geological Association, 300

NYS Invasive Plant Council, 185, 191
NYS Invasive Species Task Force, 191
NYS militia, xix*figure*
NYS Nursery and Landscape Association, 191
NYS Office of Parks, Recreation, and Historical Preservation (DEC), 325
NYS Pollution Commission, 223
NYS Power Authority, 288, 301
Nyssa sylvatica (black gum), 126*figure*, 148

oak-heath communities, 124
oaks. *See Quercus* spp.
oats, 155*table*, 253
Ocean Dumping Ban Act (1992), 227
Odocoileus virginianus. *See* deer, white-tailed
Oenothera biennis (evening primrose), 173
Oenothera perennis (little sundrops), 172
Ohio River, 252
Ohio State, 70, 188
Ojibwa tribe, 97, 106, 107, 108
Okeenokee Swamp GA, 45
Oklahoma State, 73
Olana, 300–307, 301*figure*, 302, 303, 304, 304*map*, 305. *See also specific projects*
Old Albany Post road, 258
old fields, 137, 175, 176, 177
oligochaetes (*Limnodrilus hoffmeisteri*), 236, 237, 239, 240, 241, 242
oligotrophic areas, xxi
Olmsted, Frederick Law, 295, 296, 302
omega-3 fatty acids, 106
Ondatra zibethicus (muskrats), 190
Oneonta, 69
"On the Patrons and the History of New Netherlands" (Nieuwenho), xvi*map*
opal, 70
Opalescent River, xxi
openness, 329–331, 332
"Operation Shellshock," 48
Ophioglossum spp. (adder's tongue fern), 171
Ophiostoma spp. (Dutch elm disease), 161
oral histories, 110, 112, 114
Orange County, 48, 160
orangeries, 94
orchards, 253. *See also specific fruits*
orchid, purple-fringed (*Platanthera psycodes*), 171
orchid, ragged-fringed (*Platanthera lacera*), 171, 174
Orconectes spp. (crayfishes), 239
Oregon, 157
organic carbon, 159
Organic Machine, The (R. White), 253
organic matter, xxiv
organic wastewater contaminants (OWCs), 228, 229
oriental bittersweet (*Celastrus orbiculatus*), 188
ornamental plants, 183–193. *See also specific ornamentals*
ornaments, 72

Osmerus mordax (rainbow smelt), 29*table*, 37, 265, 265*figure*
Osmunaceae, 128*figure*
Osmunda spp., 126*figure*
ostrich fern (*Matteuccia struthiopteris*), 170
Ostrya virginiana (Eastern hop-hornbeam), 138
ovens/firepits/hearths, 67–68, 69, 70, 71–72, 73
overfishing, 24, 32
overharvesting, 143, 145
Overlooks Wildfire, 149*figure*
Ovis aries (sheep), 167*figure*, 169–170, 172, 253
O&W (New York Ontario and Western Railroad), 139–140
OWCs (organic wastewater contaminants), 228, 229
ownership, 318
oxygen (dissolved oxygen): 21st century, 327; basics, xxi; generation plant studies, 265; waste treatment and, xxv, 223–224, 226, 227, 229; water quality and, 228
oysters, 42, 69, 229, 279

Packera aureus (golden ragwort), 171
PAHs (Polynuclear Aromatic Hydrocarbons), 243
painted trillium (*Trillium undulatum*), 169
painted turtle (*Chrysemys picta*), 43*table*, 44, 46*graph*
painting, landscape. *See* art; Hudson River School of landscape painting
paleoclimate, 123, 124*figure*
paleoethnobotany, 72
pale swallow-wort (*Cynanchum rossicum*), 188
paling, 29*table*, 30
Palisades, 124, 129, 258, 292, 296–298
Palisades Interstate Park (PIP), 256, 257, 259, 276, 280–282, 292, 297
Palisades Interstate Parkway, 298
Palmaghatt Ravine, 140
Palmer, Erastus Dow, 303
Panax quinquefolius (ginseng), 111
Panetta, Roger, 19
Panicum capillare (witch-grass), 169
paper birch (*Betula papyrifera*), 169
paper industry, 156–157
paper mills, 142
paper (birch) mulberry (*Broussonettia papyrifera*), 187–188
Papilio polyxenes (black swallowtail), 169
Papscanee Island, 72
parade grounds, 303
Paraquat, 159
Paris green, 160
Parker, Goeffrey, 77
parks, 203. *See also specific parks*
Parmentier, André, 184
parsley, 169
Parslow Field site, 70
parsnip, 100
Partnerships for Regional Invasive SPecies Management (PRISMs), 191

Passerculus sandwichensis (savannah sparrows), 170
pasture rose (*Rosa carolina*), 171
pastures, Columbia County, 169–172
patroonships, 24
Patterson, W. A., 72
Paul, Weiss, Rifkind Wharton and Garrison, 283
Paulding, James Kirke, 321
Pauw, Professor , 112
pavement, 130
paving stones, 297
Pavord, A., 108
Pawlonia tomentosa (princess tree), 186
PCBs (polychlorinated biphenyls): adaptation and, 243; basics, xxvi, 9–10; birds and, 328; factories and, 198, 242, 327; fisheries and, 24, 37, 242, 266, 327
peach (*Prunus persica*), 88–89*table*, 89–90, 91–93*table*, 94, 113
pearl-ash, 155
Pearl Street (Lower Manhattan), 78, 79, 80, 82–85*table*, 109–111
Pedicularis canadensis (common lousewort), 173
Peekskill bridge, 257
Peekskill Solid Waste Facility, xvi*map*
pelagic eggs, 265
peneplains, 168
Pennsylvania Railroad, 257
Pennsylvania sedge (*Carix pensylvanica*), 137, 177
Pennsylvania smartweed (*Polygonum pennsylvanicum*), 107, 169
Pennsylvania State, 68, 70, 72, 136, 168, 188, 250, 293. *See also specific locations*
Penobscot tribe, 107
Perca flavescens (yellow perch), 29*table*
perch, white (*Morone americana*), 28–29, 29*table*, 263, 265, 266, 267
perfoliated bellwort (*Uvularia perfoliata*), 168
Periplaneta americana (American cockroach), 29*table*
Perkins family, 280
Peru (South America), 95
pesticides, 159, 160–161, 326
Peter A. A. Berle Memorial Awards, 49
pharmacological pollutants, 327
Phaseolus spp. (beans), 23, 72, 95, 96, 153; *P. vulgaris* (beans), 153
Philadelphia PA, 184, 204*figure*
Phillipse, Mr., 111
Phleum pratense (timothy), 173
Phragmites (common reed), 125, 129, 130
Phytolacca spp. (pokeweed/pokeberry), 88–89*table*, 91–93*table*, 95, 96–97, 105
phytoliths, 70, 71
Phytophthora infestans (potato blight), 160
phytoplankton, 242, 262
Picea, 128*figure*
Picea spp., 126*figure*
Pickett, S. T. A., 9
Pieridae (whites) (butterfly), 169

Piermont, 256
Piermont Marsh, 123, 124, 128*figure*, 129, 131
piers, 58
Pieter Gabry and Sons, 86
Pieterszoon de Vries, David, 29
pig (*Sus spp.*), 17, 24, 44, 155
pigweed (*Chenopodiaceae*), 129
Pijpenkabinet Museum of Amsterdam, 81
pike. *See Esox* spp. (pickerel/pike)
piles, 105, 107
pilewort (lambsquarter/smearwort) (*Chenopodium* spp.), 88–89*table*, 91–93*table*, 96–97, 103, 104–105
Pinchot, Gifford, 276
Pine Hill, 142
pine resin, 70
pines (*Pinus* spp.). *See also individual species*
Pinus spp. (pines), 125, 126–128*figures*, 129, 130, 135, 139, 144, 148, 168*figure*, 169; *P. rigida* (pitch pine), 124, 137, 137*maps*, 138, 143, 146*figure*, 148; *P. strobus* (white pine), 138
PIP (Palisades Interstate Park), 256, 257, 259, 276, 280–282, 292
pipes, Native American, 316
pipes, sewer, 221, 227
pipe stem dating, 72, 81, 84*table*, 85*table*, 110
pipe-watch, 278
Pipilo erythrophthalmus (rufous-sided towhee), 176
piscivorous fishes, 9, 30*table*, 31, 36, 37. *See also individual species*
pitch pine (*Pinus rigida*), 124, 137, 137*maps*, 138, 143, 146*figure*, 148
planning, 329
Plantago spp. (plantain), 107, 125, 126–128*figures*, 129
plants. *See* flora; medicinal plants
PlaNYC, 227, 229
Platanthera lacera (ragged-fringed orchid), 171, 174
Platanthera psycodes (purple-fringed orchid), 171
Platanus occidentalis (sycamore), 170, 174
Platanus spp., 126*figure*
Plateau Path, 140
pleasure gardens, 203, 204
Pleistocene epoch, 124
Plestiodon faciatus (common five-lined skink), 43
Plethodon cinereus (Eastern red-backed salamander), 43*table*
plows, 159
plum, 90
plumes, industrial plant, 300, 302, 305, 306, 307, 328
Plumptre, G., 189
P. McCabe & Co., Inc., 213*figure*
Poaceae, 128*figure*
Poanes massasoit (mulberry wing), 171
poetry, xvi*figure*, 293, 319–321
Point Merino (Cole), 305–306
poison, 105
poison ivy, 105
poison sumac (*Toxicodendron vernix*), 170

pokeweed (pokeberry) (*Phytolacca* spp.), 88–89*table*, 91–93*table*, 95, 96–97, 105
pollen analysis: pre-contact–20th century, 125–129, 175; acorns and, 68; climate change and, 125, 131; ethnobotany and, 114; forests and, 24, 129, 130, 131; Great Lakes, 96; marshes and, 128*figure*, 131; uplands and, 123
pollution: 21 century, 328; basics, x–xi; ecosystem services and, 11; fisheries and, 32, 36–37, 266; by heat, 19; hydrological shifts and, 129; industry and, 17; invasive species and, 188; regulation of, 225; tree of heaven and, 186; visual, 300. *See also* waste and waste disposal; *specific pollutants*
Pollution Commission (NYS), 223
polychlorinated biphenyls. *See* PCBs
polycropping, 153
Polygala sanguinea (blood milkwort), 171
Polygonum spp. (bistort/knotgrass/knotweed/smartweed), 72, 88–89*table*, 91–93*table*, 95, 97, 103, 107–108; *P. arenastrum*, 107; *P. artifolium* (heart-leaved tear-thumb), 108; *P. aviculare* (prostrate knotweed), 107; *P. hydropiper* (marshpepper), 107; *P. pennsylvanicum* (Pennsylvania smartweed), 107, 169; *P. sagittatum* (arrow-leaved tear-thumb), 108
Polynuclear Aromatic Hydrocarbons (PAHs), 29
Polypodiaceae, 126*figure*, 127*figure*
Polypodiophyta, 128*figure*
Pomatomus saltatrix (bluefish), 37
Pomoxis annularis (white crappie), 30*table*
Pomoxis nigromaculatus (black crappie), 30*table*
ponds and fens, 17, 123, 129, 130
pond weed (*Potamogeto* spp.), 72
Pooecetes gramineus (vesper sparrow), 173
Popillia japonica (Japanese beetle), 189
Popular Mechanics, 282
population, human: pre-contact period, 70–71, 72; colonial period, 27; 21st century, 328; basics, xxv; Columbia County, 166; drinking water and, 218; exotic ornamental plants and, 185; Great Fire of 1835 and, 19; Hudson River, 154, 317; Lower Hudson, 154, 328; Manhattan, 220*figure*, 223*figure*, 224; Native Americans, 28, 317; New York City, 154, 206; technology and, 67; transportation and, 249; Upper Hudson, 154*table*, 328; waste and, 230
Populus deltoides (cottonwood), 170
Populus spp. (aspen), 126*figure*, 190
porcelain berry (*Ampelopsis brevipeduculata*), 189
porpoises, 42
Port of Albany, xxiii
ports, 185. *See also specific ports*
Portulaca oleracea (milk purslane), 169
Portulaca spp. (pulsey/purslane), 88–89*table*, 91–93*table*, 95, 96, 97, 106, 169
"Post-glacial Environments and Cultural Change in the Hudson River Basin" (Salwen), 67
postmolds, 67, 68, 72
Potamogeto spp. (pond weed), 72

potash, 154–155, 158
potato blight (*Phytophthora infestans*), 160
potatoes (*Solanum tuberosum*), 155*table*–156, 160
potherbs, 90*figure*, 94, 95, 96, 97, 98, 100, 106, 113
Potomac River, 252
pottery (ceramics), 69–71, 72, 80–81, 82–85*table*, 102, 109
Poughkeepsie, xxv, 197, 250, 257, 300
Poughkeepsie Railroad Bridge, 59–60*figure*, 257
poverty oatgrass (*Danthonia spicata*), 171
Power Authority (NYS), 288
power industry: 21st century, 330; fishes and, 10, 17, 19, 36, 327; hay and, 172; history, xxvi, 158, 196, 198; ice harvesting and, 207*figure*, 211, 213*figure*; regulation of, 325, 328, 331; water and, 19; wetlands and, 17. *See also specific industries; specific projects*
power take-offs (PTOs), 158
Pownell, Thomas, 248
prairies and prairie birds, 165, 169–170, 172–173, 176, 295
prairie warbler (*Dendroica discolor*), 176
precipitation, xxi–xxii, 227, 327
pre-contact period. *See* eras: pre-contact (native)
preservation, 13, 293–314, 329. *See also* conservation and restoration
"prime agricultural soils," 168
princess tree (*Pawlonia tomentosa*), 186
PRISMs (Partnerships for Regional Invasive Species Management), 191
prisons, xvi*map*
privets (*Ligustrum* spp.), 188
production, primary, 10*figure*
productivity, biological, xxi, xxii*map*, xxiv–xxv, 3, 4, 5, 11, 23, 24, 331
progress (civilization), xvi*figure*, xix, 10, 259. *See also* development
Progress (Durand), xvii*figure*, 276
Progress 2010 (panorama), xix*figure*
projectile points, 69, 70, 71, 72
prostate, 105
prostrate knotweed (*Polygonum aviculare*), 107
protein analysis, 70, 95
protests, 234, 282–283
prunts, 81, 83*table*
Prunus persica (peach), 88–89*table*, 89–90, 91–93*table*, 94, 113
Prunus spp. (cherry), 88–89*table*, 91–93*table*, 94
Pseudotriton ruber (red salamander), 44
PTOs (power take-offs), 158
public attitudes and awareness, 11, 192, 198, 255, 258, 261–262, 332. *See also* education
pulsey (purslane) (*Portulaca* spp.), 88–89*table*, 91–93*table*, 95, 96, 97, 106, 169
pumped storage, 161, 276, 277, 281, 283, 286, 288, 289, 332. *See also* Storm King Mountain ConEd plant
pumpkin. *See Cucurbita* spp.
pumpkinseed (*Lipomis gibbosus*), 30

Puritans, 317
purple-fringed orchid (*Platanthera psycodes*), 171
purple loosestrife (*Lythrum salicaria*), 189–190; Foundry Cove and, 240
purslane (pulsey) (*Portulaca* spp.), 88–89*table*, 91–93*table*, 95, 96, 97, 106, 169
pussytoe (*Antennaria sp.*), 171
Putnam County, 45, 69, 189
Pycnanthemum spp. (mountain-mint), 171
Pyne, Stephen J., 295
pyrolysis gas chromatography, 69

quail, bobwhite, 170, 172
quantification, 10–11
quarries, 304*map*, 305
Quaternary period, 123
quays, 58
Quercus spp. (oaks): *Q. alba* (white oak), 124, 137, 138, 139, 146*figure*, 148, 168; *Q. bicolor* (swamp white oak), 170; and birds, 149, 169; cropland and, 168; eras: pre-contact period, 68; fire and, 144*figure*; highlands and, 124; historical, 125, 126–128*figures*, 129, 131; *Q. iliciolia* (scrub oak), 138; industry and, 140, 142; Juet on, 317; Lower Hudson, 129; lowlands and, 67; *Q. montana* (chestnut oak), 124, 137, 139, 142, 144*figure*, 146*figure*, 148; rattlesnakes and, 43; rock walls and, 172; *Q. rubra* (red oak), 70, 124, 137, 138, 139, 141, 142, 146*figure*, 148; Shawangunk, 135, 137, 138, 139, 142, 143, 148–149; (swamp white oak), 170. *See also* acorns; *individual species*
quinces, 94
Quinn, David, 315

raccoons, 69, 72
races, river, 251
radiocarbon assays, 72, 114
radish, 98, 100, 100*table*, 101
ragged-fringed orchid (*Platanthera lacera*), 171, 174
ragweed (*Ambrosia*), 125, 126–128*figures*, 129, 169, 176
railroads: art and, 276, 293, 295, 299; basics, 197, 254–255; bridges, 257; effects of, xvi*map*, xix*figure*; ferries and, 256; fishes and, 15–16, 32, 34; Foundry Cove and, 233, 235; New York City ad, 19; Olana and, 304*map*; poetry and, 323; sediment and, 58–59; Shawangunks and, 142, 143 (*see also specific railroad companies*); shorelines and, xxiv, 327; straw paper and, 157; trap rock and, 297. *See also specific railroad companies*
rainbow smelt (*Osmerus mordax*), 37, 265, 265*figure*
Rainy Season in the Tropics (Church), 301
Rana pipiens (leopard frog/spring frog), 48, 171
Rana spp., 43. *See also Lithobates*
Ranunculus abortivus (small-flowered crowfoot), 173
Ranunculus spp. (buttercup), 72
rape, 100*table*
Rappahannock tribe, 105, 106
Raquette Lake, 296

rare species, 136
Raritan Bay, 61
raspberry (*Rubus* spp.), 68, 72, 73, 88–89*table*, 91–93*table*, 94, 103, 113, 176; *R. ideus*, 138
rattlebox (*Crotalaria sagittalis*), 172
rattlesnakes (*Crotalinae* spp.), 42–44, 48
raw materials. *See* resources
rays (fish), 28
recombinant bovine somatotropin, 157
recreation: basics, 10*figure*, 11, 54; Bear Mountain and, 257; ferries and, 256; FPC and, 284–285; mountain resorts, 144; Palisades and, 259; railroads and, 255; Shawangunks and, 135; SLC and, 307; steamboats and, 251, 321; waste and, 229. *See also specific activities*
red-backed salamander, Eastern (*Plethodon cinereus*), 43*table*
red-berried elderberry (*Sambucus pubens*), 169
redbreast sunfish (*Lepomis auritus*), 31
red cedar (*Juniperus virginiana*), 176
red clover, 153
red drum (*Sciaenops ocellatus*), 30
the Redfield, 34
redfin pickerel (*Esox americanus*), 30*table*
Red Hook, 69, 225
red maple (*Acer rubrum*), 137, 138, 141, 146*figure*, 148, 170, 174
red oak (*Quercus rubra*), 70, 124, 137, 138, 139, 141, 142, 146*figure*, 148
red raspberry (*Rubus ideus*), 138
red salamander (*Pseudotriton ruber*), 44
red-spotted newt (*Notophthalmus viridescens*), 44
red trillim, 177
red-winged blackbird (*Agelaius phoeniceus*), 173, 190
Reed, Doug, 278
reeds, 17
Reese, Frances S., 277
reforestation, 146–148, 149, 175, 177, 188, 326
refrigeration, 19, 156, 201, 208, 214
regional collaboration, 329
regulation: 21st century, 328–331, 333; basics, 277; drinking water and, 218; ecosystem services and, 11; fisheries pollution and, 31, 37; herpetofauna and, 48–49; inadequacy of, 326; industry and, xxvi; invasive plants and, 190–191; power plants and, 261–274, 325; shad and, 32; waste and, 222. *See also specific legislation; specific regulators*
Reichard, S. H., 191
Reifler, A. R., 71
religion, 11. *See also* ceremonies
remora, 30
Rensselaer County, 44
Rensselaer Polytechnic Institute, 331
Representative Important Species (RIS), 265
reptiles, 17, 24, 41–52, 43*table. See also specific reptiles*
republicanism, 250
research, xxvi

reservoirs, xxv, 220
resilience, 11, 17
resorts, 135, 144–145
"resource," 314, 315, 317–318, 325
resources, xvi*map*, 9, 11, 23–25, 67. *See also specific resources*
restoration: Clean Water Act and, 228; DEC and, 329; ecological analogies and, 165; fisheries and, 38; forest, 177; Foundry Cove and, 233, 240, 243; garden, 186; ice houses and, 209; marshes and, 131; of oysters, 229; sediment and, 62; of turtle, 48
Restoring American Gardens: An Encyclopedia of Heirloom Ornamental Plants 1640-1940 (Adams), 186
Revolution, American, xix*figure*, xxi, 156, 233, 255, 256, 318–319
rewilding, 175
Rhamnus cathartica (buckthorn), 176
rheumatism, 105, 107
Rhinebeck 2 site, 68
Rhine River, 278, 319
Rhode Island, 286
Rhododendron Swamp, 123, 124, 125, 126*figure*, 130
Rhus spp. (sumac), 72
ribbon snakes (*Thamnophis sauritus*), 171
rice, wild (*Zinzania aquatica*), 70
ridgelines, 178
Rifkind, Simon, 283
riparian zones, 32, 190
riprap, xxiv, 58, 327
"Rip Van Winkle" (Irving), 275, 276, 318–319
RIS (Representative Important Species), 265
Ritchie, William A., 67, 69, 70, 71
River City, 247–258
River Harbors Act of 1910, 56
river herring, 30, 31, 32, 34, 37, 38, 262, 263, 265, 267*table*
River House, 229
"River Indians," 73
River in the Catskills (Cole), 293
Riverkeeper, 278
River Mile 80, xxiv
River Network, 332
"River of Mountains," xv
Rivers and Harbors Act, 56
"the river that ebbs and flows," xv, 73, 315
River that Flows Both Ways, xv, 73, 315
Rivington's New York Gazetteer, 202
Rix, Julian, 296
roads, 137*maps*, 248, 257, 258, 326. *See also* highway system; turnpikes; *specific roads & highways*
Roanoke VA, 266
robber barons, 197
Robbins Swamp, 68
Roberts, William, 114
Robinia pseudoacacia (black locust), 186
Robinson Esq., Henry, 33
Rochester NY, 142, 189

rock bass (*Ambloplites rupestris*), 30*table*, 31
rock climbing, 146
Rockefeller family, 280, 282, 298
Rockland County, 44, 256, 258, 259
Rockland Lake, 207
rock walls, 172
Rocky Mountains, 295
Rocky Mountains-Landers Peak, The (Bierstadt), 295
rodents, 172
Roeliff Jansen Kill, 69, 71
Romantic movement, 183, 184, 255, 299, 320
Rondout Creek, 136
Rondout Valley, 147*figure*
Roosevelt, Theodore, 276, 280, 285, 297
Roosevelt National Park, 300
Rosa spp. (roses): *R. carolina* (pasture rose), 171; *R. multiflora* (multiflora rose), 176; *R. palustris* (swamp rose), 177
Roseton, 261, 264*figure*, 268, 287, 288
Roslund, E., 28
round worms, 105
Rubus flagellaris (dewberry), 171
Rubus spp. (brambles/raspberry/blackberry), 68, 88–89*table*, 91–93*table*; *R. ideus*, 138
Rudbeckia hirta (black-eyed Susan), 173
Rudbeckia laciniata (green-headed coneflower), 171
rufous-sided towhee (*Pipilo erythrophthalmus*), 176
Rumex spp. (dock), 125, 126*figure*, 128*figure*, 129
runoff, xxi–xxii, xxvi, 17, 61, 158, 159, 223, 326. *See also* erosion
Rupert, Colonel, 280
rural areas, 201–218, 249, 259. *See also specific features and locations*
Ruskin, John, 296
Russet Burbank, 155–156
Russia, 154, 279
rye (*Secale cereale*), 155*table*, 156–157, 167*figure*

sailing vessels, xvi*map*, 248. *See also* sloops
salamanders, 43, 43*table*, 44, 46–47, 48
"Salamanders of New York" (Bishop), 46
Salamandra coccinea (scarlet salamander), 44
Salamandra symmetrica (yellow-bellied salamander), 44
salinity, xxii*table*, xxiv*figure*, 54, 129, 130, 235, 236. *See also* salt front
Salix spp. (willow), 126*figure*, 127*figure*, 128*figure*, 177
salmon, Great Lakes, 34
Salmon, W., 107
Salmo salar (Atlantic salmon), 28, 29*table*
salt front/wedge, xxiv, xxv, 131
saltwater grass (*Spartina alterniflora*), 125
Salvelinus fontinalis (brook trout), 29*table*, 30*table*
Salwen, Bert, 67, 73
Sambucus pubens (red-berried elderberry), 169
Sampson, David, 299
Sam's Point, 47, 136
sand, 55*map*, 56

Sander vitreus (walleye), 30*table*
sand mining, 15
Sanguinaria canadensis (bloodroot), 177
Saratoga County, 71
Sargasso Sea, xxv
SASS (Scenic Areas of Statewide Significance), 305, 306, 328
sassafras (*Sassafras albidum*), 146*figure*, 148
Sassaman, K. E., 72
Satyrodes appalachia (Appalachian brown butterfly), 171
Sauer, Carl, 95
de Saussure, Nicolas-Théodore, 158
"savages," 10
savannah, 169
savannah sparrows (*Passerculus sandwichensis*), 170
sawmills, 139, 142
Saxifraga pensylvanica (swamp saxifrage), 171
SBMR & SBSA surveys, 264*figure*
Scaccia site, 71
scarlet salamander (*Salamandra coccinea*), 44
Scarry, C. M., 72
Sceloporus undulatus (brown swift (Eastern fence lizard)), 44
"Scene on the Banks of the Hudson" (Bryant), 313
scenery: 17th century, xvi*map*; 21st century and, 325, 328; history of, 198; ice houses and, 210; mountain houses and, 293, 295; Palisades, 297, 298. *See also* aesthetics; art
Scenic Areas of Statewide Significance (SASS), 305, 306, 328
Scenic Hudson, 276, 277, 278, 282–283, 284–285, 299, 325, 332
Scenic Overlooks, xix
Schackne, Patterson, 68
Schaefer, Paul, 296
Schenectady, 254
Schenectady County, 43, 160
Schizachyrium scoparium (little bluestem), 171
Schodack-Houghtaling Island, 209–210, 213*figure*
Schoharie Creek, 70, 71, 154
Scholopentria, 103
Schroon Lake, 296
Schulyerville site, 71
Schunnemunk Mountain, 48
Sciaenops ocellatus (red drum), 30
science, 13–22, 45–48, 261–269, 332, 333. *See also* biologist, role of; biology
Scirpus spp. (bulrushes), 129
Scott Brothers Ice House, 213*figure*
screens and impingement, 263, 265
scrub oak (*Quercus ilicifolia*), 138, 143, 176
Scully, Vincent, 286, 299
scurvy, 106
scurvy grass (hairy bittercress) (*Cardamine hirsuta*), 98
scythes, 173
sea levels, xxiv, 68, 132
sea-spiders, 42

370 INDEX

Secale cereale (rye), 155*table*, 156–157, 167*figure*
sedges. *See* Cyperaceae spp.
sediment: and agriculture, 13, 60, 155; basics, 53–63, 55–56; climate change and, 123–132; and dams, 60–61, 62, 129, 130; dredging and, 57; and forests, 60–61, 155; Foundry Cove and, 240–241; human impacts, 53–64, 61; ice harvesting and, 209; Jamaica Bay, 124, 130; maps, 54*map*; and railroads, 58–59; and restoration, 62; shoreline and, 58, 62; Tivoli Bay and, 69, 124; waste and, 61, 62, 221, 227–228. *See also* dredging; erosion; runoff; *specific sediments*
sediment budget, 57, 59, 60, 62
sediment distribution, 53, 54–55, 59, 60, 62
sediment supply, 60–62
seeds: pre-contact period, 70–71; contact period, 96; colonial period, 79, 81, 86, 87, 95, 101; basics, 95; invasive plants and, 189–190; sizes, 99–100*tables*. *See also* flotation technique
Seeger, Pete, 277. *See also Clearwater* (sloop)
Seeyle, John, 254
seismic profiles, 54–55
self-disclosure, 329–330, 333
Seminole tribe, 105
Semotilus corporalis (fallfish), 29
Senecio [now *Packera*] *aureus* (golden ragwort), 171
SEQRA (NYS Environmental Quality Review Act), 65, 292, 300, 306, 330, 332
sewage. *See* waste and waste disposal
sewers, 10, 220–222, 226–227, 229, 230, 327
sewing, 72
shad, xxvi, 8–9, 24, 29*table*, 31, 32, 35, 36, 70. *See also specific shad*
shadbush (*Amelanchier arborea*), 138, 172
Shakers, 170
shale, 137
Shawangunk Formation, 124, 137
Shawangunk Kill, 136, 140
Shawangunk Mountains ("Gunks"), 123, 133–152
Shawangunk Ridge, 17, 19, 46, 124, 136, 293
Shawnee-Minisink site, 68
sheep. *See also* livestock; wool
sheep (*Ovis aries*), 167*figure*, 169–170, 172, 253
sheep laurel (*Kalmia angustifolia*), 138
shell beads, 70
shellfish, 23, 24, 69, 224, 228. *See also specific shellfish*
Sherman (Augustus) Ice House, 212
Shimer, P., 106
shining clubmoss (*Lycopodium lucidulum*), 125, 126*figure*, 130
shipbuilding, 256
shipping, xvi*map*, xix*figure*, xxiii, xxiv, 168, 197, 222–223, 327. *See also* freight hauling; transportation
shipwrecks, 61, 80, 81, 82*table*
shoaling, 59, 228, 265
Shorakapkok, 315
shorelines: 17th century Manhattan, 81; 21st century, 327; data, 55; dredging and, 57; fisheries and, 37; ice harvesting and, 209–210; railroads and, xxiv, 32, 58–59, 197; sediments and, 58, 62; transportation and, 9; waste and, 219, 222, 229, 230. *See also* Hudson River, morphology; *specific locations*
shorthead redhorse (*Moxostoma macrolepidotum*), 30
shortnose sturgeon (*Acipenser brevirostrum*), xxv, 32–33
shrimps, 42, 238
shrubby pasture, 172
shrublands, 175, 176–177, 178
shrubs, 186–189, 187–188
Sicyos angulatus (bur-cucumber), 169
Sierra Club, 280, 286
"Significant Habitats and Habitat Complexes of the New York Bight Watershed," xxvi
Silent Spring (Carson), 161, 280
silos, 173
siltstones, 137
Silurian period, 124
silverfish, 29*table*, 30
silver maple (*Acer saccharinum*), 170
silversides (*Menidia* spp.), 29*table*
single-species catches, 31, 36, 37
Singleton, E., 110
Sing Sing prison, xvi*map*
Sisyrinchium spp. (blue-eyed grass), 171
Sive, David, 299
6LF2 site (Templeton), 68
skin disorders, 104, 105, 106, 107
skinks, common five-lined (*Plestiodon faciatus*), 43
skipper butterflies, 171
skyscrapers, 256
slave trade, 23, 196
SLC (St. Lawrence Cement Plant), 292, 294, 304*map*, 305, 306*figure*309, 307
slender lady's tresses (*Spiranthes lacera*), 173
sloops, xvi*map*, 53, 196–197, 248–249, 250, 277, 298*figure*, 299
sludge, 224, 225, 227
small-flowered crowfoot (*Ranunculus abortivus*), 173
smallmouth bass (*Micropterus dolomieu*), 30*table*, 36
smartweed (*Polygonum* spp.), 72
smearwort (lambsquarter/pilewort) (*Chenopodium* spp.), 88–89*table*, 91–93*table*, 96–97, 103, 104–105
Smiley, Albert ad Alfred, 144
Smiley, Daniel, 46–47, 140
Smiley Dan, 146
Smiley Sr., Daniel, 142
Smith, Anthony Wayne, 286
Smith, Bruce, 96
smoked sturgeon (Albany beef), xxv, 32–33
snakebite, 106
snakehead (*Channa argus*), 30*table*
Snake Hill, 43
Snake River Valley ID, 156
snakeroot, 108
snakes, 42–44, 45, 48, 172. *See also* herpetofauna; *specific snakes*

snake-wort, 42–43
snapping turtle (*Chelydra serpentina*), 42, 43, 48
Sneden's Landing Ferry, 256
snow, 144
soap, 154
soapstone, 70
social and natural science, 13–22
Society for Ecological Restoration Awards, 48
soft-shell turtle (*Apalone spinifera*), 44
soil: abandoned farms and, 175; agriculture and, 158–161, 167*figure*, 168, 171, 172, 174, 326; basics, 158, 159; butterflies and, 173, 176; canals and, 253; drainage and, 174; fires and, 143; flooding and, 153; forest, 168, 177; human impacts, 10–11, 131, 144, 155; invasive species and, 149, 170, 171, 186, 187, 188; management, 159, 161; pollution of, 221, 238, 242; quality, 171–172, 176*table*; Shawangunks and, 135, 137–138; wetland, 174. *See also* erosion; fertilizers; runoff
Solanaceae (nightshade), 97
Solidago spp. (goldenrod), 175, 176, 178; *S. caesia* (bluestem goldenrod), 168; *S. nemoralis* (gray goldenrod), 171; *S. tuberosum* (potatoes), 155*table*–156, 160
Somerset plant, 331
sonar, 54
song sparrow (*Melospiza melodia*), 170
Sopes Treaty, 318
sores, 105, 106
sound, 263
"source," 314, 315, 319
Sour Mountain Realty, Inc., 48
the South, 196
South America, 95, 295
South River, 315
South Seas, 106
South Street Seaport, 229
Spain, 3, 90. *See also individual Spaniards*
Spanish radish, 100
sparrows, 170, 173, 176
Spartina alterniflora (saltwater grass), 125
Spartina patens (cordgrass), 125
spatial scales of pollution, 241, 242, 264*figure*
spawning: channelizing and, 24; electric power generation and, 262, 277; eras: pre-contact period, 70; generating stations and, 262; lipfish and, 30; railroads and, 15; river herring, 31; salt front and, xxv; Storm King Mountain project and, 283, 286, 288; studies, 266, 268. *See also* larvae, aquatic
spears, 71
speciesism, 11
speed, 251–252, 254
speedwell (*Veronical officinalis*), 138
Sphagnum spp., 126*figure*, 127–128*figure*
Spicebush site, 69, 71
spice trade, 3
spiders, 46
spiked lobelia (*Lobelia spicata*), 173
spike grasses (*Eleocharis* spp.), 129

"spinach," 97, 98
spinach, wild (*Chenopodium* spp.), 68, 70, 96–97, 100*table*, 101, 105
spinning, 110–111
Spiraea alba (meadowsweet), 171, 177
Spiraea tomentosa (steeplebush), 177
Spiranthes cernua (nodding lady's tresses), 171
Spiranthes lacera (slender lady's tresses), 173
spirituality, 10*figure*
Spiza americana (dicksissel), 173
Spizella pusilla (field sparrow), 170, 176
sport fishing, xxv–xxvi, 9, 28, 36, 37
spottail shiner, 263*table*, 265, 267*table*
spotted turtles (*Clemmys guttata*), 44, 45, 171
sprains, 105
sprawl, 159. *See also* suburbs
Sprengel, Phillipp Carl, 158
spring frog (leopard frog), 48, 171
spring frog (leopard frog) (*Rana pipiens*), 48, 171
squash. *See Cucurbita* spp.
squaws, 110–111
squirrels, 172
Stachys spp. (betony/heal-all/woundwort), 88–89*table*, 91–93*table*, 108, 113
standing, legal, 284, 289, 299, 305
starchy seed plants, 88–89*table*, 91–93*table*, 94, 95, 113
starflower (*Trientalis borealis*), 177
State Environmental Quality Review Act (SEQRA), 65, 292, 300, 306
Staten Island, 225
Staten Island marsh, 124, 129
Staten Island Zoo, 44
State of the Hudson (conference), 13
State Pollutant Discharge Elimination System, 61*map*
State University of New York at Stony Brook, 239, 269, 331
Staubly site, 72
Steadman, David, 72
steamboats: basics, 197, 233, 250–252; Bear Mountain and, 257; canal barges and, 253; commerce and, 34, 320–321; ice harvesting and, 207, 210; winter and, 254
steam ferries, 255–256
steam locomotives, 155, 233
steam power: aesthetics and, xvii*figure*, 301, 302, 328; dredges, 209; generation plants, 196, 261, 277, 330; ice harvesting and, 207, 211, 213*figure*; locomotives, 155
steatite, 70
steel, 140
steel bridges, 258
Steendam, Jacob, 317–318, 319, 322
Steenwyck, 83*table*, 85*table*, 86
steeplebush (*Spiraea tomentosa*), 177
Stehling, Nancy, 114
sterile fields, 171
Stevens, John, 250
Stevens Institute of Technology, xxvi
Stewart, Margaret M., 47

stings, 107
St. Lawrence Cement Plant (SLC), 292, 294, 304*map*, 305, 306*figure*, 307, 309
St. Lawrence River, xvi*map*, 4, 70
Stockbridge-Munsee Band, 73
stocking programs, fish, 32, 33, 36
Stokes, I. N. P., 81
Stokes, John, 139, 140
Stokes Tavern, 144, 145*figure*
stomach maladies, 105, 106, 107, 108
stone, xvi*map*
Stone, Colonel, 253
Stony Brook University, 239, 269, 331
Stony Point, 256
Storm King Highway, 258
Storm King Mountain, 282*figure*, 298–299
Storm King Project, 161, 234, 276–278, 279–289, 292, 299, 332. *See also* pumped storage
storm sewers, 10
Stotts, P., 166
Strand, the, 109, 111, 113
stratigraphy, 123, 126*figure*, 129, 130
strawberry (*fragaria* spp.), 88–89*table*, 91–93*table*, 94, 105–106, 113
straw paper, 156–157
streeetcars, 222
stress, 108
strikes, 212–213
striped bass. *See Marone saxatilus*
striped mullet (*Mugil cephalus*), 28
strokes, 106
sturgeon, xxv30, 9, 29, 34, 69, 71, 72. *See also* Albany beef (smoked sturgeon)
sturgeon, shortnose (*Acipenser brevirostrum*), xxv, 32–33
Sturnella magna (meadowlark), 173, 174, 176
Sturtevvant's Edible Plants of the World (Hedrik), 98
Stuyvesant, Petrus, xix*figure*, 195
Sublime, the, 252, 276, 295, 296, 333
substantial evidence rule, 284
suburbs, 184, 185, 251, 255, 258, 259, 326
subways, 222, 257
succession after abandonment, 175–177
suckers (Catostomidae), 29*table*, 30, 34–36, 35; *Catostomus commersonii* (white sucker), 30, 31, 34
sufishes (Centrarchidae), 28
sugar maple (*Acer saccharum*), 124–125, 138, 146*figure*, 148, 190
sulfuric acid, 160
sumac (*Rhus* spp.), 72
sunfish (*Lepomis* spp.), 28, 29*table*, 30
Sunfish Pond, 206–207
sunflowers, 95, 96
Sunny Morning on the Hudson River (Cole), 293
Sunset on the Hudson (Gifford), 297
Sunset on the Palisades (Cropsey), 297
SUNY Stony Brook, 239, 269, 331
Superfund Act (1980), 238, 243

Superfund sites, 233, 238–239, 240, 242
suspension bridges, 258
Susquehanna River, 67, 69, 70, 71, 72, 73
Susquehanna Turnpike, 250
Sus spp. (pig), 17, 24, 44, 155
sustainability, xviii, 10, 11, 23, 24, 27, 37, 159, 214, 325, 327
Sutherland Fen and Pond, 124, 125
swallow-worts (*Cynanchum* spp.), 188–189
swamp forests, 170, 174
swamp rose (*Rosa palustris*), 177
swamp saxifrage (*Saxifraga pensylvanica*), 171
swamp white oak (*Quercus bicolor*), 170
swan, mute, xix*figure*, 189
sweet fern (*Comptonia peregrina*), 171
sweetflag (*Acorus americanus*), 171
swellings, 105, 109
swimming, 327
Swiss chard, 101
sycamore (*Platanus occidentalis*), 170, 174
sycamore maple (*Acer pseudoplatanus*), 186
Sylvan Lake Rockshelter site, 69
Sylvilagus transitionalis (New England cottontail), 174
Syracuse, 252

Taché, K., et al., 70
Taconic Hills, 69
Taconic Mountain Range, 67
Tafts, 258
Tall Grass Prairie, 173
tall milkweed (*Asclepias exaltata*), 169
talus, 137*maps*, 138
tanning/tanning bark, 24, 140, 142, 155, 293
tapeworm, 105
Tappan Zee, xxiii
Tappan Zee Bridge, 59, 258, 259
Tarrytown, 256, 259
Tartarian honeysuckle (*Lonicera tatarica*), 187
taxes, 326
tear-thumbs, 108
technology, x, 19–20, 250, 252, 253, 255. *See also* specific technologies
teething, 106
telegraph, xvi*map*
temperate basses. *See Morone* spp.
temperature: climate change and, 328; fisheries and, 36–37; fishes and, 10; ice harvesting and, 212; Jamaica Bay, 124; Lower Hudson, xxiii; pollution and, 19; power plants and, xxvi; sediments and, 131
Templeton site (6LF2), 68
Tennessee, 227
tension, 108
tension zone, 166
Terminus Post Quem (TPQ), 80
Terrapene carolina (Eastern box turtle), 42, 48
terrapin, northern diamond-backed (*Malaclemmys terrapin*), 42

Thamnophis sauritus (ribbon snakes), 171
thiabendazole, 160
Thlaspi arvense (field pennycress), 98
Thompson, John E., 47
Thoreau, Henry David, 201, 292
Thorin, Peter, 203, 204
Thornton, Howard, 33
3D databases, 114
three sisters, 72
throat maladies, 105, 107
Thruway, The, 259
tick-trefoil (*Desmodium spp.*), 168
tides: basics, xxiv, 54; bridges and, 59; cadmium and, 240, 242; floodplains and, 67; Foundry Cove and, 235, 236, 239; Indian Point and, 287; railroads and, 32; transportation and, 251
tiger salamanders (*Ambystoma tigrinum*), 44
Tilia, 128*figure*
Tilia americana, 126*figure*
timber/lumbering, 24, 123, 139, 141–142, 143, 155, 172, 197, 254, 296
timber rattlesnake (*Crotalus horridus*), 42, 43, 48
time, 251, 321
timothy (*Phleum pratense*), 173
Tivoli Bays, 69, 71, 72, 123, 124, 129, 130
toadflax (flax) (*Linum spp.*), 88–89*table*, 91–93*table*, 94, 103, 108, 109, 113, 155*table*, 173
toads, 43
tobacco (*Nicotiana* spp.), 88–89*table*, 90*figure*, 91–93*table*, 95, 96, 153, 317
tolls, 252
tomcod (*Microgadus tomcod*), 29*table*
tomcod (frostfish) (*Microgadus tomcod*), 29*table*, 30, 263, 265, 266, 267*table*
tonsils, 107
tools, 70, 72, 110, 142, 154. *See also* flint; projectile points
Torrey, J., 171, 173, 175
tortoises. *See Glyptemys* spp.
total susspended solids (TSS), 224
tourism, 257, 258, 296, 305, 307, 321. *See also* scenery
Tower, Fayette B., 255
towhee, rufous-sided, 176
Toxicodendron vernix (poison sumac), 170
Toxostoma rufum (brown thrasher), 176
TPQ (*Terminus Post Quem*), 80
tractors, 158–159, 161, 173, 323
trade. *See* commerce
trade-offs, 11
trailing Arbutus (*Epigaea repens*), 169
trails, 303, 329
Train, Russell, 287, 289
transmission lines, 280, 300
transportation: 21st century, 328; basics, 19, 196–197, 247–259; exotic ornamental plants and, 185; flow and, 17; food and, 161; history, 53–54; industry and, xxv–xxvi; monuments, 257; Native Americans and, 248;

passenger, 254; shorelines and, 9, 15, 19. *See also* freight hauling; *specific hubs*; *specific methods*; *specific modes*
transportation, passenger, 9, 19
Trapa natans (water chestnut), 37, 240
Trapps, the, 139, 140, 142, 146
trap rock, 297
traps, fish, 71
treaties, 11, 318
Treatise on the Theory and Practice of Landscape Gardening, Adapted to North America, A (Downing), 184
tree cover, 24, 78*figure*, 135, 148, 170, 171, 175*figure*. *See also* forests; *specific trees*
tree of heaven (*Ailanthus altissima*), 186
trees, 186, 302. *See also* forests; tree cover; *individual species*
trestles, xxiv
tributaries of the Hudson River, xxi, xxii*map*, 24, 32, 60, 70. *See also specific tributaries*
Trientalis borealis (starflower), 177
Trifolium spp. (clover), 88–89*table*, 91–93*table*, 108–109, 153, 173
Trillium undulatum (painted trillium), 169
Trinectes maculates (hogchoker), 29, 34, 265
Trinectes maculatus (flounder), 29*table*, 30
Triodanis perfoliata (Venus looking-glass), 171
Triticum spp. (wheat), 155*table*, 156, 167*figure*, 168, 169, 253, 317
Triton puctatus (crimson-spotted triton), 44
trolleys, 256
trout, 29*table*, 30
Troy: basics, xxi, xxii, xxv, 59; dams, 54, 61; fishes and, 265; forests and, 157, 196; shoreline data, 55; tides and, xxiv, 328; transportation and, 197, 254, 255; water quality and, 329; whales and, x
Truncer, J., 70
TSS (total susspended solids), 224
Tsuga canadensis (hemlock): pre-contact period -20th century, 67, 68, 126–128*figures*, 131; basics, 137*map*, 138; Columbia County, 169; decline of, 129; industry and, 140; Shawangunk, 125, 135, 139, 142, 146*figure*, 148
Tubuliflorae spp., 126*figure*, 128*figure*
Tudor, Frederick, 207
Tufano site, 71
tufted sedge (*Carex platyphylla*), 138
tunnels, 19, 256–257, 258, 299
turbidity, xxi, xxii
turkey, 69
turnip, 98, 100, 101
turnpikes, 249–250, 252
turtles, 42, 45, 48–49, 69, 71; A. *spinifera* (soft-shell turtle), 44; *C. guttata* (spotted turtles), 44, 45, 171; *C. serpentina* (common snapping turtle), 42, 43, 48; *E. blandingii* (Blanding's turtle), 47, 48; *K. subrubrum* (Eastern mud turtle), 48; (painted), 44, 46*figure*; *T. carolina* (Eastern box turtle), 42. *See also Glyptemys* spp.

twalift, 29*table*, 30
Twilight in the Adirondacks, A (Gifford), 296
Twilight in the Wilderness (Church), 301
211-1-1 site, 71
Typha spp. (cattails), 125, 127*figure*, 128*figure*, 189–190, 240; *T. augustifolia*, 29–30
typhoid, 213–214
typhus, 228
Tyrannus tyrannus (kingbird), 170

ulcers, 104
Ulmus spp. (elms), 126*figure*; *U. americana* (American elm), 138
Ulster and Delaware turnpike, 250
Ulster County, 46, 72, 142, 300
Unami language, 315
unicorns, xviii
Union Ferry Terminal, 256
United States. *See entries beginning* US
United States Centennial Exhibition (1876), 296
unity, national, 250, 252
University Heights, 315
University of Connecticut, 268
University of Massachusetts, 268
University of Michigan Datbase of Ethnobotany, 105, 107
University of New York at Stony Brook, State, 239, 269, 331
upland boneset (*Eupatorium sessilifolium*), 171
upland plovers (*Bartramia longicauda*), 173
uplands, 67, 68, 123–132, 172–173, 176, 307
uploads, 122–134
Upper Bay, 61
Upper Hudson: agriculture and, 13, 155*table*, 158*table*, 160*tables*; art and, 295; basics, xxi–xxii, xxiii, 154, 335; drainage, 154; fishes, 29–31; frogs, 46; lumbering and, 254; Native Americans and, 3, 79; PCBs and, 198, 327; population, 154*table*, 328; preservation and, 329. *See also* ice harvesting; *specific locations*
Upper Palisades, The (Moore), 297
urban areas, xxi, xxii*map*, 10, 131, 201–218, 249, 255, 259, 326. *See also specific cities*; *specific features and locations*
Urban Ecology (journal), 8
urinary tract, 105, 106, 107, 109
Ursus americana (black bear), 29, 177
US Army Corp of Engineers (USACE), xxiii, 56–57
US Atomic Energy Commission, 262
US Court of Appeals for the Second Circuit, 268–276, 284–285, 286–287, 299, 332
US Department of Agriculture Plants Database, 191
US Department of Commerce, 276
US Environmental Protection Agency (USEPA), 238, 240, 262, 268, 287, 289, 325
US Fish and Wildlife Service, 191
US Fish Commission, 33
US Geological Survey (USGS), xxvi, 295
US National Park Service, 329. *See also* National Parks

US Navy munitions storage, xvi*map*
US Nuclear Regulatory Commission (NRC), 301, 302, 303–305
US Public Health Service, 225
US Supreme Court, 251, 268
utilities, 277, 287, 325, 331. *See also specific projects*
Uva, R. H., 98
Uvularia perfoliata (perfoliated bellwort), 168

Vaccinium spp. (blueberry), 88–89*table*, 91–93*table*, 137, 138
valley floor *versus* upland model, 67
Van Corlaer, Anthony, xix*figure*
Van Cortlandt Manor, 204
Vanderbuilt, Cornelius, 197
Vanderbuilt National Park, 300
van der Donck, Adriaen, 27, 29–30, 42, 94, 98, 100–101, 103, 109–111*passim*, 110, xvi*map*
Van der Groen, Jan, 98, 100–101
Van Leuven family, 139*figure*, 140, 141*figure*, 143*figure*
Van Orden, Vanderpool, and Sherman Ice House, 70, 211, 212, 213*figure*
Van Rensselaer, Mrs. J. K., 106, 110, 111
Van Tienhoven, 86, 96, 98, 100–101
van Tienhoven "Great House," 83*table*
Van Twiller, 86
Vaux, Calvert, 295, 296, 300, 302
vegetables, 79, 90*figure*, 94, 97–101, 178. *See also specific vegetables*
vegetation. *See* flora
venereal disease, 104, 108
Venus looking-glass (*Triodanis perfoliata*), 171
Verbena hastata (common vervain), 171
verdant frame, 184
Vermont State, xxi
vernal pool amphibians, 46–47
Veronical officinalis (speedwell), 138
Verplanck, 256
Verrazano, Giuseppe da, 297, 316, 317, 322
Verrazano Bridge, 322
Verrazano Narrows, xxiii
vesper sparrow (*Pooecetes gramineus*), 173
Vespucci, Amerigo, 322
Viburnum acerifolium (maple-leaf viburnum), 138, 172
Viburnum dentatum (arrowwood), 71, 177
Viburnum lantanoides (hobble bush), 169
Viburnum lentago (nannyberry), 177
Viele, , 78*figure*
View of the Catskills — Early Autumn (Cole), 293
views. *See* scenery
vines, 188–189
Vinette site, 70–71
Viola sagittata (arrowhead violet), 171
Virginia. *See also specific locations*
Virginia pepperweed, 98
Virginia State, 3, 101, 153, 204
Virginia yellow flax (*Linum virginianum*), 172